Intellectual Mastery of Nature

Intellectual

VOLUME 2

Christa Jungnickel
and
Russell McCormmach

Mastery of Nature

Theoretical Physics from Ohm to Einstein

The Now Mighty Theoretical Physics 1870–1925

THE UNIVERSITY OF CHICAGO PRESS

Chicago and London

The University of Chicago Press, Chicago 60637
The University of Chicago Press, Ltd., London
© 1986 by The University of Chicago
All rights reserved. Published 1986
Paperback edition 1990
Printed in the United States of America

95 94 93 92 91 90   5432

Library of Congress Cataloging-in-Publication Data

Jungnickel, Christa.
  Intellectual mastery of nature.

  Includes bibliographies and indexes.
  Contents: –v. 2. The now mighty theoretical
physics 1870–1925.
  I. Mathematical physics—Germany—History.
1. McCormmach, Russell.   II. Title.
QC19.7.G3J86 1986   530.1'5   85-28920
ISBN 0-226-41584-8 (v. 2, cloth)
ISBN 0-226-41585-6 (v. 2, paper)

♾ The paper used in this publication meets the
minimum requirements of the American Standard for
Information Sciences—Permanence of Paper for Printed
Library Materials, ANSI Z39.48-1984.

FOR OUR PARENTS

Contents

Volume 2: The Now Mighty Theoretical Physics, 1870–1925

List of Illustrations

Preface

In the first volume of *Intellectual Mastery of Nature*, we analyze the development of physics in Germany from 1800 to 1870; there we show the gradual emergence of the discipline including its partially autonomous theoretical branch. In this second volume, we continue the analysis after 1870, and as before, we do this by giving as full an account as possible of the working lives of the physicists. We examine physicists individually and collectively within the institutional setting of physics throughout the several German states. We pay special attention to physicists working in the specialized institutes for theoretical physics, which came to be established in this period, and we analyze the theoretical work they did in these institutes and the relationship of this work to experimental physics. As before, we base our account largely on the extensive archival record of German physics. Here again we emphasize the unity of our book, which has been divided into two volumes only for convenience. We direct our readers to the preface of the first volume where we discuss the goal and approach of our entire study.

Although we conclude this volume with a brief discussion of major developments in atomic physics to the mid-1920s, we bring our proper study to a close around 1915. That year the Würzburg physicist Wilhelm Wien, a former student of Hermann von Helmholtz and the recent recipient of a Nobel prize for work primarily in theoretical physics, spoke of the field with pride and partiality: depending first and foremost on the creative powers of the mind, "the great masters of theoretical physics have also been the greatest scientists." Theoretical physics, he said, had now come to be recognized as a "separate science," one which was joined to experimental physics in a "higher unity." Wien spoke of "the now mighty theoretical physics."[1]

1. Wilhelm Wien, "Ziele und Methoden der theoretischen Physik," *Jahrbuch der Radioaktivität und Elektronik* 12 (1915): 241–59, on 241–43, 259.

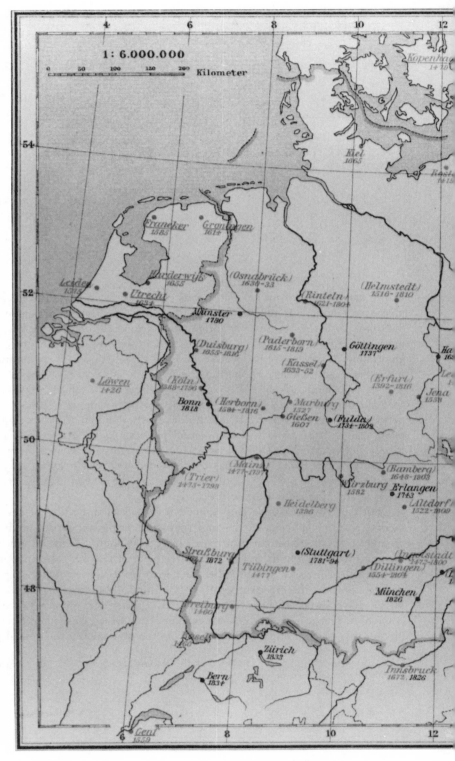

German Universities to the Beginning of the Twentieth Century.

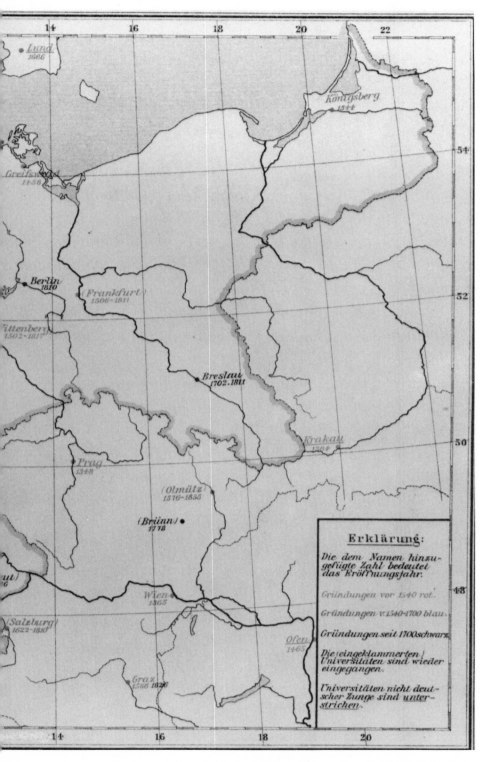

Explanation. Founded before 1700: light type. Founded since 1700: dark type. Universities in parentheses have ceased. Universities underlined are not German-language universities. Reprinted from Franz Eulenburg, "Die Frequenz der deutschen Universitäten."

VOLUME *2*

The Now Mighty Theoretical Physics 1870–1925

The Now Mighty Theoretical Physics, 1870–1925

In the last third of the nineteenth century, the political geography of Germany changed as a result of a series of wars beginning with the war with Denmark over Schleswig and Holstein. Prussia and Austria were allied against Denmark, but soon afterward they were at war with one another. Prussia emerged victorious, annexing several German states: Schleswig-Holstein, Hessen-Cassel, Hannover with its University of Göttingen, and the city of Frankfurt am Main. The old German Confederation was brought to an end as Austria was finally excluded from Germany, and most of the independent northern German states were incorporated in a new North German Confederation under Prussian leadership. The remaining difficulties standing in the way of German unification, which Prussia's Minister-President Bismarck had identified with Prussia's cause in the war with Austria, were resolved in the war between Prussia and France in 1870–71. The result was a united Germany, the Reich, proclaimed in 1871, consisting of the many formerly independent German states: four kingdoms, five grand duchies, thirteen duchies and principalities, three free cities, and Alsace-Lorraine, ceded by France and treated as a conquered territory.

After German unification the universities remained under the control of the individual states. For our study, the most obvious effect on the universities of the states annexed by Prussia in 1866 was the different spirit of administration; for example, instead of writing their former long and personal accounts to the ministries, physicists now reported on the activities of their institutes by bureaucratic forms. The war disrupted physics in minor ways; for example, the building of the new Berlin physics institute was delayed for a few years owing to financial difficulties. It also yielded tangible benefits for German physics by the incorporation into German university life of Strassburg University, which for political reasons was treated as an expensive showcase complete with a magnificent physics institute. But the war and the unification of Germany did not affect the working lives of physicists much if at all. Physics continued to be taught to an assortment of professionals in training, who were required in increasing numbers and variety by modern Germany. It continued to be taught to the small number of intending physicists as well, but with a slight difference: physics led not only to teaching jobs but to a small but growing number of jobs outside; physics was beginning to become what chemistry had long been, a science associated with industry and commerce as well as with the advancement of natural knowledge. Physics, as taught and practiced within its ever more commodious institutes, contributed its part to Germany's reputation for proficiency, which the rest of the world admired and emulated. In certain parts of physics and in the hands of certain physicists, the field was lifted above the expectations of proficient research to towering achievements.

13

Physical Research in the *Annalen* and Other Journals, 1869–1871

Contributors and Contents

With our discussion here of the *Annalen der Physik* for the years 1869 through 1871, we approach the close of J. C. Poggendorff's long editorship of Germany's leading physics journal. By 1874, the fiftieth anniversary of his editorship, he had brought out one hundred and fifty volumes.[1] *Poggendorff's Annalen*, the customary form of its citation, had become synonymous with German physics.

Inevitably, work by German authors came to occupy more space in the *Annalen* and work by foreign authors less. And inevitably, it would seem, work came to be increasingly physical, as work in chemistry, especially in organic chemistry, was channeled to other specialized journals.[2] But even as the work submitted to the journal was restricted, it increased in quantity, and Poggendorff took to bringing out supplementary volumes from time to time. In addition, toward the end of his editorship, Poggendorff began a regular supplementary series, the *Beiblätter*, for brief reports of recent work not appearing in the *Annalen*.

Over the middle half of the nineteenth century, Poggendorff's years, the *Annalen* made available throughout Germany the record of physical research at home and abroad. German readers were introduced by Poggendorff to the equivalent of nearly thirty volumes of foreign research; Michael Faraday's work alone filled over two volumes, and hundreds of pages were given over to Victor Regnault and hundreds more to other foreign experimentalists. Scarcely less important, the *Annalen* informed foreign scientists of German research; the time was long past when

---

1. The volumes Poggendorff brought out over these many years included nearly nine thousand papers and notices by over two thousand authors. Each year he published works by about thirty new authors. The numerical measure of his achievement was given by W. Baretin in the *Jubelband* for Poggendorff's fiftieth anniversary. "Ein Rückblick," *Ann.* (1874): ix–xiv, on xii–xiii. We are again making the *Annalen* the principal source for our survey.

2. W. Baretin, "Johann Christian Poggendorff," *Ann.* 160 (1877): v–xxiv, on xi.

it stood in foreign libraries, if it stood there at all, with pages uncut because it contained little but foreign research in German translation.[3]

Poggendorff opened his journal to all physicists, and although he placed greatest value on experimental work, he also welcomed theoretical work.[4] Rudolph Clausius's many researches, which were all purely theoretical and often highly mathematical, almost without exception appeared first in Poggendorff's journal. It is indicative of the attention the *Annalen* received abroad that Clausius's papers were nearly all translated within a year in the leading British physics journal, the *Philosophical Magazine*.

The *Annalen* was available to physicists in German universities and technical institutes, and it was to be found as well in many gymnasium libraries.[5] German physicists regarded the *Annalen* as an indispensable resource for their field; after Poggendorff's death in 1877, the Berlin Physical Society acknowledged an official responsibility for the journal: its title page now mentioned the collaboration of the Society and, "in particular," of Helmholtz. The new editor was again an experimental physicist, Gustav Wiedemann, but with Helmholtz to advise on theoretical physics, Germany's leading physics journal now acknowledged that experimental and theoretical research needed to be evaluated by their respective experts.

The *Annalen* for the years 1869–71 is bound in ten bulky volumes, containing some 6,500 pages. About a third of the space is given over to foreign research, in German translation where translation was called for. Fully half of the foreign work originated in Austria-Hungary, Switzerland, and Holland, where physicists had close relations with their German counterparts.[6] (There was even a certain amount of movement of physicists between teaching positions in these several countries.[7]) Britain and a handful of other European countries made up the rest of the foreign contribution to the *Annalen*.[8]

German contributors to the *Annalen* numbered well over a hundred in 1869–71. All of them, with the exception of a few government officials and a few without affiliation, were connected in one way or another with teaching institutions. They included nineteen ordinary professors of physics and directors of physics institutes from universities and technical institutes.[9] They also included about the same num-

---

3. The physicist J. F. Benzenberg reported that on his trip to Paris in 1815, he found that L. W. Gilbert's *Annalen* was not to be had at the Institute's library and that although it was at the Royal Library, its pages were uncut. In his *Ueber die Daltonsche Theorie* (Düsseldorf, 1830), preface.

4. Baretin, "Poggendorff," xii. Poggendorff's successor, Gustav Wiedemann, spoke of theory and experiment having been joined in the *Annalen*. Wiedemann, "Vorwort," *Ann.* 39 (1890): i–iv.

5. Emil Frommel, *Johann Christian Poggendorff* (Berlin, 1877), 67.

6. The total number of papers in the *Annalen* from each of those countries follows in that order: Austria-Hungary, Switzerland, and Holland.

7. For part of 1869–71, Kundt and Friedrich Kohlrausch were in Zurich; they accounted for a good share of the substantial contribution from Switzerland to the *Annalen*.

8. The other countries represented in the *Annalen* were Britain, Denmark, Sweden, (Western) Russia, France, Italy, and Belgium. Of physics published in the New World and elsewhere, Poggendorff took no notice in 1869–71. In one of these years he published a series of brief papers sent to him from Argentina by a young, world-traveling physicist from Berlin, who was an assistant to Magnus.

9. In order of the number of pages they published in the *Annalen* in 1869–71, the nineteen institute directors were: Wilhelm von Bezold (Munich), Adolph Wüllner (Bonn and Aachen), Wilhelm Hittorf (Münster), Eugen Lommel (Erlangen), O. E. Meyer (Breslau), Gustav Magnus (Berlin), Rudolph Clau-

ber of extraordinary professors, Privatdocenten, assistants, and students, whose work took up as much space in the Annalen as that of the institute directors.[10] Among the most prolific contributors to the Annalen during these years were two physicists at the Prussian Academy of Sciences, who were at the same time ordinary professors of physics without institutes at the university.[11] The rest were teachers at secondary and lower technical schools, who accounted for approximately a third of all teachers publishing in the Annalen.

The subjects of German papers in the Annalen in 1869–71 were primarily, but by

---

sius (Würzburg and Bonn), August Kundt (Würzburg), J. B. Listing (Göttingen), Eduard Reusch (Tübingen), Wilhelm Beetz (Munich), Heinrich Buff (Giessen), Johann Müller (Freiburg), Hermann Knoblauch (Halle), Heinrich Weber (Braunschweig), Gustav Kirchhoff (Heidelberg), Franz Melde (Marburg), Heinrich Wilhelm Dove (Berlin), and Wilhelm Weber (Göttingen). Dove, who was ordinary professor of physics at Berlin and director of the Prussian state meteorological institute, is included in this list because he took charge temporarily of the Berlin University physics institute for a year after Magnus's death in 1870.

University institute directors who did not publish in the Annalen in 1869–71 were: Ottokar von Feilitzsch (Greifswald), Karl Snell (Jena), Gustav Karsten (Kiel), Franz Neumann (Königsberg), Wilhelm Hankel (Leipzig), and Philipp Jolly (Munich); Rostock University did not have a physics professor. With the exception of Hankel, they published either no articles at all in any physics journals or at most an occasional article elsewhere. Hankel published thermoelectric studies of crystals regularly in the Abhandlungen of the Saxon Society of Sciences; from 1876, he republished this material in the Annalen.

10. They were mostly recent graduates, and the theoretical competence their work revealed expressed the current understanding of what a properly trained physicist should know. Through their teaching or research or both, most of them would be closely associated with theoretical physics in their subsequent careers. For physicists just beginning about 1870, theoretical physics would provide increasing opportunities.

Of the three extraordinary professors of physics publishing in the Annalen in 1869–71—Friedrich Kohlrausch, Georg Quincke, and Karl Zöppritz—the latter two taught mathematical physics in their academic positions at that time. Of the—at least—eight Privatdocenten who published in the Annalen in these years, most had close connections with theory. Wilhelm Feussner was at Marburg, where he would teach theoretical physics for fifty years, from 1880 as extraordinary professor and from 1908 as ordinary honorary professor. Friedrich Narr was at Munich, where he would teach the subject, from 1886 as extraordinary professor, to the end of his career. Eduard Ketteler was at Bonn, where, as extraordinary professor, he would teach it from 1872 to 1889, nearly the end of his career. Although Leonhard Sohncke in Königsberg and Emil Warburg in Berlin would both teach theoretical physics (Warburg as extraordinary professor for the subject), their teaching responsibility would be experimental physics over most of their careers. Yet they were associated with theoretical physics in their research: Sohncke as developer of Franz Neumann's direction in theoretical crystal physics; Warburg as one of the most theoretically competent German experimental physicists. Hermann Herwig had only a brief academic career (he died in his thirties, while professor of physics at Darmstadt); his dissertation at Göttingen had been on mathematics, and his subsequent work in experimental physics showed theoretical interest. Richard Rühlmann did not continue on in an academic career but became a gymnasium teacher of physics and mathematics. Emil Budde did not continue either, but he managed to do research all the same, much of which, like his first paper in the Annalen in 1870, was purely theoretical.

Of the—at least—six students or assistants who published in the Annalen in 1869–71, only two continued on in academic careers. Paul Glan, a Berlin graduate in 1870 and after that an assistant to Helmholtz, went on to teach theoretical physics at Berlin as Privatdocent from 1875 to the end of his life. Eduard Riecke, a Göttingen graduate in 1871, became assistant and Privatdocent there the same year; then and during appointments as extraordinary professor in 1873 and ordinary professor of experimental physics in 1881, he regularly gave lectures on theoretical physics until Woldemar Voigt was brought to Göttingen; his research was primarily theoretical throughout his career.

11. Peter Riess and the editor himself, Poggendorff, were the two physicists. Their publications in the Annalen compare in volume with those of the two most prolific institute directors, Bezold and Wüllner.

no means entirely, physical. Although the proportion of chemical to physical papers, where the two could be distinguished, had declined over the years, chemistry remained a substantial interest of the *Annalen*, as did crystallography, physiology, and other so-called border areas of physics.[12] Only pure mathematics remained outside its interests.[13]

Work by Institute Directors

Institute directors had the means and the independence to develop their own directions of research. In 1869–71, as at other times, their directions seldom coincided, and as a result their researches, when taken together, touched on most branches of physics. Wilhelm Hittorf published solely on electric conduction in gases, J. B. Listing on optical instruments, Eduard Reusch—but for a description of an instrument and of what happens when one picks up shot with tweezers—on crystals, O. E. Meyer—but for a polemic against British physicists—on the friction of air, Adolph Wüllner—but for critical remarks in support of the work of a student of his—on the spectra of gases, and Eugen Lommel—but for brief descriptions of instruments—on light absorption and fluorescence. August Kundt, who became an institute director in Germany in 1870, continued the acoustical researches he had begun abroad, though he soon took up a new topic, anomalous dispersion. Wilhelm von Bezold, the most versatile as well as the most prolific researcher, worked on many topics within electricity and on several outside.

Institute directors usually did not have to concede, as a less favored contributor to the *Annalen* did, that they could not decide which of the possible causes of a phenomenon was "reality" because of lack of "opportunity and instruments."[14] Although from their point of view they never had enough instruments of the right kind, they had control of a physical cabinet, which they made as good use of as possible in experimental research. In theoretical research, they sometimes dealt with the theory of an instrument, so that even here they might remain close to the contents of the local physical cabinet.

The work of the institute directors in 1869–71 would not have suggested to readers of the *Annalen* that research was separated into theoretical and experimental parts. Only Clausius, Gustav Kirchhoff, and Wilhelm Weber appeared there as

12. Papers by chemistry teachers took up a fifth as much space in the *Annalen* as those by physics teachers, and papers by teachers of crystallography, mineralogy, and geology together took up one-third as much space. Physiologists were well represented; less well represented were astrophysicists, mathematicians, physicians, and pharmacists. All told, the contributions to the *Annalen* from teachers outside of physics took up two thirds as much space as contributions from teachers within physics.

13. Despite the recognized connections of mathematics and physics, only two academic mathematicians published in the *Annalen* in 1869–71, and they published only one article each on mechanics, a topic that belonged as much to mathematics as to physics. Mathematicians had their own journals and did not need the *Annalen*, the journal that physicists increasingly regarded as their own.

14. Overzier, 139: 651–60, on 660. Because we refer to a large number of papers published in the *Annalen*, in the remainder of this chapter we use this highly abbreviated form of citation: author's last name, volume number: page numbers. The journal, unless otherwise specified, is always the *Annalen*, and the volumes all fall within the three years, 1869–71.

purely theoretical writers, and the latter two with only a single paper each. Other institute directors—Meyer, Bezold, Kundt, and Lommel—presented their experimental work together with some mathematical theory, and most of the rest revealed some theoretical concern. In the work of the directors in general, theoretical and experimental discussion occurred together, the balance varying from physicist to physicist and from problem to problem.

Through an experimental study of electric conduction in gases, for example, Hittorf hoped to come nearer to a mechanical explanation of electrical processes and to rid physics of its last imponderables, the two electric fluids.[15] Lommel wished to complete what he regarded as the most developed physical discipline, optics. Inflection, double refraction, circular polarization, and other complex phenomena of light had been explained by mechanical principles, but not fluorescence. Lommel now proceeded to explain it mechanically, as he did the closely related phenomena arising from the action of light on chlorophyll; then he carried out the necessary experiments for his theory.[16] Typically, in the *Annalen,* institute directors reported experiments containing measurements of some physical "constant" and its variation with other physical quantities; as constants often entered theoretically derived equations, their measured behavior could complete the theories or provide tests for them.[17] Some experiments involved the production and analysis of geometrical "figures," the lawfulness of which invited theoretical explanations.[18] Others on absorption and emission derived from—and suggested new—theoretical ideas on the relations of matter and ether.[19]

A small proportion of the work submitted to the *Annalen* by institute directors did not contain new experimental results but new theoretical reflections,[20] which had to do with particular phenomena[21] or with features of particular laws.[22] Of this purely theoretical work, Clausius's mechanical investigations of the theory of heat were the most significant for the foundations of physics.[23] Ludwig Boltzmann had worked on the same fundamental problem, and his and Clausius's exchange and, in

15. Hittorf, 136: 1–31, 197–234, on 223.

16. Lommel, 143: 26–51, 568–85, on 30–34.

17. For example, Kundt on the variation of the index of refraction with wavelength (142: 163–71), and O. E. Meyer on the variation of the constant of the internal friction of air with temperature and, possibly, with pressure (143: 14–26). Kirchhoff studied theoretically the variation of the constant of the magnetization of iron with the intensity of the magnetizing force (supplementary vol. 5: 1–15).

18. For example, Bezold on "electric dust figures" (140: 145–59) and on "Lichtenberg figures" (144: 337–63, 526–50), and Melde and Kundt on "sound figures" (Melde, 139: 485–93; Kundt, 137: 456–70, 140: 297–305).

19. For example, Magnus on radiant heat (139: 431–57, 582–93), Kundt on anomalous dispersion (142: 163–71), and Lommel on the relation of chlorophyll to light (143: 568–85).

20. These purely theoretical papers constituted only about an eighth of the institute directors' publications in the *Annalen* in 1869–71.

21. For example, the phenomena of galvanic arcs, fluorescence, and induced magnetism by, respectively, Bezold (140: 552–60), Lommel (143: 26–51), and Kirchhoff (supplementary vol. 5: 1–15).

22. Wilhelm Weber called attention to the existence of a potential for his fundamental law of electric action (136: 485–89), Bezold to analogies between the laws of photometry and the laws of gravitational attraction (141: 91–94), and Clausius to the mechanical principles underlying the second law of heat theory (142: 433–61).

23. Clausius, 141: 124–30, 142: 433–61.

general, their critical writings and priority claims lent the *Annalen* in the years around 1870 a rigorous theoretical tone, making it a risky place to publish casual statements on the nature of heat.[24] Because of the universality of the principles of heat, physicists working on a variety of topics encountered and possibly contradicted—or worse, overlooked—Boltzmann's and Clausius's statements and had to deal with their criticisms in the pages of the *Annalen* or elsewhere.[25]

Experimental Research

It was still possible to publish notices in the *Annalen* of incidental observations of nature alongside those of carefully prepared phenomena in the laboratory. During a hailstorm, for example, it occurred to Johann Müller to look at polarization in hailstones, and he reported what he saw to the *Annalen*, which was that each stone is made up of many pieces of ice oriented in various directions; but as he had not been prepared for a hailstorm, he was not in a position to make a thorough study.[26] As an institute director, Müller was assured space in the *Annalen*, to which he sent a steady stream of three-page notices on unconnected topics. Another contributor began his notice with: "On the night of . . ., I saw, as I walked to the window after finishing my work . . .," and he reported his observations of some remarkable thunder and lightning.[27] On walks in and around Frankfurt, another contributor had long listened closely to the interval, rhythm, and pitch of the two tones of the call of the cuckoo, a birdcall, he said, which every child knew and which Beethoven had included in the Pastoral Symphony.[28] Poggendorff valued the occasional uncomplicated observation; he himself extracted from a French journal an account of an "uncommon snowfall" in France, the equal of which only one elderly person could remember from 1804 or 1805.[29]

The odd or charming phenomenon was not, however, the usual substance of a paper in the *Annalen* around 1870. The unadorned description of an apparatus and its use was far more likely. Sometimes the apparatus was meant to be used in lecture demonstrations, at least in part: Kohlrausch kept the needs of "lectures" in mind; Poggendorff introduced a powerful apparatus capable of producing effects visible to

24. Clausius's work led Ludwig Boltzmann to publish a priority claim in the *Annalen* (143: 211–30). This was one of many papers by Boltzmann in the *Annalen* in these years, all in response to papers by others published there. He worked in Austria at this time and published most of his researches there.

The Stettin secondary school teacher Robert Most provoked Boltzmann by his "simple proof" of the second law of heat theory, which he regarded as "simpler" than the first law (136: 140–43). Boltzmann pointed out that Most assumed at the start that dQ/T is a complete differential, so it was "naturally not hard" for him to prove the second law (137: 495). What Most regarded as clear about his proof Boltzmann regarded as simply wrong (140: 635–44).

25. As, for example, Bezold did in connection with his study of electrical condensers (137: 223–47); Clausius responded (139: 276–81).

26. Johann Müller, 144: 333–34.

27. Hoh, 138: 496.

28. Oppel, 144: 307–9.

29. Poggendorff, 139: 510–11.

the far corners of lecture halls; Müller gave a "much simpler" experimental proof than Coulomb's of the inverse-square force of magnetism, which had more to do with teaching than with physical advance.[30] But generally the apparatus was meant to be used strictly in research. As the subjects of research came in great variety, so did the apparatus; the index of the *Annalen* for the ten years centering on 1870 listed over a hundred kinds of apparatus and instruments, which included pendulums, pipes, prisms, and a large number of measuring instruments, the point of contact between experiment and mathematical theory. The names of the instruments often ended in "meter," expressing their measuring purpose: barometer, calorimeter, electrometer, and the like. The *Annalen* was filled with the measurements Friedrich Kohlrausch made with Weber's "meters," with the bifilargalvanometer, bifilardynamometer, magnetometer, and so on.[31] The names of other instruments often ended in "scope," expressing the need of physicists not only to measure but to "see" the phenomena or the numbers on a meter: chromatoscope, chronoscope, electroscope, and erythroscope. Heavy foldout pages at the end of each issue of the *Annalen* contained elaborate drawings of the apparatus used in the researches reported there.

Like the phenomena it was designed to study, the apparatus itself had to be understood exactly; it was an interacting part of nature, operating under the same physical principles. The principles were usually assumed known, but not always; if they were not, the apparatus was, like nature, an object of theoretical study in its own right.[32]

Apparatus was often given a simple geometrical form; it was easier to build that way, and it facilitated observations and measurements, and it also corresponded to the mathematical solutions of the theoretical laws to be tested or extended. For example, with apparatus consisting of two mercury reservoirs connected by a horizontal glass capillary tube, Emil Warburg measured the relationship between the pressure and the rate of flow of mercury. According to Helmholtz's recent theory of frictional fluids, the mercury is to be understood as flowing through the circular tube in parallel, hollow cylinders; because of the friction, the cylinders should move slower the closer they are to the walls of the tube. By comparing his measurements with the theoretical formula relating flow and pressure, Warburg concluded that the mercury cylinder adjacent to the glass does not move at all, contrary to expectations based on earlier work. For a second example, Kundt derived a method for making visible as simple geometrical figures in dust the tones of strongest resonance excited in air contained within two solid, parallel plates a few millimeters apart. For these "air plates," he sketched a mathematical theory to "master" the problem. He specialized the differential equation for the propagation of sound to the two-dimensional case of a vibrating membrane, concluding that the oscillations of an air plate are opposite to those of a membrane: where the membrane is at rest, the air plate is in

30. Friedrich Kohlrausch, 136: 618–25, on 625; Poggendorff, 141: 161–205, on 203; Johann Müller, 136: 154–56.

31. Friedrich Kohlrausch, 138: 1–10, 142: 418–33, 547–59.

32. From Poggendorff's description of Wilhelm Holtz's "electromachine." He believed that the machine had "theoretical interest," since it had a property that did not allow itself to be "theoretically" explained and probably could not have been found "a priori." Poggendorff, 141: 161–205, on 204–5.

motion, and conversely. With both Warburg's and Kundt's apparatus, the simple geometry—a cylinder and a plate—made possible an immediate clarification of the phenomena through mathematical theory.[33]

Experimentalists commonly believed that certain phenomena they studied were still insufficiently understood to be explained by existing theory. Wüllner's spectra of gases, for example, were highly complex. By passing electric current through tubes containing hydrogen at different pressures, he obtained four kinds of spectra; to understand the relations between the spectra, Kirchhoff's theoretical law relating emission and absorption was of no help, and Wüllner approached the problem directly, by experiment.[34] Heinrich Buff, to take another example, thought that because of the complexity of hydraulic pressure experiments, the extensive mathematical theories of the subject did not give an accurate picture, and he set out to determine experimentally the "facts of the phenomenon."[35] Experimentalists risked criticism by claiming theoretical knowledge of phenomena that were still being ordered experimentally.[36]

Theoretical Research

As in the 1840s, in 1869–71 in the *Annalen*, German physicists often referred to British and, above all, to French theoretical work. They had a strong interest in French optical theory, especially A. L. Cauchy's and A. J. Fresnel's.[37] They recognized S. D. Poisson's contributions to nearly all parts of theoretical physics, if more often than not for their deficiencies.[38] P. S. Laplace still entered the *Annalen* for his theory of capillarity.[39] Among British theories, German physicists referred to those by Thomas Young, Michael Faraday, George Green, G. G. Stokes, and James Clerk Maxwell.[40]

What was new in 1869–71 was the extent to which German physicists now

33. Warburg, 140: 367–79; Kundt, 137: 456–70.

34. Wüllner, 137: 337–61, on 347–48.

35. Buff, 137: 497–517, on 497.

36. Magnus was criticized for making a statement that had no empirical basis but was simply a generalization from the theory of gases. Knoblauch, 139: 150–57, on 152–53.

37. In Cauchy's theories of reflection (Jochmann, 136: 561–88) and of dispersion (Ketteler, 140: 1–53, 177–219); in Fresnel's formulas for reflection (Lamansky, 143: 663–43) and for light intensity (Kurz, 141: 312–17), and in his hypothesis of ether drag (Ketteler, 144: 109–27, 287–300, 363–75, 550–63).

38. In Poisson's theories of acoustics (Kundt, 137: 456–70), of capillarity (Quincke, 139: 1–89; J. Stahl, 139: 239–61), of elasticity (Heinrich Schneebeli, who was Kundt's student in Switzerland, 140: 598–621), of pendulum motion (O. E. Meyer, 142: 481–524), of magnetism (Kirchhoff, supplementary vol. 5: 1–15), of earth temperature (Frölich, 140: 647–52), and (if the *Journal für die reine und angewandte Mathematik* is included here) of heat (Lorberg, 71: 53–90) and of hydrodynamics (O. E. Meyer, 73: 31–68).

39. For Laplace's theory of capillarity (J. Stahl, 139: 239–61; Quincke, 139: 1–89), but also for his electrodynamics (Hittorf, 136: 1–31, 197–234).

40. Young's theory of capillarity (Quincke, 139: 1–89; Paul du Bois-Reymond, 139: 262–75), Faraday's theory of charges as molecular orientation (Knochenhauer, 138: 11–26, 214–30), Stokes's theory of frictional fluids (Warburg, 139: 89–104, 140: 367–79), Green's and Stokes's theories of pendulum motion (O. E. Meyer, 142: 481–524), and Maxwell's theories of colors (J. J. Müller, 139: 411–31, 593–613) and electromagnetism (Kirchhoff, supplementary vol. 5: 1–15).

drew on German theoretical sources. They cited the earlier German theoretical work: Carl Friedrich Gauss's on capillarity,[41] Wilhelm Weber's on waves, sound, and electrodynamics,[42] and Franz Neumann's on crystallography, optics, capillarity, and magnetism.[43] But the German theoretical work they cited most often was recent work. Kirchhoff's broad and influential contributions included theories in acoustics, heat, hydrodynamics, elasticity, and electricity.[44] Clausius's included the mechanical theory of heat and the kinetic theory of gases;[45] indeed so strongly had Clausius developed this branch of physics that at times his only sources were his own previous work.[46] Of other German theoretical contributions[47] Helmholtz's were cited by far the most often, his work on acoustics most commonly, but a range of his other work as well, including that on hydrodynamics, color theory, geometry, and galvanism.[48] The German contribution to exact theory can be illustrated by a detail: an *Annalen* author observed that the electric forces entering his topic, the formation of Lichtenberg figures, involved "only electrostatic action at a distance, that is, the first term of Weber's fundamental law."[49] In other words, he regarded the law that Weber had derived and tested as the general law and Coulomb's electrostatic law as only its first

41. Paul du Bois-Reymond, 139: 262–75; J. Stahl, 139: 239–61; Boltzmann, 141: 582–90.

42. Wilhelm Weber's work on waves (Quincke, 139: 1–89; Kundt, 140: 297–305; Matthiessen, 141: 375–93), on sound (Warburg, 136: 89–102, 137: 632–40, 139: 89–104; J. J. Müller, 140: 305–8), and on electrodynamics (Wilhelm Weber, 136: 485–89, calling attention to a publication of his in the *Annalen* on electrodynamics over twenty years before).

43. Franz Neumann's theoretical work entered German physics in oblique ways in 1869–71. He had long before stopped publishing, but his work was used and cited even in cases where he had not published it himself. August Kurz cited an actual, if old, publication of 1834 on crystallography (141: 312–17), but Emil Jochmann found Neumann's formula for metallic reflection in a Swiss publication by Heinrich Wild, who had studied for a while with Neumann, and in an abstract. Jochmann thought that no derivation or statement of the suppositions of Neumann's formula had been published; he assumed that it rested on the assumption of Neumann's other optical work that the ether has the same density but different elasticity in different media (136: 561–88). Quincke, who had also studied with Neumann, stated a capillary law governing the spread of one fluid over another, which he thought Neumann was the first to express (139: 1–89). Paul du Bois-Reymond, who had spent some time with Neumann, too, called this law the "third principal law of capillarity" and supposed that the only place it was published was in his own dissertation in 1859 (139: 262–75). Riecke tested Neumann's law for the magnetism of an ellipsoid but gave no reference to any publication on it by Neumann (141: 453–56).

44. On Kirchhoff's work in acoustics (Seebeck, 139: 104–32), in heat (Wüllner, 137: 337–61; Magnus, 139: 431–57, 582–93; Lommel, 143: 26–51), in hydrodynamics (Paul du Bois-Reymond, 139: 262–75), in elasticity (Adolf Seebeck, 139: 104–32; Schneebeli, 140: 598–621), and in electricity (Knochenhauer, supplementary vol. 5: 146–66).

45. Budde, 141: 426–32, 144: 213–19; Narr, 142: 123–58; Hansemann, 144: 82–108; Bezold, 137: 223–47; Recknagel, supplementary vol. 5: 563–91.

46. Clausius, 141: 124–30, 142: 433–61.

47. For example, Riemann's work in electricity (Bezold, 137: 223–47) and in geometry (J. J. Müller, 139: 411–31, 593–613), Krönig's in kinetic theory (Hansemann, 144: 82–108; Recknagel, supplementary vol. 5: 563–91), and Carl Neumann's in optics (Ketteler, 140: 1–53, 177–219) and in electrodynamics (Wilhelm Weber, 136: 485–89). If the *Journal für die reine und angewandte Mathematik* is included here, Jochmann's and Lorberg's recent theoretical work in electrodynamics enters (Helmholtz, 72: 57–129).

48. On Helmholtz's work in acoustics (Warburg, 139: 89–104; Adolf Seebeck, 139: 104–32; Sondhauss, 140: 53–76, 219–41; Glan, 141: 58–83; Lommel, 143: 26–51; Boltzmann and Toepler, 141: 321–52), in hydrodynamics (Warburg, 140: 367–79; Paul du Bois-Reymond, 139: 262–75), in color theory and in geometry (J. J. Müller, 139: 411–31, 593–613), and in galvanism (Bernstein, 142: 54–88).

49. Bezold, 144: 337–63, 526–50, on 535.

term. German dependence on foreign theoretical work was now a good deal less complete than it had been twenty-five years before.

Research published in the *Annalen* in 1869–71 recorded a parallel drive in theory and experiment: in theory it was to understand phenomena at the level of the smallest parts and actions, in experiment to reach ever smaller quantities of space, time, and energy. Whereas the "millimeter" still satisfied the measuring needs of most of physics, with the powerful magnifications obtainable by new microscopes there was need for a smaller unit in micrography and physical optics, the "micron," a thousandth part of a millimeter.[50] With powerful magnification, previously unsolvable problems involving the fine texture and structure of bodies could now be approached.[51] In experimental mechanics, new methods of measuring extremely small intervals of time enabled the previously unsolvable problem of the duration of elastic collisions to be solved.[52] The minuscule amount of heat received on earth from an individual star[53] and the incredibly small work done by the air on the ear at the threshold of hearing[54] were among the small quantities that the *Annalen* reported regularly in these years. Molecules could not be made individually visible or tangible, but with the help of assumptions, the tiny sphere of action of their mutual forces[55] and the tiny width of their vibrations could be measured.[56]

In discussing the phenomena they were studying, German physical scientists referred regularly to "molecules," by which they sometimes meant simply the smallest particles under consideration but at other times something more precise, such as structured groups of atoms.[57] They regarded molecular reasoning as indispensable for understanding many of the intimate phenomena of matter: the motion of liquids over liquids, of liquids over solids, and of solids through gases, and the inner friction

50. Listing, 136: 467–72.
51. Listing, 136: 473–79.
52. Schneebeli explained that although the laws of collision had been known since the seventeenth century, the actual process of collision remained "rather mystical" owing to the short duration of collisions. With a new method, he determined the time of collision of steel cylinders and spheres; for a cylinder colliding with a fixed bar, the time was 0.00019 second (143: 239–50).
53. From Britain, William Huggins reported to the *Annalen* his attempt to determine the very small amount of heat the earth receives from individual stars, using an apparatus consisting of a sensitive galvanometer and various thermopiles (138: 45–48).
54. From Graz, Boltzmann and August Toepler sent the *Annalen* an experimental investigation of sound vibrations using a new optical stroboscopic method. They determined the mechanical work per second done by the air on the ear at the limits of hearing to be 1/3,000,000,000 kilogram-meter (141: 321–52, on 352).
55. Quincke determined that for glass, silver, water, and several other substances, the radius of action of molecular forces is a very small, but nonvanishing, length, of the order of 0.000050 mm, or approximately one-tenth the average wavelength of light (137: 402–14, on 413); Robert Lüdtge, also at Berlin, accepted this result (139: 620–28, on 620).
56. Boltzmann and Toepler determined that the width of an air particle's vibration at the limits of hearing is about one-tenth the wavelength of green light, showing how astonishingly sensitive the organ of hearing is (141: 321–52, on 349–52).
57. Lommel explained that he was using "molecule" in the "chemical sense": a group of atoms characterized by their nature, number, and relative positions (143: 568–85, on 573); Budde proceeded from the understanding that the "molecule" of most simple gases consists of two atoms (144: 213–19).

of gases, of liquids, and of solids. They sometimes invoked molecular actions to explain what happens when electricity, light, and radiant heat are passed through bodies and, above all, to explain the heat relations of bodies. Where an otherwise good theory did not agree with certain phenomena, molecular behavior might explain why.[58] In most physics institutes around 1870, the effects of molecular actions were studied, both experimentally and theoretically: notions of molecular behavior entered qualitatively in experimental reasoning[59] and mathematically in the assumptions of physical theory.[60]

In approaching certain theoretical problems, physicists used the continuum view of matter.[61] But to many problems they brought the discrete molecular view, which had far-reaching implications for the methods of theoretical physics. For such problems, Boltzmann, who thought deeply about these implications, believed that physicists would have to give up the language of exact science since Newton, the calculus, at least where rigorous derivations were involved.[62] Physicists usually took molecules to be bits of matter with inertia, the motion of which was governed by the laws of mechanics. The two sets of ideas, molecular and mechanical, were seen jointly to confer on physics a unity of direction: physics seeks to reduce all phenomena to "purely mechanical concepts," one author wrote, and the solution of any problem by the mechanical motions of molecules is the "final and highest task of physics."[63] Young's theory of capillarity was judged deficient for failing to give the constants entering it a "precise mechanical molecular-theoretical significance,"[64] the existence of which was assured by the understanding that the phenomena of capillarity are caused by molecular forces.[65] In optics as in capillarity, a theory was judged deficient if it failed to "give a theoretical basis" to its constants, which meant developing them from molecular and mechanical considerations. The explanation of optical absorption, dispersion, and anomalous dispersion relied on the forces be-

58. As set out by Gustav Hansemann (144: 82–108).

59. Jochmann attributed the departure of Cauchy's reflection theory from observations on reflection and refraction in thin metal sheets to "molecular" properties of the metal surface, which affect the optical constants (supplementary vol. 5: 620–35, on 632–33).

60. For example, in a theoretical investigation of the "internal constitution of gases" (Hansemann, 144: 82–108).

61. For example, the theory of a vibrating string based on the analysis of a volume element of the string rather than on molecular forces (Reinhold Hoppe, 140: 263–71); and the theory of induced magnetism based on the analysis of volume and surface elements of the magnetized body (Kirchhoff, supplementary vol. 5: 1–15).

62. Drawing on Gauss's method, J. Stahl had recently derived the fundamental equations of capillarity. According to Boltzmann, Stahl's derivation shared the defect of Gauss's method, which is to compute finite sums by integrating over all pairs of molecules. This procedure assumes that each molecule contributes a vanishingly small part of the sum, regardless of how far it is from any given molecules. Since it is likely that the contribution of a neighboring molecule is finite, Boltzmann reasoned, the use of the integral calculus is not justified here, and he showed how to develop capillarity theory using only finite sums and replacing the integration symbol $\int$ by the symbol Σ (141: 582–90).

63. Echoing Helmholtz's words here, Heinrich Schröder went on to say that it was premature to explain the action of a gas on solids or liquids by molecular motions, so that for now he had to renounce the highest task of physics (supplementary vol. 5: 87–115, on 114–15).

64. Paul du Bois-Reymond, 139: 262–75, on 267.

65. Lüdtge, 139: 620–28, on 620.

tween molecules of bodies and particles of ether; these forces and resulting motions were analyzed by the customary laws of motion of the mechanics of mass points.[66] In kinetic theory, the behavior of a gas was analyzed by mechanical collisions of molecules with the walls of the container and with other molecules and, perhaps, by the rotations and internal vibrations of the molecules.[67] Molecular considerations entered theories of heat,[68] of electricity,[69] and of other parts of physics.

To the physicists who designed theories in 1869–71, the appeal to the principle of conservation of energy had become second nature. No theory that ignored it could be taken seriously. The received authority of energy considerations led an *Annalen* author, for example, to reject all explanations of the absorption of light deriving from Cauchy's theory, since that theory violated the principle of living force.[70] The increasingly common use of potential instead of force facilitated the energetically correct formulation of problems in all parts of physics.[71] Related to these energy considerations was an interest in developing theories through the principle of least action, or some similar principle, which governed the behavior of the energies in a physical system.[72]

Other Journals

The proceedings and (for longer works) monograph series of societies and academies also frequently published the work of physicists. The proceedings were usually not specialized; although the physical-mathematical and the philosophical-historical classes of the Prussian Academy of Sciences met separately for serious scholarly work, their proceedings ran together papers on all subjects, the Vatican manuscripts side by side with electrical machines.[73] They got sorted out, if they did at all, only upon republication in specialized journals. The physics published by the Prussian Academy was almost all experimental, as was to be expected from its authors, who included Gustav Magnus, Poggendorff, Heinrich Wilhelm Dove, Peter Riess, and Reusch; even Helmholtz's one paper for the academy in this period was experimental. Kirchhoff published one theoretical paper there, which he also published in a mathematics journal. Nearly all of the other physics papers published by the acad-

66. Ketteler, 140: 1–53, 177–219, on 200; Lommel, 143: 26–51; Sellmeyer, 143: 272–82; Glan, 141: 58–83.

67. Hansemann, 144: 82–108; Recknagel, supplementary vol. 5: 563–91; Narr, 142: 123–58.

68. Clausius's central concept of disgregation was related to molecular arrangements (Budde, 141: 426–32).

69. Faraday's concept of charge as a peculiar molecular position was introduced (Knochenhauer, 138: 11–26, 214–30).

70. Glan, 141: 58–83, on 74.

71. Wilhelm Weber showed that his fundamental law of electric action could be derived from a potential (136: 485–89). Examples of other theories developed in terms of potentials are Bezold's in electricity (137: 223–47) and photometry (141: 91–94) and Kirchhoff's in magnetism (supplementary vol. 5: 1–15).

72. For example, Clausius, 142: 433–61, on 449; Boltzmann, 143: 211–30, on 220, 228.

73. The full title of the Prussian Academy of Sciences' proceedings is *Monatsberichte der königlich preussischen Akademie der Wissenschaften zu Berlin*.

emy were also published in the *Annalen*. The same was true of the many papers submitted by Beetz and Bezold to the Bavarian Academy of Sciences for publication in their proceedings.[74] The considerable quantity of physics research published in the proceedings of the Göttingen Society of Sciences covered the range of experiment and theory. Corresponding societies in Leipzig, Bonn, and elsewhere all published physics, too.

It was generally recognized that the *Annalen* was not always the appropriate place to publish a mathematical physics paper. Bezold published his experiments on optical illusions in the *Annalen,* but he published their mathematical theory elsewhere, observing that readers of the *Annalen* did not all take to mathematical deductions.[75] Leopold Pfaundler believed that Clausius's derivation of the fundamental equation of the molecular-kinetic theory of gases was too difficult for many readers of the *Annalen,* especially for chemists who did not know the integral calculus, so he published a simpler derivation of it there.[76]

Physicists often published their more mathematically rigorous work in journals they shared with mathematicians. There their work appeared alongside the mathematicians' work on physical, usually mechanical, problems, in which the interest was predominantly mathematical, not physical. The *Mathematische Annalen,* which began publishing only in 1869, contained some mathematical physics, most of it written by its co-founder Carl Neumann.[77] More mathematical physics was published in the *Zeitschrift für Mathematik und Physik,* where several German and Austrian physics institute directors published in 1869–71, though they did not publish their major work there.[78] Most of the contributors to the *Zeitschrift* were teachers at secondary schools and lower technical schools; some of their papers related directly to teaching, with criticisms of gaps in textbooks and the like, while others were concerned with advancing physics. With few exceptions, the papers were mathematical in nature.

Mathematical physics of greater significance went to the *Journal für die reine und angewandte Mathematik,* edited by Berlin mathematicians. In 1869–71, this journal published theoretical work by physicists on mechanics and electricity. Mechanics had been treated extensively by mathematicians, and electricity was being increasingly treated by them, so that physicists publishing there were assured a measure of interest as well as mathematical competence on the part of their readers. They did not, however, presume too much physical interest; Kirchhoff, for example, made no

74. The full title of the Bavarian Academy's proceedings is *Sitzungsberichte der königlich bayerischen Akademie der Wissenschaften zu München.*

75. Bezold, 138: 554–60.

76. Pfaundler, 144: 428–38.

77. Carl Neumann published a number of brief notices in his new journal, the *Mathematische Annalen,* but only one substantial work in mathematical physics, which had to do with crystal optics (1: 325–58). Karl Von der Mühll published a mathematical physics paper there (2: 643–49), as did his Giessen colleague Alexander Brill (1: 225–52).

78. Clausius and Bezold published brief notes in the *Zeitschrift,* and Lommel published an "elementary presentation" of some mathematical methods. Boltzmann published a paper on Ampère's law that was unusual for its length and its experimental nature; but it was a republication of his paper in the *Sitzungsber. Wiener Akad.* the year before.

mention of physical reality in connection with his derivation of the mathematical consequences of certain assumptions about an ideal incompressible fluid.[79] Other papers in the *Journal* discussed physical questions, but none offered new experiments.

With their work in the *Journal,* physicists acknowledged their common ground with mathematicians. One physicist, for example, called attention to the "special interest" of a calculation, the solution to a fourth-order linear partial differential equation involving roots of transcendental equations.[80] Another physicist cited sources on the series development of functions and on differential equations from the mathematical literature.[81]

Just as mathematicians constantly generalized their methods and theories, so did physicists in the work they published in the *Journal.* Kirchhoff, for example, generalized William Thomson and P. G. Tait's treatment of the motion of a body of rotation immersed in a fluid,[82] and he generalized Helmholtz's method of treating discontinuous fluid motion.[83] Boltzmann generalized Kirchhoff's derivation of the apparent forces between two rings in a moving fluid; Hermann Lorberg generalized Kirchhoff's equations for the motion of electricity in conductors; Helmholtz generalized Neumann's potential for two current elements.[84] And so it went: in physics as in mathematics, the more general the theorem, the more powerful it was understood to be.

Electrical topics dominated research in Germany in 1869–71, as they had for a long time.[85] The nature of electrical action was among these topics: the *Annalen* contained much research on the interactions of ether and matter, but little on actions in the ether itself.[86] Readers were told, and Berlin physicists were shown by demonstration, how a tuning fork disturbs nearby smoke and flame, suggesting that attractions and repulsions might be due to motions in the air or ether and not to action at a distance;[87] but physicists required more than a suggestion to be persuaded of this fundamental point. In work in the *Annalen,* German physicists took little notice yet of Maxwell's replacement of distance action by action in the ether. Kirch-

79. Kirchhoff, *Journal* 70: 289–98.
80. O. E. Meyer, *Journal* 73: 31–68, on 33.
81. Lorberg, *Journal* 71: 53–90.
82. Relaxing the restrictions that the body does not rotate around its axis and that its axis remains parallel to a fixed plane, Kirchhoff showed that the problem could still be solved, though this more general motion leads to elliptic integrals (*Journal* 71: 237–62).
83. Kirchhoff, *Journal* 70: 289–98.
84. Boltzmann relaxed Kirchhoff's assumption of the circular cross section of the rings (*Journal* 73: 111–34). Lorberg relaxed Kirchhoff's assumption that the electricity in the conductor is not acted on by outside forces (*Journal* 71: 53–90). Helmholtz's general potential reduces to Neumann's when one current is closed (*Journal* 72: 57–129).
85. The ten-year index volumes of the *Annalen* show this. Half of the topics with entries taking up an entire column or more in the index are electrical, with light and heat topics far behind. This measure applies alike to the indices for the ten years centering on 1860 and for the ten years centering on 1870.
86. Physicists still struggled with the question of whether in a light wave the oscillation of the ether is in the plane of polarization or normal to it. See Jochmann, 136: 561–88.
87. The Berlin secondary school teacher K. H. Schellbach showed his experiments to Quincke, Poggendorff, and Magnus; the latter evidently made them known to the young physicists working with him (139: 670–72).

hoff derived equations relating electric and magnetic quantities that were the same as Maxwell's electromagnetic field equations of 1865; but he was studying iron under induced magnetism and not the free ether, and he was guided by Poisson's magnetic theory and did no more than acknowledge Maxwell's derivation.[88]

It was not in the *Annalen* but in the *Journal* where German physicists in 1869–71 examined the propagation of electric action in a medium, the choice of journal reflecting the mathematical nature of the topic at this time. Here Kirchhoff developed an analogy between pressure forces acting on rings in a fluid and electrodynamic forces acting on closed currents.[89] Here, too, Boltzmann examined further Kirchhoff's analogy.[90] Helmholtz, in the most influential of the studies, examined the medium between bodies and with it Faraday and Maxwell's view of mediated electric action. Helmholtz recognized that Maxwell's theory of a polarized medium gave the same equations as distance-action theories of Poisson's type;[91] unlike Kirchhoff, he did not stop with this formal recognition but drew its physical consequences. He called attention to Maxwell's "striking result" that electrical disturbances in dielectrics travel as transverse waves with the velocity of light and to the "extraordinary significance that this result could have for the further development of physics."[92]

88. Kirchhoff, supplementary vol. 5: 1–15.
89. Kirchhoff, *Journal* 71: 263–73. Kirchhoff studied two rings of any form with infinitely small, circular cross sections placed in an infinite, frictionless, incompressible fluid. By using familiar assumptions about the motion of the fluid, he showed that the kinetic energy of the fluid has the same form as the potential of Ampère's law for the electrodynamic interaction of two electrical currents. The rings exert "apparent," or pressure, forces on one another that are the same as the forces that would act if electric currents were to flow through the rings.
90. Boltzmann pointed out that Kirchhoff's conclusion about the formal identity of apparent fluid forces and Ampèrian forces is not generally valid, and he devoted considerable discussion to the electrical side of the analogy (*Journal* 73: 111–34).
91. Helmholtz, *Journal* 72: 57–129, reprinted in Hermann von Helmholtz, *Wissenschaftliche Abhandlungen*, 3 vols. (Leipzig, 1882–95), 1: 545–628, discussion on 556–58 (hereafter cited as *Wiss. Abh.*).
92. Helmholtz, *Wiss. Abh.*, 557.

14

Helmholtz, Kirchhoff, and Physics at Berlin University

Helmholtz's Move to Prussia as a Physicist

In the 1870s and 1880s, the most influential physicist for the development of theoretical physics in Germany was Helmholtz. His influence derived from his prominent position as director of the Berlin physics institute, his research and teaching there, and his stature within German physics, which enabled him to make his view of theoretical physics bear on physics appointments. It was highly significant for theoretical physics that Germany's most famous man of science, in the middle of his career, turned primarily to teaching and research in theoretical physics and to building theoretical physics in Germany.

Helmholtz's name first appeared on a list of candidates for a Prussian chair of physics at Bonn University in 1868. This came about in part because of his reputation as a scientist who had united different parts of science and not only because of his reputation as a mathematical physicist. The Bonn mathematician Rudolf Lipschitz—who took the initiative in the effort to secure Helmholtz for Bonn at Julius Plücker's death—emphasized the "characteristic quality" of Helmholtz's "great mind." The faculty, he insisted, should make clear to the Prussian minister of culture, who would make the appointment, that "certain fundamental ideas of [Helmholtz's] researches belong to a region that is common to the sciences of inorganic and of organic nature." This had been true from the beginning of his work in science, from his paper on the conservation of force in 1847. Lipschitz's characterization of Helmholtz did not impress the scientists at Bonn, who preferred to describe Helmholtz more as a specialist who happened to have two specialties. They recommended him for his "classical investigations in the area of physical physiology" and for his outstanding achievements in "pure physics, especially in the mathematical direction," a formulation which displeased Lipschitz because he thought it was based on a wrong understanding of Helmholtz's importance.[1]

1. Rudolf Lipschitz, "Entwurf eines Votums der mathematisch-naturwissenschaftlichen Section," n.d. [1868]; and draft by the natural science faculty, also n.d. [1868]; and Lipschitz, "Separatvotum," 13 July 1868; Plücker Personalakte, Bonn UA.

Although Helmholtz had been teaching physiology for two decades, physics had always been his first love, he told the Bonn curator, and he was glad for the chance to bring his job into line with his main interest. His achievements in physiology were all based on physics, he said, but he could no longer pass on to his students what was best in his work there, since his physiology students no longer knew enough mathematics and physics to follow this part of his work. In physics, however, Helmholtz thought that he could still accomplish something, for whereas physiology in Germany was flourishing, physics was not. The few great physicists in Germany were near the end of their careers, he said, and no new generation had arisen to take their place. Physics, the "true basis of all proper science," was no longer advancing as it might, and that was true especially of "mathematical physics." If through his teaching, students were to take up the work that needed to be done in physics, he would have achieved something more important than anything he could hope for in physiology. To realize that goal, if he were to come to Bonn he would want to teach not only experimental but also "mathematical physics."[2]

Helmholtz's move to physics seemed certain, since Prussia stood to gain a "mark of glory" by attracting Helmholtz to one of its universities.[3] But the Prussian ministry of culture dragged out the negotiations, and in the end it failed to make Helmholtz a sufficiently attractive offer. Helmholtz remained at Heidelberg University as professor of physiology—for a time.[4]

In 1870 Prussia turned again to Helmholtz in its search for an illustrious physicist. Helmholtz's friend and then rector of Berlin University Emil du Bois-Reymond informed him of Magnus's death and of his own hope that Helmholtz would succeed Magnus as director of the Berlin physics institute.[5] Helmholtz answered that while he had been waiting for a decision on the appointment to Bonn, he had "thrown" himself into physical and mathematical studies again; as a consequence he now knew even better than he had at the time of the Bonn call that he had become indifferent to physiology and now was interested only in "mathematical physics." But Helmholtz did not think of himself as a trained mathematical physicist who had proven himself in the subject. Kirchhoff was such a physicist, and Helmholtz thought that Berlin needed him and that Kirchhoff would be the likely choice. Helmholtz would be satisfied to succeed Kirchhoff at Heidelberg University, which would mean simply moving from one chair to another. He urged du Bois-Reymond to do all he could to secure Kirchhoff's appointment in Berlin and, above all, to keep the Prussian min-

2. Helmholtz to Bonn U. Curator Beseler, n.d. [summer 1868], quoted in Leo Koenigsberger, *Hermann von Helmholtz*, 3 vols. (Braunschweig: F. Vieweg, 1902–3), 2: 115–16, on 116. In a letter to the physiologist Carl Ludwig, Helmholtz compared physics and physiology further: he could lecture on "all parts" of physics with "completely independent judgment," whereas the manipulations and methods of physiology had diverged so that he, or any other one person, could no longer keep up with them all. Helmholtz to Ludwig, 27 Jan. 1869, quoted in Koenigsberger, *Helmholtz* 2: 118–19, on 119.

3. Bonn University Curator Beseler to Prussian Minister of Culture Mühler, 4 Aug. 1868, quoted in Koenigsberger, *Helmholtz* 2: 116.

4. By 28 Dec. 1868, Helmholtz had been offered the Bonn physics chair at a salary of 3600 thaler; by 3 Jan. 1869, he had declined the offer. Helmholtz Personalakte, 1858/1907, Bad. GLA, 76/9939.

5. Emil du Bois-Reymond to Helmholtz, 4 Apr. 1870, quoted in Koenigsberger, *Helmholtz* 2: 178.

istry officials away from Kirchhoff, for otherwise they would certainly fail to persuade Kirchhoff to move to Berlin.[6]

The Berlin philosophical faculty regarded Helmholtz as the more "productive" and "more gifted and universal" of the two, but they preferred Kirchhoff, as Helmholtz had anticipated. Kirchhoff was a trained, experienced physicist; besides he was better at supervising beginners' exercises than Helmholtz, and his lectures were a model of "lucidity and finish"; in general, he had the greater "love of teaching."[7] Kirchhoff was offered the job. He turned it down, mainly, as he explained to du Bois-Reymond, because he was unsure of his health, but he also liked the working conditions at Heidelberg, which the Baden ministry of the interior further improved to keep him there.[8] Helmholtz suspected that Kirchhoff's decision also expressed his characteristic reluctance to subject himself to change, particularly in his work habits.[9] For his part, Kirchhoff was comforted by the knowledge that Helmholtz would be offered the job in Berlin next.

Helmholtz appreciated the immense task awaiting the physicist who would take on the Berlin professorship, and he felt ambivalent about assuming it. He was drawn to the job because he could be most effective as a teacher in Berlin, where he would find a large number of students to whom he could pass on the best that he was capable of giving. But he hesitated when he considered that his research would practically be stopped by the time-consuming chores of organizing a physics laboratory and by the many extraneous tasks expected of a Berlin professor.[10] Nevertheless, du Bois-Reymond was able to persuade Helmholtz to state the conditions under which he would take the Berlin physics professorship. This time the Prussian ministry of culture was ready to assent to them.[11]

Helmholtz's conditions sum up the collective understanding of the needs of university physics that the physicists of the previous decades had gradually reached. First, aside from an ample salary, Helmholtz wanted the ministry's assurance that a new physics institute would be built for him, equipped with everything needed for teaching, for his own researches, and for practical student exercises.[12] (To make sure that the building would be new, he considered only empty building lots; he was afraid that if a lot with a building were chosen, the old structure might not be removed but surreptitiously incorporated into the new.)[13] Second, he wanted the

6. Helmholtz to du Bois-Reymond, 7 Apr. and 17 May 1870, STPK, Darmst. Coll. F 1 a 1847. Helmholtz and du Bois-Reymond were concerned that because of penny pinching, Prussia might appoint a physicist of secondary importance, someone like Quincke.

7. Berlin Philosophical Faculty's recommendation of Kirchhoff to the Prussian Minister of Culture, n.d. [1870], quoted in Koenigsberger, *Helmholtz* 2: 179–80.

8. Kirchhoff to du Bois-Reymond, 9 June 1870, STPK, Darmst. Coll. 1924.55. To remain at Heidelberg, Kirchhoff asked for an increase in his institute budget to hire an assistant and an increase in his salary. The Baden Ministry of the Interior regarded both requests as "modest" in light of Kirchhoff's "significance" as a scholar of the first rank. Baden Ministry of the Interior, 10 June 1870, Kirchhoff Personalakte, Bad. GLA, 76/9961.

9. Helmholtz to du Bois-Reymond, 17 May 1870.

10. Helmholtz to du Bois-Reymond, 17 May 1870.

11. Helmholtz to du Bois-Reymond, 12 June, 25 June, and 3 July 1870, STPK, Darmst. Coll. F 1 a 1847.

12. Helmholtz to du Bois-Reymond, 12 June 1870.

13. Helmholtz to du Bois-Reymond, 14 Feb. 1871, STPK, Darmst. Coll. F 1 a 1847.

ministry to agree that he alone would be in charge of this institute and the instrument collection, and that it would be his decision to what extent and under what conditions the other physics teachers would be allowed to use it. He promised at the same time to be considerate of Dove, who was by then the senior physicist at Berlin. Part of the second condition was that the institute auditorium be reserved for his exclusive use, so that he could set up complicated instrumental arrangements and leave them in place. His third condition was that the institute contain the director's living quarters. In addition, Helmholtz wanted to be sure that he would be able to start work immediately, before the new institute was complete, and so he asked for rented rooms near the university for his own physical research and for the work of some students and the necessary assistants.[14]

The building of the new institute had to be deferred because of the Franco-Prussian War, which began during Helmholtz's negotiations with the Prussian government. Helmholtz was given reason to think that once the state's finances were back to normal, the building would proceed, and in December 1870 he accepted the Berlin job.[15] Something unheard-of had happened, du Bois-Reymond observed: a professor of medicine and physiology had been named to the most important physics chair in Germany.[16] Helmholtz was fifty, and from now to the end of his life his principal field was physics.

In Berlin Helmholtz soon had occasion to publicize his understanding of the proper training and practice of physics. Addressing the Prussian Academy of Sciences on his predecessor Magnus, Helmholtz explained that a good experimentalist needs thorough theoretical training and that a good theorist needs wide practical training.[17] He impressed his view of the complementary nature of experimental and theoretical approaches on his students in Berlin. Wilhelm Wien, one of these students, regarded it as a particular service of Helmholtz to strive to bring theoretical and experimental physics together into "one great science."[18] Helmholtz impressed his view, too, on faculties and ministries. Regularly consulted on appointments, Helmholtz recommended physicists for their good theoretical understanding. To be effective lecturers, he believed, even in experimental subjects, physicists had to be able to formulate clearly the concepts of theoretical physics, and to use effectively the experimental facilities of physics institutes, physicists had to command mathematical as well as experimental methods of physics. For the *Annalen der Physik*, which was under Wiedemann's editorship from 1877 on, Helmholtz refereed the

14. Helmholtz to du Bois-Reymond, 12 June 1870.

15. Helmholtz to du Bois-Reymond, 17 Oct. 1870, STPK, Darmst. Coll. F 1 a 1847; Koenigsberger, *Helmholtz* 2: 186.

16. Quoted in Koenigsberger, *Helmholtz* 2: 187.

17. Hermann von Helmholtz, "Gustav Magnus. In Memoriam," in *Popular Lectures on Scientific Subjects,* trans. E. Atkinson (London, 1881), 1–25, on 19. Helmholtz often returned to the need to avoid any divorce between theoretical and experimental physics. For example, in his foreword in 1874 to the German translation of John Tyndall's *Fragments of Science,* reprinted as "The Endeavor to Popularize Science," in *Selected Writings of Hermann von Helmholtz,* ed. R. Kahl (Middletown, Conn.: Wesleyan University Press, 1971), 330–39, on 337.

18. Wilhelm Wien, "Helmholtz als Physiker," *Naturwiss.* 9 (1921): 694–99, on 697.

papers on theoretical physics.[19] His own experimental researches and his work on the methods of precision measurements were always tied to theory.

Helmholtz's Electrodynamics

Beginning around 1870 at Heidelberg and continuing through his first several years at Berlin, Helmholtz carried out intensive researches on electrodynamics.[20] These, like his earlier researches on the conservation of force, arose from his physiological work. To interpret his work on the propagation of nervous impulses, he needed to understand the motion of electric currents in extended conductors and, in particular, the induced currents of incomplete, or open, circuits. The problem was that the theories of Franz Neumann, Weber, and Maxwell all accounted for most electrical experiments involving closed circuits, but that they had not yet been tested for open circuits for which, as Helmholtz showed, they predicted different results. By casting the laws of all three theories into a single mathematical expression, distinguished from one another by a numerical constant, Helmholtz showed how an experimental decision between them might be made.[21]

In his first paper, in 1870, on the foundations of electrodynamics, Helmholtz explained the immediate physical origins and the ultimate objectives of his theory. To understand the flow of electricity in the interior of a conductor, he had gone to Kirchhoff's formulation of the equations of motion of electric currents of variable intensity for three-dimensional conductors. Applying the equations to his problem resulted in physically inadmissible consequences. That discovery in turn led him to Weber's theory, and that theory, he convinced himself, had to be wrong; for it led to an unstable equilibrium of static electricity inside conductors and in general to a conflict with the energy principle. For small currents, Carl Neumann's theory led to Weber's, so it was vulnerable to the same criticism. Franz Neumann's induction law escaped this criticism, and Maxwell's induction law still needed to be investigated.

Helmholtz set up the most general form of the law of induction for two current elements that reduces to Franz Neumann's law in the case of closed currents. Neumann's potential between elements of length Ds and $D\sigma$ of two linear conductors s

19. Koenigsberger, *Helmholtz* 2: 233–34.

20. Koenigsberger, *Helmholtz* 2: 170.

21. Early in 1870, Helmholtz published a first announcement, "Ueber die Gesetze der inconstanten elektrischen Ströme in körperlich ausgedehnten Leitern," in *Verh. naturhist.-med. Vereins zu Heidelberg* 5: 84–89, reprinted in Helmholtz, *Wiss. Abh.* 1: 537–44, of his theory of electrodynamics; later that year he followed it by the first part of the theory, "Ueber die Bewegungsgleichungen der Elektricität für ruhende leitende Körper," *Journ. f. d. reine u. angewandte Math.* 72 (1870): 57–129, reprinted in his *Wiss. Abh.* 1: 545–628, to which our discussion refers. Helmholtz's theory of electrodynamics is discussed in A. E. Woodruff, "The Contributions of Hermann von Helmholtz to Electrodynamics," *Isis* 59 (1968): 300–311, especially on 300, 302; also in M. Norton Wise, "German Concepts of Force, Energy, and the Electromagnetic Ether: 1845–1880," in *Conceptions of Ether: Studies in the History of Ether Theories 1740–1900*, ed. G. N. Cantor and M. J. S. Hodge (Cambridge: Cambridge University Press, 1981), 269–307, on 295–301.

and σ, separated by distance r and carrying currents of intensities i and j, respectively, is:

$$- A^2 ij \, \frac{\cos (Ds, D\sigma)}{r} \, DsD\sigma,$$

where A in the electrostatic system of units is a constant numerically equal to the reciprocal of the velocity of light.[22] To form the complete potential, Helmholtz included another term that depends only on the end points of the two elements, acknowledging the unequal distribution of electricity and the forces acting on the ends of current elements:

$$-\frac{1}{2}A^2\frac{ij}{r} \, [(1 + k) \cos (Ds, D\sigma) + (1 - k) \cos (r, Ds) \cos (r, D\sigma)] \, DsD\sigma,$$

where k is a provisional constant.[23] The complete potential reduces mathematically to Franz Neumann's theory for $k = 1$, Maxwell's for $k = 0$, and Weber's and Carl Neumann's for $k = -1$. Helmholtz extended the potential for linear elements to the potential for three-dimensional currents, deriving the equation of motion of electricity for this general case.

In answer to critics in 1874, Helmholtz explained why he formulated his theory in terms of the potential law: "The potential law uses one and the same relatively simple mathematical expression to cover the whole experimentally known area of electrodynamics, ponderomotive and electromotive actions, and in the area of ponderomotive actions it produces the same great simplification and clarity that the introduction of the concept of the potential has brought to the theory of electrostatics and of magnetism. I myself can testify to this, since for thirty years I have never applied another fundamental principle than the potential law and have never needed any other to find my way in fairly labyrinthine problems of electrodynamics and at times even over previously untrod ground."[24] Because of its simplicity and clarity, the potential law had the greatest probability of being the correct law. But like the other laws, it too needed to be tested by experiment, and that was the main purpose behind Helmholtz's series of investigations, of which the 1870 paper was only the beginning installment.[25]

The first result that Helmholtz drew from his electrodynamic theory was that for negative k, which corresponds to Weber's induction law, the theory allows for a continuous increase in the motion of electricity leading to infinite velocities and infinite electrical densities. That is, Weber's law conflicts with the energy principle,

22. Helmholtz, "Bewegungsgleichungen," 562–63.
23. Helmholtz, "Bewegungsgleichungen," 567.
24. Helmholtz, "Kritisches zur Elektrodynamik," Ann. 153 (1874): 545–56, reprinted in Wiss. Abh. 1: 763–73, on 772.
25. Helmholtz, "Versuche über die im ungeschlossenen Kreise durch Bewegung inducirten elektromotorischen Kräfte," Ann. 158 (1875): 87–105, reprinted in Wiss. Abh. 1: 774–90, on 787.

a result Helmholtz had indicated earlier. Because he wanted to stay as close as possible to the "basis of facts," Helmholtz followed his theoretical discussion by an experiment, although one carried out in thought only. His experimental example was the radial motion of electricity in a conducting sphere induced by the expansion and contraction of a concentric, charged, spherical shell; for negative k, he showed that this motion is unstable. The theoretical considerations and the experimental example together supported his conclusion that a negative k is inadmissible.[26] Helmholtz attributed the difficulty of Weber's law to its dependence on motion as well as on distance. Helmholtz's own law avoided the dependence on motion.

To settle the question of which of the contending laws is correct, Helmholtz sought experiments in which k has a noticeable effect and which allow the value of k to be determined. Since for closed circuits, k disappears from the equations, the decision had to come from open circuits. Helmholtz turned to the equations of electricity spreading from a point in an infinite conductor; if the resistance is negligible, the electricity can move in longitudinal waves with a velocity of $c/\sqrt{2} \cdot 1/\sqrt{k}$, where c is the constant entering Weber's law. In Maxwell's theory, for which $k = 0$, the longitudinal waves have infinite velocity, whereas in Franz Neumann's theory, for which $k = 1$, they have the velocity $c/\sqrt{2}$, which, Helmholtz remarked, Kirchhoff had noted to be extraordinarily close to the velocity of light. Helmholtz then investigated theoretically the practical case of a long wire. By considering the course of electric waves in an infinite cylinder, Helmholtz concluded that in electrical experiments involving the velocity of longitudinal waves dependent on k, physicists would not usually need to take these velocities into account; nor would they be able to determine the value of the velocity of longitudinal waves unless they used means to measure unusually delicate time differences.[27]

Helmholtz concluded his paper of 1870 by examining Maxwell's theory and evaluating its promise in comparison with that of the other theories. From Faraday's and others' work, he knew that most physical media can be magnetized and that a state of dielectric polarization similar to magnetic polarization occurs in electric insulators, and he saw reason to assume that the light ether is magnetizable and that it is a dielectric in Faraday's sense.[28] Maxwell's theory gives the "remarkable result," he said, that electric disturbances in dielectric media propagate as transverse waves, the velocity of which in air is the velocity of light. Because of the "great significance" that this could have for the whole "further development of physics," Helmholtz investigated his general induction law for the case where magnetizable and dielectrically polarizable media are present.[29] The "remarkable analogy between the motions of electricity in a dielectric and those of the luminiferous ether," he concluded, "does not depend on the special form of Maxwell's hypotheses but arises also in an essentially similar manner if we retain the older view of electric action at a

26. Helmholtz noted that although Kirchhoff derived his system of equations from Weber's induction law, his results were not affected by Helmholtz's conclusion about this law because of the nature of Kirchhoff's applications of the equations. Helmholtz, "Bewegungsgleichungen," 551.

27. Helmholtz, "Bewegungsgleichungen," 554–56, 603–11.

28. Helmholtz, "Bewegungsgleichungen," 556–57, 612.

29. Helmholtz, "Bewegungsgleichungen," 557.

distance." The admission of electric action at a distance with finite velocity appears possible "without essential changes in the foundations of accepted electrodynamic theory."[30]

The problem of devising experiments for testing the several electrodynamic theories was formidable, and Helmholtz devoted much inconclusive effort to it after 1870. Even before leaving Heidelberg for Berlin, he had carried out the experimental investigation with which he introduced himself as a newly elected ordinary member to the Prussian Academy of Sciences in May 1871. In this work he examined the velocity of propagation of electrodynamic action, employing precise measurements for the purpose.[31] Then four years passed before Helmholtz published another experimental paper on electrodynamics. The work was demanding, and he at first lacked young experimentalists in the laboratory to take it up with him. He was also handicapped by the makeshift facilities in which he had to work until 1878, despite the ministry's assurances of a new institute. And as he had foreseen, he had little time for research owing to the many problems he faced related to construction of the institute, beginning in the fall of 1872 with the preparation of the plans.[32] "Already the Berlin bustle makes me very tired," he wrote in early 1873, "so that after the semester is over I usually wish more than anything not to have to see another person and to be able to collect my thoughts in a quiet place."[33]

Helmholtz regularly concerned himself with experimental researches during the early 1870s, even if he did not always publish them. For example, he contemplated carrying out a series of experiments early in 1873, but the preliminary experimental studies for the series took up much time, and then he found that they were not feasible after all.[34] "In the last months I have been making experiments on induction in open conducting arcs," he wrote to a colleague in early 1875. In these experiments he had observed phenomena, he said, that "I could explain to myself only by the influence of the surrounding insulators," and this led him to investigate theoretically the behavior of the insulating medium surrounding a metallic conductor.[35] He concluded that the potential theory contradicts experimental facts if it takes into account only the electrical motions in conductors and their distance actions. The law in which he had most confidence proved a disappointment. But it could still be supplemented to correspond to the experimental results if he assumed with Faraday and Maxwell that in insulators there can occur electric motions with electrodynamic actions, which are responsible for the dielectric polarization of the insulators. In the limiting case in which the potential law corresponds to Maxwell's theory, no open

30. Helmholtz, "Bewegungsgleichungen," 558, 628.

31. Hermann von Helmholtz, "Ueber die Fortpflanzungsgeschwindigkeit der elektrodynamischen Wirkungen," *Sitzungsber. preuss. Akad.*, 1871, 292–98, reprinted in *Wiss. Abh.* 1: 629–35.

32. Heinrich Rubens, "Das physikalische Institut," in Max Lenz, *Geschichte der königlichen Friedrich-Wilhelms-Universität zu Berlin*, 4 vols. in 5 (Halle a. d. S.: Buchhandlung des Waisenhauses, 1910–18) 3: 278–96, on 284.

33. Helmholtz to Knapp, 5 Jan. 1873, quoted in Koenigsberger, *Helmholtz* 2: 219.

34. Hermann von Helmholtz, "Vergleich des Ampère'schen und Neumann'schen Gesetzes für die elektrodynamischen Kräfte," *Sitzungsber. preuss. Akad.*, 1873, 91–104, reprinted in *Wiss. Abh.* 1: 688–701, on 701.

35. Helmholtz to a "colleague," 8 Feb. 1875, ETHB, Hs 87-402.

electric currents can exist at all, as Maxwell had observed; for every electric motion in a conductor that leads to an accumulation of electricity at its surface would continue into the surrounding insulator as an equivalent motion consisting of dielectric polarization.[36] Although Helmholtz published no experimental work on electrodynamics between 1871 and 1875, he regularly linked his published theoretical discussions in these years to proposals for experiments, describing in detail the experimental arrangements he envisioned and the results he expected from them.

Electrodynamic Researches in Helmholtz's Laboratory

While Magnus's chair awaited a successor, Dove had temporarily taken on the direction of the Berlin physics institute. Dove insisted that the physics laboratory, which was still in Magnus's house, be united with the physics apparatus at the university. Under pressure, the university managed to find a few rooms that could be used as an institute, which was enhanced by Magnus's private collection of instruments and by his large library, which he had left to the university. Six more rooms were added when Helmholtz arrived in 1871, and at times as many as fifteen students and assistants were doing research in the still crowded space.[37] A visiting physicist, Arthur Schuster, was impressed by the working relations between Helmholtz and his students, who were making do in the old institute, regarding it as the "ideal of what a teaching laboratory should be."[38]

Another early visitor in Helmholtz's institute was Boltzmann, who did experiments there related to Helmholtz's question about the influence of the insulators on inducing actions.[39] In his 1870 paper, Helmholtz had argued that the known influence of static electric forces on insulators, causing dielectric polarization in them, pointed to the likely influence of these insulators on the propagation of inducing actions arising from electric currents in conductors.[40] Boltzmann took up the problem of dielectric polarization from the side of static electric forces, investigating the "change in the capacity of condensers owing to insulating intermediate layers." It was a "rather extensive and time-consuming investigation," Boltzmann said, which he had undertaken "more to get into as close touch as possible with Helmholtz"— and also because he wanted to prove that a mathematical physicist could do experiments—"than because I am so interested in the subject itself."[41] Boltzmann became interested enough in the subject, however, to continue to work on it after his return

36. Helmholtz, "Versuche," 787–88.

37. Rubens, "Das physikalische Institut," 283–84.

38. Arthur Schuster, *The Progress of Physics During 33 Years (1875–1908)* (Cambridge: Cambridge University Press, 1911), 16–18, quotation on 17. Later when Schuster returned to Helmholtz's institute, no longer "a small number of badly furnished and crowded rooms" but a great many private rooms in the great new institute building, he missed the "soul and scientific spirit of the old place" (p. 17).

39. Boltzmann described his studies in Berlin in a letter to the "Director" (presumably Stefan), 2 Feb. 1872, STPK, Darmst. Coll. 30.7.

40. Helmholtz, "Bewegungsgleichungen," 612.

41. Boltzmann to the "Director," 2 Feb. 1872.

to Austria. He wanted "to decide the question of whether the change of the capacity of the condensers owing to inserted layers of different substances really derives from an electrification of the smallest parts of those substances or merely from the fact that electricity acts in a different way through them than through air." He found that the former was the case and reported the details to Helmholtz. In writing up the results of his measurements, he also found that the values he had obtained in Berlin for the dielectric constants of several substances were "all approximately equal to the square of the index of refraction"; this was just as Maxwell's theory required, and Boltzmann wrote to Helmholtz that "I must see in my experiments a confirmation of Maxwell's theory."[42]

Boltzmann was fascinated by the precision of Helmholtz's instruments, especially Thomson's new galvanometer and electrometer. He wrote to Helmholtz from Graz for more details of them so that in his papers he could give an exact description of one of the instruments and could build the other for himself.[43] Helmholtz responded to Boltzmann's accounts of his researches with suggestions for experiments that were to prove "very fruitful." Helmholtz, Boltzmann once confided, was the only person with whom he could discuss certain scientific problems.[44]

The main difficulty with the experimental testing of the different electrodynamic theories for open currents was that the electrodynamic actions to be observed occurred in time spans too brief to be measured. To circumvent the difficulty, Helmholtz devised an experiment on the interaction between a current element (of a closed current) and a current end (of an open current).[45] His student Nikolaj Schiller, who did the experiment, correspondingly used a suspended, magnetized, closed steel ring, wound with wire and suitably boxed, and a metal point through which electricity flowed out into the surrounding air. He tested the action of the metallic point on the magnetized ring by moving it toward one of the vertical sides of the ring. According to Helmholtz's potential law, this motion should have deflected the ring, but it did not. In Moscow soon after, Schiller repeated the work with better equipment and still got only a negative result. He concluded, as Helmholtz did, that either there are no electromagnetic forces at the ends of open currents, so that inferences from the potential law are incorrect, or there is no open current present in the experiment. In the latter case, the experimental apparatus would have been subject not only to the action of the current in the metallic parts of the conductor but also to the time-varying dielectric polarization of the surrounding media

42. Boltzmann to Helmholtz, 1 Nov. 1872, full letter published in Gisela Buchheim, "Zur Geschichte der Elektrodynamik: Briefe Ludwig Boltzmanns an Hermann von Helmholtz," NTM 5 (1968): 125–31, on 126–27.

43. Boltzmann to Helmholtz, 1 and 20 Nov. 1872, and 26 Feb. 1874, all in Buchheim, "Briefe Boltzmanns an Helmholtz," 126–29.

44. Boltzmann to Helmholtz, 26 Feb. 1874. Also Ludwig Boltzmann, "Experimentaluntersuchung über die elektrostatische Fernwirkung dielektrischer Körper," Sitzungsber. Wiener Akad. 68 (1873): 81–155, reprinted in Ludwig Boltzmann, Wissenschaftliche Abhandlungen, ed. Fritz Hasenöhrl, 3 vols. (Leipzig: J. A. Barth, 1909), 1: 472–536, on 480 (hereafter cited as Wiss. Abh.). Engelbert Broda, Ludwig Boltzmann. Mensch, Physiker, Philosoph (Vienna: F. Deuticke, 1955), 27.

45. Helmholtz, "Versuche," 777–80.

due to convection current.[46] Schiller's experiment did not give the desired positive proof that convection currents have the same electrodynamic action as the currents in conductors.[47]

In 1876 Helmholtz reported to the Prussian Academy that the American physicist Henry Augustus Rowland had come to his institute with a completely worked out plan for experiments that could give this "positive proof." In his experiments, Rowland showed that the "motion of electrified ponderable bodies"—the rotation of a charged gold-plated ebonite disk—produces an electromagnetic force that can be detected as oscillations of a magnetic needle. As far as the meaning of these experiments for the theory of electrodynamics is concerned, Helmholtz added, they did not decide between Maxwell's, Weber's, and his own potential theory. But like his own experiments, they showed that the potential law had to take into account the dielectric polarization of insulators.[48]

In 1878 the old cramped institute—with its scientifically productive atmosphere, according to Schuster, but not according to Heinrich Kayser, who thought it "miserable" and his days there lonely—was vacated at last. Kayser, as Helmholtz's assistant, spent his days taking books and apparatus in handcarts to the new institute, setting up instruments in cases, and cataloging them and the books in the institute library.[49] The new institute had two lecture halls, a large one accommodating over two hundred auditors and a small one accommodating sixty. One large laboratory was reserved for beginners' exercises, another for magnetic and galvanic work. In addition, there were twenty-three small laboratories or "workrooms" for students, some of which were designed for specific purposes. For example, nine rooms were reserved for optical work, four for precision measurements, and one for mechanical and acoustical exercises. The institute had a library consisting of two large halls, one of which was used as the meeting room of the Berlin Physical Society. It also contained the usual storage rooms for instruments, the technical plant, and so forth. In the new institute Helmholtz had two, later three, assistants, one to help him prepare the experimental lectures and the other two to assist in the stu-

46. Nikolaj Schiller, "Elektromagnetische Eigenschaften ungeschlossener elektrischer Ströme," *Ann.* 159 (1876): 456–73, 537–53, on 457–59. Helmholtz also gave an account of Schiller's research in "Versuche," 780–81. Schiller worked on several aspects of Helmholtz's electrodynamics, some of which he listed in his paper "Einige experimentelle Untersuchungen über elektrische Schwingungen," *Ann.* 152 (1874): 535–65. He investigated the "theoretical laws of alternating currents," using Helmholtz's method of observing "alternating electrostatic charges at the ends of an alternating current" (p. 535). He took up Maxwell's prediction of a relation between the dielectric constant and the index of refraction, determining the latter for several insulators (p. 559). And he tried to discover if the presence of an insulating substance next to an induction spiral had any effect on the electric oscillations in the spiral (p. 563); this experiment Helmholtz described to a colleague, 8 Feb. 1875, ETHB, Hs 87-402.

47. Hermann von Helmholtz, "Bericht betreffend Versuche über die elektromagnetische Wirkung elektrischer Convection, ausgeführt von Hrn. Henry A. Rowland," *Ann.* 158 (1875): 487–93, reprinted in *Wiss. Abh.* 1: 791–97, on 791.

48. Helmholtz, "Bericht," 796–97.

49. Heinrich Kayser, "Erinnerungen aus meinem Leben," 1936, 93–94, 96. Typescript in the American Philosophical Society Library, Philadelphia.

dents' laboratory practices. The assistants had their own workrooms and living quarters at the institute, as did other employees of the institute such as the mechanic and the janitors. Helmholtz's quarters, including his apartment, were in an adjoining part of the institute.[50] This spacious and well-equipped institute had advantages for independent student researchers such as Heinrich Hertz, the most important of Helmholtz's students to work on electrodynamics.

Attracted to Berlin by Helmholtz's and Kirchhoff's fame,[51] Hertz arrived just after the physics institute had moved to its new building. Noticing on the bulletin board an announcement of a prize question about the inertia of moving electricity, Hertz judged that it was "more or less" in his specialty, and he signed up for Helmholtz's laboratory practice course to work on it. Helmholtz encouraged him by suggesting the relevant literature, and a week later Hertz went back to Helmholtz ready to start work. Helmholtz took him to see his assistant, told him the best way to start and which instruments he would need, and assigned him a small private workroom, allowing him to come and go as he pleased. Hertz found everything in the new institute "beautifully furnished." His galvanometer stood securely on an iron console built into the wall; his telescope could be adjusted by screws in all directions, "more comfortable, in a way, than placing it on books." His days in the institute went as follows: "I hear an interesting lecture every morning, then I go to the laboratory where, with a short break, I stay until four o'clock, afterwards I'll work at home or in the reading room"; "Helmholtz comes in for a few minutes every day, looks at the matter, and is very friendly." As the iron console in his room was not as stable as Hertz had first thought, he moved to another room where he could use a stone post instead. The institute furnished Hertz with the things he could not provide himself, but he was expected to provide his own means as far as possible. He spent much of his time whittling cork, filing wires, and in other "not very instructive work"; he constructed apparatus for himself and used his own instruments where he could. About the experiments themselves, Helmholtz warned him at the start about the main difficulties he would encounter; within three weeks, Hertz had surmounted them. The aim of the experiments was to show that certain phenomena do not occur: "On the whole that is less fun" than if something is expected to occur, he wrote to his parents, but "such is the nature of the matter." "As far as I can carry the precision, the present theory is completely confirmed."[52]

Hertz's research was to decide if electricity has inertia by measuring the secondary current when the primary is started or stopped; his findings favored Helmholtz's

50. Albert Guttstadt, ed., *Die naturwissenschaftlichen und medicinischen Staatsanstalten Berlins* (Berlin, 1886), 140–48, on 144–47. Before and after Helmholtz moved into the new institute, he had many more students working in his laboratory than we have mentioned here; they included many working on research in areas other than electrodynamics. Guttstadt, 141–43, provides a fuller, though still incomplete, list of them.

51. Philipp Lenard, "Einleitung," in Heinrich Hertz, *Gesammelte Werke*, vol. 1, *Schriften vermischten Inhalts*, ed. Philipp Lenard (Leipzig, 1895), ix–xxix, on xii (hereafter cited as *Ges. Werke*).

52. Lenard, "Einleitung," xii–xvi. Lenard quotes Hertz's lively letters to his family from which we have taken our quotations. Hertz gives a briefer account of his work in Berlin in his "Vita," 14 Mar. 1883, LA Schleswig-Holstein, Abt. 47.7 Nr. 8.

electrodynamics over Weber's, which maintained that electricity, like ponderable matter, has inertia.[53] With this first experimental research, which Hertz finished in a third of the time allotted to it, he won the faculty prize. The work became his first publication in 1880, the year he received his doctorate and became Helmholtz's assistant. Hertz had declined Helmholtz's invitation to try another experimental problem in electricity, which Helmholtz proposed through the Prussian Academy of Sciences with Hertz expressly in mind. This problem bore on the correctness of Maxwell's theory and Helmholtz's interpretation of it, specifically on the influence of dielectrics on electrodynamic actions. The experimental difficulties seemed daunting to Hertz at the time.

Helmholtz offered his students and visiting researchers more than his experimental ideas and the facilities of his institute: he gave them the assurance that they were working on significant physical problems. Boltzmann, Schiller, and Rowland all left Germany still thinking about Helmholtz's ideas on electrodynamics. Hertz, whose initial approach was through Helmholtz's formulation of Maxwell's theory, went on to make the decisive experimental test in favor of Maxwell's theory in 1888.

Kirchhoff at Berlin

On coming to Berlin, Helmholtz continued to lecture on physiological optics and acoustics as well as give the required big experimental physics lectures, the course for practical exercises in the laboratory (for which he made the laboratory available five hours a day during the week and three hours on Saturday), and the occasional public lectures on general topics. In addition, he lectured three or four times a week on theoretical physics each semester, expecting his auditors to know mathematics at least through the differential and integral calculus.[54] It was in keeping with his emphasis on theoretical physics that he soon decided that its adequate instruction at Berlin demanded an ordinary professor for the subject, a conclusion the Prussian government accepted.[55]

Once again they looked to Kirchhoff. He had turned down Berlin in 1870, Würzburg in 1872, when Baden gave him another raise and another title,[56] and Berlin again in 1874, when he was invited there to direct the new state solar observatory.[57] In 1874 he received yet another call to Berlin, this from the Prussian Academy of Sciences and this he accepted. In the face of losing Kirchhoff, Heidelberg had designs for winning Helmholtz back as his replacement. But Prussia improved Helmholtz's conditions, and so it came about that Helmholtz and Kirchhoff

53. Heinrich Hertz, "Versuche zur Feststellung einer oberen Grenze für die kinetische Energie der elektrischen Strömung," Ann. 10 (1880): 414–48, reprinted in Ges. Werke 1: 1–36.

54. Berlin University, Index Lectionum, for the years immediately after 1871.

55. Max Planck, "Das Institut für theoretische Physik," in Lenz, Berlin 3: 276–78, on 276.

56. Baden Ministry of the Interior, 12 Feb. 1872, Kirchhoff Personalakte, Bad. GLA, 76/9961.

57. Again Baden raised Kirchhoff's salary to keep him. Baden Ministry of the Interior, 10 Mar. 1874, Kirchhoff Personalakte, Bad. GLA, 76/9961.

again taught in the same place, for ten years bringing distinction to Berlin as they had before to Heidelberg, this time both in theoretical physics.[58]

Kirchhoff's connection with the Prussian Academy was long standing, going back to 1861 when he was made a corresponding member. His researches in "mathematical physics" were judged outstanding both for their mathematics and their physics, so that the academy was unclear whether to admit him as a representative of physics or of mathematics. With the agreement of the physicists, they decided on mathematics.[59] In 1870 the academy recognized Kirchhoff again by voting him a "foreign" member, acknowledging his contributions to "mathematical physics" in the "disciplines of elasticity of solid bodies, electricity, and hydrodynamics." Even if he had made none of these contributions, his sponsors said, his law of radiant heat would have secured him a place as a physicist of the first rank.[60] In 1874 the academy offered Kirchhoff a full-time, salaried position for research with no teaching duties attached.

In the past Kirchhoff had been reluctant to leave Heidelberg, in part because with the mathematician Leo Koenigsberger he had created there a "mathematical-physical school." Now Koenigsberger had accepted an offer from Dresden, and the Baden ministry had done nothing to keep him despite Kirchhoff's warning that he too would leave Heidelberg at the next opportunity if Koenigsberger were allowed to leave.[61] Kirchhoff had explained to the ministry that without Koenigsberger, those parts of his teaching he "valued most" might lapse, since Heidelberg would no longer attract the advanced mathematics students he depended on for his lectures in mathematical physics.[62] In light of this concern, Prussia modified its offer: Kirchhoff was appointed to a newly founded ordinary professorship for "mathematical physics" at Berlin University, his salary to be paid in part by the Prussian Academy. In 1875 Kirchhoff moved to Berlin, hoping that the faculty there would receive him kindly even though he was placed among them without their initiative.[63]

In negotiations with Berlin, Kirchhoff had made it clear that he wanted to lecture on "theoretical physics" only.[64] So in the summer of 1875 he introduced his theoretical physics course at Berlin with lectures on mechanics. From fourteen students at the start, his class grew to thirty-one after a few weeks. The continuation of the course, mathematical optics, in the winter semester drew fifty-six students. With these numbers, Kirchhoff felt "very satisfied."[65]

58. Kirchhoff to Emil du Bois-Reymond, 30 Oct., 18 Nov., 13 and 17 Dec. 1874, STPK, Darmst. Coll. 1924.55.

59. "Wahlvorschlag für Gustav Robert Kirchhoff (1824–1887) zum KM," 24 June 1861, in *Physiker über Physiker,* ed. Christa Kirsten and Hans-Günther Körber (Berlin: Akademie-Verlag, 1975), 75–76.

60. "Wahlvorschlag für Gustav Robert Kirchhoff (1824–1887) zum AM," 10 Mar. 1870, in *Physiker über Physiker,* 77–79.

61. Emil Warburg, "Zur Erinnerung an Gustav Kirchhoff," *Naturwiss.* 13 (1925): 205–12, on 211.

62. Kirchhoff to the Baden Ministry of the Interior, 16 Dec. 1874, Kirchhoff Personalakte, Bad. GLA, 76/9961.

63. Kirchhoff to Eduard Zeller, 5 Jan. 1875, Tübingen UB, Md 747/373.

64. The Prussian ministry of culture told Kirchhoff that he did not have to give prescribed lectures. Kirchhoff to Emil du Bois-Reymond, 13 Dec. 1874.

65. Kirchhoff to Koenigsberger, 1 and 26 May 1875, STPK, Darmst. Coll. 1922.87. Kirchhoff to Robert Bunsen, 25 Nov. 1875, Heidelberg UB.

In the 1870s, near the end of his career, Kirchhoff held a position in the subject, theoretical physics, that he had dreamed of at the beginning of it, following his studies in the 1840s. The needs of physics instruction at Berlin now made the position possible, as Kirchhoff's poor health made it reasonable for him (as Ohm's poor health had made it reasonable for him, too, at the end of his career). Kirchhoff was only the second (or the third, if Ohm is counted here) after Listing in Göttingen to become ordinary professor for theoretical physics in Germany. Like Listing, Kirchhoff obtained the position under highly special circumstances; but unlike Listing, Kirchhoff was predominantly a theoretical physicist, which attached to his move to Berlin the additional significance of furthering the recognition of theoretical physics as an academic discipline in Germany.

Just as a discipline needs its body of knowledge and methods to define it, so it needs positions to attract young, talented scientists to ensure its continuation. Max Planck, the foremost representative of theoretical physics in Germany in the next generation, began his university studies just as Kirchhoff took up a position in theoretical physics equal in rank to a position in experimental physics. To an aspiring theoretical physicist like Planck, who attended Kirchhoff's lectures at Berlin, Kirchhoff's example pointed to the goal, if not yet to a regular route to achieving it, of a career in the emerging discipline.

15

The Creation of Extraordinary Professorships for Theoretical Physics

The dominant institutional development in theoretical physics at the German universities after 1870 was the creation of extraordinary professorships for the subject. As the activities in the physics institutes multiplied and the number of students in the experimental lectures grew in the 1870s and 1880s, the arrangement by which regular instruction in theoretical physics was left to a second, usually younger, physicist was introduced at almost every German university; it was the arrangement we found earlier in Giessen, for example. At about a third of the German universities this arrangement was put on a permanent basis: a salaried extraordinary professorship was created to which the lectures on theoretical physics were assigned. At first the positions were created solely to support the ordinary professor of physics, not to acknowledge a new specialty. They were planned as transitional positions for young physicists, whose ultimate destination was an ordinary professorship of experimental physics in the not too distant future. In pursuit of it, they needed to do experimental research; so that although the theoretical physics positions almost never came with separate institutes, they generally offered their occupants limited use of the facilities of the physics institute at the director's discretion.

New Extraordinary Professorships at Prussian Universities

Most of the new extraordinary professorships that came into being between 1870 and 1890 were at Prussian universities. Since several of them were not specifically designated for theoretical physics, it might be thought that the new positions were merely a continuation of the old Prussian pattern of multiple appointments. But that is not so. Whether so designated or not, the new extraordinary professorships were used invariably to *supplement* the work of the ordinary professor rather than to *compete* with it, and upon this understanding the ministry supported the ordinary professor in any dispute. It worked this way in Prussia as elsewhere.

33

The proper organization of university physics could best be realized in a new university, with a newly appointed faculty and new institute facilities. At Strassburg University, which Prussia acquired through its war with France, the physics institute became the model for training physicists in the early 1870s.

The first ordinary professor of physics and director of the physics institute at Strassburg was August Kundt, who had acquired great experimental skill as a student in Magnus's laboratory in Berlin. He had not, however, acquired an equally thorough training in theoretical physics at the same time, and he came to regret his one-sided education and to advocate a balance between experimental and theoretical physics. The modern experimental physicist, he said toward the end of his career, could hope for success only if he took theory as his guide in selecting a problem for research.[1]

From the beginning, the Strassburg physics institute made provision for a separate position for teaching theoretical physics. Emil Warburg, Kundt's former student and collaborator in Magnus's laboratory, was appointed extraordinary professor for theoretical physics in 1872, and Kundt left its teaching entirely to him.[2] Warburg's stay at Strassburg was the only time in his career when he came close to being exclusively a theoretical physicist, for there not only his teaching but also his research was theoretical. His reunion with Kundt in Strassburg resulted in joint investigations of the kinetic theory of gases, which had been prepared by Kundt's discovery of a simple method of measuring the velocity of sound; in their application of the method, Warburg did the theoretical work.[3] When Warburg left Strassburg in 1876 to take up the physics chair at Freiburg, he was succeeded by W. C. Röntgen, another former student and assistant of Kundt. Röntgen in turn was succeeded by the experimental physicists Ferdinand Braun and Wilhelm Kohlrausch, both of whom quickly moved on to ordinary professorships. Then in 1884, for the first time, the position went to a predominantly theoretical physicist, Emil Cohn, another former student of Kundt. Cohn taught theoretical physics for over thirty years in the Strassburg extraordinary professorship, a record that was contrary to the original conception of such a position.[4]

As at Strassburg, at Königsberg the Prussian ministry established a formal extraordinary professorship in 1876 or soon after, and it did so again at Marburg in 1877.[5] At other universities, particularly where the ordinary professor of physics was primarily a theoretical physicist or had an interest in teaching theoretical physics

1. August Kundt, "Antrittsrede," Sitzungsber. preuss. Akad., 1889, pt. 2, 679–83, on 682. Wilhelm von Bezold, "Gedächtnissrede auf August Kundt," Verh. phys. Ges. 13 (1894): 61–80.
2. Eduard Grüneisen, "Emil Warburg zum achtzigsten Geburtstage," Naturwiss. 14 (1926): 203–7. Deutscher Universitäts-Kalender (1872–76) (Berlin, 1872– [semiannual]).
3. Friedrich Paschen, "Gedächtnissrede des Hrn. Paschen auf Emil Warburg," Sitzungsber. preuss. Akad. (1932), cxv–cxxiii, on cxvii.
4. Festschrift zur Einweihung der Neubauten der Kaiser-Wilhelms-Universität Strassburg (Strassburg, 1884), 143.
5. Catalogus professorum academiae Marburgensis, ed. F. Gundlach (Marburg: Elwert, 1927), 394–95. Helmholtz to Prussian Minister Falk, 10 May 1877, STPK, Darmst. Coll. 1912.236. Prussian Ministry of Culture to Marburg U. Curator, 15 Apr. 1880, STA, Marburg, Bestand 310 Acc. 1975/42 Nr. 2037. We discuss the details of Voigt's occupancy of the Königsberg extraordinary professorship in chapter 19.

himself, the new extraordinary professorships were left unspecified; that way the younger man might be used to teach other parts of the physics curriculum as well as theoretical physics, depending on where he was needed at the time. But if at these universities the ordinary professorship passed into the hands of an experimentalist, they too followed the model of Strassburg.

Bonn is a case in point. There in 1874 Eduard Ketteler, who had been elevated from Privatdocent to extraordinary professor at his own request two years before, received the salaried extraordinary professorship the ministry had just created. Ketteler was employed, at a little additional salary, in assisting Bonn's ordinary professor Clausius in running the laboratory practice course. Clausius, a theoretical physicist by preference, lectured on theoretical as well as on experimental physics; alongside him Ketteler lectured on theoretical physics too.[6]

When Clausius died, Ketteler was temporarily assigned the lectures on experimental physics and the direction of the physics institute, which placed him in line for a possible permanent advancement. The philosophical faculty wanted to strengthen theoretical physics at Bonn, and when they confronted the problem of replacing Clausius, they decided it was time to establish an ordinary professorship for the subject. As evidence of their need, they pointed to the universities that had preceded them in establishing two positions: Berlin with Dove and Magnus and, following them, Helmholtz and Kirchhoff; Göttingen with Weber and Listing and, following them, Eduard Riecke and Woldemar Voigt; and Königsberg with Franz Neumann and Ludwig Moser. The philosophical faculty proposed a division of teaching assignments and facilities resembling the arrangement at Strassburg; namely, one of the two professors was to lecture on the general foundations of theoretical and experimental physics, give the lectures on experimental physics, and direct the exercises in the physics laboratory; the other professor was to lecture on individual branches of theoretical physics. The former professor would be in charge of the physics institute, while the latter would receive certain rooms in it for his experimental research.[7] The position the faculty had in mind for the theorist was less important than that of the experimentalist; in fact, the theorist was to be dependent on the experimentalist for space and equipment for his teaching. On the whole, neither professor could have found the arrangement to his liking. In any event, it was carried out only for a short time.

On the same day in January 1889 on which the ministry appointed Hertz to the chair of experimental physics at Bonn, it appointed Ketteler ordinary professor for theoretical physics. Hertz was told that he had Clausius's professorship, which carried the obligation "to represent experimental physics in lectures and exercises, and to direct the physics institute"; Ketteler was assigned only theoretical physics, forestalling any difficulties. By agreement between Hertz and the ministry, Ketteler was

6. Prussian Minister of Culture to Bonn U. Curator Beseler, 11 Nov. 1872 and 9 Mar. 1874, as well as many items of correspondence and faculty minutes dealing with the details of Ketteler's position from 1870 to 1889, Ketteler Personalakte, Bonn UA.

7. Prussian Ministry of Culture to Bonn U. Curator Gandtner, 24 Oct. 1888, Ketteler Personalakte, Bonn UA. Lipschitz to Dean Lubbert of the Bonn U. Philosophical Faculty, 10 Nov. 1888, Akten d. Phil. Fak., Bonn UA.

to get two rooms in the institute from Hertz. But the ministry apparently had little interest in perpetuating two ordinary professorships for physics anywhere but in Berlin or where they were already long established as at Göttingen; in October 1889 Ketteler was given the ordinary professorship of physics at Münster, and the position he had held at Bonn reverted to a properly subordinate one, that of an extraordinary professor.[8]

The vacant extraordinary professorship at Bonn was now described by Hertz as one for theoretical physics. But he did not want that designation misunderstood. To the theorist Voigt, whom he had asked for recommendations of candidates, Hertz wrote: "I saw from your letter that you understand the position a little differently than we are unfortunately forced to conceive of it. Since there are so few ordinary professors of theoretical physics, we cannot begin to claim to be proposing someone who devotes himself entirely to theoretical physics and who sees the extraordinary professorship of theoretical physics as a preliminary to an 'ordinary professorship of theoretical physics.' Instead it is basically an extraordinary professorship of physics in which the occupant is to teach theoretical physics, and so we must require that he be completely capable of this without our being prevented from giving him credit for achievements in the experimental area." To support his position, Hertz mentioned that he had heard that Breslau wanted a physicist for its job in theoretical physics who had done no work in the subject at all.[9]

The physicist whom Hertz got, Hermann Lorberg,[10] would not have expected to use the position as a stepping stone to an ordinary professorship of any kind: he was in his late fifties and had taught secondary school in Strassburg until 1889, when he became Privatdocent there. He fitted Hertz's description of someone capable of teaching the subject: he had published a number of theoretical papers on electrodynamics and a physics textbook for secondary schools. "I would like it best," Lorberg wrote deferentially to Hertz before coming to Bonn for his first semester as extraordinary professor, "if I could give a four-hour a week lecture course on 'Theory of Static Electricity and Magnetism' as well as a two-hour a week public lecture course 'On Recent Investigations in the Area of the Theory of Electricity'—if you are not reserving that for yourself—or on physical chemistry; or also a private lecture course on mechanical heat theory or mechanics—if Prof. Lipschitz does not want to lecture on these."[11] A much older man than was usually hired for the second physics position, Lorberg was at first given a temporary appointment, which was then made permanent; through Hertz's long illness and then through his own, Lorberg remained extraordinary professor for theoretical physics at Bonn to the end of his career.

At Breslau and at Halle, the extraordinary professorships were kept equally flex-

8. Prussian Ministry of Culture to Hertz, 18 Jan. 1889, Hertz Personalakte, Bonn UA. Prussian Ministry of Culture to Ketteler, 18 Jan. 1889, and to Bonn U. Curator Gandtner, 19 Oct. 1889, both in Ketteler Personalakte, Bonn UA. Hertz to Voigt, 1 and 6 Feb. 1890, Voigt Papers, Göttingen UB, Ms. Dept.

9. Hertz to Voigt, 1 and 6 Feb. 1890.

10. Bonn U. Curator to Lorberg, 20 June 1890, and other documents in Lorberg Personalakte, Bonn UA.

11. Lorberg to Hertz, 14 July 1890, Ms. Coll., DM, 2971.

ible, as the position at Bonn had been before 1890. The appointment of Philipp Lenard to the Breslau extraordinary professorship as late as 1894 still defined his duty as teaching physics in "lectures and exercises by supplementing the teaching of the ordinarius [ordinary professor] appointed for the subject and according to a more detailed agreement with him."[12] Halle had an extraordinary professor from 1878, the former Privatdocent for theoretical physics Anton Oberbeck. In 1884 Oberbeck was promoted to ordinary professor for theoretical physics because he had received an offer from the Karlsruhe Polytechnic, but like Ketteler at Bonn, Oberbeck was moved to the ordinary professorship of physics at another university the next year. His title at Halle—that of a personal ordinary professor, not of an ordinary professor—went to Ernst Dorn, probably because Dorn was already ordinary professor in his previous position and could hardly be demoted. When Dorn became the ordinary professor of experimental physics at Halle in 1895, the second position was returned to an extraordinary professorship, and its occupant, Karl Schmidt, was charged with teaching theoretical physics "in accordance with a more specific agreement with Professor Dr. Dorn."[13]

New Positions for Theoretical Physics Elsewhere in Germany

Of the non-Prussian German universities, only two acquired an extraordinary professorship for theoretical physics during this period: Munich in 1886 and Jena in 1889. These new positions were created late, perhaps because these universities and also Leipzig enjoyed regular and competent instruction in theoretical physics for some time before they made formal arrangements for it.

Leipzig attracted the theoretical physicist Karl Von der Mühll in 1867 because the mathematics faculty was then losing one of its members just when the presence of about forty students of mathematics and physics required a full complement of teachers. By offering to teach both mathematics and theoretical physics, Von der Mühll made himself acceptable to the faculty. The ordinary professor of physics, Wilhelm Hankel, taught mathematical physics as well as his regular subjects, but he was glad to have another physicist take over some of the burden.[14] In 1872 Von der Mühll was promoted to extraordinary professor of "physics," but he continued to teach theoretical physics and mathematics. He offered a fairly comprehensive program, which included "theoretical physics," elasticity theory, potential theory, topics from mechanics and from optical theory, electrodynamics, and mechanical heat theory, and he also participated in the direction of the mathematical seminar, where he took up subjects related to physics.[15] From 1878 to 1886, he was joined at Leipzig

12. *Chronik der Königlichen Universität zu Breslau* 1894/95, 6–7.

13. For this and other information on the development of theoretical physics at Halle University, we are grateful to the director of the Halle University Archive, Dr. H. Schwabe.

14. Von der Mühll to Franz Neumann, 29 Nov. 1867, Neumann Papers, Göttingen UB, Ms. Dept.

15. Saxon Ministry of Education to Leipzig U. Philosophical Faculty, 19 Dec. 1872, and Felix Klein's report on Von der Mühll's teaching, dated 15 Mar. 1886, in Von der Mühll Personalakte, Leipzig UA, Nr. 759.

by a second extraordinary professor, Eilhard Wiedemann, who also taught mathe-matical physics.[16] Von der Mühll, perhaps handicapped because he did little original research, almost none in physics, remained in his position in Leipzig for nearly twenty years without ever attaining formal recognition for his field of theoretical physics. Finally, in 1889, he returned home to Switzerland to a university post there.[17]

In the early 1880s, Munich University had two Privatdocenten for theoretical physics, Max Planck and Leo Graetz. In 1886, after Planck had left, Munich estab-lished a regular position: the Privatdocent for physics Friedrich Narr was appointed extraordinary professor with the duties of holding regular lectures on the "discipline of theoretical physics" and of directing the practice course in the institute and in the seminar.[18]

We will illustrate the process by which an informal but satisfactory arrangement for teaching theoretical physics was transformed into a regular extraordinary profes-sorship by describing the creation of the Jena position. (The process varied from place to place.) Physics acquired an up-to-date organization at Jena University fairly late, in the 1880s, largely at the initiative of Ernst Abbe. After helping to establish Jena's experimental professorship and physics institute in the early 1880s, in 1889 Abbe set in motion the negotiations for the establishment of an extraordinary pro-fessorship for theoretical physics.

Abbe's association with physics at Jena University and his awareness of its back-wardness began in his student days. He went to Jena to study in 1857, but he found that he could not get the education in mathematics and physics he wanted, and so he moved to Göttingen to continue his studies. At the suggestion of Karl Snell, Jena's ordinary professor of physics, in 1863 Abbe returned to Jena to become Pri-vatdocent, and to ensure that Abbe would not be lost to the university again, he was given a salary. In his first semester of teaching at Jena, he offered mathematics courses of the kind that had become basic to the teaching of physics: potential theory and definite integrals. For these courses, perhaps because of the students' lack of preparation, he had an enrollment of only two or three; in his next courses on mechanics and on measuring instruments, he fared better. To his first laboratory practice course, "directions for doing physical experiments," he attracted so many students, seventeen, that he had to teach it in two sections. In his second year as Privatdocent, he also took over the experimental physics lectures from Snell. What instruments Abbe needed, Snell let him assemble from the remains of old instru-

16. Saxon Ministry of Education to Leipzig U. Philosophical Faculty, 1 Feb. 1878, and the draft report of the Philosophical Faculty on Wiedemann's teaching, dated 28 Jan. 1878, in Eilhard Wiedemann Personalakte, Leipzig UA, Nr. 1060.

17. Carl Neumann reported on Von der Mühll's activities at Leipzig on 8 May 1886 for the purpose of having him promoted to ordinary honorary professor for his teaching of over twenty years. He noted Von der Mühll's lack of productivity, citing only three publications. Saxon Ministry of Education to Leipzig U. Philosophical Faculty, 5 Jan. 1889. Both in Von der Mühll Personalakte, Leipzig UA, Nr. 759.

18. Bavarian Ministry of the Interior to Munich U. Senate, 2 Aug. 1886, Munich UA, E II-N, Narr.

ments in the nearly useless physical cabinet of the university or order on Snell's authorization. For his course on measuring instruments, Abbe had most of the instruments constructed by the "local mechanic" Carl Zeiss, the beginning of an association of great importance for the future of the natural sciences, especially of physics, at Jena. In 1870 Abbe was promoted to extraordinary professor, and in 1878, in return for refusing an offer from Berlin University of a chair for optics, he was made honorary ordinary professor. (Two years earlier Abbe had entered Zeiss's firm as a scientific partner, and in the contract with Zeiss he agreed to forego any regular ordinary professorship.)[19]

Although he began by teaching both experimental and theoretical physics, Abbe came to think of his "task" at Jena University as supporting the "teaching of my colleagues in the mathematical-physical field with lectures from theoretical physics."[20] He had no formal teaching assignment for the subject, he said, but he tried to meet the students' needs as completely as he could by including in his teaching topics that were removed from his own area of interest. But as his work in optics made him increasingly specialized, the preparations for the lectures in the different parts of theoretical physics took more time than he could spare from his other responsibilities. Wishing to withdraw from teaching theoretical physics—other than optics—he suggested the establishment of a separate position for the subject in 1889.

Physics was constantly expanding, Abbe noted in his request for the position, as was the range of "purposes that physics education at the universities must serve." He feared that the means of keeping physics up to date at Jena could not keep pace, since the funds for an experimental physicist and for a physics institute a few years earlier had been difficult to acquire.[21] The loss of what little theoretical physics he had taught if he were not replaced would amount to a serious loss for physics at Jena. The ordinary professor of physics or a mathematical colleague could not substitute for him without detracting from other, equally important interests of the physics curriculum. For every university it was urgent "that besides the main professorship for physics—which is sufficiently burdened by the basic lectures and the practical exercises in the laboratory—special provisions also be made for *regular* lectures at least on the more important departments of physics in detailed, predominantly mathematical treatment." Without such lectures, not even candidates for gymnasium teaching would have an opportunity of acquiring the knowledge required of them by the state examinations. To fall behind "the demands of the times" by not offering such lectures would be detrimental to a small university such as Jena, especially at a time when the numbers of students for physics were dropping and the smaller universities were finding it difficult to compete with the larger, better equipped ones.[22]

As an honorary professor, Abbe had no say in the faculty, where requests for

19. Felix Auerbach, *Ernst Abbe, sein Leben, sein Wirken, seine Persönlichkeit* (Leipzig: Akademische Verlagsgesellschaft, 1918), 52–56, 69–72, 111, 114–24, 133–34.

20. Abbe to Jena U. Curator, 12 June 1889, Jena UA, Bestand C, Nr. 445.

21. Abbe to A. F. Weinhold, 17 Feb. 1881 and 13 Feb. 1882, Ms. Coll., DM, 1959-2.

22. Abbe to Jena U. Curator, 12 June 1889.

positions were usually initiated; but his request was favorably received all the same by a sympathetic curator, who wished to relieve Abbe of his extra teaching burden by acquiring the new position and appointment as quickly as possible. He asked the Jena philosophical faculty to suggest candidates for the position at the same time that they commented on the desirability of the extraordinary professorship.[23]

Upon seeing Abbe's proposal, the faculty agreed on the need for the position for theoretical physics, but they refused to name candidates for it until they were requested to do so by the university senate, which was the regular procedure. Their insistence on formality, the curator quickly learned, was only an excuse. In private agreement, Abbe and the ordinary professor of physics, Adolph Winkelmann, had settled on the Breslau Privatdocent Felix Auerbach as the most desirable candidate for the new position. The philosophical faculty did not want Auerbach because he was Jewish, and as a formal objection to him they brought up his ten years as Privatdocent without promotion or offer of a professorship. The curator explained that Auerbach's many years as Privatdocent did not reflect unfavorably on his ability but could "easily" be explained by his Jewish descent; it is notorious, he said, that in Prussia at present Jewish Privatdocenten are not easily promoted. This argument did not move the anti-Semitic professors. The university statute prevented the curator from ignoring the faculty's wishes, and to hurry through channels without first gathering further expert testimony in favor of Auerbach was certain to lead, in a formal vote, to the defeat of Winkelmann's candidate.[24]

The resulting delay of Auerbach's appointment allowed the faculty to use the common understanding of the nature of extraordinary professorships to try to achieve their ends. They argued that the position would probably remain a moderately salaried extraordinary professorship for a long time, and that it should be thought of as a transitional position in which younger men stayed for a few years to develop as physicists. They argued further that it would not be desirable to hire a man who might stay permanently—as was certain to be the case with a Jewish physicist such as Auerbach, given the prejudices of faculties and ministries—since he would be dependent on the director of the physics institute for the means for experimental research. Young physicists at the beginning of their careers, such as Conrad Dieterici and Wilhelm Hallwachs, whom Kundt in Berlin had suggested to Jena, were more appropriate for this transitional position than an established physicist such as Auerbach. Moreover, the connection of the young physicists with Kundt, whose assistants they had been, guaranteed that they would not stay long in Jena; they would doubtlessly benefit from Kundt's "authoritative influence" and advance soon.[25]

23. Jena U. Curator Eggeling to Dean of the Philosophical Faculty, 14 June 1889, and Eggeling's draft report on the negotiations concerning the establishment of the new extraordinary professorship of 8 July 1889, Jena UA, Bestand C, Nr. 445.

24. Dean of the Jena U. Philosophical Faculty to Curator Eggeling, 26 June 1889; H. Schroeter to Thomae, 24 June 1889 (excerpt); and Curator Eggeling's report of 8 July 1889; Jena UA, Bestand C, Nr. 445. Adolph Winkelmann to Dean of the Philosophical Faculty, 25 June 1889, "Allgemeine Fakultätsacten," Jena UA, Nr. 621a.

25. Kundt to Winkelmann, 1 July 1889 (excerpt), and Curator Eggeling's report of 8 July 1889, Jena UA, Bestand C, Nr. 445.

In the end, the question of the appointment was decided in accordance with the needs of Jena physics. Abbe and Winkelmann continued to argue for Auerbach, because they believed that the first consideration in decisions on academic appointments should be the qualifications of the candidate.[26] Their judgment of Auerbach agreed with evaluations by Helmholtz and Clausius—of which Abbe and Winkelmann had no knowledge—rendered two years earlier when the Prussian ministry considered a recommendation by the Breslau faculty that Auerbach be promoted to extraordinary professor there. Both Helmholtz and Clausius had then strongly recommended Auerbach and declared his promotion completely justified, but to no avail.[27] The evaluations of Auerbach that Winkelmann and some of his colleagues solicited were also satisfactory, particularly on Auerbach's effectiveness as a teacher, confirming Winkelmann's impression of Auerbach's skill in presenting "physical-mathematical problems" in lectures. They reported that Auerbach attracted more students than the ordinary and extraordinary professors of physics at Breslau, and they noted that Auerbach had even succeeded in interesting some students in independent scientific work.[28] On the basis of this information, Abbe and Winkelmann insisted that Auerbach had already proven himself to be far more capable and accomplished than any of the other candidates who had also been suggested by their correspondents.[29] With Winkelmann's collection of letters of recommendation for Auerbach before them, the philosophical faculty came around grudgingly to admit "no objections" to Auerbach, and in September 1889 Auerbach was appointed Jena's first official representative of theoretical physics.[30]

Qualifications for the New Theoretical Physics Positions

There were good practical reasons—as Hertz had described them to Voigt—for not making the second physics position a specialist's position at this time. In the eyes of many physicists then, there was no need to, for the best qualified young physicist could meet the demands of a theoretical position or an experimental position in equal measure. Asked his opinion about physics appointments at Greifswald University in 1884, Helmholtz wrote that he considered it desirable to have special lectures on mathematical physics; the lectures on experimental physics, which were attended by medical students, future government officials, and pharmacists, were not the place to introduce a "complete and rigorous formulation of the laws of nature, which demands a mathematical formulation and must after all be given to future teachers and mathematicians." If a university could not afford two teachers of physics, then

26. Curator Eggeling's report of 8 July 1889.
27. Clausius to Director Greiff, 15 Aug. 1887, and Helmholtz to Prussian Minister of Culture Gossler, 10 Feb. 1888, STPK, Darmst. Coll. 1913.51.
28. Notes on Winkelmann's report on the recommendations for Auerbach that he had received (possibly by the Curator), 28 June 1889; recommendation by Erdmann at Breslau, 26 June 1889 (excerpt); and Curator Eggeling's report of 8 July 1889; Jena UA, Bestand C, Nr. 445.
29. Curator Eggeling's report of 8 July 1889.
30. Dean Kalkowsky, minutes of the meeting of Jena U. Philosophical Faculty on 7 Aug. 1889, and Ministry to the "whole university," 27 Sept. 1889, "Allgemeine Fakultätsacten," Jena UA, Nr. 621a.

one physicist would do, if he were properly qualified; that is, according to Helmholtz, if he were both a physicist with "extensive knowledge of the mathematical presentation of physical theories" and an "experienced experimenter." (Helmholtz did not think that the problem should be solved by leaving the teaching of physics to mathematicians, since they "tend to treat physical problems only as paradigms for mathematical methods, without concerning themselves further with the relationship of their equations to reality.") Helmholtz did not go so far as to expect every good physicist to be capable of doing original research in mathematical theory.[31]

Helmholtz's understanding of the "complete physicist" entered his evaluations of candidates for university positions, as it entered, often implicitly, other physicists' evaluations. Whether the vacant position was for theoretical or for experimental physics, the accomplishments and the qualities of a candidate brought forward in the evaluations were much the same. In particular, candidates for theoretical positions were expected to have demonstrated experimental skill. Clausius, for example, in 1887 discussed Auerbach's researches field by field, and he noted Auerbach's "solid knowledge," his "eager scientific striving," his skill and diligence in experimenting, and his success in experimentally verifying the theories of others, such as Helmholtz's, and thereby extending and completing existing knowledge in his areas of physics.[32] For a theoretical position, Boltzmann recommended a candidate both as an experimenter and as a "theorist."[33] Georg Quincke, in considering Lenard for the theoretical job at Heidelberg in 1896, was impressed by Lenard's recent publications on cathode rays, "which have attracted attention in the widest circles," and he required of Lenard no more evidence of mathematical ability than his editorial role in the publication of Hertz's *Mechanics*.[34] Helmholtz, as we would expect, placed as much weight on experimental as on theoretical work when he considered someone for a theoretical job: evaluating Braun for one in 1877, he wrote of his accomplishments as an experimenter as well as of his "good and thorough knowledge of theory,"[35] and he recommended Auerbach for a theoretical position even though he praised him mainly for his experimental work and thought his theoretical-mathematical ability less adequate.[36]

The equal emphasis on experimental and theoretical competence in the evaluations of candidates for theoretical positions expressed more than an interest in assuring the young physicist's future employer that he was capable of moving on to an experimental professorship eventually. In the evaluations of candidates for experimental positions from this time, there is a similar emphasis on both sides of physics, which points to the existence of an ideal of the complete physicist. If a candidate for a position in experimental physics was thought to be deficient in his handling,

31. Helmholtz to Prussian Ministry official Althoff, 18 May 1884, STPK, Darmst. Coll. F 1 a 1847.
32. Clausius to Director Greiff, 15 Aug. 1887.
33. Boltzmann to Leo Koenigsberger, 3 June 1899, STPK, Darmst. Coll. 1922.93.
34. Quincke and Leo Koenigsberger to the Baden Ministry of Education, 11 June 1896, Heidelberg UA.
35. Helmholtz to Prussian Minister of Culture Falk, 10 May 1877, STPK, Darmst. Coll. 1912.236.
36. Helmholtz to Prussian Minister of Culture Gossler, 10 Feb. 1888, STPK, Darmst. Coll. 1913.51.

or even in his appreciation, of mathematical methods in physics, his chances could be hurt. Helmholtz objected to a candidate because his published "theoretical discussions" of his experimental researches "are hardly intelligible and seem ambiguous and arbitrary," leading Helmholtz to wonder if in his teaching he could give the desirable "clear and sharp statement of the conceptual formulation [of] . . . the most general laws of phenomena."[37] Knoblauch and Jolly urged Melde's promotion to ordinary professor not only for his experimental work but also for his theoretical and mathematical abilities,[38] and the Marburg faculty recommended candidates for Melde's successor for the same combined abilities.[39]

The Representative of Theoretical Physics as a Specialist

At Kiel University, the last of the Prussian universities to acquire an extraordinary professorship for theoretical physics between 1870 and 1890, the new position received a different interpretation by its initiators and by its first two occupants; namely, that of a specialized position for theory. In the fall of 1882, in connection with deficiencies of mathematical instruction at the university, Kiel considered establishing a new position for mathematical physics. The Kiel mathematician L. A. Pochhammer argued against a proposal to create an ordinary professorship for the combination of mathematics and mathematical physics, since the position would deprive mathematics ultimately of its second position. Whoever held it would have little interest in teaching elementary mathematics, which would mean a significant loss at Kiel where there were almost no Privatdocenten in mathematics. As mathematical physics was an independent and large subject, and as its representative would naturally want to stay within it, Pochhammer urged the creation of an extraordinary professorship restricted to mathematical physics.[40]

In December 1882 the Kiel philosophical faculty applied for an extraordinary professorship for "theoretical" physics, offering the following reasons for their request. They had not needed a specialist to give lectures on theoretical physics in the past, because the lectures had been given by teachers of mathematics and physics, whose own subjects were still sufficiently limited to allow them the time. Now theoretical physics was recognized as a necessary specialty at many universities, and not just at the largest, through a second professorship for physics. It was valued as a link between, and as an enrichment of, mathematics on the one hand and the natural sciences on the other. Mathematics students were protected from becoming too narrow, since theoretical physics allowed them to put their abstract mathematical knowledge to use; students of the natural sciences were motivated to complete their knowledge of mathematics when they studied theoretical physics, and they were also better prepared to study other sciences, such as those taught in the medical faculty. Finally, the philosophical faculty had a practical reason for wanting a theoretical

37. Helmholtz to Althoff, 18 May 1884.
38. "Separatvotum," 16 Feb. 1866, STA, Marburg, Bestand 305a, 1864/66 Melde.
39. Marburg Phil. Fac. to Curator, 12 Nov. 1900, STA, Marburg, Bestand 310 Acc. 1975/42.
40. Schöne to Althoff, n.d. [1882], and Pochhammer to Schöne, 14 Oct. 1882, DZA, Merseburg.

physicist: whenever the ordinary professor of physics could not attend examinations, there would now be someone else to represent his field.[41]

The Kiel curator forwarded the faculty's request together with the observation that the growth of the natural sciences since the time of C. H. Pfaff (who had taught several sciences at Kiel) made it impossible now for a professor to teach more than one field of science. He did not know if there was an urgent need to establish an extraordinary professorship for theoretical physics, but he knew that Gustav Karsten was the only one teaching physics at Kiel at the moment and that he would not be able to give extensive lectures on theoretical physics. He left the decision to the ministry. Noting that there were some forty students of mathematics at Kiel, the ministry thought that the faculty's request was justified but that a paid Privatdocent might be the answer. (In recent years, the Privatdocent Leonhard Weber had given lectures on parts of theoretical physics, but he had left.) In any case, the ministry would list the extraordinary professorship for the 1884–85 budget.[42]

The man they hired for Kiel was Heinrich Hertz. Hertz, who originally intended to become an engineer, read Wüllner's textbook on physics in his spare time while working for construction engineers in Frankfurt am Main after leaving secondary school. With his interest in the natural sciences rekindled, he attended the lectures at the Frankfurt Physical Society and soon after was briefly a student at the polytechnic school in Dresden. He regarded mathematics as his favorite subject at this time. After a year of military service, he entered the polytechnic school in Munich, where he soon decided that he liked physics even better and that he wanted to make it his profession instead of engineering. In asking his father's permission to change plans, he explained that he might have been happy as an engineer but equally so as a bookbinder or woodworker or "anything ordinary." Engineering relied on practical talents and practical knowledge, which did not interest Hertz, whereas science promised him lifelong study and the possibility of becoming an important scientist.[43]

In Munich, Hertz moved freely between the university and the polytechnic, where he took laboratory practice. Bezold, the professor for technical physics at the polytechnic, advised Hertz not to turn to physics too early but to get a good mathematics education first, which was common wisdom then.[44] On the advice of the university physics professor, Philipp Jolly, Hertz studied mechanics and mathematics

---

41. Kiel U. Philosophical Faculty to Prussian Minister of Culture Gossler, 14 Dec. 1882, DZA, Merseburg.

42. Kiel U. Curator Mommsen to Prussian Ministry of Culture, 27 Dec. 1882, DZA, Merseburg.

43. Hertz to his parents, 1, 7, and 25 Nov. 1877, in Heinrich Hertz, *Erinnerungen, Briefe, Tagebücher*, ed. J. Hertz, 2d rev. ed. by M. Hertz and C. Süsskind (San Francisco: San Francisco Press, 1977), 62–72.

44. For example, Helmholtz advised his son Robert, the future physicist, who at the time was starting his university studies, to study mathematics before studying physics. Helmholtz said of himself: "As far as mathematics is concerned, my interest developed only through its applications, especially those of mathematical physics, and I studied everything I know of mathematics only occasionally for purposes of application. But that is a method that takes a great deal of time and through which one reaches complete knowledge only very late." Anna von Helmholtz, *Anna von Helmholtz. Ein Lebensbild in Briefen*, ed. Ellen von Siemens-Helmholtz, vol. 1 (Berlin: Verlag für Kulturpolitik, 1929), 249.

on his own from the classical French works by Lagrange, Laplace, and Poisson. Finding Lagrange "horribly abstract" and more recent writers of little help, Hertz despaired of grasping the "individual parts of contemporary mathematics in their connection." He believed that in nature everything is mathematical if properly understood, but at the same time he believed that non-Euclidean geometry, geometry of four or more dimensions, elliptic functions (the subject of special lectures he attended in Munich), and, in general, the new mathematics from about 1830 on held "no great value for the physicist, however beautiful" they may be in themselves.[45]

After a year in Munich, Hertz was impatient to move on. The physics professor at the polytechnic, Wilhelm Beetz, told him that he could now expect to find a physics laboratory at any university he chose. He considered Leipzig and Bonn, "where Clausius is," then decided for Berlin. There he attended Helmholtz's lectures on mathematical acoustics and Kirchhoff's on mechanics, which held nothing new for him and which he sometimes skipped. His main interest was experimental physics, in which, as we have seen, he was successful from the start. He wrote his dissertation in three months on a purely theoretical problem concerning electrical induction in spheres rotating between magnets.[46] Following graduation, Hertz stayed on as Helmholtz's assistant for three years, supervising practice exercises connected with general physics and heat theory and doing experimental and theoretical research of his own on problems in electrodynamics and elasticity.[47] As he explained in his vita, his independent researches during his Berlin years were "in part more of a theoretical, in part of an experimental nature; they did not arise in systematic pursuit of a greater goal but from the incidental stimulus I received in rich measure from my teachers and collaborators."[48] During these years, he published a dozen or so researches, an impressive achievement, which held the promise of a better job.

About to appoint a physicist as Privatdocent for theoretical physics at Kiel, the Prussian ministry asked advice of Helmholtz, Kirchhoff, and the mathematician Karl Weierstrass. They told the ministry about Hertz, and they also told Hertz that the Kiel faculty had requested an extraordinary professorship and that there was a good chance that he might be promoted to it in two years when money for it became available. They also told Hertz about the complications of the appointment: since the Kiel faculty wanted an extraordinary professor, they might resent a Privatdocent, and the Kiel physics professor Gustav Karsten disliked the Berlin faculty and would devalue any recommendations they made. Helmholtz was lukewarm about the job, for he believed that a mathematical physicist should have means to do experimental

45. Hertz to his parents, 25 Nov. 1877, in Hertz, Erinnerungen, 68–72, on 70. Max Planck, "Gedächtnissrede auf Heinrich Hertz," Verh. phys. Ges. 13 (1894): 9–29, reprinted in Max Planck, Physikalische Abhandlungen und Vorträge, 3 vols. (Braunschweig: F. Vieweg, 1958), 3: 268–88, on 277 (hereafter cited as Phys. Abh.).

46. Heinrich Hertz, Ueber die Induction in rotierenden Kugeln (Berlin, 1880). Hertz to his parents, 4 Nov. 1878, in Hertz, Erinnerungen, 114–16.

47. Hertz, "Vita," 14 Mar. 1883, LA Schleswig-Holstein, Abt. 47.7 Nr. 8.

48. Hertz, "Vita," 14 Mar. 1883.

research, and Hertz would have none at Kiel. But Helmholtz did not advise Hertz to decline or accept the job (as Hertz wished he would); instead he advised Hertz to go to Kiel and learn firsthand the faculty's mood and the ministry's intention.[49]

Despite his belief that Berlin was the center of physics, where he could "compare his powers" with those of other researchers, Hertz was ready to move on. He wanted to begin teaching, and there were already too many Privatdocenten in Berlin;[50] Kiel's job looked to him like a solution. He found that the Kiel faculty was pleased that the gap in their curriculum was to be filled and that they would be satisfied with a Privatdocent for the purpose. And the ministry was satisfied that Hertz had, in addition to a thorough knowledge of physics, a sufficient mathematical education to meet the requirements for what its university official Friedrich Althoff called a "so-called theoretical physicist."[51]

Once installed at Kiel, Hertz lectured on theoretical physics while he did theoretical research on a variety of topics. His diary entries soon after his arrival reveal that he read Maxwell's electrodynamics and worked steadily on electrodynamic questions. The outcome of this was a critical and theoretical study in 1884 of the foundations of electrodynamics, in which Hertz concluded that Maxwell's theory does not contain within itself the proof of its incompleteness and that the opposing electrodynamics does.[52]

Like Helmholtz, Clausius, and other physicists working on electrodynamic theory, Hertz retained the customary dynamic principles where possible. To analyze the equations for closed electric and magnetic currents, he invoked the principles of the conservation of energy, the equality of action and reaction, and the superposition of actions. To these he added the two "principles" of the "unity of electric force" and the "unity of magnetic force," which he believed were implicit both in the theory of Faraday and Maxwell and in Weber's and other opposing theories of electrodynamics.

Hertz illustrated what he meant by the "unity of electric force" by this example: the force with which a friction rod attracts a charged piece of wood is the same force as that with which a changing magnet induces a current in a conductor. In both cases, the force is the electric force, from which it follows that a changing magnet attracts the charged piece of wood and that, by the principle of action and reaction, the charged wood attracts the changing magnet. It follows, too, that a changing magnet attracts another changing magnet with an electric force in addition to their mutual magnetic force; by the older electrodynamics, only the magnetic force is recognized.[53] Carrying out this analysis mathematically for ring magnets and closed

---

49. Hertz to his father, 1 Mar. 1883, in Hertz, *Erinnerungen*, 176–78.

50. Hertz to his parents, 8 Mar. 1881 and 17 Feb. 1883, in Hertz, *Erinnerungen*, 144, 172–74.

51. Hertz to Kiel U. Philosophical Faculty, 14 Mar. 1883, LA Schleswig-Holstein, Abt. 47.7 Nr. 8. Hertz to Althoff, 15 Mar. 1883, and marginal note by Althoff, DZA, Merseburg.

52. Hertz, diary entries for Jan.–July 1884, in Hertz, *Erinnerungen*, 188–94. Heinrich Hertz, "Über die Beziehungen zwischen den Maxwell'schen elektrodynamischen Grundgleichungen und den Grundgleichungen der gegnerischen Elektrodynamik," *Ann.* 23 (1884): 84–103, reprinted in *Ges. Werke* 1: 295–314, on 313–14.

53. Planck, "Hertz," 278–79.

currents of varying intensity, Hertz arrived by an iterative procedure at "Maxwell's equations."[54] "If the choice rests only between the usual system of electromagnetics and Maxwell's," he reasoned, "the latter is certainly to be preferred."[55] Like Helmholtz, Hertz tried to clear up the confusion in electrodynamics by a theoretical analysis, here by an appeal to principles standing above the conflicting theories.

Hertz's comparative study of electrodynamic theories in 1884 drew some interest—for example, from Boltzmann and Lorberg—but it did not have much impact on work in electrodynamics other than, it would seem, Hertz's own. In commenting on this study, Planck said that it had received too little attention, especially since it was a first-rate piece of theoretical work, as impressive in its way as Hertz's later experimental work, which eclipsed it.[56] Although Hertz began this later experimental work with Helmholtz's theory with its denial of the unity of the dynamic and the static electric forces, midway through it he advanced to an understanding of the unity of fields reminiscent of his principles of the unity of forces in 1884.[57]

Hertz was restless at Kiel. At his own expense he set up a laboratory in his house, but it was not the same as an institute; his means for research were meager, and it took an eternity, it seemed to him, to get a single length of platinum wire or a glass tube.[58] He had always done experimental as well as theoretical work, a pattern he wished to continue. He did not see himself as a theoretical specialist, which if he were to stay in Kiel he seemed likely to become, in teaching certainly and, by circumstance, perhaps in research as well.

In November 1884 the ministry proposed to the Kiel faculty that Hertz now be

---

54. The starting point of Hertz's mathematical analysis of electrodynamics, like Helmholtz's, was Neumann's "vector potential," which Hertz applied to both the electric current and the magnetic current to obtain the electric and magnetic potentials. In tracing the implications for these potentials of varying electric and magnetic current densities, Hertz departed from the older electrodynamics. He showed that a small correction term is needed in the expression for the ponderomotive action of the order of A^2, where $1/A$ is equal to $c/\sqrt{2}$, where c is Weber's constant, which is the same as the velocity of light. He then showed that by the principle of the conservation of energy, this term entails another correction in the inductive action, which in turn entails another correction in the ponderomotive action. Ultimately, in this way, Hertz arrived at a convergent infinite series of correction terms in increasing powers of A; the series yields electric and magnetic potentials that satisfy the standard equation for waves propagated in free space with velocity $1/A$. The electric force $\mathbf{E}$ and the magnetic force $\mathbf{H}$ satisfy the same wave equation as do the potentials, and Hertz arrived at a system of equations that he identified as Maxwell's:

$$\nabla^2 \mathbf{H} - A^2 \frac{d^2}{dt^2} \mathbf{H} = 0, \quad \text{div } \mathbf{H} = 0,$$

$$\nabla^2 \mathbf{E} - A^2 \frac{d^2}{dt^2} \mathbf{E} = 0, \quad \text{div } \mathbf{E} = 0.$$

(The vector notation is modern, not Hertz's.) Hertz also wrote the first-order differential equations mixing the electric and magnetic forces. In this paper, the equations for the electric and magnetic forces were expressed in "symmetric form" for the first time. Salvo D'Agostino, "Hertz's Researches on Electromagnetic Waves," *HSPS* 6 (1975): 261–323, on 291.

55. Hertz, "Über die Beziehungen," 313.

56. Planck, "Hertz," 278; Max Planck, "James Clerk Maxwell in seiner Bedeutung für die theoretische Physik in Deutschland," *Naturwiss.* 19 (1931): 889–94, reprinted in *Phys. Abh.* 3: 352–57, on 356.

57. D'Agostino, "Hertz's Researches," 295, 322.

58. Hertz to his parents, 27 Oct. 1883, in Hertz, *Erinnerungen*, 186.

given the planned extraordinary professorship for theoretical physics, and in December the faculty sent a report agreeing with the ministry. Hertz speculated that the faculty voted this way so that a compensating offer could be made to him if the Karlsruhe Polytechnic offered him a job, which was known to be a possibility. Hertz wanted to improve his conditions of work, and when Karlsruhe offered him a position as ordinary professor of physics and director of a decent physics institute, Hertz accepted without hesitation. He was "convinced," he told the dean of the Kiel philosophical faculty, "that there was hardly any other choice."[59]

To a hopeful colleague, Hertz wrote of his "good fortune" at getting the Karlsruhe job. In the matter of jobs, one man's success robbed another of his, Hertz reflected: an "egotistical" business. Hertz attributed his success to having caught the eye of the "authorities" such as Helmholtz and Kirchhoff.[60]

To fill the new extraordinary professorship for theoretical physics, the Kiel faculty recommended Planck, in whom they placed as much confidence as they had in Hertz. Of all the younger teachers of theoretical physics, Planck had the "longest and most successful activity" to his credit, and the faculty was as impressed by his publications as by his teaching.[61] The government approved their choice and appointed Planck to the position in May 1885. Planck agreed to teach all of mathematical physics and, if necessary, to help out in experimental physics.[62]

For Planck, the move to Kiel was more in keeping with the direction of his work than it had been for Hertz. It was also a move to a place he knew well, for he had spent his boyhood in Kiel. From there his family had moved to Munich, where his father taught law at the university and he attended the gymnasium. He early considered making a career in music, but he decided that he lacked talent for composition; he considered the humanities, too, and he later supposed he might have made a passable philologist or historian.[63] As a student at Munich University, he attended mathematical lectures, which inclined him toward the exact natural sci-

59. Hertz to his mother, 6 and 12 Dec. 1884, and diary entries for Dec. 1884, in Hertz, *Erinnerungen*, 198–200. Hertz to Dean of the Kiel U. Philosophical Faculty, 31 Dec. 1884, LA Schleswig-Holstein, Abt. 47.7 Nr. 8.

60. Hertz to Elsas, 10 and 26 Jan. 1885, Ms. Coll., DM, 3089, 3090.

61. The ministry requested another suggestion from the faculty via the curator on 2 Jan. 1885; the curator agreed with the faculty's recommendation of Planck in February. Kiel U. Curator Mommsen to Prussian Minister of Culture Gossler, 12 Feb. 1885, and Dean of Kiel U. Philosophical Faculty to Gossler, 13 Feb. 1885, DZA, Merseburg.

62. In April 1885 Kiel received an additional 3060 marks in its budget for an "extraordinary professor for theoretical physics"; it was for a salary of 2400 marks and 660 marks for rent. Planck saw the Prussian Ministry of Culture official Friedrich Althoff in Munich the same day, signing an agreement according to which Planck got 2000 marks in salary plus the 660 marks. That is, Planck received 400 marks less than the university was receiving as salary for him. Ministry document dated 10 Apr. 1885, and Prussian Minister of Culture to Planck, 2 May 1885, DZA, Merseburg. Kiel U. Curator Mommsen to Kiel U. Philosophical Faculty, 6 May 1885, LA Schleswig-Holstein, Abt. 47.7 Nr. 8.

63. Armin Hermann, *Max Planck in Selbstzeugnissen und Bilddokumenten* (Reinbek b. Hamburg: Rowohlt, 1973), 11. Max Born, "Max Karl Ernst Ludwig Planck 1858–1947," *Obituary Notices of Fellows of the Royal Society* 6 (1948): 161–88, reprinted in Max Born, *Ausgewählte Abhandlungen*, ed. Akademie der Wissenschaften in Göttingen, 2 vols. (Göttingen: Vandenhoeck und Ruprecht, 1963), 2: 626–46, on 627.

ences. He preferred physics over mathematics, he recalled, because of his "deep interest in questions of world view, which naturally cannot be solved on purely mathematical foundations."[64] Near the end of his life, he wrote that the reason for his "original decision" to devote himself to physics was his recognition that the laws of thought conform to the impressions we receive from the outer world and that by pure thought we can discover those laws.[65]

Planck had no doubt received the same advice as Hertz, for after studying physics for a time under Jolly, he too left Munich for Berlin to continue his studies. In Berlin, Planck saw Helmholtz and Kirchhoff, who impressed him for their world reputations, and he now recognized that physics in Munich had only "local significance." From Helmholtz's and Kirchhoff's lecturing, however, Planck netted "no perceptible gain." Helmholtz never prepared, constantly had to consult his notebook for facts, made frequent mistakes in calculation, and gave Planck the impression that he was as bored as the students by it all. Helmholtz's class gradually disappeared until there were only three left, Planck among them. By contrast, Kirchhoff planned his lectures to the last detail, delivered them in flawless sentences, and sounded like a memorized textbook, which made his lectures just as boring to Planck as Helmholtz's.[66]

Planck was driven to learn physics from the writings of Helmholtz and Kirchhoff and those of other recent masters. It was on a topic suggested above all by Clausius's writings that he wrote his dissertation, which he submitted in Munich in 1879. He stayed on in Munich, as we have noted, as Privatdocent for theoretical physics, waiting impatiently for a professorship. Upon receiving a call to teach physics at a forestry academy, he went to Berlin to discuss it with Helmholtz, whose prognosis about jobs in theoretical physics, Planck's preferred field, was favorable.[67] So Planck waited in Munich, and in time Kiel invited him to teach theoretical physics as Hertz's replacement. (By the time Planck came to write his scientific autobiography, he had grown pessimistic about the chances for a timely recognition of accomplishment in physics; he did not attribute his Kiel offer so much to his scientific work as to the circumstance that his father was a close friend of Gustav Karsten.)[68]

For Planck as for Hertz, Kiel was a stepping-stone to a better job. But after Kiel, their careers diverged as did their interests. Hertz moved to an experimental physics job and Planck to another theoretical one. Planck's teaching assignment at Kiel corresponded to his research interests; he was, as he later said, a theoretical physicist "sui generis."[69]

64. Planck to Josef Strasser, 14 Dec. 1930, quoted in Hermann, *Planck*, 11.

65. Max Planck, *Wissenschaftliche Selbstbiographie* (Leipzig: J. A. Barth, 1948), 7–34, on 7 (hereafter cited as *Wiss. Selbstbiog.*), reprinted in *Phys. Abh.* 3: 374–401.

66. The text was Kirchhoff's own. Planck's account, *Wiss. Selbstbiog.*, 8–9, is supported by his fellow student Leo Graetz, who wrote to a friend at the time that Kirchhoff's mechanics lectures were taken straight from his text. Graetz to Auerbach, 6 May 1877, Auerbach Papers, STPK.

67. Hermann, *Planck*, 18.

68. Planck, *Wiss. Selbstbiog.*, 13.

69. Planck, *Wiss. Selbstbiog.*, 16.

In February 1889 Planck was transferred to Berlin at the same rank, extraordinary professor for theoretical physics. The Kiel faculty had to fill the position once again, recognizing that it would not be easy. There were many Privatdocenten for them to choose from, but almost all had made their reputations in experimental physics. In part, the reason for this was that few universities had ordinary professorships for theoretical physics, so young physicists saw little chance for advancement in the field.

The Kiel faculty decided on one of the few Privatdocenten with experience in theoretical physics: Graetz, who, with Planck, had taught theoretical physics at Munich for years. In his research, Graetz had worked on purely theoretical questions and on related experimental ones, and so he seemed eminently qualified.[70] But the Prussian ministry was dissatisfied with Graetz, and the faculty then added Auerbach and, with less enthusiasm, Arthur König to their recommendation, while persisting in their request for Graetz. The ministry was still dissatisfied because they did not want to "increase the Jewish element" in Kiel, and they simply claimed that there was no suitable young physicist. Moreover, they argued, an older physicist was needed in any case, since he would have to help out in experimental physics when the aging Karsten needed help. They appointed Leonhard Weber, increasing the salary for the job to twice what Planck had got and more than extraordinary professors generally got. When in September 1892 Karsten asked to be excused from giving his lectures on experimental physics, he thought it would be no hardship on Weber if he were to take them over, as he did; there were few students for theoretical physics.[71]

Karsten wanted Weber to succeed him on a regular basis, but that idea proved unpopular in Kiel.[72] Weber was instead given a personal ordinary professorship and a few rooms for atmospheric physics at the physics institute.[73] He went on teaching theoretical physics at Kiel until 1919.

Kiel was a small university remote from the centers of German physics, and like similar Prussian universities it suffered material neglect by a ministry whose main attention was devoted to Berlin; with Karsten as its director, the Kiel physics institute was almost devoid of significance for experimental research for a half century. Yet in the establishment of theoretical physics in German universities, Kiel played

70. Prussian Minister of Culture Gossler to Kiel U. Curator Mommsen, 4 Jan. 1889, and recommendation by the Kiel U. Philosophical Faculty, 25 Feb. 1889, DZA, Merseburg.

71. Karsten said that little of Weber's time was needed for "such lectures" in theoretical physics. Prussian Ministry of Culture to Kiel U. Curator, 2 Mar. 1889; Kiel U. Philosophical Faculty to Minister Gossler, 16 Mar. 1889; Gossler to Kiel U. Curator, 16 July 1889; Gossler to Minister of State and Finance v. Scholz, 16 July 1889; Curator to Gossler, 31 July 1889; Gossler to the King, 15 Aug. 1889; and Gustav Karsten to Kiel U. Curator, 17 Sept. 1892; all in DZA, Merseburg.

72. In the examinations he gave, the mathematician L. A. Pochhammer noted the inadequate preparation of the physics students, who were taught by Weber. Ernst Hagen, the Privatdocent for technical physics, called Weber's physics laboratory a model of what a laboratory should not be today. Charlotte Schmidt-Schönbeck, 300 Jahre Physik und Astronomie an der Kieler Universität (Kiel: F. Hirt, 1965), 100.

73. Schmidt-Schönbeck, 300 Jahre Physik . . . Kieler Universität, 100.

a part: its position for theoretical physics furthered the early careers of Hertz and Planck, two major contributors to theoretical physics in Germany.

Helmholtz told his audience at Berlin University what he had repeatedly told faculties and ministries: some physicists get more satisfaction from doing theoretical work, others from doing experimental work; some physicists are better at theory, others at experiment; but whatever their inclinations, they all have to have both theoretical and experimental knowledge.

> From a purely practical standpoint, in entering more deeply into physics one will do well to decide if one wants to follow one or the other direction and correspondingly apply more effort to one or the other direction. Equally, it must be emphasized at the outset that without mathematical physics, experimental physics is a very narrowly bounded science and affords little insight into the course of physical phenomena, while the reverse, mathematical physics without experimental physics, would likewise be a rather lame and unfruitful science; one does not do well to make theories of natural processes before one has seen these processes firsthand.[74]

Kirchhoff was a good example of the kind of physicist Helmholtz described: primarily interested in constructing mathematical theories, Kirchhoff was nonetheless thoroughly knowledgeable in experimental physics. When Kirchhoff died, Helmholtz was naturally concerned that he be replaced by another representative of theoretical physics with a comparable understanding of the needs of physics. Berlin tried, in vain, to attract Boltzmann and also Hertz, who since moving to Karlsruhe from Kiel had gained renown for his experimental researches on electric waves. They settled for Planck, who stood in second place behind Hertz on the list of candidates. Planck, who had not yet done research rivaling Boltzmann's or Hertz's, was appointed extraordinary professor, which meant that for the time being the teaching of physics at Berlin reverted to the usual arrangement: beside the ordinary professor who taught experimental physics, the second physicist taught theory as an extraordinary professor.

Despite his rank, Planck had a more important position at Berlin from the start than a second physicist usually had. He was director of an institute for theoretical physics, which Helmholtz had proposed in 1889. The institute was extremely modest, and few physicists would have been satisfied with it; but it was sufficient for what Planck wanted to do, which was to teach and do research in theoretical physics, as he had done at Munich and Kiel before coming to Berlin. The institute was established with a budget of 570 marks a year and an assistant. The money went mostly to the institute's library, and the assistant was used for read-

74. Hermann von Helmholtz, *Vorlesungen über theoretische Physik*, vol. 1, pt. 1, *Einleitung zu den Vorlesungen über theoretische Physik*, ed. Arthur König and Carl Runge (Leipzig: J. A. Barth, 1903), 4.

ing and correcting the students' written exercises. Experimental work was practically precluded. Planck's principal duty was to give a cycle of lectures on theoretical physics.[75]

Planck was a specialist of the kind Helmholtz found acceptable. Although Planck did not do experiments himself, he followed closely the work of experimentalists, reproducing in his papers their tables of measurements and suggesting ways for them to develop their observations. Upon arriving in Berlin as a determined "theorist," Planck felt that the assistants in the physics institute kept him at arm's length at first. But he was welcomed by the important physicist in the institute, its director Kundt, who had just replaced Helmholtz. Kundt wrote to a colleague at the time of the appointment that "in Planck, I believe we have made an excellent acquisition; in every respect, he appears to be a splendid man."[76]

Planck was then just thirty and was in the midst of a series of researches on the thermodynamics of chemical processes. These impressed Helmholtz, who believed that physical chemistry would dominate the next development of chemistry and would lead at last to a true theory of chemistry. Helmholtz regarded Planck as the most capable young scientist working on the subject, a matter he could judge expertly, having recently worked on it himself. The Berlin philosophical faculty's report on the candidate praised Planck's researches, as exemplified in his chemical thermodynamics, for carrying through the "strong consequences of thermodynamics without interference from other hypotheses." The report also praised Planck for his "original ideas," which was what Helmholtz always looked for in candidates for physics jobs.[77]

Between 1887, when he was still at Kiel, and 1891, Planck published four papers under the title "On the Principle of Increase of Entropy," in which he applied the second law of thermodynamics to chemical problems. His goal was, as he said in the first paper of the series, to carry further the "grand generalization" of Helmholtz, Josiah Willard Gibbs, and others: like the first principle of the mechanical heat theory, the second, the "Carnot-Clausius," principle applies not only to heat phenomena but to all kinds of physical and chemical phenomena; and because the second principle applies not only to reversible processes but also to irreversible, or "natural," processes, it applies to all processes whatsoever. Planck's starting point for studying the path of chemical reactions was Clausius's definition of entropy and the associated principle: in all processes involving a change of state of bodies, the sum of their entropies increases or remains the same and never decreases. With the help of this principle and through the introduction of thermodynamic potentials, in the first paper Planck developed a general theory of chemical equilibrium, the details of which he filled in in subsequent papers of the series. His approach was carefully

75. Max Planck, "Das Institut für theoretische Physik."

76. Planck, *Wiss. Selbstbiog.*, 16. Kundt to Graetz, 26 May 1889, Ms. Coll., DM, 1933, 9/18.

77. Report by the Berlin U. Philosophical Faculty, 29 Nov. 1888, DZA, Merseburg; quoted in Hermann, *Planck*, 21–22.

considered: he would determine the laws from the facts rather than by introducing "definite ideas of the nature of molecular motions."[78]

One of the later papers of the series Einstein judged to be Planck's "first great scientific accomplishment"; what Einstein admired so about it was the generality of its formulas, which contain all that can be derived from pure thermodynamic principles.[79] In this and other later papers, Planck applied his theory to the dissociation of gases, to dilute solutions, and to electrochemical processes. Planck published other papers during these years that dealt with similar matters, especially with electrical processes in solutions, which had become an active field through the recent work by Friedrich Kohlrausch, Wilhelm Ostwald, Walther Nernst, J. H. van't Hoff, Svante Arrhenius, and others. In 1893, Planck brought together the results of his many studies in a treatise on thermochemistry.[80]

Planck gave an invited talk on recent progress in heat theory at the German Association meeting in 1891, where he discussed the dissociation of gases and other problems he had taken up in his own recent work. He recalled the daring hypotheses of August Krönig, Clausius, and Boltzmann, who had tried to answer the final questions of the mechanics of atoms through the kinetic theory of gases; after its initial successes, this theory had not fulfilled the high expectations it had aroused, in Planck's opinion, and the subsequent efforts expended on it were disproportionate to the results obtained. Planck believed that a deeper insight into the world of molecules is obtainable from the general laws of thermodynamics together with the method of "ideal processes" (reversible processes, on which conclusions from the second law depend, are "ideal"); they are a "special triumph of the human mind," a "pathfinder," which can direct us to connections between natural laws in regions closed to direct experimentation. They had led, for example, to the theoretical establishment of the dependence of the conductivity of an electrolyte—or the degree of decomposition of molecules into ions—on the dilution. The use of ideal processes was the main reason for recent progress in thermodynamics, which was all the more remarkable because they cannot be directly proven by experiment.[81]

When a few years later, in 1894, Helmholtz together with Kundt and Bezold proposed Planck as an ordinary member of the Prussian Academy of Sciences, they discussed his research: eleven papers, all of which appeared in the *Annalen der Physik* beginning in 1887 and related "preponderantly to thermochemistry." His proposers mentioned the impressive results he had derived from general principles without having to rely on hypotheses about molecular motions: these included, for example,

78. Max Planck, "Ueber das Princip der Vermehrung der Entropie. Erste Abhandlung. Gesetze des Verlaufs von Reactionen, die nach constanten Gewichtsverhältnissen vor sich gehen," *Ann.* 30 (1887): 562–82, reprinted in *Phys. Abh.* 1: 196–216, on 196–200. Born, "Planck," 629.

79. Einstein referred to the third paper of the series: Max Planck, "Ueber das Princip der Vermehrung der Entropie. Dritte Abhandlung. Gesetze des Eintritts beliebiger thermodynamischer und chemischer Reactionen," *Ann.* 32 (1887): 462–503, reprinted in *Phys. Abh.* 3: 232–73. Albert Einstein, "Max Planck als Forscher," *Naturwiss.* 1 (1913): 1077–79, on 1077.

80. Max Planck, *Grundriss der allgemeinen Thermochemie* (Breslau, 1893).

81. Max Planck, "Allgemeines zur neueren Entwicklung der Wärmetheorie," *Zs. f. phys. Chemie* 8 (1891): 647–56, reprinted in *Phys. Abh.* 1: 372–81, quotations on 380–81.

the conditions under which different states of aggregation of a given substance can coexist and the equilibrium of mixed gases. Planck's "most important and most ingenious theoretical accomplishment" was to give a theoretical basis to an empirical law according to which the freezing point and vapor pressure of dilute aqueous solutions depend only on the molecular numbers of the dissolved substances; by this application of general principles, Planck brought together two separate and independent points of departure for theoretical chemistry into a "great and comprehensive connection."[82]

Institutional Reinforcement through Technical Institutes and Technical Physics Teaching

Several leading researchers in theoretical physics lectured on the subject at polytechnics or, as they came to be renamed in the late nineteenth century, technical institutes (Technische Hochschulen). Hertz, as we have seen, both studied and taught at such institutes, and it was at one of them, Karlsruhe, that he carried out the experiments on electric waves that profoundly altered the direction of theoretical physics in Germany. These institutes have a place in our study even if it is not nearly as large as that of the universities.

In the late nineteenth century, Germany had nine technical institutes located at Aachen, Berlin, Braunschweig, Darmstadt, Dresden, Hannover, Karlsruhe, Munich, and Stuttgart; in the early years of the twentieth century two more were added at Breslau and Danzig. At these institutes, as at the universities, the size of the physics staff and the range of physics courses varied considerably from place to place. For example, at the technical institute at Berlin at the turn of the century, physics including mathematical physics and mechanics was taught by two professors, four Docenten, five Privatdocenten, and three assistants. At the technical institute at Munich, physics was also well represented. In contrast, the technical institute at Breslau relied entirely on the local university for its physics faculty; Braunschweig had only one professor of physics along with one assistant; and Aachen had only one professor and one Docent.[83]

---

82. Proposal of Planck as ordinary member of the Prussian Academy of Sciences, signed by Helmholtz, Kundt, and Bezold. Document Nr. 23 in *Physiker über Physiker*, 125–26. The proposal is undated; Planck's election was on 11 June 1894.

83. By around 1880, the German polytechnics had achieved an official position as "higher schools." The corresponding change of name from "Polytechnic" to "Technische Hochschule" occurred at different times at different schools. The full equality of these schools with universities together with the right to confer the engineering doctorate came about in the 1890s. Karl-Heinz Manegold, *Universität, Technische Hochschule und Industrie*, vol. 16, Schriften zur Wirtschafts- und Sozialgeschichte, ed. W. Fischer (Berlin: Duncker und Humblot, 1970), 72–74. Numbers of physics teachers at German technical institutes are given in Wilhelm Lexis, ed., *Das Unterrichtswesen im Deutschen Reich*, vol. 4, pt. 1: *Die technischen Hochschulen im Deutschen Reich* (Berlin: A. Asher, 1904), 217, 296; also in Paul Forman, John L. Heilbron, and Spencer Weart, "Physics *circa* 1900. Personnel, Funding, and Productivity of the Academic Establishments," *HSPS* 5 (1975): 1–185, on 10. The technical institutes used the same principal titles for teachers as the universities: ordinary and extraordinary professor and Privatdocent. They also used "Docent," which stood for a teacher who was salaried in contrast with a Privatdocent who usually was

As one would expect, the main responsibility of physicists at technical institutes was to teach basic physics, which they taught within what was often called the "general department."[84] The teaching of this department complemented that of the professional engineering schools.[85] Much like the philosophical faculties of universities, the general departments of the technical institutes, led by those of Saxony and the southern states, developed teaching goals of their own beyond providing engineers with a general education. In the 1860s, for example, the technical institute at Dresden established within its general department a professional school for training teachers of mathematics, physics, and other natural sciences. Darmstadt, Karlsruhe, Munich, and Stuttgart also offered partial or complete training for teachers of these subjects. Later the Prussian technical institutes followed their lead; Prussia's new testing ordinance of 1898 allowed teaching candidates in mathematics and the physical sciences to spend half of their time at technical institutes, and Aachen soon provided teaching for this purpose.[86]

The technical institutes did not have the right to grant doctorates in physics, which assured that they would have less significance for physics than the universities. Even so, they provided an opportunity for a good many physicists to study, teach, and practice physics. Like the universities, the technical institutes came to recognize the need for a "second" physicist; by 1891, for example, three of them— those at Aachen, Dresden, and Karlsruhe—had teachers for theoretical physics, all extraordinary professors, and in time the other technical institutes also acquired their theoretical physics lecturers.[87] Eventually, as at the universities, at these institutes the extraordinary professorships for the subject were gradually replaced by ordinary ones.

not. Otto Lehmann, for example, moved to Aachen in 1883 as Docent for physics and assistant to Adolph Wüllner, the physics professor. Lehmann's classes were small: he had two students in his lectures on the mechanical theory of heat, four in his lectures on experimental physics, three in his laboratory, and none in his class on applied physics. The space for research in the laboratory was so small that he could not even find room for the little apparatus he had brought with him from Mühlhausen, where he had been teaching in the secondary school. It took him ten times longer to do research at Aachen than at Mühlhausen, he said, but he felt compensated by the better collection of apparatus at Aachen and the increased time he had for research there. Lehmann to Warburg, 6 Nov. 1883, STPK. K. L. Weiner, "Otto Lehmann, 1855–1922," in Geschichte der Mikroskopie, vol. 3, ed. H. Freund and A. Berg (Frankfurt a. M.: Umschau, 1966), 261–71, on 262.

84. Of the fields represented by the general department, physics was regarded as the "most important natural-scientific field of instruction," since it related to all other subjects taught in the technical institutes. Robert Fricke, "Die allgemeinen Abteilungen," in Das Unterrichtswesen im Deutschen Reich, ed. W. Lexis, vol. 4, pt. 1, 49–62, on 54.

85. To the physicists Max Wien and Jonathan Zenneck at Danzig, it was "one of the most beautiful tasks of the technical institutes to introduce the exact methods of physics into the treatment of technical problems." Wien and Zenneck to the Danzig Technical Institute Senate, 15 Feb. 1906, Ms. Coll., DM.

86. Fricke, "Die allgemeinen Abteilungen," 58–61. Paul Stäckel, "Angewandte Mathematik und Physik an den deutschen Universitäten," Jahresber. d. Deutsch. Math.-Vereinigung 13 (1904): 313–41, on 323–24, 335. At Dresden, for example, where the general department had thirty-seven students in 1903–4, the teaching candidates would take the same mathematics and mechanics courses as the engineers in their first four semesters, after which they would take more advanced courses in the mathematical-physical disciplines.

87. Lehmann's 1891 survey of budgets for German physics institutes, Bad. GLA, 235/4168.

The increasing opportunities for physicists to work at technical institutes corresponded to the growth in number, size, and stature of these institutions. That growth, in turn, reflected the advanced technological needs of industrial Germany. In the face of giant industries and world trade, the conquest of space and time, and the whole drive toward the "technical mastery of lifeless nature," physics and chemistry were often viewed as driving forces of the modern age.[88] Physicists sometimes encouraged this association of their science with technological progress. Wilhelm Wien, speaking on science and German universities in 1914, observed that physics together with chemistry had "created the firm foundations on which the pillars of our industry stand, which support a great part of our economy."[89] At the fiftieth anniversary of the Berlin Physical Society in 1896, its president Emil Warburg contrasted the quiet, unnoticed condition of physics at the founding of the society with its present world renown, a difference he attributed not to physical discoveries but to the impact of physics on economic life, especially that through electrotechnology.[90] The understanding of the age as one of applied science was given historical expression by the Deutsches Museum in Munich, which was built in 1908 from public and private sources to symbolize the "mutual influence" of technical and scientific work in the past and present.[91]

The parts of physics—mechanics, heat, electricity, magnetism, and optics—had been applied earlier, but the vision of physics as a science to be systematically exploited, which in one form had flourished in the early nineteenth century only to be largely lost sight of, was a distinctive accompaniment of late nineteenth-century industrialism: it was now understood that to each branch of basic physics there corresponded a branch of applied physics.[92] From the 1860s and 1870s, technical optics, technical electricity, and other technical branches began their strong development, while institutes for conducting physical research for technical ends began to proliferate.[93] Physics was also applied in medicine.

The teaching of "applied" or "technical" physics fell to the teachers of theoretical physics at several universities and technical institutes. The aim of their teaching was to make physics useful for technology. In the introduction to his published lectures on technical mechanics, the Munich technical physicist August Föppl explained why such courses were needed. The physicist and the mathematician were of no help to the technologist, Föppl said, in their attempts to encompass all physical phenomena within a single formula, such as Hamilton's principle of least action

88. For example, by Oscar Hertwig, "Die Entwicklung der Biologie im 19. Jahrhundert," *Verh. Ges. deutsch. Naturf. u. Ärzte* 72, pt. 1 (1900): 41–58, on 41–42.

89. Wilhelm Wien, *Die neuere Entwicklung unserer Universitäten und ihre Stellung im deutschen Geistesleben* (Würzburg: Stürtz, 1915), 17.

90. Emil Warburg's address in *Verh. phys. Ges.* 15 (1896): 30–31, on 30.

91. Note on the new building for the Deutsches Museum in *Internationale Wochenschrift* 2 (1908): 608.

92. Georg Gehlhoff, Hans Rukop, and Wilhelm Hort, "Zur Einführung," *Zs. f. techn. Physik* 1 (1920): 1–4.

93. Wilhelm Hort, "Die technische Physik als Grundlage für Studium und Wissenschaft der Ingenieure," *Zs. f. techn. Physik* 2 (1921): 132–40; Friedrich Klemm, "Die Rolle der Mathematik in der Technik des 19. Jahrhunderts," *Technikgeschichte* 33 (1966): 72–91.

or Hertz's generalized law of inertia. Moreover, the rigorous demands that the physicist and the mathematician place on mechanics cannot be met in practice; the
technologist must resort to provisional, approximate theories, since he is under pressure to come up with useful results.[94] Föppl responded to criticism of his presentation
of mechanics by noting that the practical value of Hamilton's principle for technologists exists "only in the imagination" of the mathematician and that technologists
have reason to be "rather mistrustful" of it.[95] The Göttingen technical physicist
Hans Lorenz said in his published lectures on technical physics that students by and
large still learned physics in the old-fashioned experimental lectures, in associated
laboratory courses, and in theoretical lectures that were strongly mathematical and
not at all technical. In these typical theoretical lectures, Lorenz said, students
learned a "small number of the most general, sharply formulated laws," which they
were unable to apply in practice.[96] Theoretical physicists who taught technical physics were expected to recognize the practical needs of their audience and to accommodate them.

There were advantages in having theoretical and technical physics taught by the
same man, usually the extraordinary professor. To the institution at which he
taught, the combination of the two subjects offered a route to the creation of two
separate positions and perhaps of a new institute, all of which would become possible
when their need had been demonstrated by the attendance at the earlier lecture
courses.[97] For the theoretical physics lecturer, the doubling up of theoretical and
technical physics had the practical advantage of increasing his income, since his
lectures on technical physics were often well attended and brought him more fees
than his theoretical lectures. He might even take the initiative of having technical
physics added to the curriculum. Walter König, when asked about assuming the
extraordinary professorship for theoretical physics at Heidelberg University in 1899,
urged the Baden ministry of education to introduce applied physics, especially electrotechnology. He said that in recent years many universities had offered introductory lectures on these subjects, taught either by a specialist or by the "second" physicist, that is, by the theoretical physics lecturer; because of the great importance of
applied physics for practical life, there was a definite need, particularly in the training of teachers, to give students a more thorough knowledge of this subject than

94. August Föppl, *Vorlesungen über technische Mechanik*, vol. 1, *Einführung in die Mechanik*, 5th ed.
(Leipzig: B. G. Teubner, 1917), vi, 1, 9–10. The first edition (1898) was based on lectures that Föppl
gave to second semester students in technical mechanics.

95. Föppl to Arnold Sommerfeld, 29 Mar. 1902, Sommerfeld Correspondence, Ms. Coll., DM.

96. Hans Lorenz, *Technische Mechanik starrer Systeme* (Munich: Oldenbourg, 1902), v–xi. This text
is based on lectures Lorenz gave at Halle and Göttingen in 1899–1902; it is the "foundation" and the
first of several volumes covering "technical physics." Hort, "Die technische Physik," 134.

97. The Karlsruhe physics professor Lehmann, for example, in 1891 anticipated an electrotechnical
institute with an independent director, but he thought that such an institute would not be granted for
several years. He therefore proposed as a "transition phase" the creation of a separate chair for theoretical
physics, "pure as well as applied." The chairholder, he explained, "could then, as soon as the opportunity
arose, be appointed director of the new electrotechnical institute and would have the opportunity of
preparing its arrangements without rush, completely according to his wishes." Lehmann to the Director
of Karlsruhe Technical Institute, 3 June 1891, Bad. GLA, 235/4168.

they could get from the already overloaded general lectures on experimental physics.[98] When discussing the extraordinary professorship for theoretical physics at Würzburg in 1903, Friedrich Pockels said that he might like to give lecture courses meant for a larger audience, one of which might be on the foundations of electrotechnology, a subject Theodor Des Coudres had previously taught in the same position.[99] At Halle, from 1895 on, the extraordinary professor for theoretical physics Karl Schmidt increasingly taught applied physics, which as personal ordinary professor he continued to do after 1915 with the aid of a "provisional laboratory for pure and applied physics."[100] At Erlangen, to take another example, the extraordinary professor from 1904 to 1906 Arthur Wehnelt held a teaching assignment for "theoretical and applied physics." Wehnelt already had had some teaching experience in electrotechnology, and he had studied engineering before physics.[101] But his successor, Rudolf Reiger, had no such background in applied physics; as an assistant in the Erlangen physics institute and as an acquaintance of Wehnelt, he was thought to have adequate qualifications for teaching the subject, since at this time there was no regular preparation for a teacher of applied physics.[102] As late as 1931, the Hannover theoretical physicist Erwin Fues could speak of the "separation of this [theoretical physics] discipline which until recently had been united with applied physics" in many universities and technical institutes.[103]

For physicists who taught or did research in theoretical physics, technical physics did not have to mean only a source of extra income. They frequently took an interest in the scientific problems suggested by technical subjects. Clausius, as one physicist who combined theoretical and technical physics in his teaching, published on the theory of the steam engine during his researches on heat theory in the 1850s, and he published on the theory of the dynamo-electric machine during his researches on electrodynamic theory in the 1880s. He valued similar work by other physicists; for example, he recommended Auerbach for work that belonged to the earliest measuring researches on those "interesting machines," the dynamos.[104]

98. Walter König to "Geheimrat," 12 Feb. 1899, Bad. GLA, 235/3135.
99. Pockels to Wilhelm Wien, 14 Mar. 1903, Wien Papers, Ms. Coll., DM.
100. In 1921 Schmidt's laboratory was named "laboratory for applied physics," which agreed with the direction of his work. Upon Schmidt's departure, the Halle faculty in 1927 called for an ordinary professor for technical physics. Communication from Dr. H. Schwabe, Director of the Halle University Archive.
101. Wehnelt's vita submitted to Erlangen U. in 1904; recommendation by the Erlangen U. Philosophical Faculty and the Prorector that Wehnelt be appointed extraordinary professor, at a salary of 3180 marks, with the teaching assignment for "theoretical and applied physics," 3 Oct. 1904; and approval of the request by the Bavarian government, 18 Nov. 1904; Wehnelt Personalakte, Erlangen UA.
102. Reiger's vita; Erlangen U. Philosophical Faculty to Academic Senate, recommending Reiger's appointment as extraordinary professor with the same salary and teaching assignment as Wehnelt's, 7 Mar. 1907; and approval by the Bavarian government, 4 May 1907; Reiger Personalakte, Erlangen UA.
103. Quoted in Forman, Heilbron, and Weart, "Physics circa 1900," 32.
104. Clausius to Director Greiff, 15 Aug. 1887, STPK, Darmst. Coll. 1913.51.

16

Boltzmann at Graz

Early in his career, Boltzmann carried out a number of valuable experimental researches, but his work that was to have greatest influence in physics was theoretical. In a series of publications in 1868–77, he helped lay the foundations of statistical mechanics, as the subject came to be called. His probabilistic interpretation of the second fundamental law of the theory of heat belonged to this work, and the molecular methods he created to analyze the second law became standard tools of the theoretical physicist. While engaged in this research, Boltzmann held positions at Graz, at Vienna, and again at Graz, where he taught mathematical physics, mathematics, and experimental physics, respectively. The recognition his research received in Germany brought him a call in 1890 to Munich University as professor for mathematical physics; from there his teaching as well as his researches entered directly into the development of theoretical physics in Germany.

Graz Chair for Mathematical Physics

The Graz chair for mathematical physics—it was that from the start, a *Lehrkanzel*—was proposed as an extraordinary position in 1863 by the philosophical faculty, acting on the initiative of their mathematics professor. The faculty wanted to strengthen studies in the natural sciences, in their eyes a "most urgent need." They argued that mathematical and experimental physics had become so extensive that any attempt to do justice to both would exceed the strength of any single teacher.[1]

As requested, the Austrian ministry of state created a chair for mathematical physics at Graz University, but it did so mainly for reasons the Graz faculty had not

---

1. "Commissionsbericht über den in der Sitzung vom 17. Juni l. J. gestellten Antrag, die Errichtung einer ausserordentlichen Lehrkanzel der mathematischen Physik betreffend," 2 July 1863, and covering letter by the Graz U. Philosophical Faculty to Austrian Ministry of State, n.d.; Dean of Philosophical Faculty to Ministry of State, 5 July 1863; Rector of Graz U. Wagly to Ministry of State, 6 July 1863; all in Öster. STA, 5 Phil, Physik.

(and could not have) stated in their proposal. Graz was not entitled to a second physics chair since the ministry could not create chairs for all fields of research at the Austrian provincial universities. That the ministry created one all the same was due mainly to the failings of the Graz physics professor, Carl Hummel, who had been the main physicist at Graz since the physics chair had been separated from natural history back in 1850. His teaching was not regarded as being on a "scientific" level, with the result that the teaching of other sciences and medicine was hindered. To correct this problem, the ministry wanted to bring in a second man to teach both experimental and mathematical physics. Presumably, the second man would become the only teacher of physics at Graz upon Hummel's retirement; the appointment of an extraordinary professor for mathematical physics was considered a temporary measure designed to spare Hummel's feelings and the state's budget.[2]

For the next twelve years after the creation of the second physics position at Graz, there were frequent turnovers in the physics staff attended by a measure of confusion. Victor von Lang was the first to be appointed to the new extraordinary chair.[3] He promptly reported to the dean on the "complete unusability" of the physics institute and left after a year for a job at Vienna. In 1866 the new ordinary professor of mathematics at Graz Ernst Mach, who did not regard mathematics as his proper field and had been teaching the physiology of the senses under the rubric "medical physics," gave up his job in exchange for Lang's. Mach's transfer from mathematics to physics without a change in rank meant that Graz now had two ordinary professors of physics, Mach and Hummel. Mach remained at Graz for only one more year before moving to Prague as professor of physics in 1867. Hummel retired the same year, so both physics positions at Graz were vacant, and for a time all physics teaching was done by the Privatdocent Simon Subič.[4]

The Graz faculty commission asked Kirchhoff for advice about Hummel's successor as director of the physics institute. Kirchhoff told them that Subič and the other Austrians they were considering were not good enough for Graz and that, in general, if Austrian universities excluded foreign physicists on nationalistic grounds, the universities would be ruined and with them German culture in Austria. One of the physicists Kirchhoff recommended to the Graz faculty was August Toepler, a

2. Austrian Ministry of State, Section for Culture and Education, and Ministry of Finance reports, 10 July 1863, Öster. STA, 5 Phil, Physik. Carl Hummel, who was miserably paid, did little research or writing on physics: he published a secondary school text on mathematics, a text on physical geography, and one scientific paper on the electrophorus. Hans Schobesberger, "Die Geschichte des Physikalischen Institutes der Universität Graz in den Jahren von 1850–1890" (manuscript), p. 16, Graz UA.

3. Victor von Lang was the faculty's first preference, followed by Edmund Reitlinger and Ernst Mach. "Commissionsbericht," 2 July 1863, and covering letter. Lang was appointed "e. o. professor for physics," according to a letter from the Ministry to the Dean of the Graz U. Philosophical Faculty, 2 Mar. 1864, quoted in Schobesberger, "Die Geschichte . . . der Universität Graz," p. 16, and that was how he was identified in the Graz lecture catalog, *Akademische Behörden, Personalstand und Ordnung der öffentlichen Vorlesungen an der K. K. Carl-Franzens-Universität . . . zu Graz*; there he was listed as teaching "higher physics," which meant mathematical physics. This is one of the cases in which the chair, "Lehrkanzel," had one designation, here "mathematical physics," and the occupant another, here "physics."

4. Mach to Graz U. Philosophical Faculty, 10 Oct. 1865, Graz UA, N. 20 Phil. Dec. 1866. "Vortrag" of Austrian Minister of State Richard von Belcredi, 9 Apr. 1866, and Mach's appointment, 19 Apr. 1866; "Commissions-Bericht" on Simon Subič, 15 Jan. 1869; all in Öster. STA, 5 Phil, Physik. Schobesberger, "Die Geschichte . . . der Universität Graz," 24, 26, 28, 30, 34, 38.

German who was then professor at the polytechnic in Riga; Helmholtz supported Kirchhoff in this recommendation, and in 1868 Toepler was duly appointed.[5]

For the second vacant physics position—presumably the original extraordinary professorship—the Graz faculty recommended only Subič, the local man. This did not satisfy the ministry, which was advised by Josef Stefan to appoint the Vienna Privatdocent Boltzmann rather than Subič. The ministry told the Graz faculty to present a full list of candidates, which now included Boltzmann.[6] With Stefan, the ministry agreed that because of Boltzmann's "scientific accomplishments," he offered them the "best guarantee." So in June 1869 Boltzmann was appointed ordinary, not extraordinary, professor for "mathematical physics," with the additional assignment of lecturing regularly on the "elements of higher mathematics" to prepare Graz's poorly equipped students for the physics lectures.[7]

Boltzmann's Molecular-Theoretical Researches

Boltzmann lived up to Stefan's high expectations: his Graz years were characterized by "entirely uncommon accomplishments," which Stefan had promised the ministry on the basis of Boltzmann's "*extraordinary* talent" and his "thorough and *versatile* mathematical knowledge."[8] At Graz Boltzmann developed the kinetic theory of gases, his "main lifework," establishing himself as the "passionate molecular theorist" he was to remain. In going beyond the work of Krönig, Clausius, and Maxwell, among others, Boltzmann confronted increasingly difficult mathematical problems, which called into play his considerable mathematical skills.[9]

Following his initial work on the second fundamental law of the theory of heat, Boltzmann studied Maxwell's writings on the kinetic theory of gases. Recognizing their importance for his own further work on the subject, in several papers in 1868 and 1871 he examined, generalized, and applied to new problems Maxwell's law of the distribution of velocities among the molecules of a gas in thermal equilibrium, $f = Ae^{-h\varphi}$.[10] With Maxwell, Boltzmann understood that since the different gas mol-

5. Report by the Austrian Ministry of Culture and Education, 28 Dec. 1867, Öster. STA, 5 Phil, Physik. Schobesberger, "Die Geschichte . . . der Universität Graz," 50, 54.

6. "Commissions-Bericht" on Subič, 15 Jan. 1869; Josef Stefan to Austrian Ministry of Culture and Education, 8 Apr. 1869; Dean of Graz U. Philosophical Faculty D. F. Kronen to Ministry of Culture and Education, 29 July 1869; Ministry of Culture and Education to Dean of the Philosophical Faculty, 15 Apr. 1869; report by the Ministry to the Emperor, 28 June 1869; all in Öster. STA, 5 Phil, Physik. The list of candidates, which the Graz faculty supplied on request, consisted of Subič, Boltzmann, and Waizmuth, an assistant at the polytechnic in Prague, in that order.

7. Austrian Ministry of Culture and Education to the Emperor, 28 June 1869.

8. Austrian Ministry of Culture and Education to the Emperor, 28 June 1869.

9. Theodor Des Coudres, "Ludwig Boltzmann," *Verh. sächs. Ges. Wiss.* 85 (1906): 615–27, on 623. Woldemar Voigt, "Ludwig Boltzmann," *Gött. Nachr.*, 1907, 69–82, on 72.

10. In this equation, A and h are constants and φ is the sum of the living force and the force function, or the total energy, of a molecule. Whereas in Maxwell's distribution function, written in these symbols, φ is kinetic energy only, Boltzmann showed that the function could be generalized in this way. Ludwig Boltzmann, "Studien über das Gleichgewicht der lebendigen Kraft zwischen bewegten materiellen Punkten," *Sitzungsber. Wiener Akad.* 58 (1868): 517–60. Martin J. Klein, *Paul Ehrenfest*, vol. 1, *The Making of a Theoretical Physicist* (Amsterdam and London: North-Holland, 1970), 96–97. Stephen G. Brush, "Boltzmann, Ludwig," *DSB* 2 (1970): 260–68, on 261–62.

ecules pass through all possible states of motion, it is of the "highest importance" to know the probability of the various states of motion. The probability is needed to calculate, for example, the average living force, the average force function, or potential, and the mean free path of a molecule. For the case that a molecule is a single material point, as in a monatomic gas, Maxwell had derived the probability of the various states of motion. Boltzmann extended Maxwell's analysis to the more realistic case of gases composed of polyatomic molecules. A polyatomic molecule has internal atomic motions, so that its state depends on more than its translational velocity, a single variable. The probability of its state at a given moment is determined by the distribution function f, which depends on the positions and velocities, $\xi_1, \eta_1 \ldots, w_r$, of all the constituent atoms. From a knowledge of this probability, the average value of any property of the gas, as represented by a function X of the positions and velocities of the atoms, can be calculated: $\overline{X} = 1/N \int X \, dN$, where N is the number of molecules per unit volume of the gas, and $dN = f(\xi_1, \eta_1, \ldots, w_r) d\xi_1 d\eta_1 \ldots dw_r$ is the fraction of molecules whose atoms have positions and velocities falling within the intervals ξ_1 to $\xi_1 + d\xi_1$, η_1 to $\eta_1 + d\eta_1, \ldots, w_r$ to $w_r + dw_r$. Following this procedure, Boltzmann calculated the average living force of an atom, identifying it with the temperature. By the same procedure he showed that the average living force of the progressive motion of a molecule has the same value as the average living force of each of its atoms, a form of the so-called equipartition theorem for average energies, which he had stated before. He applied this theorem to the ratio of the specific heat of air at constant pressure to that at constant volume and remarked, with reference to the disparity between the theoretical and the observed values, that "each advance in overcoming the mathematical difficulties, which the calculation of the behavior of such atomic complexes [molecules] offers, appears to me to be of the highest importance for the investigation of the true constitution of gas molecules."[11]

Foremost among the problems to which Boltzmann applied his newly acquired probabilistic approach, as embodied in the distribution law, was the foundation and clarification of the second fundamental law of the mechanical theory of heat. Whereas Maxwell had recognized that the truth of the second law has a strong probability rather than absolute certainty, owing to the enormous number of molecules involved, he did not create the statistical theory that would show how a mechanical system of molecules behaves in a fashion required by the second law. That task was left to Boltzmann, who made an important step in that direction in 1871 with a new proof of the second law.[12] Five years before, he had demonstrated the existence of an entropy function by considering only the motion of atoms in closed paths, an unrealistic restriction. His new proof was based not on closed paths but on the average values of atomic motions, as determined with the aid of the distribution function. Representing the temperature T of a body by the average living force of an atom, and representing a small quantity of heat δQ added to the body

11. Ludwig Boltzmann, "Über das Wärmegleichgewicht zwischen mehratomigen Gasmolekülen," *Sitzungsber. Wiener Akad.* 63 (1871): 397–418, reprinted in *Wiss. Abh.* 1: 237–58, on 237–39, 256–58.
12. Martin J. Klein, "Maxwell, His Demon, and the Second Law of Thermodynamics," *American Scientist* 58 (1970): 84–97, on 91–92.

by the change in the average living force and force function of the atoms of the body, Boltzmann proved, as before, that δQ divided by T is a complete differential. Again he showed that the resulting equation for the entropy is useful by calculating with it the entropy of an ideal gas.[13] He referred this new analysis of the heat in a body to the problem of the specific heats of solids, from which he derived further conclusions about the real molecular world.[14]

Even with his latest proof of the existence of an entropy function, Boltzmann had not completely clarified the molecular basis of the second fundamental law. His proof had dealt with the second law only in relation to equilibrium. He still needed to examine the entropy function for the realistic, nonequilibrium case of irreversible processes, and for this he needed new methods for tracing the development of the thermal properties of a body over time. He addressed this more difficult problem the following year, in 1872, in a comprehensive work on gas molecules. The title he gave it, "Further Studies," is accurate but gives no idea of the important results it contains.[15]

To motivate these further studies, Boltzmann again elaborated on the need for using average values in exact physics, still a relatively new understanding. He pointed out that physicists do not observe molecules performing the lively motion that the mechanical theory of heat attributes to them; rather, they observe the lawful behavior of heated bodies, which they now recognize as the consequence of the "most random events" among molecules that, under the same conditions, always lead to the same average values. For calculating the necessary averages, the "probability calculus" is used, which does not mean that the mechanical theory of heat is any less certain; Boltzmann reassured his readers that the theory remains mathematically exact and confirmed by experience.[16]

For an exact theory of averages, it is important for physicists to have complete

13. The general form of Boltzmann's entropy function is:

$$-r \log h + \frac{2h}{3} \frac{\int \chi e^{-h\chi} d\sigma}{\int e^{-h\chi} d\sigma} + \frac{2}{3} \log \int e^{-h\chi} d\sigma + \text{constant},$$

where r is the number of atoms in the body, h is a constant, χ is the force function, and $d\sigma$ stands for the product of differentials of the positions and velocities of all of the atoms. Boltzmann showed that for an ideal gas this expression becomes the correct

$$\log \left(T^\lambda \, v^{\frac{2\lambda}{3}} \right) + \text{constant},$$

where λ is the number of gas molecules and v is the volume of the gas. Ludwig Boltzmann, "Analytischer Beweis des zweiten Hauptsatzes der mechanischen Wärmetheorie aus den Sätzen über das Gleichgewicht der lebendigen Kraft," *Sitzungsber. Wiener Akad.* 63 (1871): 712–32, reprinted in *Wiss. Abh.* 1: 288–308, on 303, 305.

14. Boltzmann interpreted the "Dulong-Petit or Neumann law" of the specific heats of solids by assuming that atoms are bound in place by a simple elastic restoring force. He concluded that this force holds in first approximation for bodies that obey this law and that the atoms of bodies that do not obey it must have a different law of force. Boltzmann, "Analytischer Beweis," 306–8.

15. Ludwig Boltzmann, "Weitere Studien über das Wärmegleichgewicht unter Gasmolekülen," *Sitzungsber. Wiener Akad.* 66 (1872): 275–370, reprinted in *Wiss. Abh.* 1: 316–402. Klein, *Ehrenfest*, 100.

16. Boltzmann, "Weitere Studien," 316–17.

confidence in their distribution law. In the previous year, 1871, Boltzmann had proved that *if* the distribution law has Maxwell's form, it cannot be changed by the collision of molecules or by the motion of their constituent atoms. He regarded this as the first "rigorous" proof that the distribution law meets all of the conditions that the real distribution of states of gas molecules has to satisfy. Now he set out to prove that the distribution law *must* have this form; that is, it not only yields results that agree with experience but also the equilibrium distribution of velocities it describes is the only one that does not change over time owing to collisions between molecules. In the course of proving the necessity of this distribution, Boltzmann advanced a theorem that provided the first insight into the nature of irreversible processes.[17]

To establish the uniqueness of the distribution law, Boltzmann analyzed the effect of collisions between gas molecules, two at a time, on the distribution function f at time t by studying the partial differential equation $\partial f/\partial t$ (a special case of the "Boltzmann equation" for transport phenomena). To prove that *any* distribution f always tends in time toward Maxwell's, Boltzmann introduced an auxiliary function, $E = \iint \ldots f \log f \, dx_1 dy_1 \ldots dw_r$, and showed that it behaves similarly to Clausius's entropy function. It has, namely, a directional property: E can never increase over time; it is proportional to the negative of the entropy, which can never decrease.[18] (Later Boltzmann would write H for E, and the associated theorem was to become known as the "H-theorem."[19]) To be precise, Boltzmann proved that if the initial distribution of velocities among the molecules of a gas is not described by Maxwell's law, the function E will decrease over time until temperature equilibrium is established, at which point E will have reached its constant minimum value, and the molecular velocities will be distributed according to Maxwell's law. With this result, Boltzmann provided an "analytic proof of the second fundamental law in an entirely different way," one which embraces the irreversible processes we observe in nature and not just the "ideal" reversible processes.[20] Boltzmann had completed—or so it seemed—this fundamental study: he had provided the molecular interpretation of the second law, at least in its application to thin gases, both monatomic and polyatomic.[21]

17. Boltzmann, "Über das Wärmegleichgewicht," 254–55. Klein, "Maxwell, His Demon," 92.

18. Boltzmann, "Weitere Studien," 369–402, especially on 393. Boltzmann associated the entropy with the function $E^* = N \iint \ldots f^* \log f^* \, dx_1 dy_1 \ldots dw_r$, where N is the number of molecules in the gas and $f^* = f/N$. He showed that within a constant factor and an additive constant, E^* gives the correct form for the entropy of an ideal monatomic gas, and he showed in general how E^* relates to the second law of the mechanical theory of heat in the form $\int (dQ/T) < 0$. "Weitere Studien," 399–401. The reasons for Boltzmann's choice of the letter E, in this form of writing the second law, and of the sign reversal (the entropy increases while E decreases in the passage to equilibrium) are discussed in n. 13, p. 269, as is Boltzmann's 1872 paper in general, on pp. 42–46, in Thomas S. Kuhn, *Black-Body Theory and the Quantum Discontinuity 1894–1912* (New York: Oxford University Press, 1978).

19. For a time this statement was known as "Boltzmann's minimum theory," then as "Boltzmann's H-theorem." Stephen G. Brush, *The Kind of Motion We Call Heat: A History of the Kinetic Theory of Gases in the 19th Century*, vol. 1, *Physics and the Atomists* (Amsterdam and New York: North-Holland, 1976), 238. Brush's discussion of Boltzmann's 1872 paper is on pp. 235–38.

20. Boltzmann, "Weitere Studien," 345.

21. Klein, *Ehrenfest*, 102.

While Boltzmann's work on the molecular foundations of the mechanical theory of heat took its starting point largely in Maxwell's and his own theoretical studies, he also followed the experimental work bearing on the subject, especially the experimental confirmation by his Vienna colleagues Stefan and Josef Loschmidt of Maxwell's temperature-dependent constants for diffusion and heat conduction in gases.[22] The theory behind the constants had a shaky foundation because the "intramolecular motion" was still largely unknown. Stefan's experimental determination of the heat conduction constant seemed to Boltzmann far superior to the theoretical in exactness.[23] In other words, much molecular-theoretical work remained to be done.

The New Graz Physics Institute

While teaching mathematical physics at Graz, Boltzmann was proposed as one of the candidates for a new position in mathematics at Vienna University. Boltzmann was known to be a physicist, but that did not deter the Vienna faculty, who argued that although his researches originated in physics, they were also "excellent as mathematical works, containing solutions of very difficult problems of analytical mechanics and also especially of probability calculus." Moreover, the faculty recognized in Boltzmann's use of higher analysis in the theory of heat the mark of a "decided mathematical talent."[24] Boltzmann received the appointment, and from 1873 to 1876 his position was that of ordinary professor of mathematics. His published research during these Vienna years included one paper devoted wholly to mathematical discussion,[25] but otherwise it dealt with the theory of gases and several other physical topics, including the experimental determination of the dielectric constant of insulating bodies. In his mathematical job, Boltzmann continued to work as a physicist.

As it happened, no sooner had Boltzmann left for Vienna to teach mathematics than the Graz faculty began talking of bringing him back to teach physics. Their new physics institute had just opened, which created an "entirely new state of affairs" for physics at Graz. Boltzmann did go back, as we will see, but not immediately and not in his old capacity as professor for mathematical physics but as professor for experimental physics and director of Graz's new physics institute.[26]

For now, Toepler was still the director of the institute. At the time he came to Graz, he complained that the physics cabinet contained "for the most part antiquated junk," and he immediately set about to refurbish it, increase its budget, and get extra funds for instruments. He made good use of the calls he received from

22. Ludwig Boltzmann, "Über das Wirkungsgesetz der Molekularkräfte," *Sitzungsber. Wiener Akad.* 66 (1872): 213–219, reprinted in *Wiss. Abh.* 1: 309–15.

23. Boltzmann, "Weitere Studien," 368.

24. Moth of the Vienna Philosophical Faculty to the Austrian Ministry of Culture and Education, 12 Mar. 1873, Boltzmann file, Öster. STA, 4 Phil.

25. Ludwig Boltzmann, "Zur Integration der partiellen Differentialgleichungen 1. Ordnung," *Sitzungsber. Wiener Akad.* 72 (1875): 471–83, reprinted in *Wiss. Abh.* 2: 42–53.

26. August Toepler, Johann Frischauf, and Leopold Pebal to the Graz U. faculty, 2 May 1874, Öster. STA, 5 Phil, Physik.

Zurich, Karlsruhe, and elsewhere to improve the institute in any way he could. Most important, in 1871 he got the state to commit 100,000 florins for a new institute building, which was to make Graz highly attractive to physicists. The new building was impressive, containing, for example, an arrangement of doors and windows that allowed an unobstructed horizontal observation line of over fifty meters. It had working areas free of iron for magnetic measurements, a photography laboratory, an ice cellar for constant temperature work, an astronomical observatory, and a steam engine for driving all the moving parts of the institute; it had floors isolated by separate supports to reduce vibrations for precision measurements; and all the laboratories and the big lecture hall were illuminated by sunlight. During the years Boltzmann was away in Vienna, 1873–76, Toepler bought 28,000 florins worth of apparatus for Graz. With justification, the Graz faculty boasted that their new physics institute was the "first larger institution of its kind to be founded in Austria."[27]

The intention behind the new institute was to found an "academic school for physicists." Accordingly, it was understood that the director of the institute needed to devote his time to experimental lectures and laboratory instruction and to "such extended observations and experimental investigations as correspond to the size and resources of the institution." As a corollary, the second physicist at Graz, the professor for mathematical physics, needed to teach the "purely theoretical side of physics" with "heightened care." Boltzmann had taught mathematical physics at Graz in a way that corresponded to the planned development of the new institute, and the Graz faculty commission wanted his replacement, if he were not Boltzmann himself, to be another like him, a mathematical physicist with solid scientific accomplishments. The only other Austrian physicist they could think of who fitted their description was the Prague professor Ferdinand Lippich, whom they believed they could not get. So they looked outside Austria for a mathematical physicist, recommending O. E. Meyer at Breslau.[28] The ministry said no and instructed the faculty to suggest some more Austrian names, with the result that the Vienna Privatdocent Heinrich Streintz, who had worked mainly in experimental physics, was appointed extraordinary professor for mathematical physics at Graz in 1874 (and was advanced to ordinary professor in 1885).[29]

Having spent his energies on the new institute (and his health, too, having fallen from the second story to the basement of the institute), Toepler believed that Graz needed a new physicist, and in 1876 he moved to a job at the Dresden Polytechnic. To replace him, the Graz faculty proposed Stefan, Lang, and Boltzmann, in that order. Both Stefan and Lang were attracted because of the excellence of the Graz physics institute.[30] But the ministry understood that Toepler had intended

27. Toepler et al. to Graz U. faculty, 2 May 1874; Schobesberger, "Die Geschichte . . . der Universität Graz," 76–86.

28. Toepler et al. to Graz U. faculty, 2 May 1874.

29. Schobesberger, "Die Geschichte . . . der Universität Graz," 142–44. Ministry of Culture and Education to Dean of the Graz U. Philosophical Faculty, undated but received on 4 Feb. 1885, Graz UA, Z. 168.

30. Schobesberger, "Die Geschichte . . . der Universität Graz," 102. Austrian Ministry of Culture and Education to the Emperor, 18 July 1876; Stefan to a colleague, 9 July 1876; Lang to a colleague, 9 July 1876; all in Öster. STA, 5 Phil, Physik.

Boltzmann to be his successor and, in this connection, had brought Boltzmann into the planning of the new institute. Toepler had even proposed to the ministry the specific "combination" of Boltzmann as "scientific director" of the institute and Albert von Ettingshausen, then Privatdocent at Graz, as "experimental and administrative assistant." By that arrangement, Boltzmann would be able to develop his "far-reaching theoretical ideas through the circle of his students," while Ettingshausen would insure that the "Graz physics institute, for which Austria is envied by German scholars of the first rank, also remains a model institution in the physical-*technical* sense." The minister pointed out to the emperor that the new Graz institute was fitted out for "precise measuring investigations" and that Boltzmann, in addition to teaching and doing research in mathematical physics, had "brilliantly confirmed" his gift for making such measurements, especially by his experiments on dielectrics.[31] The Graz job was specially tailored for Boltzmann, and when he was offered it he agreed to come back.

Boltzmann as Dissertation Advisor

In the winter of 1876, Boltzmann began teaching once again at Graz, this time giving the daily experimental physics lectures and, with Ettingshausen, conducting the laboratory exercises. At first he occasionally offered special lectures on his favorite subject, the mechanical heat and gas theories, but soon he left all theoretical physics lectures to Streintz.[32] The new Graz institute under Boltzmann's direction was intended for serious research as well as teaching, and Boltzmann, usually together with Ettingshausen or Streintz, advised on a good number of physics dissertations. Occasionally, he also advised on philosophy dissertations that touched on questions of physics. His formal evaluations of these dissertations provide insight into his advanced teaching at Graz.

In evaluating philosophy dissertations, Boltzmann criticized their authors' ignorance of physics, but as long as they stuck to what was "purely philosophical" and showed themselves schooled in the speculative ways of philosophers, Boltzmann approved of their work even as he disagreed with their conclusions. Of a dissertation that addressed the ultimate questions of the goal of all natural science, of the existence of the external world, and of the existence of an immortal soul and a deity, Boltzmann observed that it "does not contain much that is new [as] was to be expected by the contents."[33] The author of a dissertation on the law of causality tried

31. The ministry proposed that Boltzmann be appointed "ordinary professor of physics and director of the physics institute," with a salary of 2400 gulden plus 1440 gulden in additional pay, and that Ettingshausen be promoted to "unsalaried extraordinary professor" with the promise of an annual honorarium of 1200 gulden, both to become active on 1 Oct. 1876. Ministry of Culture and Education to the Emperor, 18 July 1876.

32. From the *Verzeichnis der Vorlesungen an der K. K. Carl-Franzens-Universität in Graz,* for the years of Boltzmann's second tenure at Graz, 1876–90.

33. Boltzmann's evaluation, 16 Nov. 1884, of Franz Lampe's dissertation "Die Causalität, ein Beitrag zur Erkenntnisstheorie," and his evaluation, 17 Jan. 1881, of Johann Svetina's dissertation, "Über Naturwissenschaften und Philosophie, ihr gegenseitiges Verhältnis und die Grenzen der durch beide erreichbaren Erkenntnis," Nr. 289 and Nr. 245, respectively, in "Rigorosenbuch 1866–1898," Graz UA. This file contains all of the dissertation evaluations discussed in this section.

to prove that every change must have a cause, which struck Boltzmann as "no better but also no worse than everything that has been said over and over again from the Eleatics to Herbart."[34]

Boltzmann advised on many experimental dissertations. As the minister had anticipated, Boltzmann made good use of Graz's facilities for physical measurements. "Very nice experimental measurements," Boltzmann observed of a dissertation on elastic aftereffects; he especially appreciated the measurements of the torsional after-effect in glass, which had not been measured before because of its softness.[35] He commended another dissertation for measurements it reported of oscillating condenser discharges, which belonged to the "most exact" in this difficult subject.[36] He acknowledged "experimental skill, caution, and perseverance," and he watched for any "independent idea" in experimental research and any "new ideas" for apparatus. He particularly liked for students to build their own apparatus, for that way they demonstrated their "thorough understanding of the theories and aims of experimental physics."[37]

The complete, well-rounded dissertation candidate in physics was Boltzmann's favorite candidate. Boltzmann would commend him for his "full understanding of the theoretical as well as of the experimental" literature of his topic, for his ability not only to build and use measuring apparatus but also to make clear and correct derivations, for his "thorough understanding of mathematical physics" as well as of "experimental physics."[38]

When a candidate revealed theoretical talent—it happened rarely—Boltzmann was happy. He praised a candidate who had arrived "entirely independently" at "important new scientific results" in theoretical physics: his "extraordinary" dissertation displayed the paths of molecules and other consequences of Maxwell's assumed inverse fifth-power law of molecular force in the kinetic theory of gases.[39] The "excellence" of another dissertation, an "entirely independent work," rested on difficult calculations that clarified the relationship of Weber's electrodynamic theory to Hertz's fundamental principle of the unity of electric forces, a "problem of the most fundamental significance for all of theoretical electricity."[40] Physics dissertations at Graz often originated in Boltzmann's own researches and therefore dealt with themes bearing on the current state of theoretical physics.[41]

34. Boltzmann's evaluation of Lampe's dissertation.

35. Boltzmann's evaluation, 13 Jan. 1879, of Ignaz Klemenčič's dissertation, "Beobachtungen über die elastische Nachwirkung am Glase," Nr. 228.

36. Boltzmann's evaluation, 11 July 1887, of Richard Hiecke's dissertation, "Ueber die Deformation electrischer Oscillationen durch die Nähe geschlossener Leiter," Nr. 324.

37. Boltzmann's evaluations of Hiecke's dissertation, and on 21 July 1882 and 1 July 1886, respectively, of Thomas Romik's dissertation, "Experimentaluntersuchung dielektrischer Körper in Bezug auf ihre dielektrische Nachwirkung," Nr. 264, and of Anton Lampel's "Über Drehschwingungen einer Kugel mit Luftwiderstand," Nr. 315.

38. Boltzmann's evaluation of Hiecke's dissertation, and on 5 Nov. 1881 of Friedrich Wrzal's dissertation, "Wärmecapacität der Wasserdämpfe bei constanter Sättigung," Nr. 256.

39. Boltzmann's evaluation, 18 Mar. 1885, of Paul Czermak's dissertation, "Der Werth der Integrale A_1 and A_2 der Maxwell'schen Gastheorie unter Zugrundelegung eines Kraftgesetzes—k^3/r^5," Nr. 298.

40. Boltzmann's evaluation, 21 Dec. 1885, of Eduard Aulinger's dissertation, "Über das Verhältniss der Weber'schen Theorie der Elektrodynamik zu dem von Hertz aufgestellten Princip der Einheit der elektrischen Kräfte," Nr. 306.

41. For example, the dissertations by Aulinger, Czermak, and Romik.

Boltzmann's New Interpretation of the Second Fundamental Law

In 1876, the year Boltzmann returned to Graz as director of the physics institute, his Vienna colleague Loschmidt presented a paper to the Vienna Academy of Sciences containing—as Boltzmann said in his response to the academy the next year—doubts about the "possibility of a purely mechanical proof of the second fundamental law." The doubts referred to any attempt to derive time-irreversible behavior, the content of the second fundamental law, from the time-reversible laws of motion of mechanics. For every process leading to an increase in entropy, the laws of mechanics allow a process with reversed molecular velocities resulting in a decrease in entropy, contradicting experience. Boltzmann took Loschmidt's objection seriously, and in showing that it did not invalidate the molecular interpretation of the second law, he deepened his and his readers' understanding of it. One had to recognize, Boltzmann argued, that the irreversibility of processes governed by the second fundamental law arises from the nature of the initial state of a molecular system and not from the equations of motion themselves, which conduct the system from that state to subsequent states in reversible ways. Boltzmann allowed that from improbable initial states, a system could evolve with decreasing entropy, but he observed that there are infinitely more initial states from which a system evolves with increasing entropy. With his discussion of initial conditions, Boltzmann underscored the probabilistic considerations in the molecular approach to the mechanical theory of heat.[42]

Later in 1877, in a paper for the Vienna Academy, Boltzmann followed up a suggestion he had made earlier that year, which was to develop a "method" for determining heat equilibrium by calculating the probabilities of the various allowed molecular states. The method followed from the understanding that a system in an improbable initial state would in time proceed through more probable states to the most probable state of all, which corresponds to heat equilibrium. Since the entropy of the system also increases as it approaches equilibrium, Boltzmann showed that entropy can be identified with probability.[43]

To make clear to his readers just what is meant by the probability of states, Boltzmann supposed that any molecule can take on only a finite number of discrete values of living force: 0, ϵ, 2ϵ, . . . , $p\epsilon$. This "fiction" facilitated the calculation of probabilities, and Boltzmann later replaced it with the realistic case of continuous energies, which molecules were assumed to allow. The detailed assignment of living force to each of the molecules constituting the system is, in Boltzmann's terminology, a "complexion": the assignment, for example, of living force 2ϵ to the first molecule, of living force 6ϵ to the second molecule, and so on, defines a specific

---

42. Ludwig Boltzmann, "Bemerkungen über einige Probleme der mechanischen Wärmetheorie," *Sitzungsber. Wiener Akad.* 75 (1877): 62–100, reprinted in *Wiss. Abh.* 2: 112–48, especially 116–22, quotation on 117. Klein, *Ehrenfest*, 102–4. Loschmidt's statement, which came to be called the "reversibility paradox," had been discussed by William Thomson in 1874. Brush, *Motion We Call Heat* 1: 238–39.

43. Ludwig Boltzmann, "Über die Beziehung zwischen dem zweiten Hauptsatze der mechanischen Wärmetheorie und der Wahrscheinlichkeitsrechnung respektive den Sätzen über das Wärmegleichgewicht," *Sitzungsber. Wiener Akad.* 76 (1877): 373–435, reprinted in *Wiss. Abh.* 2: 164–223, on 165–66.

complexion. By contrast with an assignment of living force molecule by molecule, the "distribution" specifies only the gross numbers of molecules belonging to each of the allowed values of living force: in a given distribution, ω_0 is some specific number of molecules with value o, ω_1 the number with value ϵ, ω_2 the number with value 2ϵ, and so on. By the calculus of probabilities, the number of complexions corresponding to any given distribution equals the number of permutations:

$$P = \frac{n!}{(\omega_0)!(\omega_1)! \ . \ . \ .}.$$

Boltzmann defined the probability W of a given state distribution, $\omega_0, \omega_1, \ . \ . \ .$, as the ratio P/J, where J is the sum of the permutabilities of all possible state distributions. Since J is a constant for a given system, the maximum value of P determines the state of greatest probability, which represents heat equilibrium and maximum entropy. For mathematical convenience, Boltzmann investigated the maximum value of log P instead of P, and he replaced the discrete set ω_r by the continuous velocity distribution function $f(u, v, w)$. This logarithm is, but for a constant,

$$\Omega = -\int_{-\infty}^{+\infty}\int_{-\infty}^{+\infty}\int_{-\infty}^{+\infty} f(u, v, w) \ l[og] \ f(u, v, w) \ du \ dv \ dw,$$

consistent with a constant living force and a given number of molecules of the system. Boltzmann called Ω the "measure of permutability"; it is, as we can see, the negative of the H-function, so it acquires its maximum value when f is the Maxwell distribution function, as Boltzmann had shown several years before.[44]

At the close of his 1877 paper, Boltzmann discussed the great generality of his new method. The permutability measure always increases or at most remains constant, Boltzmann supposed, which suggested a new explanation of the tendency of entropy to increase, one which extended the range of applicability of entropy: if a gas is not in thermal equilibrium before and after undergoing a change of state, the entropy cannot be calculated; but its permutability measure can be, since it is defined for every state, equilibrium or otherwise. Boltzmann believed that the new understanding of entropy applied not only to gases, which he could confirm by calculation, but also to liquids and solids, even though mathematical difficulties still prevented their exact treatment.[45]

We have discussed Boltzmann's molecular-theoretical work on heat because it was his most important work. Before closing this chapter, we will discuss briefly a theoretical work of Boltzmann's that originated in his interest in heat and extended

    44. Boltzmann, "Über die Beziehung," 168, 175–76, 190–93. Boltzmann's reasoning in this 1877 paper is analyzed in detail, for example, in René Dugas, La théorie physique au sens de Boltzmann et ses prolongements modernes (Neuchâtel-Suisse: Griffon, 1959) 192–99; Klein, Ehrenfest, 105–8; Kuhn, Black-Body Theory, 47–54.
    45. Boltzmann, "Über die Beziehung," 217–18, 223.

to another of his major interests, electricity. In 1884, in the middle of his tenure as director of the Graz physics institute, Boltzmann analyzed the relationship of the second fundamental law to the phenomena of radiant heat. In response to an observation that radiant heat offers an apparent exception to the second law, Boltzmann said that one had only to recognize the existence of a pressure due to radiation to avoid this limitation on the validity of the law.[46] He went on from there to derive theoretically, on the basis of radiation pressure, Stefan's empirical law of proportionality between the energy density of the radiant heat of a blackbody and the fourth power of the absolute temperature. In this "true pearl of theoretical physics," this "first great advance made in radiation theory since Kirchhoff," as H. A. Lorentz described it, Boltzmann showed that the fourth-power law is a consequence both of Maxwell's electromagnetic theory of light, which implies radiation pressure, and of the second fundamental law.[47] Boltzmann's theoretical support for Stefan's law stimulated experimental work on blackbody radiation, and this law together with Boltzmann's combination of Maxwell's electromagnetic theory with the mechanical theory of heat soon led to further powerful theoretical developments in the thermodynamics of radiation.[48]

46. Ludwig Boltzmann, "Über eine von Hrn. Bartoli entdeckte Beziehung der Wärmestrahlung zum zweiten Hauptsatze," *Ann.* 22 (1884): 31–39, reprinted in *Wiss. Abh.* 3: 110–17.
47. Ludwig Boltzmann, "Ableitung des Stefanschen Gesetzes, betreffend die Abhängigkeit der Wärmestrahlung von der Temperatur aus der elektromagnetischen Lichttheorie," *Ann.* 22 (1884): 291–94, reprinted in *Wiss. Abh.* 3: 118–21. H. A. Lorentz, "Ludwig Boltzmann," *Verh. phys. Ges.* 9 (1907): 206–38, reprinted in H. A. Lorentz, *Collected Papers*, 9 vols. (The Hague: M. Nijhoff, 1934–39), 9: 359–90, on 384–86. Brush, *Motion We Call Heat* 2: 517–19.
48. Brush, *Motion We Call Heat* 2: 518. Kuhn, *Black-Body Theory*, 5–6.

17

Electrical Researches

After the great experimental discoveries in electricity, magnetism, and electromagnetism in the early nineteenth century and the efforts to encompass these discoveries within a common theoretical understanding in the 1840s, electrical research underwent a strong, combined experimental and theoretical development through the second half of the century. Nearly all of the leading physicists in Germany worked extensively in the subject. In this chapter we discuss work done in the 1870s and 1880s by three of them, Weber, Clausius, and Hertz, and at the same time we discuss the institutional arrangements for research and teaching at Göttingen, Bonn, and Karlsruhe.

Weber at Göttingen

In 1866 Weber told the Göttingen curator that it was the "purpose of the [physics] institute that scientific researches be carried out there by various persons, by teachers and students."[1] It was understood that the new assistant in the institute Friedrich Kohlrausch would not only direct laboratory practice exercises and take care of the instruments but also do research of his own, and for this purpose Weber offered him space in the institute and use of its instruments. Kohlrausch was also to assist Weber in his own researches "at all times," which meant mainly carrying out assignments in the magnetic observatory. Kohlrausch had an office in the institute, where, as Weber's demanding job description specified, he was to be at nine every morning.[2]

As expected, Kohlrausch published steadily on the researches he did in Weber's institute, which helped to make him an attractive candidate for positions elsewhere. As he received job offers and inquiries repeatedly over the next several years, Weber went to much trouble to keep him at Göttingen, where Weber badly needed him.

1. Weber to Göttingen U. Curator, 29 Dec. 1866, Göttingen UA, 4/V h/10.
2. Weber to Göttingen U. Curator, 18 Oct. 1866, Göttingen UA, 4/V h/21.

72

The first offer came from an agricultural college, after Kohlrausch had been at Göttingen only a few months. To keep him, Weber proposed to the government that Göttingen be given a new physics position, which would go to Kohlrausch. To justify the new position, Weber compared the arrangements for physics with those for chemistry at Göttingen: chemistry had four extraordinary professors while physics as yet had none, despite the corresponding need in physics for laboratory instruction, which included the chemists' need for such instruction in physics. If Kohlrausch were allowed to leave Göttingen, Weber explained, physics laboratory instruction would be seriously disturbed. Weber added, appealing to the state's concern for its property, that the proper use of valuable instruments could not be guaranteed if Kohlrausch were no longer there to oversee them.[3] In Weber's opinion, the laboratory practice course had "greater importance than all the lectures on physics."[4] Weber had an even more important reason for keeping Kohlrausch, his research. In this connection—and in keeping with Weber's lifelong way of working—Weber stressed his and Kohlrausch's collaboration: their "common researches," he warned the curator, "would not merely be disturbed but completely frustrated." Weber hoped to carry out a great research with Kohlrausch as he had done with Kohlrausch's father, Rudolph, which would bring to wide notice Kohlrausch's "skill and fine scientific sense." Weber's arguments were effective; Kohlrausch was appointed extraordinary professor at Göttingen in February 1867. But Weber, certain that other job offers for Kohlrausch would soon follow, even then reminded the ministry that for their research they needed to be "assured of our collaboration for a longer time."[5]

When Kohlrausch expected a call to Würzburg as Clausius's successor and again when the Zurich Polytechnic offered him Clausius's old position there (which had since become Kundt's position), Weber urged that Kohlrausch should be made an ordinary professor at Göttingen. If that proved impossible, Kohlrausch should at least be elevated from his assistant's job to that of director of the physics practice course and co-director of the physics institute, and he should receive improved working conditions in the institute. It was "for Germany, for Prussia, and for Göttingen," Weber urged, that Kohlrausch should be given these things. But Zurich offered Kohlrausch better terms than Göttingen's, and in July 1870 Kohlrausch requested his dismissal from Göttingen.[6] Weber complained to Richard Dedekind that no one could replace Kohlrausch.[7]

Weber hoped to keep if not Kohlrausch then at least the extraordinary professorship he had obtained for Kohlrausch, but the best he could get was an assistant

3. Kohlrausch was offered the ordinary professorship for mathematics and physics at the agricultural academy in Hohenheim. As a counteroffer, Weber proposed that Kohlrausch be given an extraordinary professorship carrying a small salary in addition to his assistantship. Silcher to Kohlrausch, 24 Jan. 1867; Weber to Göttingen U. Curator, 29 Jan. 1867; Kohlrausch Personalakte, Göttingen UA, 4/V b/156.

4. Weber to an official, 17 June 1870, STPK, Darmst. Coll. 1912.236.

5. Weber to Göttingen U. Curator, 29 Jan. and 15 Feb. 1867; Prussian Minister of Culture von Mühler to Kohlrausch, 19 Feb. 1867; Kohlrausch Personalakte, Göttingen UA, 4/V b/156.

6. Göttingen U. Curator von Warnstedt to Prussian Minister of Culture von Mühler, 20 Feb. 1869 and 18 June and 2 July 1870; Kohlrausch to Göttingen U. Curator, 23 July 1870; Kohlrausch Personalakte, Göttingen UA, 4/V b/156. Weber to an official, 17 June 1870.

7. Weber to Richard Dedekind, 10 Aug. 1870, Dedekind Papers, Göttingen UB, Ms. Dept.

to take over Kohlrausch's duties.[8] He chose as his assistant another of his students, Eduard Riecke, who was then away from Göttingen, having been called up as an officer in the war with France just as he was about to present a mathematical physics dissertation to the Göttingen faculty.[9] On his return to Göttingen, Riecke took up his duties as Weber's assistant; he also became Privatdocent in 1871, and that year he took over some of the experimental lectures from Weber, who was unwell, and two years later he took over all of the experimental lectures. Also in 1873 he became extraordinary professor for the same reason that Kohlrausch had, a call to an agricultural academy and Weber's energetic efforts to counter it. In 1876 Riecke gave his assistantship to his student and part-time assistant Carl Fromme; Riecke continued to teach the advanced students in the laboratory, while Fromme taught the beginners. In this way, Riecke was gradually tested and groomed as the probable institute director when Weber should step down completely.[10]

In 1876, the fiftieth anniversary of Weber's doctorate, sixty-eight scientists, most of them German university teachers, gave Weber a present and signed themselves as his "students." In his acknowledgment, Weber wrote: "Of special value for me is your recollection of my lectures and the recognition that despite their various defects, I still succeeded in achieving the main purpose, which was to show the way of doing rigorous research in natural science and to present the connectedness [of science] that it yields."[11] Weber had, as he said, taught others how to do research for nearly fifty years, and now he wanted time to do his own research. To the Göttingen curator he explained in 1873: "You will find it . . . justified that after so many years of teaching, the longing in me grows to be partly or wholly freed from . . . lectures and business affairs during my last years, so that I can concentrate my last efforts exclusively on purely scientific problems." Twice before, the curator had persuaded Weber not to leave his teaching, but now he realized he could not persuade him again. Weber wanted to retire because, as he had often explained to the curator, he was working on a "series of large scientific questions," and his official duties had not allowed him the necessary "concentration and absorption to complete what long stood in the foreground of his thoughts." Weber's wishes were respected, but his retirement was granted only for a few years to begin with; they wanted time to see if Riecke worked out as Weber's substitute, which he did.[12]

8. Weber asked the government to try to hire the Berlin extraordinary professor Quincke, who, he supposed, might be looking for a job elsewhere after Helmholtz's move to Berlin. The government replied that there was no money to appoint someone to Kohlrausch's "professorship." Göttingen U. Curator to Prussian Minister of Culture von Mühler, 9 Sept. 1870, Göttingen UA, 4/V h/21.

9. Weber to Göttingen U. Curator, 7 Oct. 1870, Göttingen UA, 4/V h/21.

10. Göttingen U. Philosophical Faculty to Göttingen U. Curator, 29 June 1871; Weber to Curator, 6 Feb. 1873; Riecke's appointment to extraordinary professor, 26 Feb. 1873; Göttingen U. Curator von Warnstedt to Prussian Ministry of Culture, 25 Apr. 1876; Ministry to Warnstedt, 9 Sept. 1876; Riecke Personalakte, Göttingen UA, 4/V b/173. Weber to Curator, 4 Dec. 1873; Warnstedt to Prussian Minister of Culture Falk, 5 Dec. 1873; Weber Personalakte, Göttingen UA, 4/V b/95a.

11. Weber to Dedekind, 20 Oct. 1876, Dedekind Papers, Göttingen UB, Ms. Dept.

12. Weber to Göttingen U. Curator von Warnstedt, 26 Oct. 1873; Warnstedt to Prussian Minister of Culture Falk, 29 Oct. and 5 Dec. 1873; Weber Personalakte, Göttingen UA, 4/V b/95a. In productivity, Eduard Riecke lived up to Weber's example, and the nature of his research, which was alternately

The "large scientific questions" Weber worked on during his provisional retirement centered on his law of electric action and on Helmholtz's new criticisms of it. Twenty-five years before, Helmholtz had raised doubts about Weber's law because of its dependence on the motions of particles; it did not agree with Helmholtz's requirement that all forces be reducible to attractions and repulsions that depend only on the separation of particles. Helmholtz made his doubts more precise in 1870 in his first paper on the motion of electricity, where he derived unphysical consequences from Weber's law. One result of Helmholtz's renewed criticism was to help motivate Weber in the 1870s to develop a detailed molecular physics on the basis of an energy principle, which he defined in relation to his electrodynamics.[13]

In 1848 Weber had derived a potential for his force, which he returned to in 1869 and again, in a long paper belonging to the series "Determinations of Electrodynamic Measures," in 1871.[14] There he drew on his potential to answer Helmholtz's charge that his law is incompatible with the energy principle. The law of electric force has a very "complicated" character, he observed; for example, it allows an attraction between electric particles at certain separations and a repulsion at other separations. Since, by contrast, the law of electric potential is very "simple" and, to Weber, therefore more fundamental, he now chose to work with it.[15]

In deriving the motions of pairs of electric particles, Weber made assumptions about the behavior of the potential when the particles are at infinite separations. The assumptions imply that the relative velocity of the particles has an upper limit, which Weber took to be c, the ratio of electrical units that enters his law of force or potential. He then applied the limiting velocity to answer Helmholtz's criticism of his law: to demonstrate that his law permits unlimited velocities and therefore

---

theoretical and experimental, was reminiscent of Weber's practice. Although his method was predominantly theoretical, he regularly did experiments, which were usually to make measurements using theory as a guide. Woldemar Voigt, "Eduard Riecke als Physiker," *Phys. Zs.* 16 (1915): 219–21, on 219; Emil Wiechert, "Eduard Riecke," *Gött. Nachr.*, 1916, 45–56, on 47–48. Both Riecke's research and his handling of the physics institute were praised by Weber, Listing, and the rest of the Göttingen science professors when they recommended his promotion to ordinary professor. Göttingen U. Curator to Prussian Ministry of Culture, 6 Sept. 1881, Riecke Personalakte, Göttingen UA, 4/V b/173. The minister warned the Göttingen curator that Prussia would not support three physics professors at Göttingen, so that if Riecke were made professor he would have to replace Weber. The curator replied that Weber, who had given up hope of getting Friedrich Kohlrausch as his successor, wanted to be replaced by Riecke and that it was what the faculty wanted, too. He added that after Riecke had run the institute for so many years, no one could now take it away from him in any case. So in December 1881 Riecke formally succeeded Weber. Prussian Minister of Culture Gossler to Göttingen U. Curator von Warnstedt, 4 Nov. 1881; Warnstedt to Gossler, 8 Nov. 1881; Minister to Warnstedt, 14 Dec. 1881; Riecke Personalakte, Göttingen UA, 4/V b/173. Riecke's salary was 3500 marks—raised 1000 marks in 1883 and another 2000 marks in 1886—plus 540 marks for rent.

13. Edmund Hoppe, *Geschichte der Elektrizität* (Leipzig, 1884), 511–12; Eduard Riecke, "Wilhelm Weber," *Abh. Ges. Wiss. Göttingen* 38 (1892): 1–44, on 26–27.

14. Wilhelm Weber, "Ueber einen einfachen Ausspruch des allgemeinen Grundgesetzes der elektrischen Wirkung," *Ann.* 136 (1869): 485–89; "Elektrodynamische Maassbestimmungen insbesondere über das Princip der Erhaltung der Energie," *Abh. sächs. Ges. d. Wiss.* 10 (1871): 1–61, reprinted in *Wilhelm Weber's Werke*, vol. 4, *Galvanismus und Elektrodynamik, zweiter Theil*, ed. Heinrich Weber (Berlin, 1894), 243–46 and 247–99.

15. Weber, "Elektrodynamische Maassbestimmungen insbesondere über das Princip der Erhaltung der Energie," 254–55.

unlimited work, in violation of the energy principle, Weber noted, Helmholtz had to assume an initial relative velocity of two electric particles greater than c; Weber objected that nowhere in nature do we observe velocities that great. Weber ruled out the possibility that two electric particles with initial finite relative velocity could achieve infinite relative velocity at infinitely close separations by assigning a finite size to electric particles, just as ponderable particles are assigned a finite size to avoid the infinite accelerations the law of gravitation otherwise predicts. In general, Weber argued, Helmholtz was unjustified in making assumptions about molecular motions on the basis of macroscopic physics. Until there existed a well-developed molecular dynamics, a notion—however remarkable—such as a maximum relative velocity in nature could not be dismissed on a priori grounds.[16]

Helmholtz, Weber, and their respective supporters sustained a controversy through the 1870s. It proved inconclusive, and even today it is not obvious which position, Weber's or Helmholtz's, was the strongest then.[17] Helmholtz's supporters included Hertz, Planck (who believed that his part in the controversy had a favorable influence on his career[18]), and the British physicists Maxwell, Thomson, and Tait. Among Weber's many supporters were his Göttingen colleagues: his replacement at the earth-magnetic observatory, Ernst Schering,[19] and his replacement in experimental physics, Riecke. It is not surprising that Schering took an interest in Weber's electrodynamics, as his work was generally rooted in that of his Göttingen predecessors and colleagues. He was Gauss's student, editor of Gauss's collected works, extender of Gauss's and Bernhard Riemann's mathematics, and inheritor and employer of Gauss and Weber's earth-magnetic instruments. In his first publication, a prize essay at Göttingen in 1857, Schering derived Neumann's electrodynamic equations from Weber's, proving their identity; and he showed that all interactions between linear currents can be explained by forces that depend on the motion as well as on the instantaneous position of the electric particles. At Göttingen, Schering lectured on Weber's electrical law, and in 1873 he published a completely general theory of forces of Weber's kind.[20]

16. Weber, "Elektrodynamische Maassbestimmungen insbesondere über das Princip der Erhaltung der Energie," 296–99.

17. K. H. Wiederkehr, *Wilhelm Eduard Weber. Erforscher der Wellenbewegung und der Elektrizität 1804–1891*, vol. 32 of Grosse Naturforscher (Stuttgart: Wissenschaftliche Verlagsgesellschaft, 1967), 106.

18. Planck, *Wiss. Selbstbiog.*, reprinted in Planck, *Phys. Abh.* 3: 380–81.

19. When Weber stepped down as director of the Göttingen observatory in 1868, he was replaced by Ernst Schering and the astronomer Wilhelm Klinkerfues, who divided it between them. Schering at first called himself director of the "Department for Geodesy and Mathematical Physics," but after a disagreement, which Weber was called in to moderate, he became director of the "Department for Theoretical Astronomy and Higher Geodesy." Klinkerfues directed the equally new "Department for Practical Astronomy." Ernst Schering to the Curator, 26 Jan. 1869, Göttingen UA, 4/V f/58. Felix Klein, "Ernst Schering," *Jahresber. d. Deutsch. Math.-Vereinigung* 6 (1899): 25–27. Emil Wiechert, "Das Institut für Geophysik," in *Die physikalischen Institute der Universität Göttingen*, ed. Göttinger Vereinigung zur Förderung der angewandten Physik und Mathematik (Leipzig and Berlin: B. G. Teubner, 1906), 119–88, on 132–34. As Privatdocent at Göttingen in 1879–83, Ernst Schering's younger brother Karl lectured on mathematical physics and worked in the earth-magnetic observatory. Hans Baerwald, "Karl Schering," *Phys. Zs.* 26 (1925): 633–35.

20. Ernst Schering, "Zur mathematischen Theorie elektrischer Ströme," *Abh. Ges. Wiss. Göttingen* 2 (1857), also in *Ann.* 104 (1858): 266–79; "Hamilton-Jacobische Theorie für Kräfte, deren Maass von der Bewegung der Körper abhängt," *Abh. Ges. Wiss. Göttingen* 18 (1873), also in condensed version in *Gött. Nachr.*, 1873, 744–53; Weber's and Helmholtz's electrodynamic laws are discussed on 751–52.

Like Schering, Riecke from the start of his career took a lively interest in Weber's electrodynamics. He recognized that although Weber's fundamental law provides a solution to the general problem of how to proceed from the known laws of interaction of closed currents to the laws of "elementary" actions, it is not the only solution; indeed, there are an infinite number of possible solutions, and in his first paper on the subject, which Weber communicated to the Göttingen Society of Sciences, Riecke examined Helmholtz's new, alternative electrodynamic elementary law. Riecke next studied Weber's law in connection with the question of whether there are one or two electricities. Then he used Weber's law to examine the orbit of an electric particle, which with another particle forms a persisting molecular system.[21] Throughout the 1870s, Riecke did a good deal of theoretical work on electrical laws, paralleling Weber's own continuing work.[22]

With the time he gained by detaching himself from official duties at the institute, Weber worked not only on his electrodynamic theory to insure its compatibility with energy considerations but also on the application of his theory to phenomena lying outside of electricity proper. He showed that his law allows bounded motions of electric particles, which enabled him to make a "first *reconnaissance*" of chemical phenomena. By supposing that an electric particle adheres to every ponderable atom, he could account for the permanence of the atomic aggregates that chemistry studies, and by supposing that around a ponderable atom an electric particle moves in a stable orbit, forming an elementary Ampèrian current, he could explain the thermal properties of electric conductors. Weber acknowledged that his equations of motion of electric particles could not be tested directly by experiment, but he believed that they afforded theoretical clues for investigating the "still obscure regions" of the molecular world.[23]

Weber now worked indefatigably on molecular physics. He showed, for example, that the reflection and scattering of electric particles agree with the motions described by gas theory, with the difference that gas particles are ponderable and electric ones are not. But even this difference could be removed, he showed, by appealing to F. O. Mossotti's theory, which reduces gravitation to electric forces. If each ponderable gas particle is thought of as a positive and negative pair, like a double star, the laws of collision can be determined by the laws of electric interaction, which eliminates the need for the untested assumptions about molecular forces of the kinetic theory of gases. By this approach, Weber showed that positive electric particles can form an imponderable ether, the medium of light, and so on. In outline, Weber had constructed a nearly complete electrical view of physical nature:

21. Eduard Riecke, "Ueber das von Helmholtz vorgeschlagene Gesetz der electrodynamischen Wechselwirkungen," *Gött. Nachr.*, 1872, 394–402; "Ueber das Weber'sche Grundgesetz der electrischen Wechselwirkung in seiner Anwendung auf die unitarische Hypothese," *Gött. Nachr.*, 1873, 536–43; "Ueber Molecularbewegung zweier Theilchen, deren Wechselwirkung durch das Webersche Gesetz der electrischen Kraft bestimmt wird," *Gött. Nachr.*, 1874, 665–72.

22. Riecke published an extended comparison of electrical laws in vols. 20 and 25 of the Göttingen Society's *Abhandlungen*, extracting from it a substantial paper, "Ueber die electrischen Elementargesetze," which he published in the *Ann.* 11 (1880): 278–315.

23. Weber, "Elektrodynamische Maassbestimmungen insbesondere über das Princip der Erhaltung der Energie," 249.

to explain the phenomena of electrodynamics, gravitation, light, heat, chemistry, and other molecular processes, he needed only positive and negative electric particles, the laws of their motion (which mechanics provides), the fundamental law of electric action (which is derived from the fundamental law of electrostatics), and an appropriate version of the energy principle. In his last writing on the subject, Weber said that in principle all physical processes in nature could be calculated from the fundamental law of electric action once the "position and motion of all electric molecules . . . were given at any time." In simplifying physics to the study of the laws of only one substance, imponderable electricity, Weber worked within the tradition of research in which the fluids of heat and magnetism had been eliminated and the ether of radiant heat had been reduced to the ether of light. It was within that tradition that he expressly placed his work.[24]

In 1880 Weber published his last experimental work, a collaboration on electrodynamic measures with Friedrich Zöllner, an ardent champion of Weber's side in his dispute with Helmholtz.[25] Weber's last publication, in 1883, was an account of a precision instrument,[26] a fitting conclusion to a career that had been inseparably bound to exact measurements. Weber's (and Gauss's) instruments and plans remained active at Göttingen even as Weber withdrew from them into his final theoretical studies, as is evident from Karl Schering's report in 1884:

> In the year 1878 we, my brother Ernst Schering, director of the earth-magnetic observatory in Göttingen, and I had constructed an instrument by a new application of the principle underlying Weber's earth inductor. For the inclination of the earth-magnetic force, it permitted an equally exact determination as the present form of Gauss's magnetometer gave

24. Wilhelm Weber, "Elektrodynamische Maassbestimmungen insbesondere über die Energie der Wechselwirkung," Abh. sächs. Ges. d. Wiss. 11 (1878): 641–96, reprinted in Werke, vol. 4, pt. 2, pp. 361–412, on 394–95; "Elektrodynamische Maassbestimmungen insbesondere über den Zusammenhang des elektrischen Grundgesetzes mit dem Gravitationsgesetze," handwritten manuscript, published posthumously in 1894 in Werke, vol. 4, pt. 2, 479–525, on 479–81. Weber's biographer writes of Weber's enduring "vision that all natural phenomena are governed by a single law: his fundamental law of electric action." Wiederkehr, Weber, 181. Weber's views of the ether—as formed of the two electric fluids in their neutral state or as formed of pairs of positive particles—are discussed in Wise, "German Concepts," 276–83. Weber referred his law of interaction to more fundamental features of nature. He showed that the interaction between pairs of particles was completely determined by the principle of the conservation of energy, suitably formulated, and the law of electrostatic interaction, which for him had the requisite simplicity of a fundamental law. He saw his accomplishment as reducing his general law of interaction to a "theorem," a deductive consequence of what was truly fundamental. Wilhelm Weber, "Ueber das Aequivalent lebendiger Kräfte," Ann., Jubelband (1874): 199–213; "Ueber die Bewegungen der Elektricität in Körpern von molekularer Konstitution," Ann. 156 (1875): 1–61, reprinted in Werke, vol. 4, pt. 2, pp. 300–311 and 312–57; "Elektrodynamische Maassbestimmungen insbesondere über die Energie der Wechselwirkung," 372.

25. Wilhelm Weber and Friedrich Zöllner, "Ueber Einrichtungen zum Gebrauch absoluter Maasse in der Elektrodynamik mit praktischer Anwendung," Verh. sächs. Ges. Wiss. 32 (1880): 77–143, reprinted in Werke, vol. 4, pt. 2, 420–76.

26. Wilhelm Weber, "Ueber Construction des Bohnenberger'schen Reversionspendels zur Bestimmung der Pendellänge für eine bestimmte Schwingungsdauer im Verhältniss zu einem gegebenen Längenmaass," Verh. sächs. Ges. Wiss., 1883, reprinted in Ann. 22 (1884): 439–49.

for the declination and horizontal intensity. Since then, we have been occupied with the problem of making an equally reliable measuring apparatus for variations of the vertical intensity, which one already has for variations of the direction and strength of the horizontal earth-magnetic force, using Gauss's unifilar and bifilar in connection with Weber's needle.

We believe we have solved this problem through a new instrument, the "quadrifilar-magnetometer," which has been set up in the basement observation room of the Göttingen observatory since the fall of 1882. . . .[27]

Clausius at Bonn

Clausius spent the last twenty years of his career as director of the Bonn physics institute, doing purely theoretical research as before. His teaching was both theoretical and experimental, and he took seriously the need for laboratory opportunity for his students.

When in 1869 Clausius accepted the job at Bonn, he understood that one of his obligations was the "establishment and direction of a physical laboratory." (His predecessor at Bonn, Plücker, had already introduced laboratory practices in 1867.[28]) He "most definitely" intended to be a director of the Bonn seminar for the natural sciences, since he hoped to be in touch with the abler students: "The one decisive reason that caused me to exchange my position here with that at Bonn," Clausius wrote to the Bonn curator from Würzburg, "was that I counted on finding better conditions for advanced studies among the students at Bonn than in Würzburg."[29] As for lectures at Bonn, he said he wanted to devote roughly equal time to the two parts of physics, theoretical and experimental, but the Prussian requirement that an ordinary professor give a series of public lectures each semester upset his plans somewhat.[30]

Clausius increased the Bonn physics institute budget and got large, one-time grants for instruments.[31] He tried to get a new institute building as well, but he only succeeded in getting the institute moved—and then not until 1885—into a space

27. Karl Schering, "Das Quadrifilar-Magnetometer, ein neues Instrument zur Bestimmung der Variationen der verticalen erdmagnetischen Kraft," *Ann.* 23 (1884): 686–92, quotation on 686–87.

28. Heinrich Konen, "Das physikalische Institut," in *Geschichte der Rheinischen Friedrich-Wilhelm-Universität zu Bonn am Rhein,* vol. 2, *Institute und Seminare, 1818–1933,* ed. A. Dyroff (Bonn: F. Cohen, 1933), 345–55, on 346–48.

29. Clausius to Bonn U. Curator Beseler, 12 Mar. 1869, Clausius Personalakte, Bonn UA.

30. For his first semester at Bonn, in the summer of 1869, Clausius originally intended to teach "optics, electricity, and magnetism treated experimentally, 5 hours per week" and "heat theory treated mathematically, 4 hours per week." He changed his schedule to an elementary public two-hour lecture course on the mechanical heat theory, a two-hour lecture course on elasticity theory and the theory of elastic oscillations treated mathematically, and the five-hour experimental lecture course he had proposed before. Clausius to Bonn U. Curator Beseler, 27 and 29 Jan. 1869, Clausius Personalakte, Bonn UA.

31. Konen, "Das physikalische Institut," 348–49. Clausius improved the budget in 1871 and 1873; with extraordinary grants, he bought, for example, a goniometer, an air pump, a microscope, and various electric and magnetic apparatus.

vacated by the surgical clinic in a wing of the main university building. The new quarters were awkward and uncomfortable, especially the cellar-like laboratory rooms; Hertz, who followed Clausius at Bonn, reported that after a hard rain, water ran down the institute walls all day and one worked with an umbrella. The maximum number of students in the laboratory practice course, fourteen, was reduced to six because there really was no more room.[32]

When Clausius was offered a job at Strassburg University in 1871, he described the conditions under which he would remain at Bonn. His main wish was to be relieved of some of the burden of teaching, which until then had included lecturing in both experimental and mathematical physics and directing both the physical seminar and the laboratory exercises. If all of that work is done as it should be, he told the Bonn curator, it is too much for one person. In Strassburg, two ordinary professorships for physics were planned, but Clausius did not want a corresponding division of his job at Bonn. He asked only for a small sum of 500 thaler to pay an extraordinary professor or Privatdocent to conduct the physics exercises. He was convinced that a young physicist could be found to do this, since he would get an income and an opportunity to do scientific research with "beautiful instruments." Clausius "could then, beside [attending to] experimental physics, devote more time to mathematical physics, which would be useful for the university."[33]

Clausius did not want to hire a physicist from the outside for this work. Instead he wanted to assign the supervision of the exercises to one of the younger physicists already at Bonn, one semester at a time, which would keep him dependent on Clausius for his position in the institute. Clausius was to retain complete control over the apparatus because of its "great value" and the "many expenses that must occur with practical work." (He gave out apparatus to others in the institute only sparingly, which assured that the apparatus remained in good condition; but it did not make Clausius a particularly successful institute director.)[34]

In 1883 Clausius had another chance to ask that his teaching be brought more into line with his research. He was offered Listing's professorship for theoretical physics at Göttingen, and he let it be known at Bonn that he would prefer a professorship that would require him to lecture on theoretical physics only, since he saw his "real scientific calling in the exclusive occupation with this subject." The prospect that Clausius might move to Göttingen was used at Bonn to argue for a second physics professorship, one more to Clausius's liking. It would be for theoretical physics, so that in effect Clausius would then receive a new call to Bonn to fill it. But the Prussian government did not create this second professorship for Clausius, nor did Clausius leave Bonn. There was a persuasive argument against Clausius making a move of this kind: if he taught from Göttingen's (or Bonn's) chair for theoretical

32. Barbara Jaeckel and Wolfgang Paul, "Die Entwicklung der Physik in Bonn 1818–1968," in *150 Jahre Rheinische Friedrich-Wilhelms-Universität zu Bonn 1818–1968*. (Bonn: H. Bouvier, Ludwig Röhrscheid, 1970), 91–100, on 93. Konen, "Das physikalische Institut," 349. Hertz to his parents, 5 Apr. 1889, quoted in Hertz, *Erinnerungen*, 288.

33. Clausius to Bonn U. Curator Beseler, 24 Dec. 1871, Clausius Personalakte, Bonn UA.

34. Clausius to Beseler, 24 Dec. 1871. Konen, "Das physikalische Institut," 349.

physics, he would get far less in student fees than he now got at Bonn, where he taught the popular experimental course. For staying at Bonn, Clausius was even given a raise, which made him the Bonn professor with the highest salary.[35]

When Clausius arrived at Bonn, the Privatdocent Eduard Ketteler had already been teaching physics there for several years, since 1865. Ketteler was a capable theoretical physicist who specialized in optics, and he did some experimental research as well. His lectures, like Clausius's, included both theoretical and experimental topics. Even after Clausius took up regular theoretical lecturing, Ketteler offered a full cycle of "theoretical physics" lectures, and he conducted laboratory classes, which Clausius did not care to do.[36]

In his teaching at Bonn, Ketteler was probably as successful as circumstances permitted: his required, free "public" lectures each semester drew five, ten, and as many as twenty-one students, but his "private" lectures, for which he could collect fees, drew very few students and frequently none at all. To obtain better conditions, in 1870 he requested promotion to "professor," pointing out the "difficulties . . . that face a physicist inclining more to the experimental direction under existing [teaching] conditions."[37] Clausius appreciated that Ketteler "greatly longed to be able to have a physical cabinet at his disposal."[38] With Clausius's approval, the Bonn faculty proposed Ketteler for extraordinary professor, and, after a delay due to the war, the promotion was confirmed in 1872.[39]

After twelve years of service in this capacity, which entailed some of the work, if not the responsibility, of an institute director, Ketteler tried to obtain an independent position. In 1884 he applied to the minister to be made a second ordinary professor for physics at Bonn. The minister declined on the grounds that there was no second salaried ordinary professorship for physics at Bonn and no thought of establishing one. On the other hand, the minister appreciated Ketteler's work and went along with the faculty's recommendation to raise Ketteler's salary and to "raise" his position, which meant to assign him more work. The curator consulted Clausius about the change in Ketteler's position, with the result that Ketteler, who until then had acted as assistant in the physics institute, was appointed "practice director in the physics laboratory." He was now bound to spend three hours every weekday

35. Lipschitz to Bonn U. Curator, 12 May 1883, Clausius Personalakte, Bonn UA. Clausius's salary was 2700 thaler, the equivalent of 8100 marks; after his call to Göttingen, his salary was raised to 9000 marks.

36. Entries for the years around 1870 in *Vorlesungen auf der Rheinischen Friedrich-Wilhelms-Universität zu Bonn.*

37. Ketteler to the Bonn U. Philosophical Faculty, 27 June 1870, Ketteler Personalakte, Bonn UA.

38. Clausius's recommendation of Ketteler for a position at Karlsruhe, 29 Oct. 1870, Bad. GLA, 448/2355.

39. Clausius's recommendation, 29 Oct. 1870. Ketteler was not given a salary at first. Then, in the spring of 1874, he was given one of 900 thaler, though he received it without an accompanying teaching assignment. Bonn U. Philosophical Faculty discussion of Ketteler's request for promotion, 29 June 1870, and their recommendation drafted 13 July 1870 and submitted to Bonn U. Curator Beseler, 22 Nov. 1870; Prussian Minister of Culture to Beseler, 11 Nov. 1872; Minister to Ketteler, 17 Mar. 1874; Ketteler Personalakte, Bonn UA.

afternoon at the laboratory in accordance with Clausius's wishes.[40] Moreover, the position in the laboratory was not permanent but was for two years only, an arrangement Ketteler tried to change in 1887, pointing out that his work was hardly that of a mere assistant. The curator agreed with Ketteler and tried to get him a permanent appointment. But Clausius would not hear of it; believing that a certain constancy in the conduct of the affairs of the institute was useful, Clausius wanted to see Ketteler reappointed, but he was afraid that if the appointment ever became undesirable it would be harder to change if it were permanent.[41] Ketteler was, of course, not told of Clausius's opposition to his request; it was refused on the basis of budgetary arrangements that did not allow any such change, or so he was told.[42]

Upon his move to Bonn, Clausius pursued a deepening interest in the foundations of electrodynamic theory. He had long been interested in a variety of electrical topics, approaching the subject within the context of his researches on the mechanical theory of heat. Now, at Bonn, he entered a "second electrodynamic epoch" in his research, as Riecke described it;[43] in 1875 he put forward a new "fundamental law of electrodynamics," and for several years after he elaborated its theory.[44]

At the time of Clausius's work on electrodynamic theory, Weber's theory was being extended, modified, and challenged. Riemann's attempt to introduce finitely propagated electric action into Weber's electrodynamics—and parallel attempts to refer "electrodynamic forces to known electrostatic forces" by Carl Neumann and E. Betti—prompted Clausius to discuss the foundations of electrodynamics in 1868.[45] When he introduced his own law several years later, he referred to Helmholtz's objections to Weber's law and to reasons of his own, independent of Helmholtz's, for coming to believe that Weber's law "does not correspond to the reality."[46]

In part, Clausius constructed his own electrodynamic theory in connection with his disagreements with Weber's. With Carl Neumann and Riemann, he doubted that a current consists of Weber's opposing streams of electricities. He doubted, too, Weber's assumption that the force between moving electric particles acts only along

40. As an assistant, Ketteler had been paid 300 marks; now as "practice director," he was to receive 600 marks along with the new title. Prussian Ministry of Culture to Bonn U. Curator Beseler, 30 Aug. 1884; Curator to Minister, 11 Sept. 1884; and Ministry to Curator, 30 Jan. 1885; Ketteler Personalakte, Bonn UA.

41. Bonn U. Curator Gandtner to Clausius, 16 Feb. 1887, and Clausius to Gandtner, 18 Feb. 1887, Ketteler Personalakte, Bonn UA.

42. Bonn U. Curator Gandtner to Ketteler, 23 Feb. 1887, Ketteler Personalakte, Bonn UA.

43. In 1879 Clausius reworked his electrical researches to include electrodynamic phenomena in *Die mechanische Behandlung der Electricität* (Braunschweig, 1879), which he presented as the second volume of *Die mechanische Wärmetheorie*. This work contained, for example, his theories of dielectrics, electrolytic conduction, thermoelectricity, and his fundamental theory of electrodynamics. Eduard Riecke, "Rudolf Clausius," *Abh. Ges. Wiss. Göttingen* 35 (1888): appendix, 1–39, on 24. Walter Kaufmann, "Physik," *Naturwiss.* 7 (1919): 542–48, on 546.

44. Rudolph Clausius, "Ueber ein neues Grundgesetz der Elektrodynamik," *Sitzungsber. niederrhein. Ges.*, 1875, 306–9; translated as "On a New Fundamental Law of Electrodynamics," *Philosophical Magazine* 1 (1876): 69–71.

45. Rudolph Clausius, "Ueber die von Gauss angeregte neue Auffassung der electrodynamischen Erscheinungen," *Ann.* 135 (1868): 606–21. Riecke, "Clausius," 24–28.

46. Clausius, "On a New Fundamental Law," 69.

the line joining them. He argued that the force between two electric particles may also be influenced by the direction of motion of the particles. If Newton had considered electrodynamic forces, Clausius speculated, he would have agreed that Weber was unjustified in his assumption that the force between particles is independent of the directions of their motions. Riemann, too, did not accept Weber's restriction to forces acting along the line between particles. In the derivation of his law, Riemann assumed the forces acting on the two particles to be equal and opposite, that is, parallel but not necessarily in a line; but even Riemann's assumption, Clausius reasoned, was too restrictive. In electrodynamics, Clausius was prepared to proceed without Newton's third law, the law of equality of action and reaction, in its familiar statement; this law is founded on our experience with static forces only, Clausius observed, and so the analogy with the static gravitational force is misleading in electrodynamics.[47]

At the start of his derivation, Clausius did not impose preconditions on the mathematical form of the fundamental electrodynamic law, which he preferred to establish by appeal to experience with electric actions and to the principle of conservation of energy. In constructing the law, Clausius endowed its static part with the well-confirmed inverse-square dependency on the separation between electric particles, and its dynamic part he expanded in quantities linear and bilinear in the absolute velocities of the two particles and linear in their absolute accelerations. The coefficients in this expansion are undetermined functions of the separation of the particles. Clausius fixed all but one of the undetermined coefficients by adopting Carl Neumann's assumption of unitary currents, rather than Weber's dualistic currents, by applying the known force between closed currents, and by requiring that the law satisfy the principle of conservation of energy. He then set the remaining coefficient equal to zero on the grounds that in this way the electric potential receives its "simplest and therefore most probable form." Comparing his potential with the potentials of the rival electrodynamics, which were, in his opinion, Weber's and Riemann's, Clausius found his the simpler, which argued in its favor. The main advantage of his potential was, for Clausius, its greater generality; although he had derived it by assuming that only one electricity moves, he showed that it was still valid if both electricities move and, indeed, if they move with different velocities, as they do in electrolytic conduction.[48]

Clausius's law is expressed in terms of absolute motions of electric particles

47. This account of Clausius's theory is from his derivation of the "new electrodynamic fundamental law" in Rudolph Clausius, *Die mechanische Wärmetheorie*, vol. 2, *Die mechanische Behandlung der Electricität* (Braunschweig, 1879), 227–81.

48. The dynamic part of Clausius's potential is:

$$V = \frac{kee'}{r}\left(\frac{dxdx'}{dtdt} + \frac{dydy'}{dtdt} + \frac{dzdz'}{dtdt}\right),$$

where k is a constant depending on units, e and e' are the electric masses, r their separation, and dx/dt, dy/dt, dz/dt and dx'/dt, dy'/dt, dz'/dt their absolute velocities. If ϵ is the angle between the absolute velocities v and v', the potential acquires the simpler form: $V = (kee'/r)\, vv' \cos \epsilon$. Upon applying his

rather than of relative motions, as Weber's law is. Absolute motions presuppose a stationary medium, an ether, which Clausius invoked to explain the apparent violation of Newton's third law by the motion of electric particles in his theory: if the medium takes part in electrodynamic actions, it can contain the necessary momentum to preserve Newton's third law. For the same reason, not even energy has to be conserved by the motions of electric particles, but Clausius's law does conserve it. Clausius did not attribute specific properties to the medium nor attempt to locate its momentum nor make its mode of propagation of electric action a central concern. As Weber's, Clausius's fundamental law describes only the motions of electric particles.[49]

In his work on electrodynamic theory, Clausius revealed the same inclination toward general formulations of laws that he revealed in his work on heat theory. To construct his fundamental law of electrodynamics, he found it sufficient to call upon a limited number of experimentally confirmed results in electricity, the principle of conservation of energy, and internal theoretical considerations such as the simplicity and symmetry of the mathematical expressions. His writings on electrodynamics offered their close readers a lesson on how to do theoretical physics, Clausius's way.

In 1884, in his inaugural address as rector at Bonn, Clausius spoke on the subject of the "inner connection" between the "natural forces," or the "great agents of nature." He dwelt at length on the connection of heat and electricity, the two agents to which he had devoted the most research. Despite frequent claims to the contrary, heat and electricity had not, to Clausius's satisfaction, been reduced to a single agent: electricity was still elusive, despite its connections with heat. It is true that electric currents set atoms into motions constituting heat, and, conversely, that thermal motions of atoms give rise to electric currents, but from this familiar transformation of one kind of motion into another we cannot, Clausius cautioned, draw inferences about the nature of electricity.[50]

The nature of electricity is illuminated, Clausius said, by its connection with radiant heat and light through Weber's law and Weber and Kohlrausch's measurements of the relationship between electrostatic and electrodynamic forces. He explained what he meant: according to Weber's theory, if two like electric particles move in parallel directions and with uniform velocities, they mutually attract by their electrostatic force and mutually repel by their electrodynamic one, and the repulsion and attraction become equal and cancel at a definite relative velocity,

law to the interaction of two current elements, Clausius arrived at the same force that the mathematician Hermann Grassmann had published many years before, "Neue Theorie der Elektrodynamik," Ann. 64 (1845): 1–18; in light of their "entirely different" starting points, Clausius considered their agreement an "encouraging corroboration." Clausius, Die mechanische Behandlung, 276–77.

49. Clausius, "On a New Fundamental Law," 69; "Ueber das Verhalten des electrodynamischen Grundgesetzes zum Princip von der Erhaltung der Energie und über eine noch weitere Vereinfachung des ersteren," Sitzungsber. niederrhein. Ges., 1876, 18–22, translated as "On the Bearing of the Fundamental Law of Electrodynamics toward the Principle of the Conservation of Energy, and on a Further Simplification of the Former," Philosophical Magazine 1 (1876): 218–21, on 218–19.

50. Rudolph Clausius, Ueber den Zusammenhang zwischen den grossen Agenten der Natur, Rectoratsantritt, 18 Oct. 1884 (Bonn, 1885), 20–23.

which according to Weber and Kohlrausch is the velocity of radiant heat and light in empty space. The agreement of a magnitude belonging to electricity with a magnitude belonging to heat and light "cannot be without an inner reason," Clausius remarked. This and related agreements "leave no doubt that in the propagation of light or, what is the same, in the propagation of radiant heat, electric forces must be acting." Indeed, physicists had already begun to connect the agents of light and electricity: whereas before, physicists had derived the equations for the propagation of light from the elastic forces of the ether, Maxwell had recently derived the same equations from electric forces, founding an "*electrodynamic* or, as he calls it, an *electromagnetic* theory of light." Clausius did not know if Maxwell's assumptions were correct, observing that they had not been established by mechanical considerations, but he allowed that if it could be shown that radiant heat and light are explained by electric forces, as Maxwell claims they are, the ether would have to be viewed as nothing but electricity. Today, Clausius said, only two substances are assumed to exist, electricity and matter; everything else is explained by motion. This great internal connection had come about through the perfection of the theoretical conception of nature. This much Clausius was confident of.[51]

Hertz at Karlsruhe and Bonn

As early as the 1850s, Wilhelm Eisenlohr had taught theoretical physics at the Karlsruhe polytechnic, and for a time so did his successors. But beginning with Ferdinand Braun's tenure in 1882, the expanding field of electrotechnology temporarily dominated physics instruction. Braun started an electrotechnical practice course in the physics institute to supplement lectures on electrotechnology given by an applied physicist, and in his teaching of physical theory Braun stressed the theoretical foundations of electrotechnology. The committee charged with recommending candidates for Braun's successor in 1884 decided that they must consider above all physicists who had done scientific work in electricity. They did not yet want a division for electrotechnology separate from the physics institute; electrotechnology and physics were to remain together, and the new professor of physics was to continue giving lectures on "electrotechnology on a mathematical basis" and the corresponding practical exercises. For the time being, theoretical physics was "pushed aside."[52]

With advice from Braun and from outside physicists, the Karlsruhe faculty recommended three candidates for Braun's replacement. The three they recommended all did research in electricity and all did both theoretical and experimental research. And all were recommended by Helmholtz, who said of Hertz, the youngest of the

---

51. Clausius, *Ueber den Zusammenhang*, 24–27.

52. "Bericht" by Grashof and Engler, members of the Karlsruhe faculty, to the Baden Ministry of Justice, Culture, and Education, 30 Nov. 1884, Bad. GLA, 448/2355. Karlsruhe Technical Institute, *Festgabe zum Jubiläum der vierzigjährigen Regierung Seiner Königlichen Hoheit des Grossherzogs Friedrich von Baden* (Karlsruhe, 1892), xxviii, xxx, 261. Friedrich Kurylo and Charles Süsskind, *Ferdinand Braun: A Life of the Nobel Prizewinner and Inventor of the Cathode-Ray Oscilloscope* (Cambridge, Mass.: MIT Press, 1981), 48–59, discuss the scientific orientation toward electrical engineering that characterized the Karlsruhe physics institute under Braun.

three, that he raised the "greatest hopes for the future."[53] Hertz, who was offered the position, had other prospects at the time, but a visit to the Karlsruhe physics institute decided him in its favor.[54]

As ordinary professor of physics at Karlsruhe, Hertz assumed the combination of subjects the faculty intended for him, experimental physics and electrotechnology. He read up on the theory of electrodynamic machines for his lectures, set up an electrotechnology laboratory, and, in general, as his predecessor Braun had done, devoted much of his attention to electrotechnology.[55]

To a colleague, Hertz wrote that at Karlsruhe he had, on the whole, "very beautiful rooms and very beautiful means."[56] But for a time after his move there, he was without direction in his research. For intellectual stimulation, he joined a group of mathematicians with whom he discussed his work, and he lectured to the Karlsruhe Natural Scientific Society on telegraph and telephone transmission and other current topics.[57] Then, toward the end of 1886, he found his direction, and for nearly three years he worked steadily on researches related to electric waves and not, as he usually did, on a variety of topics. These researches led to his experimental confirmation and subsequent theoretical reformulation of Maxwell's electromagnetic theory of light.

Hertz's researches depended on two related experimental arrangements, which he arrived at partly by theoretical understanding and partly by experimental insight, if in Hertz's case these two can be meaningfully separated. To detect electric waves, he used the resonance principle, tuning a secondary circuit to receive waves radiated by a primary one; to produce waves of manageably short length in the primary circuit, he discharged an induction coil across a spark gap between two spheres.[58]

Working in the laboratory of the Karlsruhe physics institute, by November 1886 Hertz had propagated an electric induction between two open current loops across a space of a meter and a half. By December he had produced resonance between two electric oscillations along with related phenomena, which encouraged him to write to Helmholtz to tell him of his new work.[59] With an assembly of primary and sec-

53. The other two candidates were Anton Oberbeck and Franz Himstedt, both of whom were extraordinary professors at the time. Helmholtz recommended Oberbeck and Hertz to the Karlsruhe faculty directly and Himstedt indirectly. In addition to Helmholtz, three of Hertz's colleagues at Kiel also wrote recommendations for him. Grashof and Engler, "Bericht," 30 Nov. 1884.

54. Hertz went to Berlin to discuss Karlsruhe with Helmholtz, Kirchhoff, Heinrich Kayser, and Althoff, who also spoke of a possibility for him at Greifswald. After Hertz went to Karlsruhe, Oberbeck, the competing candidate for that job, took the job at Greifswald. From Hertz's diary in his *Erinnerungen*, 200.

55. From Hertz's diary in his *Erinnerungen*, 210, 222. In April 1886, funds were approved for Hertz's electrotechnology laboratory. Otto Lehmann, "Geschichte des physikalischen Instituts der technischen Hochschule Karlsruhe," in Karlsruhe Technical Institute, *Festgabe*, 207–65, on 262. A year later, Hertz could report to a colleague that the machines were already there. Hertz to Elsas, 7 Mar. 1887, Ms. Coll., DM, 3091.

56. Hertz to Elsas, 7 Mar. 1887.

57. From Hertz's diary in his *Erinnerungen*, 208, 210, 214, 218.

58. Planck, "Hertz," 281–82.

59. Hertz to Helmholtz, 5 Dec. 1886, quoted in Koenigsberger, *Helmholtz* 2: 344. Hertz's diary in his *Erinnerungen*, 212–16.

ondary circuits, spark gaps, and blocks of paraffin and other dielectrics, Hertz now demonstrated the existence of inductive effects of a polarization current in dielectrics. In 1887 the Prussian Academy awarded him its prize for solving Helmholtz's problem of 1879. His proof, Hertz said, was of an action that "everyone has long assumed," but it had long seemed hopeless and its discovery seemed for that reason "a kind of personal triumph."[60]

At first Hertz interpreted his experiments in light of Helmholtz's general theory of dielectric polarization, according to which electrostatic and electrodynamic waves propagate with different velocities. Later in the course of his experiments, he turned from Helmholtz's to Maxwell's theory, which recognizes only one electric force, and from the electromagnetic effects of changing polarizations in material dielectrics to the free propagation of electric waves. He came to see that the finite propagation of electric waves in space, or air, is the central point of Faraday's and Maxwell's understanding of electricity, and that its proof would solve the rest of Helmholtz's problem. (Earlier the Prussian Academy had judged the proof that air and empty space behave like dielectrics as too demanding and had suppressed it from the original problem.)[61]

To his parents at the time, Hertz wrote that the problem of electric waves had interested him for years without his ever believing in its solution. Now he believed in it. His greatest moment came at the end of 1887, when he measured the velocity of electric waves in air. His first measures suggested that the velocity is greater than that of light, perhaps infinite, which disappointed him because it contradicts Maxwell's theory, and for a time he discontinued his experiments. Then he resumed them on the grounds that a disproof of Maxwell's theory would be as important as a proof. He moved his experiments from the laboratory to the big lecture hall where there was more room.[62]

As quickly as Hertz completed his experiments, he wrote them up, usually sending them to Helmholtz for publication with the Prussian Academy. On Wiedemann's request, he published papers in the Annalen der Physik, too. Early in 1888 he reported in the Annalen an indication of a "finite velocity of propagation of electric distance actions," which gave him renewed confidence in his work.[63] In subsequent papers, Hertz elaborated experimentally and theoretically on the fact of finite propagation. His experiments made electric waves in air "almost tangible," promising a "decision between the conflicting theories" of electrodynamics and their differing predictions for open currents.[64]

In one of the later papers, "The Forces of Electric Oscillations, Treated according to Maxwell's Theory," Hertz concluded that Maxwell's theory explains all the

60. Hertz to his parents, 30 Oct. 1887, in Hertz, Erinnerungen, 232.
61. Planck, "Hertz," 282.
62. Hertz to his parents, 13 Nov. and 23 Dec. 1887, 1 Jan. 1888, in Hertz, Erinnerungen, 236–48.
63. Heinrich Hertz, "Ueber die Einwirkung einer gradlinigen elektrischen Schwingung auf eine benachbarte Strombahn," Ann. 34 (1888): 155–70, on 169.
64. Heinrich Hertz, "Ueber die Ausbreitungsgeschwindigkeit der elektrodynamischen Wirkungen," Ann. 34 (1888): 551–69, on 568–69; "Ueber elektrodynamische Wellen im Luftraume und deren Reflexion," Ann. 34 (1888): 610–23, on 610.

facts he had investigated and is superior to all of the other theories.[65] He exhibited Maxwell's equations in the same symbols and form as he had in his earlier theoretical work on the subject at Kiel. His theoretical orientation was now firmly Maxwell's, for he viewed the production of electric waves as owing not only to the source but also to the condition of the surrounding space, the seat of the electromagnetic energy. By means of great parabolic mirrors, lenses, gratings, and prisms, he showed that the electric force exhibits all of the main properties of light waves. In the paper "On Electric Radiation," which Hertz regarded as a "natural end" of this series of researches, he reported that his experiments had removed "any doubt as to the identity of light, radiant heat, and electromagnetic wave motion": electric rays are light rays of long wavelength.[66]

In the course of his experiments, Hertz inadvertently discovered that discharges across an air gap are facilitated if the negative electrode is illuminated with ultraviolet light. This discovery revealed a new "relation between two entirely different forces," light and electricity. He was particularly impressed, since the known relations between these two forces were few: there was, for example, Faraday's rotation of the plane of polarization of light by electric currents, but instances of the reverse influence of light on electricity were almost unknown. Hertz hoped that his discovery—his hope was borne out, for he had discovered the photoelectric effect—would one day yield important understanding.[67]

Throughout Hertz's experiments on electric waves, he made frequent reference in his diary to Maxwell's theory, increasingly the center of his theoretical concern. In the performance of the experiments, he even went beyond what Maxwell himself had worked out, as he did, for example, with his theoretical understanding of the relationship between the radiating source and the resulting electromagnetic field. Hertz's success, Helmholtz observed, required an "intellect capable of the greatest acuteness and clearness in logical thought, as well as of the closest attention in observing apparently insignificant phenomena." Helmholtz added that "in the whole investigation [of electric waves] one scarcely knows which to admire most, his experimental skill or the acuteness of his reasoning, so happily are the two combined."[68] In the introduction to *Untersuchungen über die Ausbreitung der elektrischen Kraft (Electric Waves)*, the collection of Hertz's papers that the publishers of the *Annalen der Physik* asked him to prepare, Hertz gave his view of the need for both theory and experiment in his demonstration of electric waves in air. If he had not given the demonstration, he said, it probably would have been given by Oliver Lodge, who was doing experiments on the discharge of condensers guided by Maxwell's theory; G. F. FitzGerald's theoretical discussion of electric waves was not in

---

65. Heinrich Hertz, "Die Kräfte elektrischer Schwingungen behandelt nach der Maxwell'schen Theorie," *Ann.* 36 (1888): 1–22, on 1.

66. Heinrich Hertz, "Ueber Strahlen elektrischer Kraft," *Ann.* 36 (1889): 769–83, on 781.

67. Heinrich Hertz, "Ueber einen Einfluss des ultravioletten Lichtes auf die elektrische Entladung," *Ann.* 31 (1887): 983–1000; Hertz to his father, 7 July 1887, in Hertz, *Erinnerungen*, 224–28, on 226.

68. Helmholtz's preface to Heinrich Hertz, *Die Prinzipien der Mechanik, in neuem Zusammenhange dargestellt*, ed. Philipp Lenard (Leipzig, 1894), translated as *The Principles of Mechanics Presented in a New Form* by D. E. Jones and J. T. Walley (1899; reprint, New York: Dover, 1956). The role of theory in Hertz's experiments is analyzed in D'Agostino, "Hertz's Researches."

itself enough, since it was not possible to proceed by theory alone. To arrive at a knowledge of electric waves, the peculiar property of the electric spark had to be understood, something which "could not be foreseen by any theory."[69]

With his experiments, Hertz had gone far toward realizing his goal of testing Maxwell's theory decisively. Helmholtz informed the Berlin Physical Society of Hertz's demonstration of electric waves with these words: "Gentlemen! I have to communicate to you today the most important physical discovery of the century."[70] In summing up the significance of Hertz's experiments, Helmholtz said that they showed that light and electricity are "most closely connected" and, even more important from the theoretical standpoint, that "apparent actions-at-a-distance really consist of a propagation of an action from one layer of an intervening medium to the next."[71]

Kundt spoke of Hertz as the "new *great attraction*" of physics,"[72] and indeed Hertz's experiments on Maxwell's theory had made him something of a celebrity. That meant he was asked to lecture on his experiments and to repeat them in Berlin and elsewhere.[73] That meant, too, that almost everyone who lectured on them and repeated them, or wanted to repeat them, wrote to him to tell him so.[74] Franz Himstedt wrote from Giessen, for example, to say that Hertz's work had "really enraptured" the local physicists.[75] From America, a correspondent wrote that Hertz's experiments had "attracted the whole world."[76] From Britain, FitzGerald wrote that "no more important experiment has been made in this century" and that "your experiment will be called 'Hertz's classical experiment that decided between theories of electromagnetic action at a distance and by means of the ether.' "[77]

Also from Britain, Oliver Heaviside wrote to Hertz that he had learned some-

69. Heinrich Hertz, "Introduction" to his *Electric Waves, Being Researches on the Propagation of Electric Action with Finite Velocity through Space*, trans. D. E. Jones (1893; reprint, New York: Dover, 1962), 1–28, on 3. Originally published as *Untersuchungen über die Ausbreitung der elektrischen Kraft* (Leipzig, 1892).

70. Eugen Goldstein, "Aus vergangenen Tagen der Berliner Physikalischen Gesellschaft," *Naturwiss.* 13 (1925): 39–45, on 44.

71. Helmholtz's preface to Hertz, *Principles of Mechanics*.

72. Kundt wrote *"great attraction"* in English. Letter to Graetz, n.d. [Dec. 1888], Ms. Coll., DM, 1933, 9/18.

73. To overflow audiences in the spring of 1889, Hertz lectured on his work at the local Natural Scientific Society and at the Karlsruhe Technical Institute. Hertz to his parents, 24 Feb. 1889; Elisabeth Hertz to his parents, 10 Mar. 1889; in Hertz, *Erinnerungen*, 282, 284. Hertz was invited to show his experiments at the tenth International Medical Congress in Berlin, but he could not attend. He proposed Kundt in his place, explaining to the Prussian Minister of Culture von Gossler that the Berlin physics institute had already constructed excellent apparatus for this purpose. Gossler to Hertz, 13 June 1890; Ms. Coll., DM, 2907. Hertz to Gossler, 17 June 1890, STPK, Darmst. Coll. 1913.51.

74. From Sofia the physics professor P. Bachmetjew wrote to Hertz on 15 Mar. 1890 to ask where he could order Hertz's apparatus for electric waves. From Groningen, H. Haga wrote on 5 Feb. 1890 that he had repeated Hertz's experiments. Letters with similar contents include: G. F. FitzGerald to Hertz, 14 and 23 Jan. 1889; Arthur Oettingen to Hertz, 31 Mar. 1888; James Moser to Hertz, 21 Mar. 1890; Himstedt to Hertz, 31 May 1889; Walter König to Hertz, 25 Apr. 1890; Lehmann to Hertz, 12 July 1890. These letters are in Ms. Coll., DM, 2865, 2941, 2887–88, 2989, 2982, 2934, 2955, and 2968, respectively.

75. Himstedt to Hertz, 31 May 1889.

76. Lucien Blake to Hertz, 11 Dec. 1889, Ms. Coll., DM, 2876.

77. FitzGerald to Hertz, 8 June 1888, Ms. Coll., DM, 2886.

thing about resonators for detecting electric waves from Hertz's experiments and that he thought the experiments served a purpose in persuading people to give up untenable theories, especially the German electrodynamic theories with their "absurd speculations." For his part, he had expounded Maxwell's theory since 1882 and had long been theoretically convinced of the existence of electric waves in dielectrics "in spite of the absence of experimental evidence." He explained that the "man who goes by hardheaded reasoning of a legitimate nature, on the basis of laws known with great exactness, does not want an experimental proof." He assured Hertz that his experiments were "highly appreciated in England," even if he did not need them himself.[78] Nor, as it happened, did a number of other leading British physicists need them to be convinced of Maxwell's theory.[79] Hertz understood this point. "You may believe that I was fully in earnest when I said you could not learn very much of my experiments," Hertz replied to Heaviside in 1889: for whoever "was fully convinced of the truth of Maxwell's equations and was able to interpret them, did know as much about these things before my experiments as after them." Hertz went on to say that he did not mean to belittle his experiments in any way, since "there were many people not convinced of those equations or not able at all to see what they meant."[80]

German physicists, as Hertz knew, were not so inclined as their British colleagues to accept Maxwell's theory without proof. One reason was the long development in Germany of distinctive optical and, especially, electrodynamic theories. As Hertz said of Weber's fundamental electrodynamic law: "Whatever one may think about the correctness of this [law], the totality of efforts of this kind constitutes a closed system full of scientific appeal; once you had wandered into its magic circle you remained imprisoned in it."[81] Another, more general reason was the experimental emphasis in physical research in Germany accompanying the development of physics institutes in the late nineteenth century. As the young Leipzig physicist Walter König wrote to Hertz, the agreement of "facts" was much more interesting than the agreement of "equations" for proving that light is electrical: "Decisive for the electromagnetic theory of light are naturally only quantitative determinations of the velocity of propagation, of the index of refraction, etc., of the electric rays."[82]

Personally gratifying as all of this notice was—a colleague called his research that followed Hertz's his "apostledom"—Hertz felt oppressed by the load of correspondence and complained about it repeatedly in his letters to his parents and col-

78. Oliver Heaviside to Hertz, 14 Feb., 1 Apr., 13 July, and 14 Aug. 1889, Ms. Coll., DM, 2922–25.

79. A good number of British physicists believed in electric waves before Hertz's demonstration of them, and its greatest impact in Britain was going to be on engineers. Bruce Hunt, "Theory Invades Practice: The British Response to Hertz," *Isis* 74 (1983): 341–55.

80. Hertz to Heaviside, 3 Sept. 1889, quoted in Rollo Appleyard, *Pioneers of Electrical Communication* (London: Macmillan, 1930), 239.

81. Heinrich Hertz, *Ueber die Beziehungen zwischen Licht und Elektricität* (Bonn, 1889), address at the 1889 meeting of the German Association, reprinted in Hertz, *Ges. Werke* 1: 339–54, on 342.

82. Walter König to Hertz, 27 May 1889, Ms. Coll., DM, 2957.

leagues.[83] Most of the correspondence his "external successes" brought him was "superfluous, in part absurd." But there was some correspondence with worthwhile scientific content, such as that with Heaviside, and Hertz joined in an extended discussion of research in Maxwellian electrodynamics.[84]

In addition to correspondence, wanted and unwanted, Hertz's celebrity brought him invitations; one, which he accepted, was to speak at the German Association meeting in Heidelberg in 1889. His talk there was scheduled for a general session rather than for a physics session because of the widespread interest in his experiments. It was a distinction for someone so young, but Hertz half regretted that he had agreed to it. He believed that physicists learned nothing from such general talks; he struggled to prepare a good one all the same, determined that it should not be "entirely unintelligible to the laity."[85] The result was one of the great nineteenth-century popular talks on science, "On the Relations between Light and Electricity." It dealt with Hertz's proven theme of the growing unity of physics, which his own researches had recently furthered.

In his Heidelberg talk, Hertz explained that the relations between the two great classes of phenomena are more intimate than was expected: light turns out to be nothing but an electrical phenomenon. "Remove electricity from the world, and light disappears; remove the light ether from the world, and electric and magnetic forces could no longer traverse space." The relations between light and electricity are not affirmed by the direct testimony of the senses, he went on to explain; indeed, they are "false" from the standpoint of our senses, inaccessible to our hand, ear, and eye. They are accessible only to our inner intuition through mathematical physics, the key to the hidden unities of nature. Maxwell's theory shows the power of mathematical physics: when one studies that theory, Hertz said, one feels "as if the mathematical formulas have an independent life and intelligence, as if they are wiser than we, even wiser than their discoverer, as if they give us more than he put into them." Hertz made it clear that in his researches, he first established the correctness of Maxwell's electromagnetic theory and then gave a direct experimental proof of the connection between light, or rather long waves in the ether, and electricity.[86]

Hertz, in the conclusion of his Heidelberg talk, singled out three "ultimate" problems of physics. These, he suggested, would be solved by starting from the views of Faraday and Maxwell and the understanding that light and electricity share a common ether. The first problem is gravitation, the one remaining action at a distance, which Hertz believed would be shown to be finitely propagated, too. The second is the nature of electricity, which was now understood to extend "over all of

83. Ernst Lecher to Hertz, 11 June 1890, Ms. Coll., DM, 2970. Hertz to his parents, 20 Jan. 1889, in Hertz, *Erinnerungen*, 278.

84. Hertz to Emil Cohn, 31 Dec. 1890, Ms. Coll., DM, 3204. Hertz's serious scientific correspondents included, besides Heaviside and FitzGerald in Britain, Heinrich Rubens and Paul Drude in Germany and Lucien de la Rive and Édouard Sarasin in Switzerland. Much of this correspondence, in incomplete form, is in Ms. Coll., DM.

85. Hertz to Cohn, 15 Sept. 1889, Ms. Coll., DM, 3202. Hertz to his parents, 8 Sept. 1889, in Hertz, *Erinnerungen*, 292–94.

86. Hertz, *Ueber die Beziehungen*, reprinted in his *Ges. Werke*, 339–40, 344, 352–53.

nature." The third is the "nature, the properties of the space-filling medium, of the ether, its structure, its rest or motion, its infinite or bounded extent." This is the "all-important question," the solution of which will reveal the nature of electricity and matter: "Today's physics is inclined to ask the question if all that exists has not been created from the ether." The vision with which Hertz closed his talk was of a unified physical theory of the ether, the substratum of the entire phenomenal world, of light, electricity and magnetism, gravitation, heat, and all the rest.[87]

Despite his comparative youth, Hertz had become a spokesman for physics, one qualified by his researches to point to the tasks of the future. At the Heidelberg meeting, he associated with Helmholtz, Kundt, and Wiedemann, the leaders of German physics. "It flattered me," he wrote to his parents of the meeting, "that the older and better known always drew me into their circle and that because of this, my authority with the younger quickly and so-to-speak visibly grew."[88]

When Hertz learned of Kirchhoff's death in 1887, he wrote to his parents that he would miss Kirchhoff, one of the few great physicists.[89] A year later, in the wake of his experimental success, Hertz was called into the office of the Prussian ministry of culture to discuss Kirchhoff's position among other positions. Hertz protested that he was "not really a mathematical physicist," as Kirchhoff had been. His true calling was experimental physics, which he would have to abandon, or at least dissociate his teaching from, if he went to Berlin as Kirchhoff's successor. The ministry brought up Königsberg, but Hertz had no interest in that place. The ministry also brought up Bonn, among other possibilities, and in December 1888 it offered Hertz the choice between Berlin and Bonn.[90] Hertz did not hesitate: he chose Bonn,[91] and he did not give Baden a chance to try to keep him, since Karlsruhe could not have matched his Bonn income.[92]

Hertz's successor Otto Lehmann wrote to Hertz in 1890 that his substantial budget request at Karlsruhe was approved, a fortunate outcome he attributed to Hertz's researches, which had made the "significance of physics appear in a brighter light."[93] The acclaim Hertz's researches received called attention to the use of physics institutes as places for research.

Karlsruhe did not yet have a chair for theoretical physics when Hertz worked there, and Lehmann proposed one in 1891. Lehmann did not then ask for an institute for theoretical physics, and in fact he wanted the professor for theoretical physics to represent electrotechnology as well and eventually to direct an institute for that subject.[94] A year later, in rivalry with the chemist at Karlsruhe, Lehmann

87. Hertz, *Ueber die Beziehungen,* 353–54.

88. Hertz to his parents, 26 Sept. 1889, in Hertz, *Erinnerungen,* 294–96, on 296.

89. Hertz to his parents, 17 Oct. 1887, in Hertz, *Erinnerungen,* 230.

90. Hertz to his parents, 5 Oct. 1888, in Hertz, *Erinnerungen,* 1st ed. of 1927, 195–98.

91. Hertz to Director Schuberg, Hertz Personalakte 1885/1894, Bad. GLA, Diener 76/9942. As Clausius's successor, the Bonn faculty proposed Hertz in first place. Konen, "Das physikalische Institut," 350.

92. Even if Hertz's salary at Karlsruhe had been raised to Bonn's, his total income there would still have been considerably less than at Bonn owing to the difference in lecture money. Hertz to Director of Karlsruhe Technical Institute, 24 Dec. 1888, Hertz Personalakte 1885/1894, Bad. GLA, Diener 76/9942.

93. Lehmann to Hertz, 12 July 1890, Ms. Coll., DM, 2968.

94. Otto Lehmann to Director of Karlsruhe Technical Institute, 3 June 1891, Bad. GLA, 235/4168.

became bolder in his demands for physics. With time, physics had grown so much, he argued, that one teacher could no longer represent it all. If, nevertheless, this growth had not brought major changes in physics instruction at Karlsruhe as elsewhere, it was because physics institutes could not be divided, and the establishment of a new chair of physics necessarily required the setting up of a corresponding new institute, which cost a great deal of money. In the past, the general public had had little understanding of the needs of the exact sciences, so that it was hopeless to make large requests; but now, Lehmann noted, modest beginnings of an expansion of physics instruction making use of existing facilities had a chance. He therefore proposed for Karlsruhe *eight* new physics chairs, one each for theoretical physics, electrotechnology, physical chemistry, technical physics, meteorology, history of physics, physical measurements, and optics and photography. There might even be a ninth chair, since theoretical physics was already too large for one representative and might best be divided right away. On the other hand, some parts of theoretical physics such as theoretical mechanics, mechanical heat theory, and hydrodynamics were already included in lectures by teachers in the school of mechanical engineering, which made the subject easier to handle. Some of the other subjects were also being taught already, but they all needed institutes of their own.[95] With a few exceptions, Lehmann's proposals were realized in the next decade. In particular, theoretical physics was represented at Karlsruhe by an ordinary professorship in 1896.[96]

Hertz understood that the Prussian government valued him above all as a researcher. Whereas the Bonn faculty wanted Hertz to teach this and that, the ministerial official Althoff let Hertz know that beyond giving the required experimental physics lectures and directing laboratory work, he was to spend his time on his own research.[97] Althoff's advice agreed with Hertz's inclination, which was to avoid all local entanglements that might interfere with his research. Even after taking the best precautions, however, Hertz soon complained of the endless account forms, lost keys, and all the other distractions of running a physics institute.[98]

At Bonn Hertz took charge of an institute that was uncomfortable and run down and an instrument collection that was quite limited. He got Althoff to promise means for instruments and for outfitting additional rooms from the unused director's residence.[99] At best his arrangements fell far short of those of new physics institutes,

95. Lehmann to Director of Karlsruhe Technical Institute, 22 July 1892, Bad. GLA, 448/2355.
96. The first ordinary professor for theoretical physics at Karlsruhe was August Schleiermacher. *Minerva. Jahrbuch der gelehrten Welt* (Berlin, 1891–   ).
97. In private, Althoff told Hertz not to load himself down with lectures; Hertz was not to take seriously the "public" lectures customarily expected of Prussian professors. Hertz to his parents, 25 Dec. 1888, in Hertz, *Erinnerungen,* 272–76, on 276.
98. Hertz to A. Schulte, 12 Jan. 1889, Bonn UB. Hertz to his parents, 8 Nov. 1890, in Hertz, *Erinnerungen,* 304–6. Hertz complained soon after moving to Bonn: "To my great grief, I have no time to go further on in these things [experiments on electromagnetism] for a year or so, having to spend too much time with my lectures, laboratory, examinations, etc." Hertz to Heaviside, 10 Aug. 1889, quoted in Appleyard, *Pioneers,* 239.
99. Althoff told Hertz that his salary was to be 6400 marks with 660 marks for living quarters, and Althoff mentioned some figures for the budget and extraordinary grants. Hertz to his parents, 25 Dec. 1888, in Hertz, *Erinnerungen,* 272–76, on 276. According to Konen, who saw Hertz's correspondence about the budget, Hertz received an extraordinary grant of 11,000 marks for instruments and machines and a regular budget of 4570 marks. Konen, "Das physikalische Institut," 350.

but he decided that his would do for a time. With extra funds for the purpose, he bought apparatus for elementary laboratory instruction and costly instruments for research, which included an Appun sonometer, a Rowland optical grating, several telescopes for reading scales, and, close to his own interests, fine galvanometers and a series of other electrical measuring instruments. As he was skilled with his hands, he could build whatever apparatus he needed beyond the metering devices.[100] With laboratory rooms for beginners, other laboratory rooms for advanced students who still required direction, and still others for fully independent researchers, and with the help of his assistant, Hertz brought to life a modest experimental activity in the Bonn physics institute.[101]

Responding to an inquiry from the Norwegian Vilhelm Bjerknes, Hertz explained that his Bonn physics institute was not rich. That did not discourage Bjerknes from coming, since one of his reasons for wanting to work with Hertz was to learn how to do experiments with simple means, which was all that most physicists could look forward to anyway.[102] Other young physicists, often from abroad, wanted to come to Bonn to work with or study under Hertz.[103] If they came, they carried out experiments deriving from Hertz's researches, while Hertz did experiments of his own, for example, stringing parallel wires along a passageway of the institute to clear up a contradiction in his earlier work.[104]

When Hertz came to Bonn, he had great hopes for progress in understanding electromagnetic phenomena through more experimental research. "Theory goes much further than the experiments," he told Heaviside. "But I think in due time there will come from experiment many new things which are not now in theory, and I have even now complaint against theory, which I think cannot be overcome until further experimental help." Hertz had specific problems in mind: "I hope for many new things to come from the experiments. . . . The motion of the ether relatively to matter—this indeed is a great mystery. I thought about it often but did not get an inch in advance. I hope for experimental help; all that has been done till now has given negative results."[105]

Hertz's considerable experimental work at Bonn dealt with a miscellany of top-

100. Hertz's annual report on the physics institute for 1890–91, *Chronik der Rheinischen Friedrich-Wilhelms-Universität zu Bonn* 16 (1890–91): 43–44. Konen, "Das physikalische Institut," 351.

101. In 1890–91, in the rearranged and expanded institute quarters, the beginner's practice course drew eight students in the summer semester and seven in the winter; in 1891–92, it drew eleven and thirteen; two or three students did scientific research each semester. Until December 1890, Hertz's assistant was Carl Pulfrich; after him a student assisted Hertz until Philipp Lenard took over in April 1891. Hertz's annual reports on the physics institute in *Chronik . . . Bonn* 16 (1890–91): 43–44, and 17 (1891–92): 46–47.

102. Bjerknes to Hertz, 2 Jan. 1890, Ms. Coll., DM, 2871.

103. As an example, Vilhelm Bjerknes's friend K. Birkeland from the University of Christiania wanted to complete his studies with Hertz. Letter to Hertz, 3 May 1893, Ms. Coll., DM, 2879. As another, the Italian Antonio Garbasso, who had a grant from his government to continue his studies abroad, wanted to spend two semesters with Hertz. Letter to Hertz, 17 Oct. 1893, Ms. Coll., DM.

104. For example, in Hertz's institute Bjerknes did research on the damping of electric waves, and F. Breisig did research on the influence of ultraviolet light on electric discharge. Hertz's report on his institute in *Chronik . . . Bonn* 16 (1890–91): 44. Konen, "Das physikalische Institut," 350.

105. Hertz to Heaviside, 10 Aug. and 3 Sept. 1889, in Appleyard, *Pioneers*, 238–39.

ics; it was not a sustained series of experiments, as his work at Karlsruhe had been. Rather his Bonn years were to be distinguished not by experiments but by theoretical research. In the spring of 1889, just before moving to Bonn, Hertz began reflecting on the fundamental equations of electrodynamics,[106] the outcome of which was two papers in 1890 on the theory of electrodynamics. In these, Hertz proceeded from the understanding, which he had reached in his experimental study, that Maxwell's theory was "richer and more comprehensive" than any other. But if the theory was "perfect" in content, it was not in "form," and Hertz set about to simplify it, expose the "essential ideas," and, above all, remove all traces of reasoning from action at a distance, which Hertz believed Maxwell's formulation of his theory still retained.[107]

In the Introduction to *Electric Waves*, Hertz explained that at the time of his experimental work, he had admired Maxwell's mathematical statements, but he had not always been sure that he grasped their physical significance, which was why he had approached his work through Helmholtz's theory. Now that he was convinced of Maxwell's theory, he knew of no shorter or better answer to the question "What is Maxwell's theory?" than this: "Maxwell's theory is Maxwell's system of equations." It followed that Maxwell's, Helmholtz's, and Hertz's own representations of electrodynamics were all forms of Maxwell's theory, since they all yielded the same system of equations.[108]

In his theoretical work in 1890, Hertz postulated the mathematical connections between the main observable quantities of Maxwell's theory: the electric and magnetic forces. He made no attempt to derive the equations from a physical conjecture about the constitution of the ether and the nature of the forces, as these things were still "entirely unknown"; he regarded it as "expedient to start from these equations in search of such further conjectures respecting the constitution of the ether."[109] Writing the mathematical connection between the electric and magnetic forces to express clearly the logical structure of Maxwell's theory, Hertz gave the equations for the free ether, in Gauss's absolute units:[110]

$$A\frac{dL}{dt} = \frac{dZ}{dy} - \frac{dY}{dz}, \qquad A\frac{dX}{dt} = \frac{dM}{dz} - \frac{dN}{dy},$$

106. Hertz, diary entries for 23 Feb. and 1 and 29 Mar. 1889, in Hertz, *Erinnerungen*, 282, 286.

107. Hertz observed that Maxwell distinguished between polarization, or dielectric displacement, in the free ether and the electric force that produces it, which implies that if the ether were removed from a portion of space, the electric force would remain. It is a distinction, Hertz said, that makes sense only within the theory of action at a distance. Heinrich Hertz, "Ueber die Grundgleichungen der Elektrodynamik für ruhende Körper," *Gött. Nachr.* 19 (1890): 106–49, reprinted in *Ann.* 40 (1890): 577–624, and in Hertz, *Ges. Werke*, vol. 2, *Untersuchungen über die Ausbreitung der elektrischen Kraft*, 2d ed. (Leipzig, 1894), 208–55; translated as "On the Fundamental Equations of Electromagnetics for Bodies at Rest," in *Electric Waves*, 195–240, quotations on 195–96.

108. Hertz, "Introduction," *Electric Waves*, 20–21.

109. Hertz, "Fundamental Equations . . . at Rest," 201.

110. Hertz, "Fundamental Equations . . . at Rest," 201. These became a standard form of the equations of Maxwell's theory; Hertz's left-handed coordinate system determines the sign. Hertz's introduction of these equations is analyzed in Tetu Hirosige, "Electrodynamics before the Theory of Relativity, 1890–1905," *Japanese Studies in the History of Science*, no. 5 (1966), 1–49, on 2–6; and in P. M. Heimann, "Maxwell, Hertz and the Nature of Electricity," *Isis* 62 (1970): 149–57.

with corresponding equations for the y and z components, where X, Y, Z are the components of the electric force and L, M, N those of the magnetic force, and A is the reciprocal of the velocity of light. Hertz added the supplementary equations to this connection, which distinguish the ether from ponderable matter:

$$\frac{dL}{dx} + \frac{dM}{dy} + \frac{dN}{dz} = 0, \qquad \frac{dX}{dx} + \frac{dY}{dy} + \frac{dZ}{dz} = 0.$$

From these equations, Hertz deduced certain principal electric, magnetic, and optical phenomena: Ohm's law, Kirchhoff's laws of circuits, Neumann's potential for the ponderomotive force between currents, the law of electrostatic force, which was the starting point for theories like Weber's but was a "remote final result" in Maxwell's theory, induction in open circuits, which was the "richest region of all" and one still little explored experimentally (Hertz here cited his papers on electric waves), and more.

In his effort to "sift Maxwell's formulae and to separate their essential significance from the particular form in which they first happened to appear," Hertz had been anticipated by Heaviside. When Heaviside's work came to his attention, Hertz recognized that Heaviside had "gone further on than Maxwell" by doing away with "unnecessary potentials," namely, the vector potentials, making the "electric and magnetic forces the objects of direct attention."[111] But Heaviside's work was mathematically obscure, and it was to Hertz's work that German physicists turned to understand Maxwell's theory. From Erlangen, the young physicist Hermann Ebert wrote to Hertz of his gratitude for his paper on the electrodynamics of bodies at rest. In preparing lectures on the electromagnetic theory of light, Ebert had been struggling unsuccessfully with Maxwell's theory; upon reading Hertz's paper the difficulties vanished at a stroke.[112]

Hertz's reformulation of Maxwell's theory held an interest of its own, quite apart from the clarity it brought to the subject. By starting from bare differential equations describing experimental results rather than from detailed physical pictures, Hertz offered physicists a model of, as Boltzmann called it, "mathematical phenomenology." Mach, for example, admired this way of doing physics: he read Hertz's 1890 work with "special interest," he told Hertz, since in it Hertz followed the "ideal of a physics free of mythology" that he had been advocating.[113]

---

111. Hertz to Heaviside, 21 Mar. 1889, in Appleyard, *Pioneers*, 238. Hertz acknowledged Heaviside's priority in his paper "Fundamental Equations . . . at Rest," 196–97. Heaviside sent Hertz his work on Maxwell's theory and discussed it in a letter to Hertz, 14 Feb. 1889, Ms. Coll., DM, 2923. Earlier, Cohn had drawn Hertz's attention to Heaviside's parallel work, in particular to Heaviside's elimination of the potential. "Thus ψ and A are murdered," Cohn reported, and he thought it would be good if certain other quantities, in Heaviside's spirit, were also "ge-'murdered.' " Cohn to Hertz, 15 Apr. and 8 June 1889, Ms. Coll., DM, 2880–81.

112. Ebert to Hertz, 2 June 1890, Ms. Coll., DM, 2899.

113. Ludwig Boltzmann, "Über die Entwicklung der Methoden der theoretischen Physik in neuerer Zeit," in *Populäre Schriften* (Leipzig: J. A. Barth, 1905), 198–227, on 221. Mach to Hertz, 25 Sept. 1890, Ms. Coll., DM, 2976. Hertz acknowledged that in his presentation, Maxwell's theory looked abstract and

To give his theoretical work on Maxwell's theory a wide distribution, Hertz sent a large number of reprints to colleagues and one to Wiedemann for the *Annalen*.[114] He had written the paper quickly and published it before it was complete, he told Cohn, because the subject was "especially timely." Alongside Hertz's paper in the *Annalen*, Cohn published one of his own on the same subject; in it Cohn referred to Hertz's recent "brilliant experimental success," which had inspired in many physicists the wish for a "systematic presentation of electrical theory"; only in this way could they feel "completely happy with the newly won possession." Hertz had now given them such a presentation, and Cohn offered another by developing electrical theory from concepts defined "purely mechanically."[115]

Hertz wrote to Cohn that he was inserting moving conductors into the system of equations and to Wiedemann that he would soon send him a sequel paper for the *Annalen* on the electrodynamics of moving bodies.[116] In this, his last paper on electrodynamic theory, Hertz assumed that the ether within bodies moves with them, an assumption that allowed him to formulate Maxwell's equations for moving charged bodies. He thought it was, in fact, more likely that the ether and bodies move independently of one another, but to work with that idea required making assumptions about the unknown motion of the ether. He was content to seek a "systematic arrangement" of the theory, regarding it as premature to seek a "correct" theory.[117]

In both of his theoretical studies in 1890, Hertz proceeded from the fundamental physical concept of contiguous action: according to "Maxwell's standpoint," electric and magnetic forces at a point arise from the state of the medium at that point. Hertz cited the writings by Maxwell and his British followers who worked with that concept; otherwise, he cited principally Helmholtz's work in 1870, which presented a "limiting case" of Maxwell's theory and which had directed Hertz toward his experimental decision in favor of that theory.[118]

---

colorless: "If we wish to lend more colour to the theory, there is nothing to prevent us from supplementing all this and aiding our powers of imagination by concrete representations of the various conceptions as to the nature of electric polarisation, the electric current, etc. But scientific accuracy requires of us that we should in no wise confuse the simple and homely figure, as it is presented to us by nature, with the gay garment which we use to clothe it." Hertz, "Introduction," *Electric Waves*, 28.

114. Hertz sent off 110 reprints, "rather a lot," from the original place of publication, the *Göttinger Nachrichten*. Letter to Gustav Wiedemann, 1 June 1890, Ms. Coll., DM, 3234.

115. Hertz to Cohn, 9 June 1890 and 25 Feb. 1891, Ms. Coll., DM, 3203, 3205. Emil Cohn, "Zur Systematik der Electricitätslehre," *Ann.* 40 (1890): 625–39, quotation on 625. There was a fair amount of German theoretical work on Maxwellian electrodynamics, but it came largely after Hertz's. In the *Annalen* for the three years 1888–90, as we will see in the next chapter, Hertz's and Cohn's were the only theoretical studies by German physicists on the subject; in addition, the Austrian physicist František Koláček published on it there.

116. Hertz to Cohn, 9 June 1890, and to Gustav Wiedemann, 1 June 1890.

117. Hertz restricted his considerations narrowly to electromagnetic processes, which do not suggest the independence of ether and matter, as optical processes do. Heinrich Hertz, "Ueber die Grundgleichungen der Elektrodynamik für bewegte Körper," *Ann.* 41 (1890): 369–99; translated as "On the Fundamental Equations of Electromagnetics for Bodies in Motion," in *Electric Waves*, 241–68, quotation on 268.

118. Hertz also cited subsequent papers by Helmholtz, an early paper by Emil Jochmann, Kirchhoff's mechanical lectures, and papers by two more German authors, W. C. Röntgen and Cohn, appearing in these same volumes of the *Annalen* with his own.

18

Physical Research in the *Annalen* and in the *Fortschritte*

Hertz's work on Maxwell's electromagnetic theory in 1888–90 facilitated its acceptance in Germany and stimulated physicists there to examine its ramifications throughout physics. Because of the importance of this work, we select those years as the basis for our next partial survey of research in the *Annalen der Physik*. In addition to the electromagnetic and optical work surrounding Hertz's, we also look at work in molecular physics. For the years following 1890 were to see important developments originating in Maxwell's electromagnetic theory and in molecular physics, leading to—as it came to be called—"modern" theoretical physics.

Contributors and Contents

In the twenty years between the time of this survey and that of the previous one, the number of contributors to the *Annalen der Physik* in Germany increased substantially, by a third. Within that increased number, certain categories of contributors grew while certain others declined, a redistribution which reflected two trends. One was that the journal under Poggendorff's successors was becoming a journal sustained almost exclusively by physicists.[1] While it continued to publish, for example, occasional discussions of the eye and the ear and of nature outside the laboratory, of rainbows, of waves on the sea, and the like,[2] and even of the history of physics (making room for Eilhard Wiedemann's historical studies of early physics, which

---

1. In 1888–90, the *Annalen* contained contributions by twenty-six ordinary physics professors, including one emeritus, together with seven extraordinary physics professors, and at least ninety Privatdocenten, assistants, students, and recent graduates in physics. Only seven scientists in universities and technical institutes who did not have positions in physics numbered among the contributors.

2. On the eye: Ebert, 33: 136–55; Wolf, 33: 548–54; Geigel, 34: 347–61; Brodhun, 34: 897–918. On the ear: Voigt, 40: 652–60; Preyer, 38: 131–36. On the weather: Helmholtz, 41: 641–62; Pulfrich, 33: 194–208.

began in good physics fashion, "Our knowledge . . . still is in a bad way"),[3] the contributors to these border areas of physics were often themselves physicists. The other trend was an increasing predominance in the *Annalen* of work by physicists in universities. The technical institutes increased their contribution, too, as is to be expected from their rapid development in this period. From outside these two kinds of institutions, there were only about half as many contributors as before, now mostly secondary school teachers with a handful from other occupations.[4]

The earlier outstanding contributors of purely theoretical studies to the *Annalen* were Kirchhoff and Clausius, both of whom died just before the time of this survey. Wilhelm Weber was no longer active, and Bezold had moved from teaching physics to directing the meteorological institute in Berlin. Of the institute directors from the earlier time, only Lommel was still publishing on theory in the *Annalen*. Hertz, Warburg, and other newer institute directors who were publishing on theory were in the early part of their careers and were outstanding for their combined theoretical and experimental talents.[5] To these should be added two physicists who published on theory and who were near the end of their careers: Hankel, who was emeritus at Leipzig, and Helmholtz, who had just moved to an administrative position at the Imperial Institute of Physics and Technology, the Physikalisch-Technische Reichsanstalt, in Berlin.

Since 1869–71, the time of our previous survey, many extraordinary professors for theoretical physics had been appointed, most of whom published in the *Annalen* in 1888–90. Their research included some purely theoretical papers, Planck's above all, which had considerable significance for physics.[6] Below them were Privatdocenten, assistants, and others who published purely theoretical papers in the *Annalen*, among them several experimental physicists with strong theoretical ability.[7] In general, with the major exception of Planck, physicists publishing in the *Annalen* in 1888–90 showed little tendency to specialize in pure theory.[8]

As before, we should note, examples of the best theoretical work by German physicists continued at this time to be published in journals other than the *Annalen*, for instance, in the mathematicians' *Journal für die reine und angewandte Mathematik*.[9]

3. Eilhard Wiedemann, 39: 110–30, and more following on the Arabic predecessors of the founders of modern physics.

4. Along with twenty universities—only Münster Academy was unrepresented—eight technical institutes were represented by authors in the *Annalen* in 1888–90. There were at least twenty secondary school teachers and administrators, and there were three technical school teachers. There were two employees from the Physikalisch-Technische Reichsanstalt in Berlin and two more from the Hamburg state physics laboratory. The remaining dozen contributors no doubt included some unaffiliated persons.

5. Besides Hertz and Warburg, they included Voigt, Braun, and Oberbeck.

6. Besides Planck, they included Cohn, Paul Volkmann, Lorberg, and, for the theory of instruments, Felix Auerbach. Other extraordinary professors for theoretical physics, such as Ketteler and Narr, published experimental work with theoretical discussions.

7. Drude was foremost among them, but also Otto Wiener, Walter König, and—though he was soon to turn to technical physics—Föppl.

8. Lorberg, as well as Planck, was doing purely theoretical research at this time.

9. The editors of the *Journal*, Leopold Kronecker and Karl Weierstrass, on the occasion of the hundredth volume of the journal in 1887, noted correctly that throughout its history it had published work by "many of the most significant . . . mathematicians and mathematical physicists." In the past

A contributor to the *Annalen* in 1888–90 observed that physics brought the theoretical and the exact observational work characteristic of mathematics and astronomy together with the experimental work characteristic of chemistry.[10] It was an accurate observation: most physical research in the *Annalen* was experimental and was often accompanied by theoretical and mathematical discussion. Now as in the past, experimental papers commonly contained separate parts entitled "Mathematical Treatment," "Comparison between Theory and Observation," "Theoretical Investigation," "Theoretical Observations," "Theoretical Part," or simply "Theory."[11] These parts were given over to the never ending work of examining, criticizing, and developing laws and concepts in light of experimental results. This work included the elucidation of physical constants to be measured, the inexhaustible common ground of theory and experiment: with improved instrument collections, physicists measured and remeasured indices of refraction and the theoretical molecular constants determining them, elastic constants of solids, light absorption constants of crystals, magnetic constants of isotropic and crystalline bodies, metal optical constants, the constant expressing the relationship between electromagnetic and electrostatic units, among other constants.[12]

Occasionally, theoretical discussion in the *Annalen* was purely qualitative; it was in Braun's theory of change of state and in Lehmann's theory of crystals, which was illustrated by a hundred or so polarization pictures and was without equations or numbers.[13] Occasionally, too, it was presented with the aid of geometrical or graphical methods to make the calculations more visual.[14] But most often it was presented in analytical form, reminiscent of the early French mathematical physics.[15]

Molecular Work

Much physical research appearing in the *Annalen* in 1888–90, as in the time of the previous survey, was guided by molecular reasoning. From Maxwell's theory of gases

---

they had included Gauss, Dirichlet, and Riemann. This, the hundredth, volume contained work by Helmholtz and Boltzmann as well as by Kronecker, Kummer, and other leading mathematicians, both German and foreign. "Vorwort zum hundertsten Bande," *Journ. f. d. reine u. angewandte Math.* 100 (1887): v–vi.

10. Eilhard Wiedemann, 39: 110–30, on 110–11.

11. Wiedeburg, 41: 675–711; Messerschmitt, 34: 867–96; Galitzine, 41: 770–800; Eilhard Wiedemann and Ebert, 35: 209–64; Pockels, 37: 144–72, 269–305; Karl E. Franz Schmidt, 33: 534–48; and others.

12. Ketteler, 33: 353–81, 506–34; Wesendonck, 35: 121–25; Voigt, 36: 743–59; Drude, 40: 665–80; Henri du Bois, 35: 137–67; Voigt, 39: 412–31, and many more places; Kundt, 34: 469–89; Himstedt, 33: 1–12, 35: 126–36.

13. Braun, 33: 337–53; Lehmann, 40: 401–23, gave a few temperatures, but nothing more in the way of numbers.

14. Following Kundt's example, Wiener treated the combined phenomena of double refraction and circular polarization by "geometrical" methods, imparting to the theory greater "Anschaulichkeit" (35: 1–24). Similarly, by a "graphical representation," Auerbach made a calculation in the theory of an instrument, an air pump, "anschaulich" (41: 364–68).

15. German physicists in 1890 still discussed, criticized, and developed the work of Fresnel, Poisson, Cauchy, and other French mathematical physicists.

German physicists adapted the concept of "relaxation time" to their electrical studies,[16] and in general they drew on work on the theory of gases by Maxwell, Clausius, Boltzmann, O. E. Meyer, and others.[17]

Hertz along with Hittorf and several others had made quantitative studies of electric conduction in gases, but the law of conduction continued to elude them. Since gaseous behavior is relatively simple, they were puzzled why they had not been able to describe conduction in thin gases with the same exactness with which Ohm's law describes conduction in metals and electrolytes. One approach was to go after more experimental facts, which Narr saw as proper as long as physicists were still far from an "approximately clear mechanical idea of the passage of electricity through gases."[18] Some physicists offered solutions to the problem—"still full of riddles"—in the course of their experimental search for the exact law: these included "mechanical theories," "ether theories," and, in this set of *Annalen* volumes, an explanation based on the kinetic theory of gases together with the laws of electrostatics.[19] Molecular ideas entered the explanations in one way or another.

Other instances of electric conduction received their explanations and laws. Eilhard Wiedemann explained Hertz's discovery of the action of ultraviolet light on an electrode by the passage of energy from the light to the molecules of the electrode and from them to the molecules of the surrounding gas, setting up a convective current. Hankel explained the main laws of galvanic currents—Ohm's law, Kirchhoff's laws of currents, and Joule's law of the heat of a current—by the electric state of material molecules, which consists of a rotary oscillation on the surfaces of the molecules. Graetz extended to electrolytically conducting solids Clausius's explanation of electrolytically conducting solutions based on the assumption of freely moving charged molecules or partial molecules. Working on the same problems as van't Hoff, Arrhenius, and Nernst, Planck developed the theory of chemical equilibrium and of the excitation of electricity and heat in dilute solutions, but he made no hypotheses about the motions of the molecules that entered his discussion.[20]

Physicists working in 1888–90 did not believe that the sharply distinguishable forms of bodies in the phenomenal world necessarily have a correspondence in the molecular world. Bodies that appear to be solid, for example, may contain Graetz's "fluid" molecules. Or they may contain Lehmann's "liquid" crystals: the essence of crystals, for Lehmann, is not the geometrical arrangement of molecules determined

16. Cohn and Arons, 33: 13–31; Graetz, 34: 25–39.

17. Schleiermacher, 34: 623–46; Föppl, 34: 222–40; Galitzine, 41: 770–800; Eilhard Wiedemann, 37: 177–248; Wüllner, 34: 647–61; Ebert, 36: 466–73. The most sustained work on the kinetic theory of gases came from abroad: Ladislaus Natanson's series of purely theoretical, highly mathematical studies (33: 683–701, 34: 970–80, etc.).

18. Narr, 33: 295–301, on 298.

19. Narr, 33: 298. To explain electric conduction in gases, Föppl preferred the earlier mechanical theories of Gustav Wiedemann and Rühlmann to the ether theories of Erik Edlund, who identified the "extra-molecular ether" as the substratum of the current, and of Eilhard Wiedemann, who explained the current by the electrical deformation of ether shells surrounding molecules. The difficulty with Föppl's kinetic theory was that it did not say how two electrified molecules behave upon collision (34: 222–40).

20. Eilhard Wiedemann's "Theoretical Observations" at the end of his and Ebert's study of electric discharges (35: 209–64, on 255–64); Wilhelm Hankel, 39: 369–89; Graetz, 40: 18–35; Planck, 34: 139–54, 39: 161–86, 40: 561–76.

by elastic forces, for that arrangement can be destroyed by external pressure without in the least affecting the crystalline nature of the bodies; rather it is the structure of the physical molecule that explains crystal properties such as double refraction. For Voigt, crystals with directional properties and solids with identical properties in all directions may have a common molecular basis; to explain the empirical failure of Poisson's theory of isotropic bodies, Voigt extended to the molecules of such bodies his understanding of the molecules of crystals: molecules have a certain polarity, so that their mutual forces depend not only on their separation but also on the direction of their connecting line.[21]

In their optical researches, German physicists continued to use molecular reasoning extensively. One of their basic ideas was that the oscillations of ether particles cause the molecules of bodies to oscillate in the same direction and with the same frequency, a consequence, as Wiener put it, of the "striving of physics to learn to conceive of all phenomena as processes of motion." In their optical explanations, physicists used auxiliary molecular ideas such as shells of ether particles surrounding molecules of bodies; after experimentally deciding between conflicting laws of the index of refraction of compressed water, Ludwig Zehnder interpreted the correct law by a physical picture: when water is compressed, some of the less dense ether outside the shells of ether surrounding water molecules is forced out, which changes the index of refraction in the right direction.[22] To take another example from optics: the explanation of spectra, according to Eilhard Wiedemann, requires consideration of the various internal and external motions attributed to gas molecules by the "new views" of the constitution of matter.[23] Since various types of motion are associated with various types of differential equations, Ebert argued, the equations—even if they cannot describe spectra quantitatively—can suggest the molecular motions responsible for the observed line, band, and continuous spectra.[24] Physicists called upon different assumptions within the "molecular theory of radiation" to defend different views of the origin of the spectra of gases.[25] In general, the motion of light in bodies, the production of light in fluorescence and phosphorescence, and the production of light by thermal, electrical, chemical, frictional, and other processes all found interpretations within this broad molecular conception of matter and ether.[26]

Physicists, of course, could not see molecules directly, but they could draw conclusions about the laws that govern them. They could, for example, conclude that

21. Graetz, 40: 18–35; Lehmann, 40: 401–23; Voigt, 38: 573–87.

22. Wiener, 40: 203–43, on 241–42; Zehnder, 34: 91–121, on 117–21.

23. Eilhard Wiedemann, 37: 177–248, on 178–79, decided that it is the oscillatory motions within gas molecules and not their rotations and translations that produce light.

24. Ebert, 34: 39–90.

25. Ebert, 34: 39–90, on 89. Ebert, 33: 155–58, and Wüllner, 34: 647–61, criticized one another's molecular interpretations. Wüllner criticized Kayser's view that certain molecules produce line spectra and certain others produce band spectra; his own view was that the transition between the two kinds of spectra is gradual (38: 619–40).

26. Voigt, 35: 370–96, 524–51; Eilhard Wiedemann, 34: 446–63; Wüllner, 34: 647–61; Ketteler, 33: 353–81, 506–34.

the same Doppler principle that applies to the greatest bodies in the universe applies as well to individual luminous molecules at the opposite end of the scale of things.[27] By analogy with the forces that move the greatest bodies, they could reason about the forces that move the smallest. They could assume that molecular forces together with the laws describing their effects were more or less similar to those they were familiar with, and they could measure their extraordinarily short range.[28]

Optical Theories

When physicists discussed the foundations of optical theory in the *Annalen* in 1888–90, it was often in connection with Franz Neumann's theory from the 1830s and Kirchhoff's and Voigt's extensions of it. It was to the advantage of Neumann's theory, Voigt said, that it alone was supported "by the consistent connection to the laws of elasticity theory and by the use of unassailable fundamental laws of general mechanics," and it therefore merited attention.[29] That is, to Voigt, Neumann's theory was an instance of a well-constructed theory. Several other German physicists who published on optics in the *Annalen* in these years—Paul Drude, Pockels, and Karl Schmidt, among them—also approached the subject with a theoretical viewpoint they saw as descending from Neumann's.

These physicists responded to new challenges to the optical theories they worked with, for example, to an objection based on J. C. Jamin's experiments with polarized light that applied to both Neumann's and Fresnel's theories.[30] They responded, too, to the old conflict between these two theories: in their work on the reflection of light, they constantly had to confront the different assumptions of Neumann's and Fresnel's theories concerning the density and elasticity of the ether and the associated difference in the relationship of the plane of polarization to the direction of light oscillation.[31] New experiments on standing light waves were being put forward by Otto Wiener as evidence for Fresnel's assumptions and against Neumann's. Moreover, some German physicists assumed that the decision in favor of

27. Ebert, 36: 466–73.

28. From experiments on the thickness of oil on water, Sohncke calculated the radius of action of molecular forces to be equal to or greater than 55.75 $\mu\mu$ (40: 345–55, on 354). From Dorpat, the *Annalen* received an article containing the law of force between pairs of molecules, m_1m_2/r^2, which showed that the "same law that rules the macrocosm has also proven valid for the microcosm" (Bohl, 36: 334–46, on 346). It was usual to assume a more complex and uncertain law in analyzing processes on the molecular level. Hermann Ebert, for example, wrote the force producing motions responsible for line spectra as $d^2x/dt^2 = f(x, a, b, c, \ldots)$, where $a, b, c, \ldots$ are constants that depend on the properties of the radiating molecule (34: 39–90, on 89–90).

29. Voigt, 35: 76–100, on 100.

30. It had been argued that Jamin's experiments on the reflection of polarized light made invalid the theories of light of Neumann, Voigt, and Fresnel; Voigt explained the departure from theory by a surface layer on the reflecting body (31 [1887]: 326–31), and Drude (36: 532–60, 865–97) and Karl Schmidt (37: 353–71) examined this explanation.

31. For Fresnel, we recall, the density of the ether changes in bodies, and for Neumann its elasticity changes; for Fresnel the oscillation of polarized light is perpendicular to the plane of polarization, and for Neumann it is parallel to the plane.

Fresnel had already been made, adducing as support a variety of phenomena: Newton's colored rings, Fresnel's experiment with three mirrors, phase changes with reflection of polarized light, crystal fluorescence, optical properties of moving bodies, and more; these were all rejected as supports for Fresnel's superiority by Drude and Voigt. In fairness, Voigt said, both theories should be used and both named, and Paul Volkmann and Pockels spoke of the "Fresnel-Neumann" laws for the propagation of light.[32]

The foundations of optical theory were constantly disputed. In light of developments since Neumann's theory, Volkmann, for example, examined the mechanical foundations of optics: he reduced one part of the subject, the motion of light in unbounded, transparent media, to elasticity theory "rigorously in the sense of mechanics." He could not do the same for the problem of reflection and refraction, where it was necessary to leave the "mathematically rigorous basis" of "pure elasticity theory." If all of optical theory were to be developed from "pure mechanics," another basis had to be found, and Volkmann noted that Kirchhoff had sought it in the interaction of the particles of ether and matter.[33]

As the electromagnetic theory posed a challenge to all mechanical theories of light, physicists had begun to examine the relationship between the two sets of foundations.[34] Old questions of the optics of moving bodies, such as the question of the dragging of ether by moving bodies, Voigt treated within the framework of the mechanical theories and their disagreements,[35] while others now treated them from the electromagnetic side.[36] The seemingly irresolvable controversy between Neumann's and Fresnel's theories over the plane of polarization appeared resolvable within the electromagnetic theory: if light consists of an electric oscillation and a magnetic oscillation at right angles to one another, the question of the direction of the light oscillation loses its meaning.[37] This implicit advantage of the electromagnetic theory could be appreciated even by developers of the mechanical theory such as Drude. Drude was open-minded; he did not count recent experiments on metal optics reported in the *Annalen,* for example, as refutation of the electromagnetic theory of light, since the optical constants for waves of the length of electric waves were all too uncertain. As Maxwell's theory demanded that these constants pass over continuously, with decreasing wavelength, to those for heat and light, Drude looked for a decision on the theory from new experiments with new apparatus: he hoped for a single apparatus capable of testing metals for all wavelengths, for those of light, heat, and electric waves.[38]

32. Wiener, 40: 203–43, on 240; Drude, 41: 154–60; Voigt, 35: 76–100, on 100; Volkmann, 35: 354–60, on 355; Pockels, 37: 144–72, 269–305, on 151.
33. Volkmann, 35: 354–60, on 354–55, 359–60.
34. Wiener, 40: 203–43; König, 37: 651–65; Geigel, 38: 587–618; Drude, 39: 481–554; and, from abroad, Koláček, 34: 673–711.
35. Voigt, 35: 370–96, 524–51.
36. Des Coudres, 38: 71–79; Röntgen, 35: 264–70.
37. Koláček, 34: 673–711; Drude, 41: 154–60, on 154.
38. Drude, 39: 481–554, on 553–54.

Electrodynamics

Because of the exactness of its measurements and the accessible mathematical regularities of its phenomena, optics remained the ideal testing ground for theory construction. A good number of German physicists worked theoretically in the subject before Hertz's experiments, as they did after; Maxwell's electromagnetic theory of light only enhanced the challenge of optical problems or, as some of them viewed it, extended the domain of optics. But optics did not have the foundational significance that some other parts of physics had around 1890; it resembled acoustics in that its foundations had been sought, and found, in mechanical principles, and if they were also sought in electromagnetic principles, the dependence of optics remained. For physics, the establishment of Maxwell's theory within electricity was the main task, and to this end optics could provide confirmation, or refutation, of the assumptions of the theory.

The pages of the *Annalen* in 1888–90 continued the controversies of electrodynamics, which seemed as irresolvable as Neumann's and Fresnel's in optics.[39] The same issue of the journal contained both Hankel's proof that the electrodynamic fundamental law is a "point law" like gravitation and Hertz's recognition that his experiments were best explained by the Faraday-Maxwell view of electric forces.[40] A physicist reading the *Annalen* from cover to cover in the years 1888–90 might well have been unclear about the direction electrodynamics would take in the near future.[41] But if he had followed the next stage of physics, he would have recognized that the most significant development of electrodynamics in Germany, following Hertz's work, lay with Maxwell's theory and that much of the most significant development in theoretical physics in general lay in that direction.

Besides the then familiar tests of Maxwell's theory—comparisons of the ratio of electrical units with the velocity of light and tests of the relationship between the dielectric constant and the index of refraction for insulators—other tests were tried in Germany. The results were inconclusive; in varying degrees, they confirmed the theory, disconfirmed it, and found its predictions to be the same as those of other theories. According to one result, optical observations of crystals were "a proof of Maxwell's theory";[42] according to another, experiments on the transparency of thin metal sheets contradicted Maxwell's theoretical relationship between optical trans-

39. For example, in connection with the long-studied phenomena of induction by the rotation of a magnet around its axis with respect to a conductor, Lorberg discussed many contending laws: Weber's, Edlund's, Clausius's, Riemann's, and Maxwell's (36: 671–92).

40. Hankel discussed the ether theory he had developed since the 1860s, according to which all electrical phenomena are oscillations (36: 73–93).

41. In any case, the older methods might still be seen to have advantages. Following Hertz, the Austrian physicist Stefan studied high-frequency oscillations in straight conductors, this time presenting the problem of the distribution of a varying current in a straight conductor of circular cross section "not from Maxwell's theory of the electromagnetic field but from the formulas that F. Neumann and W. Weber have laid down for the electrodynamic potential of two current elements" (41: 400–420, quotation on 406).

42. Geigel, 38: 587–618, on 610–11.

parency and electrical conductivity of a body;[43] according to yet another, the theoretical relationship between the dielectric constant and the index of refraction was confirmed for insulators and poor conductors but not even approximately for conducting fluids.[44] Experiments on electric residues were inconclusive,[45] as were experiments on the capacity of a condenser, and it was not always possible to derive different mathematical conclusions from Maxwell's and others' theories, even if their different physical assumptions might suggest them.[46] Moreover, Maxwell's was not the only electromagnetic theory of light.[47]

Incompletely tested as it was, by the late 1880s in Germany, Maxwell's theory had drawn wide notice, as it had not twenty years earlier when Maxwell's own papers were its difficult source and Maxwell's *Treatise* and its German translation did not yet exist. Most of the notice in the *Annalen* came from physicists who were at the start of their careers, still Privatdocenten or assistants or even only advanced students. The few established physicists among them were in their early careers, too: Cohn was an extraordinary professor for theoretical physics and engaged in a systematic formulation of Maxwell's theory; Röntgen and Himstedt were institute directors who were working on experiments to confirm the existence of an electrodynamic action of a dielectric moving in an electric field, an effect anticipated by both Maxwell's theory of dielectric displacement and the theory of the dielectric as polarized particles; and, in addition, Röntgen was trying experiments from the point of view that the light ether is the medium of electric forces.[48] But it was only Hertz who applied the resources of his institute single-mindedly to a set of experimental problems central to the claims of Maxwell's theory. In place of the confusing evidence for and against these claims in the *Annalen,* Hertz's experimental work was seen by his colleagues as decisive proof of the finite propagation of electric forces in air, and his theoretical work was seen as clarifying these forces and the equations describing their action.

Responses to Hertz's experiments were themselves predominantly experimental. They divided into two sorts. One was a widespread interest in the discovery he made in 1887 incidental to his researches on electric waves; physicists repeated and varied Hertz's experiments, explaining as best they could the action of the ultraviolet light on the electrode or on the surrounding air that facilitated the discharge.[49] The other

43. Wien, 35: 48–62.
44. Cohn and Arons, 33: 13–31, on 23.
45. Wüllner, in his study of electric residues, 32 (1887): 19–53, doubted that the values of the dielectric constants obtained so far agreed with the Faraday-Maxwell theory. Leo Arons regarded the accurate determination of these constants as difficult but not impossible (35: 291–311, on 307).
46. At first, Himstedt believed that his experiments on a "Schutzring-Condensator" showed that Maxwell's capacity formula was correct and Kirchhoff's "false" (35: 126–36, on 129–30). Soon he acknowledged his error: Kirchhoff's formula was as good as Maxwell's (36: 759–61).
47. A work on the refractive indices of fluids discussed the formula relating the index and the density from the side of the "electromagnetic theory of light," mentioning not Maxwell but Ludwig Lorenz, who independently developed an electrical theory of light, and H. A. Lorentz and Koláček (Ketteler, 33: 353–81, 506–34, on 354–55).
48. Cohn, 40: 625–39; Röntgen, 35: 264–70; Himstedt, 38: 560–73, 40: 720–26.
49. Eilhard Wiedemann and Ebert, 33: 241–64, 35: 209–64; Hallwachs, 33: 301–12; Narr, 34:

and more widespread response was to the central direction of Hertz's experiments. Boltzmann demonstrated electric waves in a great auditorium before two hundred viewers; Emil Wiechert demonstrated them by a similar method; and on Kundt's suggestion, Robert Ritter demonstrated them in a long passageway in the cellar of the Berlin physics institute, using an ingenious modification for making them visible to a large number of viewers.[50] All sorts of variants of the experiments were tried and reported in the *Annalen*.[51]

Hertz's experiments inspired others to try original ones. Since his experiments made the "identity of electric waves and light waves" appear no longer as merely the "result of mathematical developments but as the object of the most direct perception," one of his readers explained, the contradiction between the optical transparency and the good electrical conductivity of electrolytes appeared serious, and he did experiments to clarify it.[52] Hertz's proof of the finite propagation of electric actions inspired another reader to revive experiments of his own on the detection, by electrical means, of the motion of the earth through the ether. To yet another reader, Hertz's production of standing electric waves in air by bringing incident and reflected waves together suggested a method for producing standing light waves for the first time.[53] All in all, Hertz's work quickly stimulated a good deal of varied experimental activity among physicists in Germany.

By following Hertz's researches as they appeared and, in many cases, constructing and handling apparatus like his and reproducing his results, German physicists became intimately familiar with electrical phenomena according to Maxwell's teachings. Even if they had studied Maxwell independently, after Hertz they had a guide through the complexities of the theory and a persuasive demonstration of its usefulness in aiding experimenters in exploring new phenomena.

The lesson of Hertz's work was more than the indispensability of Maxwell's theory, it was as well the indispensability of good theoretical understanding for good experimental work and conversely. Hertz's experimental researches began with problems posed within the older action-at-a-distance electrodynamics, and they ended with Hertz's denial of action at a distance; in the Introduction to his *Electric Waves*, Hertz told the history of the researches the book contained, giving ample recognition to the interplay of the theoretical and experimental reasoning behind them.

To many German physicists, especially those at the start of their careers who were defining their orientation in research, Hertz's work suggested Maxwell's theory as a source of rewarding problems. Among those writing on Maxwell and Hertz in the *Annalen* in 1888–90 were several future institute directors, including Cohn,

---

712–19; Lenard and Wolf, 37: 443–56; Elster and Geitel, 38: 497–514; and, from abroad, Arrhenius, 33: 638–43.

50. Boltzmann, 40: 399–400; Wiechert, 40: 640–41; Ritter, 40: 53–54, demonstrated the oscillations in the detector of the electric waves by convulsions of a frog's leg.

51. Classen, 39: 647–48; Rubens and Ritter, 40: 55–73; Waitz, 41: 435–47; Elsas, 41: 833–49.

52. Cohn, 38: 217–22, on 218.

53. Des Coudres, 38: 71–79, on 72; Wiener, 40: 203–43.

Drude, Föppl, and the Austrian and later German physics institute director Boltzmann.[54] These four were the authors of the major texts that, in addition to Maxwell's own, introduced Maxwell's theory into Germany. Drude, for example, who discussed his favorite subject, optics, in the *Annalen* in 1890 with an awareness of Maxwell's work, made an early reputation through his experimental study of the optical properties of Hertzian electric waves; later, as director of Germany's largest physics institute, Berlin's, he continued his research on Hertzian waves. His successor at Berlin, Heinrich Rubens, who in 1890 reported in the *Annalen* quantitative experiments on Hertzian waves, devoted himself from the start of his career to the study of heat radiation from the electromagnetic point of view, experimentally bridging the gap between optical and electric waves, which Hertz's experiments had made a central problem for physics. Besides these, there were several more future institute directors among those responding to Maxwell and Hertz in the *Annalen* in 1888–90.[55]

Divisions of Physical Knowledge: *Fortschritte der Physik*

Through the nineteenth century, new physical disciplines were organized, representing both a furthering of specialization and, by bridging the more established disciplines, a corrective to it. Universities offered courses in and created positions and institutes for meteorology, geophysics, and astrophysics, and a good many physicists worked on problems originating in these fields.[56] Helmholtz, for instance, published fundamental studies in meteorology, examining the earth's atmosphere from the standpoint of hydrodynamics. Meteorology was characterized as the "physics of the atmosphere," and Bezold, who moved from a position in physics to the first professorship in meteorology in Germany, sought to make it an exact science through the application of thermodynamics. Several men who taught theoretical physics in universities did research in, and sometimes also taught, subjects belonging to the earth sciences. For years Dorn collected observations made in Königsberg on the temperature of the earth. Pockels studied the magnetization of basalt, the accumulation of rain in mountains, and other problems of interest to geophysicists and mineralogists. Johann Koenigsberger worked extensively in mineralogy. Leonhard Weber built meteorological instruments, collected weather information with them, did research on the upper atmosphere, trained meteorologists, and, in general, directed his research toward meteorology and geophysics. Karl Zöppritz worked almost entirely in geophysics, eventually moving from a position in theoretical physics to one in geography. Wiechert made a similar move to become director of a geophysics institute and an authority on seismology.[57] Physical chemistry, the most important of the neigh-

54. Boltzmann had already long been interested in Maxwell's theory.
55. They included Lenard, Des Coudres, König, Wiechert, and Wiener.
56. Courses on astrophysics, geophysics, meteorology, and "cosmic physics," which encompassed them all, are included in the semester-by-semester physics course announcements for all German-language universities and technical institutes in the *Physikalische Zeitschrift* from its founding in 1899.
57. Paul Volkmann, "Hermann von Helmholtz," *Schriften der Physikalisch-ökonomischen Gesellschaft*

boring sciences for physicists, acquired a fundamental physical theory through the application of thermodynamics. Certain physicists and chemists even anticipated that physical chemistry would dominate chemistry in its next stage, perhaps reducing it to a chapter of physical theory.[58] One such physicist was Helmholtz, who did much influential work on the thermodynamics of chemical reactions, and many of the leading theoretical physicists working in the 1880s and after—Clausius, Kirchhoff, Boltzmann, and Planck, among others—worked on problems in physical chemistry. For their part, chemists who worked in it required a knowledge of heat theory. Nernst, for example, studied theoretical physics under Boltzmann, for him "always the first model" of a scientist; from the start, Nernst believed that the proper task of physical chemistry was to apply the "methods of theoretical physics" to chemical problems, and after moving to physical chemistry he maintained close working relations with theoretical physicists.[59] Physical chemistry had other characterizations than that given by Nernst. In any case, physical chemistry was by no means thought to be an appendage to physics, theoretical or otherwise; but given their connections, it is not surprising that the teaching of theoretical physics and physical chemistry was sometimes combined.[60]

From its beginning in 1845, the Berlin Physical Society's abstracting journal the *Fortschritte der Physik* included the border regions of physics, such as physiological optics and electrochemistry. In 1847 the *Fortschritte* gave over an entire section to meteorology, in 1849 another to physical geography, combining the two in 1850, and in 1852 combining with them earth magnetism to constitute a section entitled "physics of the earth." The *Fortschritte* for 1880 published a separate volume on the "physics of the earth," and in 1890 the volume was renamed "cosmic physics," the parts of which, astrophysics, meteorology, and geophysics,[61] its editor said, "belong without doubt to physics," even if they do not belong in all respects to the working region of "pure physics." That cosmic physics belonged to physics was evident, he said, from a few examples. It was in spectroscopy and stellar photography that optics had had its most outstanding successes. In the theory of winds and other weather

zu Königsberg 35 (1894): 73–81, on 77. C. Voit, "Wilhelm von Bezold," *Sitzungsber. bay. Akad.* 37 (1907): 268–71, on 270–71. Albert Wigand, "Ernst Dorn," *Phys. Zs.* 17 (1916): 297–99, on 297. Johann Koenigsberger, "F. Pockels," *Centralblatt für Mineralogie, Geologie und Paläontologie*, 1914, 19–21, on 20. Joachim Schroeter, "Johann Georg Koenigsberger (1874–1946)," *Schweizerische Mineralogische und Petrographische Mitteilungen* 27 (1947): 236–46. Schmidt-Schönbeck, *300 Jahre Physik . . . Kieler Universität*, 99–101. Gustav Angenheister, "Emil Wiechert," *Gött. Nachr., Geschäftliche Mitteilungen aus dem Berichtsjahr 1927/28*, 53–62.

58. Eduard Riecke, "Rede," in *Die physikalischen Institute . . . Göttingen*, 20–37, on 35.

59. Walther Nernst, "Antrittsrede," *Sitzungsber. preuss. Akad*, 1906, 549–52.

60. For example, Gerhard Schmidt, who was trained primarily in chemistry and who was Gustav Wiedemann's assistant, alternated lectures on theoretical physics with lectures on physical chemistry and astronomy as Privatdocent at Erlangen in 1896–1900. In 1901, he was appointed extraordinary professor of physics with an assignment for both theoretical physics and physical chemistry at Erlangen. Gerhard Schmidt Personalakte, Erlangen UA.

61. The term "geophysics" was also introduced in 1890. The term "astrophysics" was introduced earlier, in 1873; in that year, astrophysical topics, which before had been included in the chapter "Meteorological Optics" were given prominence in a new chapter "Astrophysics and Meterological Optics," and five years later, in 1878, they received their own chapter, "Astrophysics."

topics, the laws of the mechanical theory of heat had played a major part. In geodesy and in the determination of gravity, the laws of mechanics and the theory of measures had proven indispensable. In sum, the editor said, in the sciences comprising cosmic physics, pure physics had found not only extensive applications but perhaps proof of its conclusions.[62] In 1903 the *Fortschritte* brought together in a single section the scattered chapters on physical chemistry, since it had grown into a large research area and the "representatives of this discipline" did not yet have their own yearly report on the literature. At the same time, the Physical Society decided to eliminate from its *Fortschritte* all reports on purely chemical and purely technical publications.[63]

The body of physical knowledge, narrowly speaking, was divided into partially autonomous branches such as mechanics and electricity. In the teaching of physics, in the words of a leading textbook, it was necessary to invoke a "principle of division" to deal with the enormous body of knowledge. The principle could not, however, any longer be justified by the distinction between, say, the physics of ponderable bodies and the physics of imponderable bodies or by the association of the parts of physics with the different sense organs. The parts of physics had "close relations" with one another, and the retention for teaching purposes of the old division of physics into mechanics, acoustics, optics, heat, electricity, and magnetism was justified on purely "practical grounds."[64] It was not, that is, thought to be based on the nature of things. In textbooks and in lectures on general physics, the older practical divisions persisted.

In teaching and in research, the principal divisions of physics imposed a degree of specialization. In addition to what each division shared with the rest of physics, it had its general and special theories, its concepts, measures, and apparatus specific to its needs. Each had a tradition of work on detailed problems, which led to new methods, concepts, and facts, to new theoretical and experimental problems, and to disputes, dissertations, and the rest of what made up the practice of physics. Each of the principal divisions of physics in turn contained narrower specialties; in addition there were hyphenated specialties such as "thermo-electricity," which acknowledged theoretical and experimental bridges across the divisions. Occasionally an entire principal division of physics was subsumed under another, or it was judged no longer to belong to physics proper. The classifying scheme of physics was fluid; it changed as physical understanding changed, which the categories of the *Fortschritte* repeatedly show.

62. Richard Assmann, et al., ed., "Vollendung des 50. Jahrganges der 'Fortschritte,' " *Fortschritte der Physik des Aethers im Jahre 1894* 50, pt. 2 (1896): i–xi. The remarks by Assmann, the editor of the cosmic physics department, are on viii–xi. Although the title "cosmic physics" appeared in the 46th volume, 1890, the volumes for the years 1893 and 1894 actually came out first, which is why the editor discussed it there. Richard Assmann's "Vorwort," 3d vol., *Fortschritte der Physik* 46 (1890).

63. "Tagesereignisse," *Phys. Zs.* 3 (1902): 559–60. Karl Scheel, "Vorwort," *Fortschritte der Physik* 59, pt. 1 (1903): iii–iv. The elimination of pure chemistry and technology is announced in *Verh. phys. Ges.* 3 (1901): 130.

64. Leopold Pfaundler, ed., *Müller-Pouillet's Lehrbuch der Physik und Meteorologie*, 10th rev. ed., vol. 1 (Braunschweig: F. Vieweg, 1905), 10–11.

The *Fortschritte* started out in 1845 with six main divisions, each with its sub-division for "theory": general physics, which included mechanics; acoustics; optics; heat, which included radiant heat; electricity, which included magnetism; and applied physics, which was distributed among the other categories in the next report. Over the years these divisions underwent some modification. The report for 1882 recognized certain long-standing connections between them by combining optics, heat, and electricity under the designation "physics of the ether" and acoustics and general physics under that of "physics of matter."[65] In 1902 the Physical Society decided that the *Fortschritte*'s ordering no longer corresponded to the state of physics; they thought it was now appropriate to place electricity and magnetism before, instead of after, optics and heat, and to extend optics to the optics of the total spectrum.[66] That organizational change corresponded to the major change in physical understanding associated with the names of Maxwell and Hertz.

65. E. Rosochatius, "Vorwort," *Fortschritte der Physik* 38 (1882): v–vi. Emil Warburg, "Zur Geschichte der Physikalischen Gesellschaft," *Naturwiss.* 13 (1925): 35–39, on 37.
66. Scheel, "Vorwort," iii. The new ordering appeared in the *Fortschritte* for 1903.

19

Göttingen Institute for Theoretical Physics

Voigt at Königsberg and Göttingen

Beginning in 1875 and for eight years after, Woldemar Voigt was the extraordinary professor for mathematical physics at Königsberg.[1] Voigt had begun his university studies at Leipzig; he had taken the standard experimental physics lecture course and helped out the professor in the lecture preparation room, but he had not gone far in his studies when they were interrupted by the Franco-Prussian War. After his tour of active duty he did not return to Leipzig but went to Königsberg to continue his studies in the "school of Fr. Neumann and Fr. Richelot."[2] Although he was drawn to music as well as to physics, he doubted that he had the talent to make a career in music and decided for physics instead.[3]

Voigt persisted in his studies at Königsberg, even though this university did not offer him as much physics as he wanted. Moser, who was ill by then, directed his lectures to the interests of medical students. Neumann, the only "exact physicist" there, did not always lecture, and for two out of the six semesters that Voigt studied at Königsberg, no "physically exact lecture" was given at all.[4] In 1874 he completed his dissertation on the elastic relations of rock salt, a natural topic for a student of Neumann's. Voigt then taught briefly at a gymnasium in Leipzig, and after habilitating at Leipzig University he moved back to Königsberg in accord with Neumann's wishes.

Neumann gave his last lectures in theoretical physics in the winter semester of

---

1. Although Voigt received no salary when he was first appointed extraordinary professor at Königsberg, he later received one of 2800 marks. Prussian Ministry of Culture to Voigt, 2 Oct. 1875, Voigt Papers, Göttingen UB, Ms. Dept.

2. Letter by Voigt from the field, 21 Nov. 1871, quoted in Woldemar Voigt, *Erinnerungsblätter aus dem deutsch-französischen Kriege 1870/71* (Göttingen: Dietrich, 1914), 211. Voigt's interrupted studies at Leipzig are referred to on pp. 6, 9. The records of his university studies are in Voigt Papers, Göttingen UB, Ms. Dept.

3. Carl Runge, "Woldemar Voigt," *Gött. Nachr.*, 1920, 46–52, on 47.

4. Voigt to (probably) Göppert, 4 Sept. 1875, STPK, Darmst. Coll. 1913.51.

1875–76.[5] Voigt then took them over along with Neumann's seminar, and he set up a mathematical-physical laboratory of his own, which he supported with his own very limited means. He tried hard to offer a complete program of physics instruction at Königsberg without much encouragement. He had expected in time to be promoted to ordinary professor for mathematical physics, but Neumann did not make way for him as his successor. Voigt's position was anomalous; for example, during the years Volkmann, Neumann's eventual successor, studied at Königsberg, Voigt did all of Neumann's teaching, and yet Volkmann's dissertation was formally approved not by Voigt but by Neumann (who, because of advanced age, excused himself from attending Volkmann's examination).[6] In effect, Neumann set back the cause of theoretical physics at Königsberg, since by his holding on to the ordinary professorship for the subject and forcing it to be taught by an extraordinary professor, theoretical physics would disappear from the examinations and along with it potential students. Moreover, as long as Voigt remained outside the faculty as extraordinary professor, the likelihood that Königsberg would at last get an official institute for theoretical physics seemed nil. (The institute was built only after Voigt had left Königsberg and Neumann, who had long wanted it, was too old to use it.[7])

As few universities had two ordinary professors for physics, let alone two ordinary professors representing theoretical physics, Voigt's expectations for a promotion at Königsberg were not based so much on precedent as on the government's earlier acceptance, in principle, of a double institute at Königsberg, one for experimental physics and the other for mathematical physics. The experimental one was intended mainly for teaching with the aid of demonstration experiments, the mathematical one for advancing physics through exact measurements in the laboratory. Mathematical physics, that is, was supposed to be more or less equal to experimental physics at Königsberg, which it decidedly was not during Voigt's tenure. Voigt was only the director of a makeshift mathematical physics institute, or "laboratory," and he was not an ordinary professor; and his institute received far less financial support than the experimental institute. The better part of what support Voigt did receive went for rent and heat, which left him with scanty means for instruments, materials, and assistance; so that only by petitioning for every purchase or by paying out of his own pocket could Voigt keep the institute going at all. By contrast, the experimen-

5. In 1871, Neumann was still lecturing on mineralogy, the other part of his teaching responsibility. Albert Wangerin, *Franz Neumann und sein Wirken als Forscher und Lehrer* (Braunschweig: F. Vieweg, 1907), 56–58.

6. Luise Neumann, *Franz Neumann, Erinnerungsblätter von seiner Tochter*, 2d ed. (Tübingen: J. C. B. Mohr [P. Siebeck], 1907), 395–96. Wilhelm Lorey, *Das Studium der Mathematik an den deutschen Universitäten seit Anfang des 19. Jahrhunderts* (Leipzig and Berlin: B. G. Teubner, 1916), 97–98. Runge, "Voigt," 49. Recommendation of Voigt by the Göttingen Philosophical Faculty, 12 Jan. 1883, Voigt Personalakte, Göttingen UA, 4/V b/203. Like Neumann before him, Voigt organized his laboratory in his own house. Kathryn Mary Olesko, "The Emergence of Theoretical Physics in Germany: Franz Neumann and the Königsberg School of Physics, 1830–1890" (Ph.D. Diss., Cornell University, 1980), 464. Draft of Neumann's evaluation of Volkmann's dissertation, n.d., Neumann Papers, Göttingen UB, Ms. Dept.

7. The institute, without living quarters, was built in the mid 1880s, when Neumann was almost ninety. Wangerin, *Neumann*, 185.

tal physics institute, which Carl Pape directed after Moser's death in 1880, received a relatively ample budget, and it already had a collection of apparatus. Moser and Pape, moreover, could count on large numbers of students while Voigt had to limit his.[8] Even if Voigt had got the money he asked for, mathematical physics would have remained poorly regarded by students. That was because it was represented by an extraordinary professor instead of a chairholder.

By 1883 Voigt had been teaching at Königsberg for a good many years under circumstances that had failed to develop as he had originally hoped.[9] He remained an extraordinary professor without an institute. Neumann and the Königsberg faculty feared, correctly, that he would not stay on if another university approached him; when in 1883 Göttingen offered him Listing's ordinary professorship, it strongly appealed to him even though the Königsberg faculty took steps to keep him and even though Listing's mathematical physics institute was in a sorry state of neglect.[10]

At Listing's death in 1882, Riecke temporarily took over as director of the Göttingen mathematical physics institute.[11] In the interest of instruction in physics, Riecke and the faculty urged the government to replace Listing promptly. He had in mind a proper theoretical physicist for the job, not another experimental physicist to share teaching duties. This view was accepted by the Göttingen faculty, who recommended that the government try to hire Clausius, a theoretical physicist who would continue the Göttingen "tradition" established by Gauss, Gustav Lejeune Dirichlet, and Riemann.[12] Realistically they did not have much hope of getting Clausius: he was expensive, Bonn would take steps to keep him, and Göttingen's mathematical physics institute had "completely inadequate quarters."[13] At one point in the negotiations, it seemed that Clausius was actually going to come, but he refused in the end.[14] The government then turned to the faculty's second choice, Voigt, who did not pose the financial problem Clausius did.

Letters from the faculty commission to Voigt characterized the position of theoretical physics at Göttingen. The mathematician H. A. Schwarz stressed the advantage for theoretical physics of the traditional mathematical strength at Göttingen: there physics was counted as mathematics, he explained, and the physicists

8. Olesko, "Emergence," 464–65. Otto Lehmann, "Zusammenstellung der Etatsverhältnisse der physikalischen Institute an den deutschen Hochschulen," 26 June 1891, Bad. GLA, 235/4168. Voigt to (probably) Göppert, 4 Sep. 1875.

9. Initially Voigt had been encouraged because chemistry had three ordinary professors and physics only two, one of whom was very ill. Voigt to (probably) Göppert, 4 Sep. 1875.

10. Neumann to a Prussian government official (draft), n.d. [soon after 20 Apr. 1883], Neumann Papers, Göttingen UB, Ms. Dept.

11. Göttingen U. Curator to Riecke, 10 Jan. 1883, Listing Personalakte, Göttingen UA, 4/V b/108.

12. Riecke to Voigt, 11 Jan. 1883, Voigt Papers, Göttingen, UB, Ms. Dept. Philosophical Faculty recommendation, drafted by Riecke, for the commission appointed to decide on Listing's replacement, 12 Jan. 1883, and Göttingen U. Curator to Prussian Minister of Culture von Gossler, 15 Jan. 1883, Voigt Personalakte, Göttingen UA, 4/V b/203.

13. Göttingen U. Curator to Prussian Minister of Culture von Gossler, 15 Jan. 1883, Voigt Personalakte, Göttingen UA, 4/V b/203.

14. Riecke to Voigt, 4 Apr. 1883, Voigt Papers, Göttingen UB, Ms. Dept.

Weber, Riecke, and earlier Listing belonged to the mathematical, not the physical, class of the Göttingen Royal Society of Sciences. The practical advantage was that at Göttingen Voigt would find over 130 well-prepared, hardworking students enrolled in mathematics.[15] From the side of physics Riecke wrote that he especially welcomed Voigt since they complemented one another so well. Riecke had been lecturing only on experimental physics, so that Voigt would have the "whole area of mathematical physics" to himself, which corresponded to Voigt's interests.[16] The difference between the experimental and the theoretical institutes was to be real, no longer "merely nominal," as it had been in the past, when Listing only formally, not actually, represented the entire field of mathematical physics. Riecke would retain control of the beginners' exercises and the examinations for students of medicine, pharmacy, and agriculture, but he would divide with Voigt the advanced exercises and the examinations for secondary teachers and doctoral candidates. The budget of the experimental physics institute was more than twice that of the theoretical, he told Voigt, but the experimental one paid for heat and lighting for the whole building that the two institutes shared. He warned Voigt that the physics institute was on a busy street and not solidly built, but there was reason to expect a new institute building in a few years. Since Listing had had no assistant, Voigt should ask the minister for one. Finally, Riecke assured Voigt that if they worked together, their institute would compare in scientific significance with the better equipped ones.[17]

The Göttingen faculty had a good idea of the man they were hiring. They valued Voigt as an "independent researcher in the area of mathematical physics," and they noted, with approval, that he was familiar with the methods and means of experimental research. He had attained "beautiful results" through the "connection of the theoretical and the experimental sides of physical research," which reassured the faculty that he would not abandon himself one-sidely to "abstract mathematical problems." Furthermore, because Voigt's research had centered on elasticity theory and optics, his direction would complete that of physicists already working at Göttingen, and in this respect "none of the physicists who might be named could be regarded as equal to him."[18]

In August 1883 Voigt became Göttingen's "ordinary professor for theoretical (mathematical) physics" and director of the mathematical physics institute. Soon after this at his request, he was also appointed co-director of the physical department of the mathematical-physical seminar.[19] Unlike his predecessor Listing, Voigt was trained in theoretical physics and regarded it as his field, so that his appointment at Göttingen was an important step in the establishment of theoretical physics in German universities.

15. H. A. Schwarz to Voigt, 7 Jan. 1883, Voigt Papers, Göttingen UB, Ms. Dept.
16. Riecke to Voigt, 9 Jan. 1883, Voigt Papers, Göttingen UB, Ms. Dept.
17. Riecke to Voigt, 11 and 22 Jan. 1883, Voigt Papers, Göttingen UB, Ms. Dept.
18. Göttingen Philosophical Faculty recommendation, 12 Jan. 1883.
19. Voigt's salary was 4200 marks with an additional 540 marks for rent. Prussian Ministry to Göttingen U. Curator von Warnstedt, 3 Sept. 1883; Voigt to Curator, 4 Oct. 1883; Curator to Voigt, 9 Oct. 1883; Voigt Personalakte, Göttingen UA, 4/V h/203.

Voigt's Researches in the Elastic Theory of Light

In April 1883, as he was negotiating for the Göttingen job, Voigt published his first major contribution to the elastic theory of light.[20] With it, he linked the two subjects, elasticity and optics, to which the Göttingen faculty had drawn attention in their evaluation of his work. This publication provides a good example of Voigt's preferred method of theoretical research, which was to draw mathematical consequences from a few general principles based on experience, above all the principles of mechanics and heat theory, rather than to draw them from special pictures or mechanisms.[21]

Voigt's optical work belonged to an active research area in Germany, as one can see at a glance from the quantity of pages in the *Annalen der Physik* devoted to it. By Voigt's time, the early nineteenth-century theories of Franz Neumann, James MacCullagh, Green, Cauchy, and others were seen to be unsatisfactory in certain ways. It was now recognized, for example, that the optical relations of matter could not be understood by a simple elastic theory in which the only consideration is the influence of matter on the density or elasticity of the ether: Neumann, for one, from about 1865 began to consider simultaneous equations of motion for the ether and matter in his Königsberg optical lectures, and there were a few other early recognitions of the need to consider the reaction of ether on matter as well as that of matter on ether. But it was only from the 1870s that the theory of the mutual reaction of ether and matter was systematically developed and criticized. This theory came to be especially cultivated in Germany; Helmholtz, W. Sellmeyer, Ketteler, Lommel, and others all devised simultaneous systems of equations of motion for the interacting particles of ether and matter, each system differing in its expression for the interaction.[22] They all worked with mechanical conceptions, but their results were not always in full agreement with mechanical laws.[23] With this general type of elastic theory, they were able to explain normal and anomalous dispersion as well as the customary optical phenomena of reflection, refraction, and the like.

Because existing theories of light were built upon systems of equations that contradicted the usual equations of motion of elastic matter, Voigt believed it was desirable to prove that the laws of the motion of light could be derived from a suitably

---

20. Karl Försterling, "Woldemar Voigt zum hundertsten Geburtstage," *Naturwiss.* 38 (1951): 217–21, on 219. Woldemar Voigt, "Theorie des Lichtes für vollkommen durchsichtige Media," *Ann.* 19 (1883): 837–908.

21. Försterling, "Voigt," 217–18.

22. Glazebrook discusses primarily German work in the part of his report dealing with theories of the mutual reaction of ether and matter. R. T. Glazebrook, "Report on Optical Theories," in *Report of the Fifty-Fifth Meeting of the British Association for the Advancement of Science* (London, 1886), 157–261, on 212–33.

23. Glazebrook, "Report," 228. Voigt criticized Lommel's theory, for example, for introducing a frictional force between ether and matter that contradicts the principle of equality of action and reaction or the principle of the conservation of the motion of the center of gravity. Lommel denied that his theory contradicts the first of these principles but accepted that it does, and should, contradict the second. Woldemar Voigt, "Bemerkungen zu Hrn. E. Lommel's Theorie der Doppelbrechung, der Drehung der Polarisationsebene und der elliptischen Doppelbrechung," *Ann.* 17 (1882): 468–76; "Zur Theorie des Lichtes," *Ann.* 20 (1883): 444–52. Eugen Lommel, "Zur Theorie des Lichtes," *Ann.* 19 (1883): 908–14.

reformulated elasticity theory, and in 1883 he set out to give the theory its greatest possible extension and to establish rigorously the connection between optical and elastic phenomena.[24] He did not attempt to formulate an elastic theory for the whole of optics. He excluded for now dispersion and other optical phenomena accompanied by the absorption of light, since the theory of elasticity did not yet yield oscillations that could be identified with heat; the fundamental physical ideas for an exact theory of these phenomena were lacking, and for this reason Voigt saw the immediate task of elastic theory as one of building an optical theory for completely transparent bodies only. For these bodies, he could apply the principle of conservation of energy to the oscillations.

To preserve the fundamental connection between optical and elastic theories, Voigt invoked a major use of the energy principle in theoretical physics. With its help, he identified all energetically possible mathematical expressions for the law of interaction between the elastic ether and matter; he then examined the expressions to see if any of them allowed the elastic theory to be brought into agreement with optical laws. In this way, the energy principle served Voigt as a guide to the construction of mathematical expressions, but it did not at the same time serve him as a guide to new ideas about the nature of the interaction of ether and matter; Voigt pointed to the remaining task of forming physical ideas for deriving the interaction and for measuring its constants in the laboratory.

The central interest of Voigt's paper of 1883 is not its mathematics, though it is dense with formulas, but its mode of explanation. Voigt's starting point was two sets of mechanical equations of motion, one for the ether and one for matter; the two sets contain the confirmed inner elastic forces of matter and ether and, as well, expressions standing for their as yet unknown interaction. For the equations of motion of the ether inside an element of volume of the medium containing matter and ether, Voigt wrote:

$$m\partial^2 u/\partial t^2 = X' + X + A,$$

with corresponding equations for the y and z components of the motion. And for the equations for matter, he wrote:

$$\mu\partial^2 U/\partial t^2 = \Xi^1 + \Xi + A,$$

again with corresponding equations for the other two components.[25] By the principle of action and reaction, the two terms for the interaction, A, the action of matter on ether, and A, that of ether on matter, are the negative of one another, and by requiring the equations to satisfy the energy principle, Voigt decreased the number of physically possible expressions for the interaction A to eight. Then by comparing

24. Voigt, "Theorie des Lichtes," 873.
25. Here u and U are components of the displacements of ether and matter, m and μ their densities, X' and Ξ' the components of the external forces, which vanish for optical phenomena, and X and Ξ their inner elastic forces. Voigt, "Theorie des Lichtes," 874.

them with optical laws and by making physical assumptions about the density and elasticity of the ether, he further reduced the possibilities. Applying the resulting elastic equations to a wide range of optical phenomena, he concluded that for all optical phenomena in transparent, or nearly transparent, bodies, the elastic theory agreed completely with observation. He called the theory an "explanation" of non-absorptive optical phenomena, even if the task remained of deriving the interactions from a physical idea and determining the constants.[26]

In 1884, the year after publishing his theory, Voigt decided that he had been wrong after all in thinking that it was impossible to treat absorptive optical processes without making "precarious assumptions about the motion of ponderable molecules." It was only necessary to assume that any such motion is disappearingly small, which enabled him now to develop an optical theory of absorbing media, based "strongly on mechanical foundations." He then took up the experimental problem of determining the optical properties of such media.[27]

Voigt's Arrangements

Voigt came to the Göttingen mathematical physics institute after eight years of teaching in Germany's other institute of this kind, Königsberg's, and he brought with him firm ideas about what such an institute should accomplish. Now as an ordinary professor at Göttingen, Voigt fully expected more support for his teaching than he had received as an extraordinary professor at Königsberg. Having received official assurances of financial support during negotiations about the job, he confidently opened a laboratory practice course in the new institute to draw immediate attention to his teaching, and Riecke encouraged him in this and lent him apparatus for the purpose. Voigt intended the course for the more mathematically trained students, conceiving of it in part as a laboratory extension of his lecture course on theoretical physics and modeling it after the course he had given at Königsberg. It proved a miscalculation.

While the Göttingen curator cooperated with Voigt by installing ovens in some rooms in his institute, which otherwise would have been too cold in the winter to hold classes in, the Prussian ministry did not respond to Voigt's requests at Göttingen anymore than it had to his requests at Königsberg. The first time Voigt taught the laboratory practice course at Göttingen he paid a mathematics student to assist him, managing to give the course at considerable expense to himself. The government eventually reimbursed him for the student's salary, but it denied his request for extraordinary funds to bring Listing's institute up to date, recommending instead that he try again later. Voigt made one more attempt; but when in the second year

26. Voigt, "Theorie des Lichtes," 907.

27. Woldemar Voigt, "Theorie der absorbirenden isotropen Medien, insbesondere Theorie der optischen Eigenschaften der Metalle," Ann. 23 (1884): 104–47; "Ueber die Theorie der Dispersion und Absorption, speciell über die optischen Eigenschaften des festen Fuchsins," Ann. 23 (1884): 554–77; "Zur Theorie der Absorption des Lichtes in Krystallen," Ann. 23 (1884): 577–606; "Ueber die Bestimmung der Brechungsindices absorbirender Medien," Ann. 24 (1885): 144–56.

the government put up only enough money for another student assistant for the laboratory practice course and promised nothing more, Voigt rejected the money and declared himself ready to close the course. After warning that the institute would amount to no more than Listing's institute, an institute for demonstrations only, unless the ministry came through, Voigt gradually began to get some of the things he needed for his teaching.[28] In all of this, he did not bring up his own research; for this purpose he provided his own instruments.[29]

Voigt's and Riecke's institutes—formally two "departments" of a single physics institute—needed extra money to bring their instrument collections up to standard. The instrument collection Voigt inherited from Listing was limited to optics, Listing's special interest, since Listing's budget had not reached beyond that. Riecke's instrument collection had been maintained by Weber on relatively little money because Weber had built most of what he needed himself, a practice neither Riecke nor any institute director could follow any longer.[30] In general, the arrangements for physics in Göttingen were behind the times, the Göttingen curator told the minister. Recent developments in physics demanded costly instruments, which Göttingen's two new, young institute directors, Voigt and Riecke, rightly wanted for their teaching. Bonn and even Kiel had spent more in equipping their physics institutes than Göttingen had. The ministry eventually complied with an extraordinary grant to be divided between the two institutes at their directors' discretion.[31] Voigt applied his share to filling the gaps in his institute's collection of "measuring instruments." With the request for a second extraordinary grant to complete the collection, Voigt listed the most essential instruments to be bought, all of them instruments for measuring the basic quantities of physics—time, length, and weight; with his first extraordinary grant he had also bought mainly "meters" of one kind or another: piezometer, spectrometer, photometer, galvanometer, rheometer, and *Kathetometer*.[32] The list confirmed what Voigt always maintained: measurements are the proper work of a mathematical physics institute.

28. Soon after taking up his duties at Göttingen, Voigt requested a regular position for an assistant with a salary of 1200 marks; he also requested an increase of his institute budget from 1000 to 1800 marks annually and an extraordinary grant of 5000 marks, later 7500 marks, for instruments, 3000 of which was to be made available immediately. He was offered 600 marks for a student assistant in 1884, and when he rejected the money because he had been denied the money requested for instruments, the ministry let him keep it to use for instruments instead and added another 1500 marks. Göttingen U. Curator to Prussian Minister of Culture von Gossler, 16 Oct. 1883 and 22 Feb. 1884; Minister to Curator, 24 Nov. 1884; Curator to Voigt, 8 July 1884; Voigt to Minister, 3 Sept. 1884; Curator to Minister, 18 May 1885; Ministry to Curator, 20 Nov. 1885; and Riecke and Voigt to Curator, 2 Dec. 1885; Göttingen UA, Universitäts-Kuratorium XVI. IV. C. v.

29. Voigt's budget was taken up in running his institute and completing its collection. Voigt to Göttingen U. Curator Ernst von Meier, 26 Jan. 1892, STPK, Darmst. Coll. 1915.2.

30. Listing's budget was 960 marks; Voigt's first budget of 1000 marks was not so much more as to make a difference. Göttingen U. Curator to Prussian Minister of Culture von Gossler, 16 Oct. 1883 and 22 Feb. 1884.

31. The ministry provided an extraordinary grant of 8000 marks for 1886–87; Riecke took 5000 marks, Voigt 3000. Göttingen U. Curator to Prussian Minister of Culture von Gossler, 18 May 1885; Minister to Curator von Warnstedt, 20 Nov. 1885; Riecke and Voigt to Curator, 2 Dec. 1885.

32. The second extraordinary grant Voigt requested was for 3000 marks in 1887–88. Voigt to Göttingen U. Curator von Warnstedt, 12 May 1887, Göttingen UA, Universitäts-Kuratorium XVI. IV. C. v.

Measurement versus Experiment

The desire by physicists such as Voigt for close cooperation between theory and experiment was partly based on their understanding that the experimentalist worked from or toward a theory. They distinguished between the two laboratory activities of "experiment" and "measurement," and they typed themselves accordingly, Kundt as the most important German "experimental physicist" and Friedrich Kohlrausch as the most important "measuring physicist." The two activities related to theoretical physics somewhat differently, they believed. The experimental physicist often explored unknown territory, in which case his work preceded a completed theory. The work of the measuring physicist too could precede theory and lead to discovery; the measurements that Kirchhoff and Bunsen made of spectra were recognized as an instance of this way of discovery. But even if it did not, measurement was valued for the detailed knowledge of nature it provided; it made accessible to us parts of the world where our unaided senses would have failed us. With the increasing completeness of the knowledge of nature that measurement made possible, the theory of physics became increasingly unified. In turn the unified theory indicated new measurements to be made using precision instruments. The first aim of the interaction of measurement and theory, Wiener said, was a theory that was at once unified, comprehensive, and independent of the nature of our senses.[33]

Measurement stood in particularly close relationship to theory, since the measuring physicist often worked from an already completed theory. The most important quantities in theoretical physics were the universal constants and functions, for example, the velocity of light and the energy distribution function of blackbody radiation, and these quantities were accordingly among the most important objects of precision measurements. Other precision measurements concerned arbitrary standards, such as the ohm, the unit of resistance, and constants and functions that were not universal but characteristic of particular substances, for example, the electrical conductivity of iron. The latter kind of precision measurement was carried out extensively in Voigt's laboratory.

Voigt discussed the problem of determining physical constants with the greatest possible accuracy by referring to a discussion by Hertz. In the determination of a physical constant, according to Hertz, an error of 1/100 of the true value is the borderline for the "desirable" accuracy and an error of 1/1000 the borderline for "possible" accuracy, while hardly any constant could be "defined" more accurately than to within 1/10000 of its true value. This seems remarkable, Voigt said, since any carpenter with a measuring rule can be more accurate than one millimeter per meter. But Hertz was right. The accuracy of the physical constant is not the same

---

33. Otto Wiener, "Nachruf auf Wilhelm Hallwachs," *Verh. sächs. Ges. Wiss.* 74 (1922): 293–313, on 294. Emil Warburg, "Verhältnis der Präzisionsmessungen zu den allgemeinen Zielen der Physik," in *Physik*, ed. Emil Warburg, Kultur der Gegenwart, ser. 3, vol. 3, pt. 1 (Berlin: B. G. Teubner, 1915), 653–60, on 654. Paul Volkmann, *Einführung in das Studium der theoretischen Physik insbesondere in das der analytischen Mechanik mit einer Einleitung in die Theorie der physikalischen Erkenntniss* (Leipzig: B. G. Teubner, 1900), 174–75. Otto Wiener, "Die Erweiterung unsrer Sinne," *Deutsche Revue* 25 (1900): 25–41, on 27–36, 38–39.

as the accuracy of the measuring instrument, and the reason is that physical constants are seldom measured directly but rather through combinations of measurements aided by theoretical formulas. The errors of individual measurements multiply rapidly. More important, the phenomena of interest usually cannot be totally isolated from other phenomena, so that extraneous causes can seriously distort the measurements. To arrange the measurements in such a way that the disturbances can be ignored or calculated, the physicist has to call upon his knowledge of theoretical physics: "the task arising out of this for theory is hardly less important for the advance of physics than the deduction of general laws, which one might consider the only task of theory."[34] To bring about an increase in the accuracy of observation, without which new problems in physics often could not even be tackled, required a good grounding in theoretical physics. That was provided for in Voigt's Göttingen institute for theoretical physics.

The scale of Voigt's subject can be appreciated from the efforts by the measuring physicist Friedrich Kohlrausch to keep up with it. Kohlrausch regularly brought out new editions of his laboratory manual, but by the beginning of this century the subject had grown too big for one book. Originally he had tried to include "all physical measuring methods," but now he had to be content with listing the specialized literature for electrostatics, electrochemistry, and the like. The subject was too big for one man as well, and he had to take much on the authority of others. It was frustrating for him to bring out a new edition knowing that daily it could be enlarged. "Measuring physics increases like an avalanche," Kohlrausch told Wilhelm Wien in 1909: "This is the last time I undertake the sifting."[35]

Voigt's Seminar Teaching and Textbooks

After 1870, as before, the Göttingen mathematical-physical seminar was well attended. Because of growing numbers of students, practical exercises in the seminar tended to give way to experimental lectures with demonstrations by staff and students. Listing, who limited his seminar teaching to optical topics, limited it even more in the summer of 1875 when he replaced "exercises in experiments and measurements" with lectures and written assignments. His institute was too small for the twenty-nine seminar members that semester, an enrollment twice that of any recent semester. Riecke, with his larger institute, continued to offer exercises on absolute measurements that semester.[36] Listing tried to keep up laboratory exercises, but in the winter of 1879–80, he had to cancel them again.[37] In 1881–82, the year of

34. Woldemar Voigt, "Der Kampf um die Dezimale in der Physik," *Deutsche Revue* 34 (1909): 71–85, on 73–74.

35. Kohlrausch to Wien, 5 May 1905, 23 Dec. 1908, and 26 Apr. 1909, Ms. Coll., DM, 2450, 2455–56. Corresponding to the flourishing state of measuring physics in Germany at the turn of this century, the German precision instrument industry also flourished: German Association, "German Scientific Apparatus," *Science* 12 (1900): 777–85.

36. Annual report of the Göttingen mathematical-physical seminar for 1874–75, Göttingen UA, 4/V h/24a.

37. Annual report for 1879–80, Göttingen UA, 4/V h/24a.

maximum enrollment, sixty-seven in the summer and seventy-two in the winter, Listing admitted only a few students to his optical laboratory and had to use the institute's large auditorium for his lectures. That year the mathematical director Schwarz observed that the seminar's large enrollment—it had doubled in the last five years—had disadvantages, but that he and the other directors did not want to change the liberal rules of admission to the seminar, since it might turn away some students unfairly.[38]

After Voigt replaced Listing in the seminar, the enrollment remained high through the 1880s, only somewhat more manageable, varying between thirty and forty. Voigt offered lectures and written exercises on topics in theoretical physics, beginning with Laplace's theory of capillarity and, in following semesters, proceeding through topics in electricity, optics, mechanics, heat, and elasticity, in that way covering all branches of theoretical physics and not just optics, Listing's subject. In this regard, Voigt's teaching in the Göttingen seminar was much as it had been in Königsberg's. But the caliber of students he now taught was different; his early seminar class at Göttingen included Paul Drude and Friedrich Pockels, who won stipends in the seminar and whom he counted among the best students he ever had.[39]

Voigt extended his teaching at Göttingen to a wider German audience, becoming known as a writer of textbooks, which he based largely on his lectures. His first text, in 1889, developed the fundamental laws of mechanics as an "introduction to the study of theoretical physics." Its need arose from the "constantly growing significance that theoretical-physical, especially mechanical, considerations" had acquired throughout the natural sciences. Voigt approached mechanics not as a branch of mathematics but as a "branch of that exact natural science the final goal of which is to derive numerical laws of actual phenomena." Because human understanding requires that new phenomena be explained by familiar ones, especially by the simplest, which are mechanical, and because almost all exact instruments rest on mechanical principles, mechanics is the "foundation of the entire theoretical physics."[40]

Voigt opened his presentation of mechanics with a discussion of units, since the basic assumption behind the application of theory to phenomena is their measurability. The absolute system of units, to which all other units in physics can be reduced, Voigt said, lends physics its great simplicity and unity. In keeping with the emphasis on measurability, Voigt introduced the physical quantities that physicists confront in nature: in order of complexity, scalars, vectors, and tensors, each of which is associated with a physical "field." It is again a sign of the simplicity and unity of physics that its many concepts can be expressed by so few kinds of quantities.

38. Annual report for 1881–82, Göttingen UA, 4/V h/24a.

39. Drude and Pockels both won seminar stipends in 1884–85. Report for that year, Göttingen UA, 4/V h/24a. Runge, "Voigt," 49–50. Försterling, "Voigt," 217.

40. Woldemar Voigt, *Elementare Mechanik als Einleitung in das Studium der theoretischen Physik*, 2d rev. ed. (Leipzig: Veit, 1901), quotations on iii, 1. This edition differed from the first edition of 1889 mainly in form, not in approach or content.

To derive the laws of motion of inertial bodies, the "task of mechanics," Voigt introduced the necessary concepts one by one rather than laying down Newton's axioms or a general principle such as Lagrange's at the start. He gradually built up the science by considering single mass points, then pairs of mass points, then many, then, by replacing mass points with volume elements, rigid bodies, and finally non-rigid bodies, which correspond to the bodies we know through experience. He discussed the theory of measuring instruments and analyzed the sources of error in observation. Since, in his approach, results in mechanics are important insofar as they relate to experiment, he emphasized the importance of approximate solutions, for example, for hydrodynamic problems, which seldom have strong and complete solutions of any importance for experiment. Voigt's students and readers were left in no doubt that mechanics treats the real world and that it is the proper introduction to theoretical physics.

From his text on mechanics as an introduction to theoretical physics, Voigt turned to a text on all of theoretical physics. By this time, the 1890s, it had become customary for lecturers on theoretical physics—or their students—to bring out an edition of their lectures; Franz Neumann's came out in the 1880s, Kirchhoff's in the early 1890s, and Helmholtz's from the late 1890s. They were regarded as standard reference works and were cited in research publications, and professors sometimes directed their students to them. But they were bulky, several volumes each, uneven, sometimes indifferently edited, and out of date in parts. In any case, they were intended mainly as a monument to their authors. Other literature on theoretical physics was less comprehensive, taking the form of handbooks and published special lectures treating one or another branch of the subject. Voigt saw a clear need for a convenient, comprehensive text on theoretical physics, which would provide students with a view of the entire discipline and would supplement—or substitute for—lectures on theoretical physics. In 1895 and 1896, Voigt published in two volumes his *Kompendium der theoretischen Physik*, the express purpose of which was to present a unified view of the whole field of theoretical physics with a physical, as opposed to a mathematical, emphasis throughout. There was no German textbook like it at the time.[41]

The first volume of Voigt's compendium dealt with mechanics and heat. Mechanics took on a strange form there, Voigt explained; he passed lightly over the large areas of the subject that had become almost pure mathematics to dwell on areas that did not belong to mechanics proper. He treated at length the nonmechanical parts of physics as special problems of general mechanics. Some of these mechanical theories were no more than mechanical "analogies," Voigt said; he believed, however, that such analogies were useful for the visualization they allowed; and since they ascribed to physical systems only the most general properties of mechanical systems without regard to their real possibility, they were superior to the older mechanical explanations, which provided complete mechanisms for nonmechanical processes.[42] Under mechanics, then, Voigt treated not only mechanics it-

41. Woldemar Voigt, *Kompendium der theoretischen Physik*, 2 vols. (Leipzig, 1895–96), 1: iii.
42. Voigt, *Kompendium* 1: v, 91–92; 2: 63.

self but Coulomb's and Weber's electrical laws, Maxwell's electrodynamics, optics, the kinetic theory of gases and solutions, electric and thermal conduction, capillarity, elasticity, crystals, and more.

By ignoring special mechanisms, Voigt impressed upon theoretical physics a characteristic direction. In his treatment of optics in the second volume of his text, for example, he did not proceed from Maxwell's electromagnetic theory of light because it required the mixed method of hypothetical views and experimental facts, which he everywhere sought to avoid. Instead, he based the fundamental equations of the motion of light on experience, introducing a light vector that could stand as well for elastic displacement or magnetic field intensity. He called attention to both the elastic and the electromagnetic theories as he went along.[43]

Voigt's compendium was difficult, dense, and, in Voigt's eyes, something of a failure. He came to see that he had tried to put too much in two volumes, to present and connect the whole of theoretical physics *"at one stroke."* He had composed the book too fast, which resulted in errors and in a case of nervous exhaustion. Then the timing could not have been worse; he soon considered revising the text, for it had come out just before the penetration into physics of the electron theory, vector analysis, and the Zeeman effect, which influenced great parts of optics. "So the book was obsolete almost with its appearance!—and the great work and joy I have invested in it almost make me sorry now."[44]

43. Voigt, *Kompendium* 2: v, 89. The optical theory Voigt presented in this textbook is discussed in Försterling, "Voigt," 219.
44. Voigt to Göttingen U. Curator Höpfner, 1 Oct. 1895, Voigt Personalakte, Göttingen UA, 4/V b/203. Voigt to Sommerfeld, 26 Nov. 1899 and 25 Nov. 1909, Sommerfeld Correspondence, Ms. Coll., DM.

20

Mechanical Researches and Lectures

Through the last quarter of the nineteenth century, the period we are concerned with here, and beyond, German physicists took a lively interest in the choice between, and the correct formulation of, the general principles of mechanics. They were equally interested in the use of mechanics as a source of theories or "pictures" or "analogies" for understanding physical phenomena falling outside of mechanics proper. They were interested above all in the question of the foundation of physics as a whole, and in light of recent work in electricity and heat, it was not clear to them that mechanics was adequate to the task of providing this foundation. To our account of Voigt's mechanical work of the previous chapter, we here add an account of Kirchhoff's and Helmholtz's mechanical work at Berlin, Hertz's at Bonn, and Volkmann's at Königsberg.

Kirchhoff at Berlin

Unlike Voigt, Kirchhoff lacked an institute for theoretical physics equipped for experimental work. He did some work in the private laboratory of a friend, but most of his work in Berlin was theoretical, a narrowing forced on him in part by poor health. He published a paper or two a year, but his main achievement at Berlin was his lectures on theoretical physics, which were published.[1]

Before coming to Berlin, Kirchhoff had already begun to prepare an edition of his Heidelberg lectures on mathematical physics. One reason he gave for declining the job at the solar observatory in Berlin in 1874 was that it would interrupt his work of several years on the edition. When he moved to Berlin in 1875 as member

1. Gustav Kirchhoff, *Vorlesungen über mathematische Physik*, vol. 1, *Mechanik*, 3d ed. (Leipzig, 1883), quotation from the preface to the first edition in 1876. The volume is based on lectures Kirchhoff gave at Heidelberg just before moving to Berlin. Ludwig Boltzmann, *Gustav Robert Kirchhoff* (Leipzig, 1888), 22. Robert Helmholtz, "A Memoir of Gustav Robert Kirchhoff," trans. J. de Perott, in *Annual Report of the . . . Smithsonian Institution . . . to July, 1889*, 1890, 527–40, on 531.

of the Prussian Academy and professor at the university, he continued to work on it. At Heidelberg, where he had been the experimental professor, he had only lectured on those parts of mathematical physics that most interested him then; at Berlin he was the mathematical physicist, and he felt the need to give his lectures on the subject a "certain completeness," which meant new preparations. In 1876 his lectures on mechanics came out, and within a year his publisher Teubner informed him that a second edition was called for.[2]

What appealed to the readers of Kirchhoff's mechanical lectures was their compelling logic, according to Boltzmann. By bringing out the logical structure of mechanics, Kirchhoff resolved differences of opinion on such questions as the origin of the law of inertia; in his presentation it was unnecessary to decide if this law is a summary of experience or an axiom or a law deducible from something else. The task of mechanics, Kirchhoff said, is not to explain phenomena, but to "describe completely and in the simplest way the motions occurring in nature." Boltzmann recalled the initial astonishment over this characterization of mechanics.[3]

Kirchhoff constructed the general equations of mechanics from the fundamental concepts of space, time, and matter. He introduced force and mass, primitive concepts in other presentations, as derivative concepts useful only for simplifying the equations. The reason is that, for Kirchhoff, mechanics has nothing to do with metaphysical notions such as the cause of motions but only with the equations describing them. There is nothing more to solving the problem of the motion of a material point than this: integrate the three differential equations $d^2x/dt^2 = X$, $d^2y/dt^2 = Y$, $d^2z/dt^2 = Z$ given the initial position and velocity of the point and given fact that these equations contain second-order derivatives and not third- or higher-order ones; experience has shown that these derivatives allow a simple description, which is sufficient.[4]

Kirchhoff's published lectures on mechanics contain much material he reworked for the purpose from his own theoretical papers on continuum mechanics. He developed mechanics from the assumption that matter fills space continuously, which describes how things appear to be and which agrees with his characterization of mechanics. He did not take up molecular theories of matter.

Kirchhoff developed his lectures on the theory of heat from the same standpoint as he did his lectures on mechanics, beginning with the dictum: "to order [physical phenomena] clearly and to present them as simply as possible." He followed this with the observation: "Of all physical phenomena, the simplest, that is, those that lie closest to the understanding, are the *phenomena of motion*, which comprise the subject of *mechanics*. The fewest fundamental views occur here; namely, only space, time, and matter." The well-known hypothesis that all physical phenomena arise from motion leads to the goal of reducing all of physics to mechanics: "Had that [reduction] succeeded, then with regard to the simplicity of presentation, the highest

2. Kirchhoff to Koenigsberger, 15 Feb. 1876, STPK, 1922.87.
3. Kirchhoff, *Mechanik*, preface. Boltzmann, *Kirchhoff*, 25.
4. Kirchhoff, *Mechanik*, chap. 1.

conceivable [goal] would be accomplished; the reduction in question is therefore a goal that is in the fullest measure worth aiming at." But if the reduction of heat phenomena to motion satisfies the criterion of simplicity, it does not satisfy the other criterion, that of clarity. The idea of heat motion is "unclear," Kirchhoff said, and even in its most developed form, in the theory of gases, the assumed molecular collisions remain "obscure." In the first of the heat lectures, and again in later lectures, Kirchhoff went to lengths to show the extent to which heat is—and is not yet—ordered under the concepts of mechanics.[5]

In the early heat lectures, which deal solely with the distribution and variation of temperature, Kirchhoff treated matter as continuous in the space of bodies. This viewpoint did not allow him to reduce temperature changes to mechanics, and he introduced "pure mechanics" here only as a guide in developing the formulas of "pure heat theory." It was not until the fifth lecture that Kirchhoff introduced the "mechanical theory of heat," or "thermodynamics," which deals with motions as well as with temperature changes. But even here Kirchhoff held to his initial assumption that temperature is a property of matter that does not "need to be reduced to motion." Only with his introduction of the molecular view of matter in connection with the kinetic theory of gases did Kirchhoff interpret heat and temperature in terms of motion. Kirchhoff cautioned that the "leap" from the phenomena that molecules are intended to explain to the molecules themselves is so great that it is hard to decide on the right assumptions for drawing strong conclusions; yet it is possible, he said, from molecular assumptions to build a theory that "represents many properties of *gases* in a satisfactory way and offers a valuable guide to the further investigation of these properties." Since the properties depend on average values and not on individual molecules, Kirchhoff introduced Maxwell's velocity distribution along with the "concept of probability" to deal with the "statistical" nature of this subject, and he introduced the appropriate branch of mathematics, the "probability calculus," for carrying out the calculations.[6]

In an aside in the heat lectures, Kirchhoff remarked that specifically optical concepts such as light intensity, color, and the state of polarization are reducible to mechanical concepts; the reason for this is the agreement with experience of the hypothesis that light consists of oscillations, the living force of which stipulates the intensity of light, the duration of which stipulates the color, and the direction of which stipulates the state of polarization. In the optical lectures proper, Kirchhoff referred the oscillations to an ether filling empty space. Assuming that the free ether is a homogeneous, isotropic body on which no forces act except those that account

5. Gustav Kirchhoff, *Vorlesungen über mathematische Physik*, vol. 4, *Vorlesungen über die Theorie der Wärme*, ed. Max Planck (Leipzig, 1894), quotations on 1–3. Kirchhoff gave these lectures in 1876, 1878, 1880, 1882, and 1884 at Berlin University. Planck worked from notebooks that Kirchhoff wrote and edited, and he also referred to lecture notes by a student to fill in the gaps. Planck assured the reader that the lectures, as edited, were a true picture of Kirchhoff's lectures (pp. v–vi).

6. Kirchhoff, *Theorie der Wärme*, 51, 135. Kirchhoff gave eighteen lectures on heat: after the first four lectures, in which he treated pure heat theory, he devoted eight lectures to the mechanical theory of heat from the standpoint of the continuum theory of matter and the final six chapters to the mechanical theory supplemented by the molecular theory of matter of the kinetic theory of gases.

for its elasticity, he based his investigation on the equations of infinitely small motions, which he had derived in his mechanics lectures. Later, in his treatment of absorption and dispersion of light, he allowed an influence of the ponderable parts of bodies on the motion of the ether, a more intricate mechanical problem.[7]

In the lectures on electricity and magnetism, which complete the published lectures on mathematical physics, Kirchhoff assumed the imponderable fluids of electricity and magnetism and their associated distance forces instead of mechanical motions in an ether. He worked from the potential of the distance actions, frequently calling attention to the analogous expressions for the potential here and in other parts of physics. For example, to clarify the discussion of the change in direction and magnitude of an electric force, he drew an analogy with the velocity potential in hydrodynamics: according to this analogy, each electrical problem corresponds to a stationary motion of an incompressible fluid, as lines of electric force correspond to lines of fluid flow.[8] Only near the end of the lectures did Kirchhoff acknowledge that not everyone accepts the distance forces with which he began. According to Faraday and Maxwell, he said, electric and magnetic forces are transmitted by an insulating dielectric and arise from pressures or stresses, and he then developed the equations for the displacements in such a medium. By assuming that the electric waves in the medium correspond to the "transverse waves of an elastic medium according to our equations," he showed that the equations lead to the optical laws of Fresnel and Neumann. Planck, who edited these lectures in 1891, noted that Kirchhoff had treated only in passing Faraday and Maxwell's theory, "which at present perhaps offers most hope for a fruitful development." Planck explained that Kirchhoff did not treat this or any other theory extensively because of the "expressly established unity of conception," which was characteristic of all of Kirchhoff's writings.[9]

In his lectures, Kirchhoff made an effort to exclude everything that was unnecessary. They were highly mathematical lectures, directed toward establishing strong conclusions from clearly defined concepts and physical assumptions. Kirchhoff coupled his emphasis on rigor with frequent reference to experimental facts, especially where they were in conflict with the theory, as they were in the theory of an "ideal" gas.

Kirchhoff's method in physics, according to Boltzmann, was to avoid "bold hypotheses" and to "build equations that correspond to the phenomenal world as truly as possible and quantitatively correctly, unconcerned with the essence of things and

7. Kirchhoff, Theorie der Wärme, 1. Gustav Kirchhoff, Vorlesungen über mathematische Physik, vol. 2, Vorlesungen über mathematische Optik, ed. K. Hensel (Leipzig, 1891), 4. The editor wanted to present Kirchhoff's lectures on optics in the form of his last lectures in Berlin, acknowledging in that way Kirchhoff's "continuously renewed, thoroughgoing occupation" with the subject (p. v).

8. Gustav Kirchhoff, Vorlesungen über mathematische Physik, vol. 3, Vorlesungen über Electricität und Magnetismus, ed. Max Planck (Leipzig, 1891), 11. This volume is based on lectures Kirchhoff gave five times at Berlin between 1876–77 and 1885–86. Kirchhoff reworked them each time he gave them; taking the resulting changes into account, Planck refrained from extending Kirchhoff's theories, since it was impossible to find a stopping place in the "mighty, daily expanding field" of electricity and magnetism. Planck, "Vorwort des Herausgebers," v–vii, on vi.

9. Kirchhoff, Electricität und Magnetismus, 180, 228; Planck, "Vorwort," vi–vii.

forces."[10] While Kirchhoff accepted the goal of the mechanical reduction of physics, he did not base his lectures on the identification of heat with motion or of electric and magnetic forces with stresses in a dielectric medium, hypotheses which invited mechanical development but which seemed to him complicated and problematic.

Helmholtz at Berlin

Helmholtz's later researches in mechanics spanned his last years as director of the Berlin physics institute and his entire directorship of the Reichsanstalt. This latter directorship was a newly created job intended for Helmholtz, as Emil du Bois-Reymond explained: "There came a time when our great friend Werner von Siemens, with a huge donation that only he could afford, prepared the way for the founding of an imperial institute for physics and technology in Charlottenburg. Now we were not unaware that Siemens always regretted that Helmholtz had to devote a large share of his time and energy to his teaching duties instead of to the continuation of his incomparable researches, and we were also not unaware that he had intended the position of president of the institute for Helmholtz. His intention was for it to free him from all but scientific work, a situation that only a pure academic could imagine as ideal."[11] In the spring of 1888 Helmholtz was duly appointed president of the Reichsanstalt, realizing Siemens's intention and Helmholtz's, as well as the involved officials', wishes.

With his new job, Helmholtz did not give up all teaching at the university. He accepted a less demanding teaching assignment, which was financially tied to his retention of a salary from the Prussian Academy of Sciences. The request for it, made formally by the Prussian minister of culture, was justified in part by Kirchhoff's continuing serious illness, which left physics inadequately represented at the university. The plan was for Helmholtz to lecture from one to three hours a week in the "area of theoretical physics" and otherwise to be free of all academic duties. To this request, the imperial government gave its "revocable" permission after being assured that Helmholtz's teaching would not interfere with his duties at the Reichsanstalt. The plan was acceptable to the Berlin philosophical faculty, and since it was acceptable to Helmholtz, he remained affiliated with Berlin University through a professorship carrying a circumscribed teaching responsibility.[12] During the negotiations concerning Helmholtz's move from the university and the Prussian payroll to the Reichsanstalt and the imperial payroll, officials were concerned about Helmholtz's student fees from his lectures on theoretical physics. If the imperial government were

10. Boltzmann, *Kirchhoff*, 25.

11. Emil du Bois-Reymond quoted in Koenigsberger, *Helmholtz* 2: 346.

12. Prussian Minister of Culture Gossler to Otto von Bismarck, 20 May 1887, quoted in Koenigsberger, *Helmholtz* 2: 352–53. To make the job of presidency of the Reichsanstalt acceptable to Helmholtz, the imperial government had to offer him an income comparable to the one he had at the university. This meant that, as usual, the desired sum had to be brought together from here and there. How this was done and how Helmholtz's teaching assignment entered into it are discussed in David Cahan, "The Physikalisch-Technische Reichsanstalt: A Study in the Relations of Science, Technology and Industry in Imperial Germany (Ph.D. diss., Johns Hopkins University, 1980), 200–204.

to match exactly Helmholtz's income at the university, which would presumably include student fees, he would then be paid twice for the same thing. The officials did not realize that the fees would be entirely inconsiderable because the enrollment would be so small.

While he was director of the Berlin physics institute, Helmholtz took up a new research subject, physical chemistry. This he developed with the help of the "free energy," a potential useful for measuring the work done in a reversible process with constant temperature. From the thermodynamics of chemical processes, Helmholtz turned—or returned—to the problem of the foundations of thermodynamics. In his 1847 memoir on the conservation of force, he had given a mechanical explanation of the principle that came to be known, in one of its applications, as the first law of thermodynamics; now later, in several papers in 1884, he examined the second law of thermodynamics, again from the standpoint of mechanics. Specifically he showed that the limited transformability of heat energy has an analog in the behavior of mechanical systems containing an inner motion. To form an analogy with reversible heat processes, he developed the theory of hidden "inner cyclic motions," a class of motions which preserve the living force and the total energy of the system while the individual parts of the system rapidly change their positions. The coordinates defining these positions cannot, then, enter the expressions for the energy of the system as do the corresponding cyclic velocities. Helmholtz assumed that any changes in the cyclic velocities produced by external forces are brought about relatively slowly. In the analogy he intended, the velocities of the cyclic coordinates correspond to molecular heat motions of a gas and are related to the temperature. The remaining noncyclic, slowly varying coordinates of the mechanical system, such as those that define the volume of the gas, enter into equations that are identical to the equations relating heat to work. As an example of such motion, Helmholtz mentioned a spinning top: the rotation of the top corresponds to the motion identified with heat; this hidden motion is rapid compared with, say, the observable slow precession of the axis of the top.[13]

To formulate the analogy exactly, Helmholtz analyzed the simplest case, that of a "monocyclic" system. For it only one cyclic velocity q enters; the system may contain many inner cyclic motions, but if it is monocyclic, they all must depend on the one parameter. In this system, the analog of the introduction of heat from the outside is the performance of work dQ by an external force tending to increase the cyclic velocity. For this work, Helmholtz derived from Lagrange's equations of mo-

13. The properties of monocyclic systems and Helmholtz's reasons for studying them, along with his conclusions, are given in Martin J. Klein, "Mechanical Explanation at the End of the Nineteenth Century," *Centaurus* 17 (1972): 58–82, on 63–67; Leo Koenigsberger, "The Investigations of Hermann von Helmholtz on the Fundamental Principles of Mathematics and Mechanics," *Annual Report of the . . . Smithsonian Institution . . . to July, 1896*, 1898, 93–124, on 120–23; and, the main source of our discussion, Helmholtz's Prussian Academy papers on the subject and his recapitulation of them in his Berlin lectures on the theory of heat: "Studien zur Statik monocyklischer Systeme," *Sitzungsber. preuss. Akad.*, 1884, 159–77, 311–18, 755–59, and *Vorlesungen über theoretische Physik*, vol. 6, *Vorlesungen über die Theorie der Wärme*, ed. Franz Richarz (Leipzig: J. A. Barth, 1903), 338–70.

tion the formula $dQ = qds$. The symbol s is shorthand for $-\partial H/\partial q$, where H enters Lagrange's equations as the difference between the potential energy and the living force of the system. (H is the negative of the familiar Lagrangian function.) Assuming that the measure of the temperature is the living force, Helmholtz transformed the equation to read $dQ = LdS$, where the integrating factor L is the living force and S, a function of s, is the measure of entropy. With this familiar result, Helmholtz observed, the "Carnot-Clausius principle" is no longer a principle derived from experience but appears as a special case of a law derived from the "general principles of mechanics." At the same time, Helmholtz did not think that heat motion is, in reality, monocyclic "in the strict sense," since each atom probably changes its motion and a great number of atoms probably represent all phases of motion. But he regarded the motion as stationary in the "looser sense": on the average, the motion remains the same. In response to a criticism by Clausius of his work on monocyclic systems, Helmholtz clarified his intention: it was to draw attention to "analogies" between monocyclic and thermal motions rather than to claim "'an explanation' of the second principle of the mechanical theory of heat." The analogies do not take the form of a detailed mechanical model but concern only the "most general conditions" under which the "most general physical characteristics of heat motion" can be represented by mechanical motions.[14]

Helmholtz always sought causal laws rather than statistical regularities, and his work on the second law of thermodynamics was no exception. Boltzmann was interested, and in several papers in 1884 and 1886 he examined the relationship of monocyclic systems to his own work on the second law; as we will see, he used monocyclic systems again, and extensively, in his Munich lectures on Maxwell's electromagnetic theory soon after. And as we will see, Hertz also found Helmholtz's monocyclic studies useful as the first general treatment of hidden motions, upon which he based his new principles of mechanics.[15]

In his monocyclic studies, Helmholtz showed that the laws of reversible heat processes, including the law of entropy, can be expressed in the form of Lagrange's equations of motion and therefore in the form of a certain "minimal law," which he referred to as "Hamilton's principle of least action." In 1886 he began a systematic study of the least-action principle itself. He concluded from a comparison of the diverse formulations of the fundamental principles of mechanics in the past that the principle of least action is to be preferred, and that his formulation of that principle encompasses the other formulations as special cases. He anticipated that the principle would allow a uniform dynamical representation of the laws of all parts of physics. It was a continuation of the general direction of his research in theoretical

14. Helmholtz, "Studien zur Statik monocyklischer Systeme," 159–60, 169–70, 755–59; *Theorie der Wärme*, 351, 364–65. Klein, "Mechanical Explanation," 67, 70–71. In this analogy, the second law of thermodynamics is written $dQ = \theta\, dS$, where S is the entropy and θ is the absolute temperature.

We again wish to acknowledge discussions with Stephen M. Winters, at work on a major study of Helmholtz's physics, including his monocyclic systems and least-action principle below.

15. Wien, "Helmholtz als Physiker," 696. Klein, "Mechanical Explanation," 70–75.

physics, and it corresponded to that of certain British physicists whose work he followed and admired, Maxwell's above all.

Helmholtz undertook this study of the least-action principle as a direct result of his interest in the form of the kinetic potential demanded by Maxwell's electromagnetic equations. "Kinetic potential" is the name Helmholtz used for the function H entering the principle of least action and the Lagrangian equations of motion resulting from it. He had introduced it in his monocyclic studies, where he gave it a generalized form; for thermodynamic systems it is identical with the free energy. In the analytical formulation of the least-action principle, it appears in the variational equation determining the entire behavior of the system:[16]

$$\delta\Phi = 0, \qquad\qquad\qquad \text{where}$$

$$\Phi = \int_{t_0}^{t_1} dt \left\{ H + \sum_a (P_a \cdot p_a) \right\}.$$

Of the consequences of this generalized principle of least action and Lagrange's equations of motion, Helmholtz regarded the "exchange relations," or "reciprocity laws," between forces as especially useful. Exchange relations are a property of any system that has a kinetic potential; one set of such relations obtains between forces and accelerations, another between forces and velocities, and another between forces and coordinates. The relations between forces and velocities, for example, are:

$$\frac{\partial P_a}{\partial q_b} = - \frac{\partial P_b}{\partial q_a},$$

which assert, in words, that if the force P_b increases with increasing velocity q_a, the force P_a decreases with a corresponding increase in velocity q_b. Helmholtz referred

16. C. G. J. Jacobi had shown that Hamilton's principle still holds if H depends explicitly not only on positions and velocities but also on time, and Helmholtz included the sum $\sum_a(P_a p_a)$, where the P_a are forces that depend on time and p_a are the corresponding coordinates. In this generalized form, the principle applies to systems acted on by friction, galvanic resistance, and other nonconservative forces. In words, the principle states that for equal time intervals, the negative average value of the kinetic potential, as supplemented by the terms involving time-dependent forces, is a minimum for the actual path in comparison with all neighboring paths leading from the starting point to the end point of the action. The calculus of variations yields the Lagrangian equations of motion, which are, in Helmholtz's notation,

$$0 = P_a + \frac{\partial H}{\partial p_a} - \frac{d}{dt}\left[\frac{\partial H}{\partial q_a}\right],$$

where $q_a = dp_a/dt$ are the velocities of the system, and P_a are the forces exerted by the system on its surroundings. Proceeding from these equations, Helmholtz examined the behavior of mechanical, thermodynamic, and electrodynamic systems. Hermann von Helmholtz, "Über die physikalische Bedeutung des Princips der kleinsten Wirkung," *Journ. f. d. reine u. angewandte Math.* 100 (1887): 137–66, 213–22, on 139–40, 145.

to the following thermodynamic law as an example of exchange relations: if with an increase in the temperature of a given system, the pressure increases, then compression of the system will increase the temperature. Helmholtz showed that with different physical meanings attached to the generalized forces, velocities, and accelerations, these abstract exchange relations express confirmed connections between all kinds of phenomena: mechanical, electrodynamic, electrochemical, thermodynamic, or thermoelectric.[17] In this way, through the least-action principle, he connected the laws of the different parts of physics, much as he had done before through the principle of conservation of force, or energy.

In an address to the Prussian Academy of Sciences in 1887, Helmholtz compared the least-action principle with the conservation of energy principle: the former completes the latter, since the former yields the latter as a consequence. The two principles had similar histories: both originated within the mechanics of ponderable bodies and were subsequently extended to the imponderables, to heat and electricity. Clausius, Boltzmann, and recently Helmholtz himself, in his studies of monocyclic systems, had applied the principle to the second law of thermodynamics, and Franz Neumann had applied it to electrodynamics and after him so had Weber, Riemann, Carl Neumann, Clausius, and, abroad, Maxwell and others. The only limitation to the validity of the least-action principle, as Helmholtz saw it, is that it does not apply to irreversible processes. He did not see that limitation as fundamental, since he thought that irreversibility is not in the nature of things but is only a consequence of our inability to trace the irregular motions of individual atoms. Helmholtz valued the least-action principle because it compresses in the narrowest space, that of a single equation, all of the conditions of a physical problem; it yields more than the energy principle, since it also determines the path of a process. He valued it also because it applies throughout physics, providing a guide for formulating the laws of new classes of phenomena wherever they occur.[18]

In his researches from 1892 until his death in 1894, Helmholtz studied electrodynamics from the standpoint of the generalized least-action principle. Earlier he had discussed its applicability to Franz Neumann's potential law and other limited

17. Helmholtz, "Über die physikalische Bedeutung," 161–66. The kinetic potential entering the comprehensive least-action principle is no longer limited to its original form, the difference between potential energy as a function of position and living force as a quadratic homogeneous function of velocity; the potential energy now may be a function of velocity as well as of position, and the living force need not be a homogeneous function and may depend on position and on terms linear in velocity. The function H is required only to have finite first and second derivatives with respect to the coordinates and velocities. For each area of physics there is a kinetic potential, which in general consists of two series of variables, p and q; among the p are parameters that determine the position of the masses, the acceleration of which is referred to ponderomotive forces; the q need not refer to the velocities of ponderable masses but may refer to functions of the temperature, intensities of electric currents, and changes of dielectric and magnetic polarizations, among other physical quantities. The kinetic potential is constructed from variables of state and has the dimensions of energy, or work, but the individual parts of the kinetic potential are not specified as kinetic or potential energy.

18. Helmholtz, "Über die physikalische Bedeutung," 142–43; "Rede über die Entdeckungsgeschichte des Princips der kleinsten Action," in Adolf Harnack, ed., Geschichte der Königlich preussischen Akademie der Wissenschaften zu Berlin, 3 vols. (Berlin: Reichsdruckerei, 1900), 2: 282–96, on 283–87. Helmholtz did not publish this address, which he gave on 27 January 1887 at the Prussian Academy of Sciences, because he learned that the theme had already been treated by the mathematician Adolf Mayer.

electrodynamic processes. Now he extended it to the more complete electrodynamic equations of Maxwell and, especially, of Hertz, who had recently introduced terms that depend explicitly on the motion of the medium. Assuming that the pure ether is a frictionless, incompressible fluid without inertia, Helmholtz derived from the variational principle Hertz's fundamental electromagnetic equations for bodies in motion. He derived as well the ponderomotive forces of the theory, but these posed complications that he did not fully resolve.[19]

Helmholtz emphasized in these later researches the general principles of physics, the principles of energy and least action, governing the "flux of the eternally inde-structible and uncreatable energy supply of the world." He preferred to work from the differential equations embodying these general principles rather than from hy-pothetical models, though he fully appreciated the value of the latter.[20]

In 1891, on the occasion of Helmholtz's seventieth birthday, Hertz wrote that it was not well known that Helmholtz had recently returned to his earlier interests: in 1847 he had extended the energy principle to all forces, and now he was extend-ing the least-action principle to all of nature. Helmholtz was pursuing a lonely path, Hertz said, and he speculated that it would be years before a follower appeared to continue Helmholtz's work of tracing all phenomena to the least-action principle.[21]

The most important, if not also the first, of those followers was Max Planck. He saw the question of the significance of the least-action principle for the whole of physics as one worthy of his, as of Helmholtz's, best efforts. Helmholtz had pointed the "way to a unified conception of all natural forces," Planck said of Helmholtz's accomplishments in theoretical physics, adding that the "future must bring the re-alization of his ideas."[22]

It was not intended that Helmholtz should be the principal lecturer in theoret-ical physics at Berlin University after his move to the Reichsanstalt. Kirchhoff's replacement Planck had the official job, and he and Helmholtz were not in compe-tition.

Even so, Helmholtz gave a complete lecture course in theoretical physics at the university, in successive semesters covering all of its main parts. Having begun teaching physics with the conviction that theoretical physics was badly needed in Germany, he ended his teaching by lecturing on it exclusively. In 1892 he let it be known that he wanted to see his lectures published, and over the next dozen years

19. Hermann von Helmholtz, "Das Princip der kleinsten Wirkung in der Elektrodynamik," *Ann.* 47 (1892): 1–26, reprinted in *Wiss. Abh.* 3: 476–504; "Folgerungen aus Maxwell's Theorie über die Bewe-gungen des reinen Aethers," *Ann.* 53 (1893): 135–43, reprinted in *Wiss. Abh.* 3: 526–35; "Nachtrag zu dem Aufsatze: Ueber das Princip der kleinsten Wirkung in der Elektrodynamik" (1894), in *Wiss. Abh.* 3: 597–603.

20. Helmholtz, "Rede über die Entdeckungsgeschichte," 287. Helmholtz's preface to Hertz, *Princi-ples of Mechanics.*

21. Heinrich Hertz, "Hermann von Helmholtz," in supplement to *Münchener Allgemeine Zeitung,* 31 Aug. 1891, reprinted and translated by D. E. Jones and G. A. Schott in Hertz's *Miscellaneous Papers* (*Schriften vermischten Inhalts*) (London, 1896), 340.

22. Hertz, "Helmholtz," 340. Max Planck, "Helmholtz's Leistungen auf dem Gebiete der theore-tischen Physik," *ADB* 51 (1906): 470–72, reprinted in Planck, *Phys. Abh.* 3: 321–23, on 323.

they came out in a series of handsomely produced volumes, one for each part of physics. The publication was based mainly on stenographic notes of the lectures supplemented by Helmholtz's lecture notebooks and students' notes. The editors, mainly former students of his, wanted the published lectures to be a faithful account of the lectures as Berlin audiences heard them. They were at once a tribute to their author and a textbook for future physicists.

During the several years between the delivery of the lectures and their publication, physics underwent great changes. To a reviewer it appeared that a revolution in physics was well under way by the time Helmholtz's introductory lectures appeared in 1903, and they were welcomed as a steadying hand from the past.[23] These introductory lectures, *Einleitung zu den Vorlesungen über theoretische Physik*, consisted not of the expected lectures on mechanics but of even more fundamental matters; namely, methodological principles that apply to all parts of physics. Helmholtz justified their inclusion on the grounds that "we must investigate the instrument we work with," arguing that the construction of concepts, hypotheses, and laws and their quantitative formulation in differential equations and integrals have to be discussed if the work of theoretical physics is to be understood.[24] One of his students, Wilhelm Wien, regarded the Berlin lectures on theoretical physics as a great accomplishment not least because they were so completely Helmholtz's creation, free from previous models.[25]

Helmholtz's introductory lectures recalled the introduction to his memoir on the conservation of force, with its discussion of forces as constant causes. In the lectures he emphasized that our "ideas" and "wishes," our "consciousness" and "will," cannot influence the phenomena we know through experience. They exist independently of us. We seek their laws. These laws we sometimes call "forces," he said, since "force" means no more than that the "law will show itself in each case where the conditions for the phenomena are given." Forces belong to the "outer world," always present and ready to act under the right conditions. We assume that all changes in bodies obey laws and have causes intelligible to us. That is the justification of physics.[26]

To get on with the task of physics, Helmholtz told his audience, one needs a "certain penetration into the greater depths of mathematical analysis." He discussed the axioms of arithmetic, pointing out that their applicability to physical magnitudes has to be decided by experiment. He discussed the various sorts of numbers that enter the statement of physical laws, and he discussed the conceptual basis for the quantitative description of physical phenomena. In the *Einleitung* he did not present mathematical methods useful in doing theoretical physics; he did that in later lectures, in which he developed the laws of physical phenomena.[27]

Helmholtz's later lectures are filled with mathematical formulas, and in the *Ein-*

23. Karl Böhm, "H. von Helmholtz, *Einleitung zu den Vorlesungen über theoretische Physik*," *Phys. Zs.* 5 (1904): 140–43.

24. Helmholtz, *Einleitung*, quotation on 1. These lectures, constituting the first part of volume 1 of Helmholtz's *Vorlesungen über theoretische Physik*, were given in 1893.

25. Wien, "Helmholtz als Physiker," 697.

26. Helmholtz, *Einleitung*, 7, 10–11, 14–16.

27. Helmholtz, *Einleitung*, quotation on 25.

leitung he explained why this must be so. To express the laws of changes of bodies, we need differential equations; they contain what is essential and common to all cases of a class of changes, and they omit everything that is accidental, such as the form, size, time, and place of the individual bodies we observe in individual experiments. We construct the differential equations by analyzing the mutual actions of the elementary parts of bodies. To obtain from them the actions of the larger bodies of our experience, we need to add the elementary actions, to integrate the differential equations, which is the reason why theoretical physics is predominantly mathematical. The construction and integration of differential equations involve a "thought process" that reappears throughout all parts of theoretical physics.[28]

In the introductory lectures, Helmholtz discussed in general terms the two ways physicists have of treating bodies. Depending on the problem at hand, they regard bodies either as aggregates of material points or as volume elements filled with matter. The differences between the two ways are far-reaching, and Helmholtz devoted a separate semester—and volume of lectures—to each. In the first set of lectures following the introduction, Helmholtz treated the dynamics of discrete material points, or "mass points." The mass point is a useful concept for solving problems in which the real form and extension of a body can be neglected, as they can be when two bodies act on one another over a long distance. For the description of the motion of a mass point, which is one of the "first and most important tasks of theoretical physics," the position of the point must be defined by continuous and differentiable functions of time; otherwise, the point could be in two places at the same time, disappearing here and instantly reappearing there. Such a discontinuity would violate the identity of the mass point, which derives from a "fundamental law of experience of all natural phenomena": matter cannot be created or destroyed. Within this "picture" of indestructible and discrete mass points, Helmholtz introduced Newton's three axioms of motion together with supplementary assumptions about the acting forces; from these he derived the "general fundamental principles of dynamics."[29]

In the second set of dynamical lectures, Helmholtz elucidated the central concept of continuously distributed masses, comparing it with the concept of discrete

28. Helmholtz, *Einleitung,* 22–24. In the later dynamical lectures, Helmholtz further discussed the role of differential equations in theoretical physics. He explained how physicists derive general laws by differentiating with respect to time the mathematical formulation of a particular experimental observation. When physicists do not have such a mathematical formulation to begin with, they set up differential equations from assumptions or analogies and compare the resulting integrals with experimental observations. Hermann von Helmholtz, *Vorlesungen über theoretische Physik,* vol. 1, pt. 2, *Vorlesungen über die Dynamik discreter Massenpunkte,* ed. Otto Krigar-Menzel (Leipzig, 1898), 42–44.

29. Helmholtz, *Einleitung,* 22, 38, 41; *Dynamik discreter Massenpunkte,* 2, 7. These latter lectures are based on a transcript of Helmholtz's Berlin lectures from December 1893 to March 1894. Commissioned by Helmholtz, the transcript was supplemented by the editor Krigar-Menzel from his own notes of Helmholtz's theoretical physics lectures from earlier years and, for material bearing on the validity of the energy principle, from Helmholtz's experimental physics lectures on the subject. Because time ran out, Helmholtz often did not get through everything he intended to in his lectures; that can be seen, Krigar-Menzel said, from the notebook Helmholtz used, but seldom looked at, during his lectures. Krigar-Menzel, "Vorwort," v–vi. This volume of lectures is divided into four parts: kinematics of a material point, dynamics of a material point, dynamics of a system of masses, and comprehensive principles of dynamics.

mass points. The difference between the two is evident from the calculation of density, or the ratio of mass to volume: within the picture of continuous masses, we can imagine a closed volume of diminishing smallness in which the density approaches a limiting value at a given position; by contrast, within the picture of discrete masses, we can imagine a sufficiently small closed volume containing only a few mass points, so that in this case it makes no sense to speak of a limiting value of the density.[30]

Although the concept of continuously distributed masses corresponds to our direct sense impressions of sight and touch, Helmholtz cautioned that we cannot conclude that matter is actually continuous. Of the final division of matter we know nothing; all we can do is form hypotheses about it to explain the properties of bodies, mechanical, thermal, and chemical. We generally work with the hypothesis that the final division is into atoms and molecular groupings of atoms, the main properties of which correspond to the picture of discrete mass points interacting through characteristic central forces. But when we treat phenomena by mathematically dividing bodies into volume elements that are large compared with molecular separations, as we do, for example, in hydrodynamics and elasticity theory, we then work within the picture of continuously distributed masses rather than of atomistically structured ones. Even here, Helmholtz noted, there are problems such as the dispersion of light in which the simple picture of continuously distributed masses is insufficient and the molecular hypothesis must be invoked.[31]

After introducing the fundamental concepts for treating the dynamics of continuously distributed masses, Helmholtz described how we apply mathematics to them. We introduce partial differential equations, the mathematical language of the dynamics of systems with more than one independent variable. We imagine cell walls distributed throughout a continuous body, dividing it into masses that retain their identity as the body moves as a whole. Without misunderstanding, we may speak of a "mass point" in the present case just as we can in that of the dynamics of discrete masses. Only the meaning is different here: the "mass point" now has no definite mass; it is a shorthand expression for the corner point of a volume element filled with mass. Exactly as in the molecular picture, in the picture of continuous masses the positions of the mass points can be differentiated with respect to time; since, however, these mass points form a continuum, as they do not in the molecular picture, their velocities can be differentiated not only with respect to time, to construct their acceleration, but also with respect to the three spatial coordinates. It is this mathematical distinction—the presence here of spatial coordinates as variables

30. Hermann von Helmholtz, *Vorlesungen über theoretische Physik*, vol. 2, *Vorlesungen über die Dynamik continuirlich verbreiteter Massen*, ed. Otto Krigar-Menzel (Leipzig: J. A. Barth, 1902), 2. These lectures, given in Berlin in 1894, were Helmholtz's last, and they were broken off due to his illness. The major topic is the elastic theory of solids, but Helmholtz also treated hydrodynamics; he intended to conclude the lectures with his work on vortex motion, but he never got around to it. The book—a "torso" of the lectures—is divided into four parts: the first two treat the kinematics and the dynamics of continuously distributed masses, and the last two treat the problem of determining the forces given the deformation and the converse problem of determining the deformation given the forces. Krigar-Menzel, "Vorwort," v–vi.

31. Helmholtz, *Dynamik continuirlich verbreiteter Massen*, 2–3.

with which the velocities can be differentiated—that is the "essential characteristic of continuously distributed masses in contrast to systems of discrete mass points, in which time is the only primitive variable."[32] Within the picture of continuously distributed masses, Helmholtz developed the general mathematical theory of the deformations and forces of bodies, of "strain" and "stress," and applied it to problems involving dilation, bending, sheer, pressure, tension, torsion, and the like.

Helmholtz underscored the distinction between the concepts of discrete mass points and of continuously distributed masses by distinguishing between the concepts of "unordered" and "ordered" motions. In unordered motion, each molecule carries out its motion independently of that of its neighbors. In the case of continuously distributed masses, adjacent volume elements press against one another and so cannot move independently, and the concept of unordered motion cannot apply here. To illustrate the distinction, Helmholtz invoked an image: he likened unordered motion to the motions of individual flies within a swarm and ordered motion to the change in shape and position of the whole swarm. In the lectures on elasticity theory, he was interested in the swarm only.[33]

The concept of unordered motion, which Helmholtz introduced in the lectures on the dynamics of continuously distributed masses but did not use there, is central to his lectures on the mechanical theory of heat. For heat is the "unordered motion" of the smallest particles of bodies; the motion, that is, of the flies, in which the velocity and displacement of any individual have no relation to those of its neighbors. From the hypothesis that the molecules of a gas are, like flies, in unordered motion, Helmholtz deduced most of the properties of gases. Since we cannot describe the motions of individual molecules or mass points, he explained, we can only calculate their average values.[34]

With both ways of picturing the constitution of matter, the changes of bodies are brought about by forces, the proper subject of "dynamics." Continuously distributed masses introduce forces—for example, surface forces—that are not encountered in systems of mass points, but they are nonetheless forces, an extension of the concept of force from the case of mass points. In the lectures on heat theory, Helmholtz offered a further extension of the concept of force, and at the same time he extended the number of thermodynamic variables of state beyond the usual two of temperature and volume. In general, for each change in a state variable, or generalized "coordinate," v_a, there is an opposing generalized "force," P_a, defined such that the product $P_a dv_a$ is work. It follows that the analog of the first law of thermodynamics $dQ = dU + pdv$ for more complex processes is $dQ = dU + \sum(P_a dv_a)$. Helmholtz spoke of a true advance of the "newer physics" in the common representation and measure of work of all forces, whether they are mechanical, electrical, chemical, or what-

32. Helmholtz, Dynamik continuirlich verbreiteter Massen, 8–9.
33. Helmholtz, Dynamik continuirlich verbreiteter Massen, 7–8.
34. Helmholtz, Theorie der Wärme, 256–58. This volume was compiled from Helmholtz's notebook for the summer semester of 1890, from stenographic notes taken of his lectures in the summer semester of 1893, and from notes taken by the editor in the early 1880s. Richarz, "Vorwort." The volume is divided into three parts: pure heat theory, thermodynamics or the mechanical theory of heat, and theories of molecular heat motion.

ever. Helmholtz was responsible for much of this advance himself, and his lectures, as we see here and below, reported his own recent researches extensively.[35]

In the lectures on the dynamics of discrete masses, Helmholtz deduced the conservation of energy principle from mechanical laws and from the assumption of conservative forces. He pointed out that this derivation does not imply that the consequences of the principle belong only to mechanics. There exist various "energy forms," and although efforts have been made to "reduce the nonmechanical energy forms through hypothetical ideas to the mechanical as the original forms," the hypothetical ideas do not belong to the content of the energy principle. The principle stands as an independent fact of experience: it is a completely general principle that "governs all phenomena of nature."[36]

Although the energy principle is highly useful in solving dynamical problems, Helmholtz pointed out, it is often an insufficient basis for carrying through the calculations. Other "comprehensive principles" must also be considered, to which Helmholtz devoted the last part of the lectures on the dynamics of discrete masses. The advantage of principles such as the principle of virtual displacements, d'Alembert's principle, and Hamilton's principle of least action is that they provide a general view of the behavior of systems without requiring the special consideration of each mass point. Of the comprehensive principles, Helmholtz naturally gave most attention to Hamilton's; in several places in the lectures, he derived the principle, either starting directly from Newton's axioms and the assumption of conservative forces or, as in the lectures on the dynamics of discrete mass points, from d'Alembert's principle together with the same assumption about the forces. The variational equation embodying Hamilton's principle yields Newton's equations of motion, and it contains nothing that is not already expressed in these equations; its "great significance" is its independence from the choice of coordinate systems. Introducing generalized forces and coordinates and Lagrange's equations of motion, Helmholtz extended the least-action principle, as he had extended the energy principle, beyond the domain of mechanics in which it originated.[37] In particular, Helmholtz extended Hamilton's principle to thermodynamics at the end of his lectures on heat theory. There, as an accompaniment to the kinetic theory of gases, he derived the mechanical analogy of heat as cyclic motions from the extended Hamilton's principle, employing the kinetic potential, and from the associated Lagrangian equations of motion.[38] He observed that no natural phenomena are known to contradict Hamilton's principle or to show its restriction on natural forces as being too narrow; we are justified, he concluded, in assuming that this principle too has "universal validity."[39]

35. Helmholtz, *Theorie der Wärme*, 277–80; *Dynamik discreter Massenpunkte*, 22; *Dynamik continuirlich verbreiteter Massen*, 55, 72.

36. Helmholtz, *Dynamik discreter Massenpunkte*, 231.

37. Helmholtz derived Hamilton's principle of dynamics in several places; for example, in *Dynamik discreter Massenpunkte*, 309–14; *Theorie der Wärme*, 338–41; and *Vorlesungen über theoretische Physik*, vol. 3, *Vorlesungen über die mathematischen Principien der Akustik*, ed. Arthur König and Carl Runge (Leipzig, 1898), 57–62, quotation on 62.

38. Helmholtz, *Theorie der Wärme*, 353–70. Helmholtz developed the kinetic potential in *Dynamik discreter Massenpunkte*, 359–73.

39. Helmholtz, *Dynamik discreter Massenpunkte*, 368–69, 373.

The remaining volumes of Helmholtz's lectures on theoretical physics treat acoustics, optics, and electrodynamics and magnetism. Helmholtz presented acoustics as an application of ordinary mechanics, since it is the study of small oscillations in bodies under the action of conservative forces.[40] Although he began his optical lectures with the mechanical optical theories because of their "great historical interest" and the "enormous quantity of factual knowledge" expressed in their concepts and language, he used as title for these lectures "electromagnetic theory of light," not the customary "optics." Helmholtz explained that Hertz's experiments left no doubt of the reality of a medium for electric oscillations and of their finite velocity of propagation through it: they "possess all objective properties of light oscillations," justifying the emphasis in these lectures on the electromagnetic theory.[41] Helmholtz's lectures on the theory of electrodynamics and magnetism were the last to appear. The fifteen years or so since Helmholtz had delivered the lectures had seen an enormous development of the subject; the volume of lectures was not even given a proper review in the *Physikalische Zeitschrift* but was acknowledged as the completion of a "lasting memorial of the great master."[42]

Helmholtz's lectures on theoretical physics presupposed some familiarity with physics, which his students would have acquired from a course of general physics lectures. They presupposed as well some familiarity with analysis; for example, the theory of equations, the differential and integral calculus, the variational calculus, ordinary and partial differential equations, and vector analysis. Helmholtz made occasional mathematical "digressions" when he thought that the students did not have enough preparation to follow his derivations.

It was not for mathematical reasons that Helmholtz's lectures were difficult, as they were found to be. Kirchhoff, for example, in his lectures, emphasized the mathematical side of theoretical physics more than Helmholtz did. Helmholtz's were difficult largely because of their emphasis on conceptual development, which required of students a physical way of thinking.[43]

Helmholtz's audience at Berlin came away from his course of lectures with a view of physics as a connected science, not as so many isolated explanations of things. The connectedness depends above all on dynamics: dynamics provides the concepts, pictures, laws, comprehensive principles, and invariant magnitudes by which physics expresses its most general viewpoints. It is the source of the universal in physics: Hamilton's principle has universal validity, the energy principle has universal validity, Newton's third axiom of motion has universal validity. "Universal,"

40. Helmholtz, *Akustik*, 1.

41. Hermann von Helmholtz, *Vorlesungen über theoretische Physik*, vol. 5, *Vorlesungen über die elektromagnetische Theorie des Lichtes*, ed. Arthur König and Carl Runge (Hamburg and Leipzig, 1897), 14–16.

42. Hermann von Helmholtz, *Vorlesungen über theoretische Physik*, vol. 4, *Vorlesungen über Elektrodynamik und Theorie des Magnetismus*, ed. Otto Krigar-Menzel and Max Laue (Leipzig: J. A. Barth, 1907). Acknowledged by Emil Bose, in *Phys. Zs.* 9 (1908): 141.

43. Reviewers of Helmholtz's lectures on theoretical physics compared them with Kirchhoff's; for example, Hans Lorenz's review of vol. 6 in *Phys. Zs.* 4 (1903): 684–85, and Ernst Zermelo's review of vol. 2 in *Phys. Zs.* 5 (1904): 475.

"comprehensive," "general," and "invariant," these are words that stand out in Helmholtz's presentation of theoretical physics and give it its individual stamp. In this respect, Helmholtz's lectures and researches constitute a whole, as we have seen in this discussion.

Hertz at Bonn

"As to the structure of the ether," Hertz wrote to Heaviside in 1889, "the structure of all the models imagined until now is certainly not the structure of the ether."[44] In keeping with the spirit of his Heidelberg talk, through a variety of approaches Hertz worked toward a solution of the fundamental problem of physics, the properties of the ether. He initiated an astronomical correspondence, for example, in search of evidence for a possible finite velocity of propagation of the gravitational force through the ether. It was a question of the "highest interest" to him and to "all physicists," since it bore on the prospect of a consistent development of physical theory in terms of contiguous action instead of distance action.[45]

Hertz was at the same time in correspondence with a number of physicists working on optical questions, to which, he insisted, Maxwell's electrical theory could be applied. He told Drude that it would be "ridiculous" to doubt the electrical theory because of difficulties in applying it to the reflection of light from metals. He told Voigt that the latter's optical theory was convincing more for its results than for the clarity of its principles and that for this reason Voigt should support his equations by the electrical theory.[46] The optical problem, he told Wiener, interested him "mainly from the electrical standpoint," and he advised Wiener to apply his discovery of standing light waves, the optical counterpart of Hertz's standing electrical waves, to the problem of detecting the absolute motion of the earth through the ether. The ether remained puzzling to him: "I think that in our ideas of the ether there are still a good many contradictions that are hard to understand, but perhaps we must still wait a long time until we understand them."[47]

At the end of 1890, Hertz had finished arranging his laboratory at Bonn and now had more time for his research. He tried an experiment on polarization through gravitational action, and he tried a variety of other experiments, grew tired of their repeated failures, and looked for a new direction. In March 1891 he noted in his diary that he was now working on mechanics.[48] From then on, he gave little thought to experimental work; he wrote to a colleague that he was "entirely occupied" with

44. Hertz to Heaviside, 3 Sept. 1889, in Appleyard, *Pioneers*, 239.
45. Hertz to Rudolf Lehmann-Filkés, 10 and 17 Nov. 1889; Lehmann-Filkés to Hertz, 12 Nov. 1889, Ms. Coll., DM, 3178–79, 2975.
46. Hertz to Drude, 15 May 1892, Ms. Coll., DM, 3209. Hertz to Voigt, 6 Feb. 1890, Voigt Papers, Göttingen UB, Ms. Dept.
47. Hertz came to recognize that the experiment with standing waves looked hopeless, but he could not believe that the enormous velocity of the ether would not have some detectable optical result. Hertz to Wiener, 25 Dec. 1890 and 6 Jan. 1891, Ms. Coll., DM.
48. Hertz to Cohn, 31 Dec. 1890, Ms. Coll., DM, 3204. Hertz's diary in Hertz, *Erinnerungen*, 314.

a work on mechanics, which "unfortunately has a purely theoretical interest and no practical interest at all."[49]

The work was to become a book. For three years, Hertz tried to write "something good and above all lasting," changing each word three or four times and re-writing the whole as often. His highest aim was to bring "absolute clarity" to mechanics. Following the example of the "ancients," he presented it by definitions and propositions.[50]

The book *Die Principien der Mechanik* (*Principles of Mechanics*), which came out in 1894, opened with the observation that "all physicists agree that the problem of physics consists in tracing the phenomena of nature back to the simple laws of mechanics." But since all physicists did not agree on what those simple laws are, it was "premature to attempt to base the equations of motion of the ether upon the laws of mechanics." That would have to come later. Hertz set out to reformulate the laws of mechanics so that they would embrace all known motions and exclude all unknown ones. His work, he said, owed a good deal to Mach's *Science of Mechanics* and to Thomson and Tait's and other treatises on mechanics, but it owed most to Helmholtz's mechanical investigations of the 1880s, which Hertz called the "furthest advance in physics."[51]

The first part of the *Principles* treats geometry and kinematics and is "completely independent of experience," concerned only with statements about the paths and connections of material particles that satisfy the demands of thought. The second part treats mechanics proper and appeals to experience through the "fundamental law" of mechanics, which is a statement about the path followed by a free system.[52] The concept of force does not appear as one of the fundamental concepts, since Hertz regarded it as logically obscure and belonging to the superseded physics of action at a distance. Hertz replaced force—this invisible something responsible for the visible motions of masses in the world—with masses and motions. The only difference between the latter and the moving masses physicists observe is that they are too fine for our senses. They are related to the "concealed masses" and "concealed motions" that Helmholtz had introduced in his mechanical studies, and their actions on visible bodies are what physicists in the past had ascribed to forces. Their presence clearly excludes distance forces from mechanics.

Hertz believed that his formulation of mechanics would be suited for "tracing back the supposed actions-at-a-distance to motions in an all-pervading medium whose smallest parts are subjected to rigid connections."[53] It was in this area that he expected the decision between the different formulations of mechanics to take

49. Hertz to Sarasin, 19 May 1893, Ms. Coll., DM, 3149.

50. Hertz to Arthur Meiner, 12 Nov. 1893, Ms. Coll., DM, 3245.

51. Hertz, *Principles of Mechanics*, author's preface and 17.

52. Hertz's fundamental law reads: "Every free system persists in its state of rest or of uniform motion in a straightest path." *Principles of Mechanics*, 144. Here the expression "free system" refers to a material system with only internal and normal, or time-independent, connections; and "straightest path" refers to a path of smallest curvature.

53. Hertz, *Principles of Mechanics*, 41.

place, and just as he had recently helped decide between rival electrodynamic theories, he now wanted to do the same for the principles of mechanics.

On reading Hertz's book, Helmholtz thought that Hertz's mechanical principles might lead to a new understanding of forces, and he admired the logic and the mathematics. But he regretted that Hertz did not give concrete examples to show how the mechanism of hidden masses works.[54] Hertz's principles were applied by others to some problems, but physicists, by and large, looked at them as Helmholtz did, with respect and suspended judgment. Hertz's viewpoint was compared with other viewpoints in discussions of the foundations of mechanics around the turn of this century, but it was his epistemological and methodological discussion in the introduction to the book that was to have the greatest impact. There, in his criticism of the competing representations of mechanics, Hertz closely analyzed physicists' "pictures" of nature and the criteria by which they were to be judged.

In 1892, while working on his mechanics, Hertz contracted a head infection, which became so serious that he had to cancel lectures. Rest, surgery, and spas brought only temporary relief; just after sending off most of the manuscript of *Principles of Mechanics*, in January 1894 Hertz died.[55] Helmholtz wrote in a preface to the posthumous book that Hertz's death brought "deep sorrow"; of all of his students, Helmholtz said, Hertz was the "one who had penetrated furthest into my own circle of scientific thought, and it was to him that I looked with the greatest confidence for the further development and extension of my work."[56]

After Hertz's death, the scientific section of the Bonn philosophical faculty proposed as his successor several experimentalists: Friedrich Kohlrausch, Warburg, and in third place, alongside Leonhard Sohncke, Heinrich Kayser. The job went to Kayser, a student of Kundt and Helmholtz, who was recommended above all for his "extremely exact investigations of the spectra of chemical elements, which belong to the most outstanding accomplishments of physics today."[57]

Kayser found the Bonn physics institute damp and unhealthy and "almost impossible" to work in because of its nearness to trains. He stopped trying to work there in time, and the Privatdocent Alexander Pflüger worked there only at night from one to four, after placing a flag outside to signal traffic not to pass by the building. Because spectroscopic research required the kind of exact measurements that usually could not be carried out in the institute, Kayser devoted himself to

54. Helmholtz's preface to Hertz, *Principles of Mechanics*.

55. Hertz's doctor apparently believed he died of the institute. Hertz's replacement Kayser eventually got the ministry to make changes to dry out the rooms. Kayser, "Erinnerungen aus meinem Leben" (1936), 181–83. This unpublished autobiography is in the American Philosophical Society Library, Philadelphia.

56. Helmholtz's preface to Hertz, *Principles of Mechanics*.

57. After Kayser's researches were praised, his teaching was mentioned; it included lecturing, directing a laboratory, and writing textbooks on experimental physics and spectral analysis. Draft of recommendations of candidates to succeed Hertz, 31 Jan. 1894, Akten der phil. Fak. betr. Hertz, Bonn UA.

writing a handbook on spectroscopy, which eventually ran to six volumes and which associated Kayser's name forever with this literary monument. From the time he arrived, Kayser appealed for a new institute building, but it was only in 1910 that the necessary money became available and 1913 that the new building was ready. It was well equipped for spectroscopic research, which was Bonn's specialty.[58]

From the standpoint of theoretical physics, Bonn was not the same again after Clausius and Hertz. Kayser showed little interest in the subject. In his spectroscopic research, he barely mentioned the lawful regularities of the wavelengths of spectral lines; for him, the precise measurement of the wavelengths had paramount importance, not the theoretical insight that might be attached to them.[59] He did not lecture on theoretical physics but left it to the extraordinary professor and a Privatdocent or two.[60]

Volkmann at Königsberg

In the new Königsberg physics institute, built in 1884–86, the mathematical and the experimental physics professors were each given a wing. Each had all the parts of a complete institute, so that the plan of the building expressed something of an equality between the representatives of the two halves of physics. For a small university, Königsberg now provided reasonable space for its new theoretical physicist, Paul Volkmann.[61]

Volkmann's enrollments increased gradually. In the physics section of the Königsberg mathematical-physical seminar, Volkmann had no students in 1892, one student the following year, and two students the year after, which encouraged him to offer regular seminar exercises again after a long interruption. In 1901, owing to an influx of students, the ministry granted the institute a sizable extra allowance for new instruments and an increase in its budget for the laboratory. By 1902, the teaching demands of the laboratory had grown sufficiently for Volkmann to request a position for a second assistant.[62]

58. Kayser, "Erinnerungen," 183–87. Heinrich Kayser and Paul Eversheim, "Das physikalische Institut der Universität Bonn," *Phys. Zs.* 14 (1913): 1001–8.

59. Friedrich Paschen, "Heinrich Kayser," *Phys. Zs.* 41 (1940): 429–33, on 432.

60. Lorberg, the extraordinary professor for theoretical physics, had difficult relations with Kayser, which were exacerbated by Lorberg's illness in his last years and Kayser's impatience. The Privatdocenten Alexander Pflüger and Alfred Bucherer took over the theoretical lectures at first, but Kayser wanted an extraordinary professor who could treat all parts of theoretical physics together with their connections. He got what he wanted in 1903 in the experimentalist Walter Kaufmann. But Lorberg did not want to be replaced, which complicated Kayser's plans nearly until Lorberg's death in 1906. Kayser to the Bonn U. Curator, 10 Dec. 1901 and 4 Oct. 1902, and many more letters in the same file, Bonn UA, IV E II b, Lorberg.

61. This discussion of Volkmann's institute is based in part on the annual report, *Chronik der Königlichen Albertus-Universität zu Königsberg i. Pr.* The years considered here are 1892–93 through 1901–2. After Voigt left Königsberg for Göttingen, Volkmann inherited responsibility for the mathematical-physics institute. A graduate of Königsberg in 1880, he became Privatdocent there in 1882, extraordinary professor in 1886, and ordinary professor in 1894. "Neubau des physikalischen Instituts in Königsberg i. Pr.," *Centralblatt der Bauverwaltung* 7(1887): 13–14.

62. Volkmann to Königsberg U. Curator, 14 July 1902, STPK, Darmst. Coll. 1923.16.

From twelve students in 1899, in three years the number of students attending Volkmann's laboratory course increased to twenty-four. This doubling of attendance entailed a good deal of extra work, Volkmann explained—or rather complained—to the curator. Allowing for an understandable measure of exaggeration, Volkmann's account provides a good idea of the work of laboratory supervision. Every student who came to the laboratory, Volkmann told the curator, was assigned a problem in "physical precision measurements"; before beginning it, the student had to prepare, and for that he came to the laboratory at the hours scheduled for laboratory practice; he did the actual measurements usually after hours. If the student continued to come to the laboratory in subsequent semesters, he began "physical research." Physically and mentally, Volkmann said, the direction of student laboratory work belonged to the most demanding teaching in the university. Problems had to be assigned to the members of the class in such a way that the solution of each problem required only instruments that could be chosen and arranged so as not to disturb the experiments of other students. The director could not concentrate his attention as he could in a lecture but had to attend to the students' various needs all at the same time. Because the class was divided among a series of rooms, including the corridors and stairway shafts, the director and his assistant had to run back and forth to see that everything was going all right and to answer students' questions from every side of physics. They had to see to the proper use of the instruments, since damaged ones might require weeks to be repaired. Finally, they had to examine and, in many cases, recalculate the work the students did on their observations at home. All of this demanding labor in the laboratory course, Volkmann explained, had its basis in "the nature of the discipline of mathematical physics."[63]

From Volkmann's writings, we can learn more about his understanding of the nature of this discipline. In 1896 he published a book on the epistemology of the natural sciences; it is a subject, he said, which has recently come into prominence as a result of Faraday and Maxwell's direction of physics, and as evidence he cited writings by Helmholtz, Mach, Boltzmann, Ostwald, and Hertz. Physics, according to Volkmann, experiences the strongest epistemological drive of all the natural sciences: no other natural science is as little satisfied as physics with the mere arrangement of facts; none is so determined to go beyond them to their connection through basic laws; none is so committed to carefully constructed exact theories. Volkmann made a lively and provocative case for physics—and the other natural sciences—as the "driving force" of all current intellectual life.[64] This book Volkmann wrote for the educated public. He wrote much the same thing when he wrote for students and colleagues, only he included more technical matter; in his lectures on theoretical physics, in particular, he stressed the great, general intellectual significance of the science.

The great question physics had to decide then, whether light is an elastic or an

63. Volkmann to Curator, 14 July 1902.

64. Paul Volkmann, *Erkenntnistheoretische Grundzüge der Naturwissenschaften und ihre Beziehungen zum Geistesleben der Gegenwart. Allgemein wissenschaftliche Vorträge* (Leipzig, 1896), iii, 4, 13.

electromagnetic phenomenon, led Volkmann in 1891 to publish his lectures on theoretical optics. He did not give his students a straight exposition of the accepted
understanding of the subject but a comparison between the two great and equal
theoretical conceptions leading to a decision between them. He began his optical
lectures by developing the theory of light from pure mechanics, examining the assumption that light consists of transverse waves in an elastic substance filling the
universe. He then examined the electromagnetic theory, first the older one based
on action at a distance and then the newer one of Faraday and Maxwell, and he
showed that each yields differential equations identical with those for transverse
elastic waves. He discussed Hertz's experiments, which were then only three years
old; and he discussed in addition to Maxwell's and Helmholtz's work on the electromagnetic theory Hertz's and Cohn's very recent theoretical work. Because of the
natural way the velocity of light enters the electromagnetic theory, Volkmann concluded that there can no longer be any doubt about the superiority of the electromagnetic view of light. Yet the elastic view retains an advantage: optics teaches us
that transitory stages of science hold a permanent value for "physical *Anschauung,*"
on which rests the "progress of the science." Comparisons between theories extend
the students' understanding of physics; they have epistemological value.[65]

In 1900 Volkmann published his mechanics lectures as an introduction to theoretical physics, which he prefaced by nearly fifty pages of physical epistemology,
and he made frequent reference to this epistemological discussion throughout the
rest of the text. With regard to the question of which "special physical discipline we
should choose as the foundation for all other physical disciplines," he acknowledged
that electricity might qualify. But to treat it as such was to disregard the historical
development of the physical sciences, in which mechanics had been fundamental.
Besides the historical argument, the logical position of mechanics within physics also
argued for its choice. So did the widespread use of "mechanical analogies." Volkmann regarded his task in his lectures on theoretical physics as showing that physics
is nothing but mechanics.[66]

In his presentation of mechanics proper, Volkmann proceeded from Newton's
three axioms of motion, citing Helmholtz's and William Thomson's presentations as
precedents. He disliked Lagrange's formulation of mechanics, which as a "closed
system" was appropriate for the study of mathematics but not for that of theoretical
physics. The flaw in most subsequent presentations of mechanics was that in their
mathematical one-sidedness they lost sight of the "subjective element of research."
Hertz and Boltzmann were exceptions in that they took into consideration the subjective as well as the objective side: they viewed mechanics as a "picture of reality,"
which relates the subjective to the objective as a picture is related to its object.
Still, Volkmann thought, they paid too much attention to the picture and not
enough to the object, which made their pictures closed, again chiefly of interest to

65. Paul Volkmann, *Vorlesungen über die Theorie des Lichtes. Unter Rücksicht auf die elastische und die
elektromagnetische Anschauung* (Leipzig, 1891), iii–vii, ix, 17–30.
66. Volkmann, *Einführung*, 9, 41–42, 349.

mathematicians and not a fitting introduction to the study of theoretical physics. Volkmann avoided their mistakes by treating thoroughly the relationship of mechanics to reality, paying attention to the needs of physics. His presentation, from beginning to end, was an argument for the "position of mechanics as the fundamental discipline within the physical system."[67]

In his lectures Volkmann carefully discussed the sorts of mental constructions—concepts, laws, hypotheses, and postulates—that enter the structure of theoretical physics. He also discussed at length precision measurements, and he frequently displayed data, the record of physical reality. He made clear the suppositions of his science and subjected them to criticism. When addressing his "students" directly, he always spoke of "the study of" theoretical physics; he was teaching them how to think about physical theories, which required them to reflect on the relationship between theory and reality, on physical epistemology.[68]

As electrical power became conveniently available in Königsberg, Volkmann with the help of an extra allowance from the curator connected his "mathematical-physical laboratory" to the central station. In addition to bringing advantages to the institute, convenient electrical power also created a problem for it. Because of the likely disturbance of precision measurements in the institute by electric trams passing by, the authorities ordered determinations of the magnetic field of the institute. Volkmann and a "large number" of students set to work measuring the "vagabond currents" day and night; Volkmann complained to a colleague that he worked "like a horse" on the problem.[69] The threat to the laboratory of electrical disturbances was evidently averted by the curator's decisive action. But little research was done in Volkmann's institute, which was not primarily owing to any inadequacies of the institute.[70] Volkmann did almost no research himself, and he did not attract physicists. He was mainly a physics teacher and popularizer.

When Franz Neumann's heirs left to the university a collection of his apparatus and instruments, which Volkmann regarded as having historical value, particularly those Neumann had used in his work, he made an inventory of them and with a grant for the purpose set them up in a separate room. Volkmann believed that great works of physics like Neumann's could not be done any longer, and although he introduced recent developments in physics into his lectures, his admiration of the past and his penchant for directing students to the "classics" of physics lent physics instruction in his institute aspects of a museum tour.[71] The future theoretical physicist Arnold Sommerfeld attended classes in the Königsberg theoretical physics insti-

67. Volkmann, *Einführung*, iii–vii, 349. In the last part of the text, Volkmann treated the general principles of mechanics introduced since Newton—those of d'Alembert and Lagrange through Gauss and Hamilton—and discussed their agreement with Newton's laws.

68. Volkmann, *Einführung*, 3–4.

69. Volkmann to Sommerfeld, 17 Oct. 1899, Sommerfeld Correspondence, Ms. Coll., DM.

70. In the ten years from 1892, Volkmann reported in the Königsberg *Chronik* only one publication from his institute.

71. Paul Volkmann, *Franz Neumann. 11. September 1798, 23. Mai 1895* (Leipzig, 1896), 27.

tute, but he looked not to Volkmann but to Volkmann's assistant of many years, Wiechert, as his "highest model."[72]

The level of Volkmann's teaching suffered a setback when the professor of mathematics Hilbert and then his successor, Minkowski, left Königsberg in the 1890s. Under their successor, Franz Meyer, mathematics instruction became embarrassingly elementary. Königsberg's reputation had been above all as a "school" of mathematics. Mathematical physics had for a time flourished there along with mathematics, as it could not have without it.

72. Arnold Sommerfeld, "Autobiographische Skizze," AHQP.

21

Munich Chair for Theoretical Physics

When Kirchhoff died in 1887, the theoretical physicist of comparable standing who might replace him at Berlin was Boltzmann, then teaching physics at Graz in Austria—with growing dissatisfaction. Boltzmann's lecture course on elementary physics attracted mainly students of medicine and pharmacy, providing him no "stimulation or leisure for theoretical lectures," and Graz had few students in any case who were prepared for theoretical physics. Boltzmann also felt dissatisfaction with the progress of his research owing to the "burden" of directing the physics institute; in contrast to the typical institute director, who tended to complain of the limitations of his institute, Boltzmann complained that his institute was too big and, for that reason, not really suited for conditions at Graz.[1] In short, Boltzmann was in a mood to leave Graz and to leave his position in experimental physics for one in theoretical physics. So when Berlin offered him Kirchhoff's job, he accepted, or so it seemed. Boltzmann told the Prussian ministry he was coming to Berlin, asked about his teaching assignment and when it was to begin, and even provisionally selected his room, with Kundt's consent, in the Berlin physics institute. But then he began to talk to Berlin of his problems and limitations; he had eye trouble, he had a nervous problem, and he was master of only parts of theoretical physics, not the whole, as Kirchhoff had been. He had private reservations about going to work in Berlin, too, about living among Prussians, who seemed dour to him. In the end, he asked to be released from his commitment to Berlin, even though it meant giving up a salary nearly twice his at Graz.[2]

---

1. Boltzmann to a colleague [Lommel or Bauer] in Munich, 3 Nov. 1889, Munich UA, E II-N, Boltzmann.

2. Boltzmann to Althoff, 6 and 24 June 1888, and Boltzmann to Prussian Ministry of Culture, 24 June 1888, STPK, Darmst. Coll. 1913.51. In Berlin, Boltzmann's salary would have been 13,700 marks, or 8220 fl. In Graz, his salary was 4,240 fl., which the Austrian government raised by 1,000 fl. Schobesberger, "Die Geschichte . . . der Universität Graz," 102.

Establishment of the Munich Chair and Boltzmann's Appointment

Hardly had Boltzmann informed Berlin that he was not coming than Munich approached him, offering him a job like Berlin's, an ordinary professorship for theoretical physics. As it was a new position at Munich, and a nearly unheard-of position
elsewhere, the Munich philosophical faculty had to justify it carefully. The first of
their arguments, which were written up by the physics professor Lommel and the
mathematics professor Gustav Bauer, was that a "progressive separation of theoretical from experimental physics" was becoming noticeable as a natural outcome of
their "difference in methods," calling for a division of labor: "while experimental
physics in its inductive work requires the knowledge and practice of an experimental
art that becomes more and more intricate, theoretical physics uses mathematics in
its deductive process as its main tool and demands intimate familiarity with all
means of this quickly advancing science." With the rapid growth of physics, the
argument continued, there would be fewer and fewer physicists who could master
both methods to the same perfection, which left physicists no choice but to specialize in one or the other and which left universities no choice but to provide specialized chairs. The Munich faculty did not expect, however, that physics would split
into completely separate branches but that experimental and theoretical physics
would "supplement and penetrate one another." Their second argument for the establishment of a chair for theoretical physics was that it would bring to Munich a
physicist whose research belonged to theoretical physics. Several men already taught
theoretical physics at Munich, including an extraordinary professor who gave a year's
course covering the whole subject, which meant that in theoretical physics, students
at Munich were served as well as students at other German universities. But if the
theoretical physics teacher was to do more than simply transmit knowledge of the
subject to students, he needed to be an independent researcher who contributed
"new truths" to theoretical physics; Munich's present lecturers had not done that
and were in any case more interested in experimental physics. The philosophical
faculty wanted Boltzmann.[3]

Boltzmann's researches dealt almost exclusively with theoretical physics, the
Munich faculty noted. Because of his outstanding talent for theoretical researches
together with his "most thorough mathematical education," he was able "to develop
further and to supplement the theories of Maxwell, Clausius, and Helmholtz." His
work in elasticity, hydrodynamics, electricity, and, above all, the mechanical theory
of heat and the kinetic theory of gases had earned him a reputation as one of the
best theoretical physicists. The faculty singled out as examples Boltzmann's papers
on the mechanical meaning of the second law of thermodynamics and on the relationship of that law to probability theory, on the theory of gas friction and diffusion,
on heat conduction and equilibrium in gases, and on the nature and the velocity of
gas molecules.[4]

3. Dean von Baeyer of Section II of the Munich U. Philosophical Faculty to Munich U. Senate,
24 Nov. 1889, Munich UA, E II-N, Boltzmann.
4. Dean von Baeyer to Munich U. Senate, 24 Nov. 1889.

Boltzmann was eager for the new Munich job and for the same reason that the Munich faculty wanted him: as he told them, it brought into "agreement the areas of my teaching and research."[5] By combining the salaries of two lapsed professorships, Munich could afford Boltzmann, and in August 1890 he became their "ordinary professor of theoretical physics."[6]

Physics Teaching at Munich

For some years before Boltzmann joined them, the Munich faculty had tried to improve their facilities for instruction in physics. When Jolly died in 1884, they could not replace him by a renowned physicist because of the poor working conditions at Munich. They first recommended Mach, who not only had published much experimental research but also had mastered "theory in its full compass"; but Prague made an effort to keep Mach, and he stayed. Next Kundt was approached, but with his good institute at Strassburg, he was not tempted to move. So they were reduced to considering physics professors at other Bavarian institutions: Beetz at the Munich Polytechnic, whom the ministry regarded as too expensive and too old, and Friedrich Kohlrausch at Würzburg, who would consider the Munich job only if he were promised an adequate physics institute.[7]

Invited by the ministry to comment again, the Munich faculty explained that students of all faculties had attended Jolly's lectures on experimental physics, the "most frequented" of any lectures at the university. Primarily to insure the continuation of Jolly's teaching success, the faculty wanted a first-rate experimental physicist, and to attract him, they realized, the rooms and instrument collection available for physics instruction would have to be expanded. Since what was really needed was a new physics institute, which required a large sum of money, the faculty appealed to the general usefulness of physics as a cultural good: "physics is neither merely an auxiliary science of medicine nor merely a technical science for natural scientists, but it is a science that belongs to a general education as no other of the so-called natural sciences does to the same degree."[8]

As the faculty's appeal did not generate immediate prospects of a new physics institute, the best Munich could do was to replace Jolly by Lommel, the physics professor at Erlangen, a small provincial Bavarian university with a poor physics institute. By promoting Lommel from Erlangen to Munich after he had received no

5. Boltzmann to a colleague [Lommel or Bauer], 3 Nov. 1889.

6. Boltzmann's salary was 7800 marks. Munich U. Senate to Bavarian Ministry of the Interior, 11 June 1890; letter of appointment by Luitpold, Prince of Bavaria, 6 July 1890, Munich UA, E II-N, Boltzmann.

7. Munich U. Senate to Bavarian Ministry of the Interior, 28 Jan., 20 Apr., and 20 May 1885; Ministry to Senate, 5 May 1885; Mach to Hugo Seeliger, 28 Feb. 1885; Munich UA, E II-N, Boltzmann.

8. There were some specifically Bavarian reasons for this argument for university physics. Whereas elsewhere physics was generally taught in the gymnasium, in Bavaria it was excluded from there, but only because Bavarian university students were required to take eight lecture courses belonging to general education, by which "physics above all is understood." Munich U. Philosophical Faculty to Academic Senate, 3 Dec. 1885, Munich UA. Munich U. Senate to Bavarian Ministry of the Interior, 20 Apr. 1885.

calls from other universities in over a decade, the Bavarian government had resigned itself, at least temporarily, to dropping out of the competition for the first-rate talent it could not afford. After taking on the Munich job, Lommel spent several years pushing through the proposal of a new building for physics and several more years supervising its construction.[9]

Narr, Planck, and Graetz, whom we have mentioned in this connection, had been teaching theoretical physics at Munich before Boltzmann's arrival. The oldest of the three was the Munich graduate Narr, who began teaching the subject in 1870.[10] As a student, he had attended Clausius's lectures at Würzburg on the theories of heat and electricity; his Munich experimental habilitation thesis on the thermal properties of gases contained a "theoretical conclusion" pointing to his indebtedness to Clausius's work on heat theory. Narr regularly lectured on theoretical physics at Munich, and he sometimes taught experimental and laboratory physics and helped out in the seminar.[11] In the 1880s Planck and Graetz, two more Privatdocenten, joined him in regularly offering theoretical physics lectures. Graetz had studied at Berlin, where he was inspired by Helmholtz and Kirchhoff to devote himself mainly to "theoretical physics," at Strassburg, where he learned experimental physics in Kundt's laboratory, and at Breslau, where he graduated in 1879.[12] Planck, after graduating at Munich in 1879 with a dissertation relating to Clausius's mechanical heat theory, qualified as a Privatdocent there the following year with another theoretical work on heat.[13] In a progressive course extending over four or five semesters, Planck lectured on all of the "main subjects of theoretical physics," and he lectured repeatedly on analytical mechanics. He and Graetz together conducted the "physics colloquium" with about ten members. When Planck left for Kiel in 1885, his nine semesters of lectures at Munich, which had drawn an attendance varying between ten and thirty-odd, were regarded as an "excellent success."[14] The Munich

9. Ludwig Boltzmann, "Eugen von Lommel," *Jahresber. d. Deutsch. Math.-Vereinigung* 8 (1900): 47–53, on 49–50. Wilhelm Wien, "Das physikalische Institut und das physikalische Seminar," in *Die wissenschaftlichen Anstalten der Ludwig-Maximilians-Universität zu München*, ed. K. A. von Müller (Munich: R. Oldenbourg und Dr. C. Wolf, 1926), 207–11, on 208–10.

10. Bavarian Ministry of the Interior to Munich U. Senate, 25 Dec. 1870, Munich UA, E II-N, Narr.

11. Narr's "Curriculum vitae" in connection with his Habilitationsschrift, *Ueber die Erkaltung und Wärmeleitung in Gasen* (Munich, 1870), Munich UA, E II-N, Narr. Dean von Baeyer to Munich U. Senate, 24 Nov. 1889.

12. Bavarian Ministry of the Interior to Munich U. Senate, 8 Aug. 1881, and Graetz's "Curriculum vitae," Munich UA, E II-N, Graetz.

13. Planck to Dean of Section II of the Munich U. Philosophical Faculty, 12 Feb. 1879; Dean Seidel of Section II to members of the section, 14 Feb. 1879, with Jolly's and Bauer's evaluations of Planck's doctoral dissertation; Munich UA, OCI-5p. "Max Planck, Promotion. . . . 28 Juni 1879," including Planck's "Curriculum vitae," Munich UA. Planck to Dean of Section II, 28 Apr. 1880, with "Curriculum vitae"; Dean L. Radlkofer of Section II to members of the section, 29 Apr. 1880, with Jolly's evaluation of Planck's Habilitationsschrift and remarks by Bauer; "Protocoll über die Habilitation des Herren Dr. Max Planck als Privatdozent für Physik am 14. Juni 1880"; Munich UA, OCI-6–7. Dean Radlkofer to Munich U. Senate, 14 June 1880; Munich U. Senate to Bavarian Ministry of the Interior, 15 June 1880; Munich UA, E II-N, Planck.
Munich UA, E II-N, Boltzmann.

14. Dean A. Stimming of the Kiel Philosophical Faculty to the Prussian Minister of Culture Gossler, 13 Feb. 1885, DZA, Merseburg.

faculty took the occasion to request an extraordinary professorship for theoretical physics to be given to Planck's replacement. Their request was granted, and Narr was appointed to the position in 1886.[15]

So when Boltzmann arrived at Munich in 1890, Narr and Graetz were already giving lectures on theoretical physics (and Lommel also was giving special lectures on topics in theoretical physics in addition to his main experimental physics lectures).[16] The difference between Boltzmann's position and that of the other theoretical physics lecturers at Munich was that his was independent. He was the director of a state institute and one of the directors of the mathematical-physical seminar.[17] He was Lommel's equal, offering as complete a program in theoretical physics as Lommel did in experimental physics.

The state institute that Boltzmann directed from 1891 was the "mathematical-physical collection." It consisted of apparatus for research, which had earlier served Ohm among other Munich physicists, and which Boltzmann enlarged. His assistant made an inventory of the collection, listing more than a thousand pieces;[18] this well-equipped laboratory enabled Boltzmann to direct "experimental researches" as well as the theoretical ones he could direct in the seminar.[19]

15. The dean of the second section of the Munich Philosophical Faculty recalled that its members had repeatedly argued that the establishment of an extraordinary professorship for theoretical physics was an "undeniable desideratum, which must be granted sooner or later." In support of this appeal, the dean referred to Planck as Munich's "only representative of this discipline," which made the question of a professorship "very urgent." Dean H. Seeliger of Section II to Munich U. Senate, 12 Apr. 1885, Munich UA, E II-N, Planck. Narr was appointed to the newly established extraordinary professorship with the assignment of lecturing regularly on the "discipline of theoretical physics" and directing practice exercises in the institute and in the seminar. Bavarian Ministry of the Interior to the Munich U. Senate, 2 Aug. 1886, Munich UA, E II-N, Narr.

16. Because of illness, Narr was almost constantly on leave from about the time of Boltzmann's appointment. Lommel to Section II of the Munich U. Philosophical Faculty, 23 Oct. 1892, Munich UA, E II-N, Graetz. As he had done before, in 1892 Lommel proposed that Graetz be made extraordinary professor without salary, with the assignment of conducting the laboratory practice course. In 1893, when Narr died, Graetz was promoted; he was made extraordinary professor with the assignment of lecturing regularly on theoretical physics as well as directing the practical exercises in the physics laboratory. He received a salary, too, of 3180 marks. Long before this assignment, as a Privatdocent, Graetz had already lectured and conducted exercises on the "whole area of theoretical physics." Kiel U. Philosophical Faculty recommendation, Feb. 1889, DZA, Merseburg. Boltzmann, "Lommel," 50.

17. Arnold Sommerfeld, "Das Institut für Theoretische Physik," in *Die wissenschaftlichen Anstalten . . . zu München,* 290–91. Bavarian Ministry of the Interior to the Munich U. Senate, 15 May 1891, Munich UA, E II-N, Boltzmann.

18. Boltzmann, report on Ignaz Schütz, 19 Dec. 1893, Munich UA, Acta . . . phys. Cab., Nr. 289. Boltzmann's inventory book was forty-five pages long, listing 1254 different instruments. Of the fourteen classes, optics, Munich's specialty, accounted for the greatest number of instruments, then mechanics, then weights and measures. Many of the instruments, the inventory noted, had been built by famous Munich instrument makers such as Carl August Steinheil and Georg von Reichenbach. Boltzmann, "Inventar der mathematisch-physikalischen Sammlung des Kgl. bayerischen Staates," 1894, Ms. Coll., DM, 1954–52/8.

19. Boltzmann's report on Schütz, 19 Dec. 1893. Schütz assisted Boltzmann in the seminar and in directing experimental researches by students. Broda, *Boltzmann,* 5. The collection did not cause Boltzmann to vary from his practice of having only a few special students and directing only a few researches. Voigt, "Boltzmann," 81.

Boltzmann's Writings and Lectures on Maxwell's Electromagnetic Theory

As conservator of the mathematical-physical collection, Boltzmann added some pieces of electrical apparatus, which included his own so-called bicycle, a model for demonstrating the mutual induction of electric currents.[20] When in 1892 the German Mathematical Society published a catalog of instruments and mathematical and physical models, including Boltzmann's, Boltzmann used the occasion to write up his thoughts on the methods of theoretical physics; models, a recent supplement to lectures in physics, symbolized for him the most promising new method of the field. The older, supplanted method of theoretical physics Boltzmann associated above all with the French, who believed that physical theory "explains" the phenomena and that the electric and material points and central forces on which they built their theories correspond to reality. Boltzmann agreed with Kirchhoff, who corrected the French by pointing out that a theory only "describes" the phenomena, not explains them. By this newer understanding, it was still proper to develop physical theory from mechanical conceptions, as the French had done, but now the mechanisms had to be recognized as "pictures," or "analogies," not as reality. Maxwell had early grasped this point and, by appealing to mechanical analogies, had developed electromagnetic equations of "unbelievable magical power." Maxwell had been impressed by the analogies of nature, by the reappearance of the same plan throughout nature: the same laws or differential equations apply to heat conduction as to the distribution of electricity in conductors, to vortices as to electromagnetic processes in the ether. Maxwell had taught physicists that to understand is to see analogies, and his method became the new method of theoretical physics, one to which, Boltzmann believed, the immediate future belonged. Demonstration apparatus could visually display mechanical analogies, which was why Boltzmann set so much store by them.[21]

At the time of his essay on the new method of theoretical physics, Boltzmann gave great attention to its leading illustration, Maxwell's electromagnetic theory. At the German Association meeting in 1891, Boltzmann reported on it, and at the 1893 meeting he reported generally on the new theories of electricity and magnetism. There, in addition to classifying several ether theories developed by German and British physicists and sketching a new theory of his own, he cautioned against following Hertz's example exclusively. Hertz's presentation in 1890 of the equations of electromagnetism as the experimentally given is useful, Boltzmann acknowledged, but physics would be held back if physicists relaxed their zeal for mechanical theories. Recalling Maxwell's observation that an infinite number of mechanical conceptions are possible, Boltzmann said that a good number had already been tried and still more should be; for mechanical theories have great value for illustrating and for discovering facts, as Maxwell's own career of discovery had shown.[22]

20. Sommerfeld, "Das Institut für Theoretische Physik," 290.

21. Ludwig Boltzmann, "Über die Methoden der theoretischen Physik" (1892), in *Populäre Schriften*, 1–10.

22. Ludwig Boltzmann, "Ueber die neueren Theorien der Elektrizität und des Magnetismus," *Verh. Ges. deutsch. Naturf. u. Ärzte* 65 (1893): 34–35, reprinted in *Wiss. Abh.* 3: 502–3.

To the Bavarian Academy in 1892, Boltzmann presented a paper on the mechanical properties of an ether characterized by Maxwell's electromagnetic equations. He represented the ether by two mechanical analogies, the first as a continuous, incompressible fluid with mass and inertia, but without weight, and the second as a homogeneous, isotropic, elastic solid. By use of the laws of continuum mechanics and the energy principle, Boltzmann showed that the forces acting within the fluid and the solid give rise to equations that, when the symbols are appropriately interpreted, are identical with Maxwell's. There is, of course, no suggestion in Boltzmann's analysis that the ether really is an incompressible fluid or elastic solid.[23]

At Munich Boltzmann lectured on Maxwell's electromagnetic theory, publishing, at the request of his students, excerpts from his lectures in two texts in 1891 and 1893.[24] Like Hertz, Boltzmann had no quarrel with the "content" of Maxwell's theory but only with its "form," which in Maxwell's presentation remained "obscure." Like Hertz again, he saw in Maxwell's *Treatise on Electricity and Magnetism* a confusing mixture of old and new electrical concepts. He recommended reading Maxwell's earlier papers on electromagnetism to grasp the meaning Maxwell intended for "electricity" and "dielectric displacement" and to see that he did not proceed from distance forces. Boltzmann spoke of himself as an "interpreter" of Maxwell.[25]

Boltzmann had already worked with mechanical analogies in the mechanical theory of heat and the kinetic theory of gases; moreover, as we have noted, he had recently become interested in Helmholtz's monocyclic analogy of the second law of heat theory, which he developed further in several papers and compared with his own earlier attempt at a mechanical reduction of the second law.[26] In the first part of his Munich lectures on Maxwell's electromagnetic theory, Boltzmann interpreted as Helmholtz's "cyclic" motion the motion in the ether and in ponderable bodies giving rise to the phenomena of electric currents. Boltzmann studied this motion by means of Lagrange's equations, an approach that reflected Boltzmann's great admiration for Maxwell's Lagrangian mechanical derivation of actions between electric currents; that derivation did not depend on any detailed hypotheses about the mechanisms responsible.[27] For a "bicycle," a system of masses for which the motion and

---

23. Ludwig Boltzmann, "Über ein Medium, dessen mechanische Eigenschaften auf die von Maxwell für den Electromagnetismus aufgestellten Gleichungen führen," *Sitzungsber. bay. Akad.* 22 (1892): 279–301, reprinted in *Wiss. Abh.* 3: 406–27.

24. Ludwig Boltzmann, *Vorlesungen über Maxwells Theorie der Elektricität und des Lichtes*, vol. 1, *Ableitung der Grundgleichungen für ruhende, homogene, isotrope Körper* (Leipzig, 1891); vol. 2, *Verhältniss zur Fernwirkungstheorie; specielle Fälle der Elektrostatik, stationären Strömung und Induction* (Leipzig, 1893). Boltzmann explained the genesis of his published lectures in their expanded English version: Boltzmann, "Preface" to Charles Emerson Curry, *Theory of Electricity and Magnetism* (London, 1897).

25. Boltzmann, *Maxwells Theorie* 2: iii–iv.

26. Klein, "Mechanical Explanation," 70–71.

27. In the motion that Helmholtz called "cyclic," the kinetic energy of a system does not depend on any cyclic coordinate l, but it does depend on the velocity of the cyclic motion l'. It may depend, too, on slowly varying coordinates k, whose time derivatives are negligible. For a "cycle," defined as a system of masses whose velocities at any point are determined solely by the slowly varying coordinates k and the single velocity l' of the driving point, Lagrange's general equations of motion may be applied even though the real mechanism of the system is unknown: $L = d/dt \cdot \partial T/\partial l' + W$, where T is the

position are determined by two "cyclic" coordinates, Boltzmann wrote the Langran-
gian equations of motion. He then illustrated these equations by a model, which
was essentially two independent cranks linked by a movable middle shaft; representing the two cyclic coordinates by the angular positions of the cranks, he showed
that the motion of the bicycle forms a complete analogy with two interacting electric
currents. But lest the reader confuse the apparatus with reality, Boltzmann returned
to the methodological lesson with which he introduced the mechanical analogy: the
mechanism of the electric current is completely different from that of the apparatus
and, moreover, is completely unknown to us. That the apparatus has only a gross
analogy with nature does not detract from its heuristic value, however; it offers the
advantage of working with a well-defined mechanical system, which for Boltzmann
is a considerable advantage.[28]

In the second volume of his lectures on Maxwell's theory, Boltzmann proceeded
from the hypothesis that the ether communicates electric and magnetic actions from
volume element to neighboring volume element. Each element supposedly contains
an unknown motion causing a displacement, which Boltzmann called the "electro-
tonic state," following Faraday's terminology. As a starting point for the imagination, Boltzmann conceived of the displacement as the turning angle of a nucleus
within each element, any change of which gives rise to a kinetic energy. To represent the potential energy of an element, Boltzmann fell back on Maxwell's idea that
between two rotating nuclei are particles, which serve as friction rollers; the potential energy is then proportional to the work done by the displacement of these particles. By substituting the terms for the kinetic and potential energies into the equation that expresses Hamilton's principle, Boltzmann derived one set of Maxwell's
equations; the other set simply states the assumptions of the derivation.[29]

It was part of Boltzmann's intention in his lectures to show how ideas from the
distance-action theories relate to Maxwell's theory and to treat special problems by
Maxwell's theory as thoroughly as they were treated by the other theories, making
it unnecessary for students to go to the older textbooks any longer. In the language
of potentials, Boltzmann compared Maxwell's equations with those of the older theories. As the potentials are expressed as integrals over all space rather than in terms
of states of neighboring volume elements, Boltzmann referred to the integral equations as "Maxwell's distance-action equations" in contrast to his "contact-action"
partial differential equations. Boltzmann went on to derive Neumann's induction
law, Helmholtz's general theory, and other parts of the older electrodynamics.[30]

kinetic energy, W is the resistance to the cyclic motion, and L is the force tending to increase l. In the
analogy, the electric current is a cycle, so that l' increases with current intensity and L with the electromotive force, and W measures the electric resistance. The form and position of the conductor and the
relative positions of nearby iron masses are determined by k, and if they change, the motion l' changes.
Boltzmann, Maxwells Theorie, vol. 1, lecture 3.

28. Boltzmann approved of Maxwell's expression "dynamical illustration" for the mechanisms that
could be imagined to represent the electromagnetic field. He discussed this and related points in Maxwells
Theorie 1: 13, 35, and elsewhere in his lectures.

29. Boltzmann, Maxwells Theorie, vol. 2, lecture 1.

30. Boltzmann, Maxwells Theorie, vol. 2, foreword and lectures 12 and 13.

To develop the consequences of Maxwell's equations, Boltzmann introduced new "mechanical pictures," which he distinguished from the "mechanical foundations" of the theory. In general, throughout his Munich lectures, Boltzmann constantly reminded his audience that he was doing theoretical physics according to the preferred new method, not describing reality. He showed, that is, how to build theories of physics—Boltzmann's way.[31]

End of Boltzmann's Position

Early in 1893 Stefan at Vienna University died, and Austria tried to get Boltzmann to return to Vienna as his successor in the "ordinary chair of theoretical physics." To hold him in Munich, the Bavarian government hastened to improve his position, giving him a substantial raise, a title, and an assistant. Boltzmann, claiming to be indebted to Bavaria for this generosity, declined the Austrian offer, but only for the present. He wanted the chair of theoretical physics in Vienna, but not until the fall of 1894. Meanwhile he quietly continued to negotiate with Austria, naming terms that were quite in keeping with his high reputation, even in the eyes of the Austrian ministry of culture. It was a "question of honor" to them to bring back a native of Austria who, not only in Austria but also in Germany, was considered one of the "most outstanding representatives of the subject of theoretical physics." If they did not grant his request for a salary of 9000 florins (plus another 2000 florins in additional income), they did offer him the highest salary then paid to any Austrian university professor, 6000 florins, and eventually they added enough income from other sources to bring the total up to just over 9000 florins anyway. During the negotiations with Boltzmann, Stefan's teaching duties had been assigned to an extraordinary professor for theoretical physics, Gottlieb Adler. When Adler died in the spring of 1894, Boltzmann and the Austrian government reached a final agreement, and Boltzmann took the occasion to break the news to the Bavarian government that he would be leaving. In July he obtained the official release from his duties at Munich, where he had taught theoretical physics for four years.[32]

Boltzmann's departure again left Munich without a theoretical physicist on its faculty. Among the existing "small number of representatives of theoretical physics,"

31. Boltzmann, *Maxwells Theorie* 2: 22, 50. Boltzmann discussed other new methods of doing theoretical physics and their limitations, too. He acknowledged the value of Hertz's acceptance of Maxwell's equations as the empirically given, but he pointed to Hertz's failure to give a clear insight into the "inner connection"; by contrast, Maxwell's mechanical analogy gives a "deep insight into the inner connection" of the facts of electromagnetism. Boltzmann also appreciated the value of Kirchhoff's way of doing theoretical physics, which was to find particular integrals of general differential equations and then to look for their physical significance; but Boltzmann preferred his own way, which was to look for the physical reasons why certain particular integrals are significant and others not, which offered advantages to the understanding. Boltzmann, *Maxwells Theorie* 2: 42, 113–14.

32. Reports to the Austrian Ministry of Culture and Education by ministry officials on negotiations with Boltzmann, 14 July 1893, 3 Dec. 1893, and 30 May 1894, including drafts of letters written to Boltzmann, and Boltzmann's reply, 26 Dec. 1893, Öster. STA, 4 Phil, Physik, 1375/1849. Boltzmann to the Munich U. Rector, 9 May 1894, Munich UA, E II-N, Boltzmann. Dean Zittel to the Munich U. Philosophical Faculty, 21 May 1894, Munich UA, OCI 20. Boltzmann received his official dismissal on 14 July 1894.

the Munich faculty first considered Voigt or Planck as the potential successor to Boltzmann. Planck they especially wanted but believed they had little hope of getting. (On Boltzmann's suggestion, they considered Lorentz, too.) Because of the difficulty of finding a suitable theoretical physicist, Lommel suggested that they convert the position from theoretical physics to physical chemistry, which was not yet represented at Munich, and offer it to Nernst.[33]

If they were going to hire a theoretical physicist at all, he had to be someone of Boltzmann's stature, since the Munich faculty had defined their position as one for an outstanding researcher who would teach the methods of theoretical research. For a time, their problem appeared to be solved in 1896, since Boltzmann wrote to them from Vienna that he would gladly return if he could receive the salary offered him in 1894 to keep him in Munich. The faculty commission for the appointment accordingly made up a new list of two candidates, Boltzmann and Voigt, but since the philosophical faculty wanted Boltzmann, they formally asked only for him. Boltzmann, they reasoned, had obeyed a "patriotic duty" to return to Vienna, but now having reconsidered, he preferred Munich after all. They urged that every effort be made to recall him, since Boltzmann was the "uncontested first representative" of theoretical physics and was "recognized as such by all nations."[34] The Bavarian ministry of the interior, however, refused to offer the same job again to Boltzmann.[35] The faculty persisted; when Lommel died in 1899, they proposed as his successor Boltzmann and Lorentz, in that order. If they could not have Boltzmann or Lorentz as their theoretical physicist, they would have one or the other as their experimental physicist. Boltzmann, they said, was the first European physicist "in general," known not only as a "theorist" but also as a "teacher of experimental physics." Their third candidate was Röntgen, who got the job in the end.[36]

Boltzmann's position in theoretical physics at Munich remained vacant until well after the turn of this century. The ministry allowed Boltzmann's mathematical-physical collection to be used for research and maintained its budget and assistant's salary and even gave the collection an extraordinary grant.[37] Theoretical physics lectures continued to be given at Munich by Graetz and, beginning in 1895, the year after Boltzmann left, by the Privatdocent for "theoretical physics" Arthur Korn.[38]

33. The Munich faculty proposed H. A. Lorentz in first place, Planck in second, and Walther Nernst in third. Nernst was actually called first, and he declined. Minutes of the meeting of Section II of the Munich U. Philosophical Faculty on 6 and 13 June 1894, 4 May 1895, and 5 Mar. 1896, Munich UA, OCI 20, 21, and 22, respectively.

34. Minutes of the meeting of Section II of the Munich U. Philosophical Faculty, 30 Apr. 1896, Munich UA, OCI 22.

35. Bavarian Ministry of the Interior to the Munich U. Senate, 21 May 1896, Munich UA, OCI 22.

36. Letter containing the faculty's list of candidates for the experimental physics chair at Munich: Dean Lindemann of the Munich U. Philosophical Faculty to Academic Senate, 19 July 1899, Röntgen Personalakte, Munich UA.

37. General Conservatory of the Scientific Collections of the State to the Munich U. Senate, 31 July 1894, Munich UA. Bavarian Ministry of the Interior to the General Conservatory, 21 Mar. 1896, Munich UA, OCI 22.

38. Dean to Munich U. Philosophical Faculty, 1 July 1895, with Lommel's evaluation of Korn's work, 11 July 1895, Munich UA, OCI 21.

The Place of Theoretical Physics within German Physics in the 1890s

Boltzmann's top salary reflected the value placed on top researchers in physics, their rarity, and their reluctance to accept the working conditions of theoretical physics jobs. They could, after all, command institutes that went with chairs in experimental physics. Boltzmann was unusual in being more than content with institutional arrangements his colleagues would consider inadequate.

The rarity of physicists, especially good physicists, who specialized in theoretical research, which the Munich faculty noted, was seen by physicists as part of a general neglect of theoretical physics. The reasons, as they saw them, had to do with the preference of physicists for working in experimental physics. From at least the mid-1890s, Voigt had complained that theoretical physics in Germany was not valued as highly as experimental physics, as if, he said, brilliant experimental results were possible without theoretical preparation. Asked for recommendations of young theoretical physicists for a position in 1899, Voigt could not come up with many names and fewer yet that he could feel enthusiastic about. The reason he gave was that "Kundt's purely experimental direction" had become the "standard in Germany." Young physicists scarcely valued "exact theory"; among them there were no theorists as good as Planck, Wien, Drude, and Sommerfeld.[39] Voigt wrote to Sommerfeld in 1899 that he was afraid Sommerfeld, who was about to begin teaching engineers at the technical institute in Aachen, would be lost to theoretical physics. He wrote again in 1902 to say that he rejoiced that Sommerfeld was still working in theoretical physics and that he expected him to "help reconquer for *us* the world position lost since Kirchhoff etc."[40] In a letter to Sommerfeld in 1898, Wien analyzed the state of German theoretical physics in terms similar to Voigt's:

> Theoretical physics in Germany lies as good as completely fallow. That ought to be a reason to help revive it, but the situation has already gotten to the point that even the need for theoretical physics disappears more and more. The reasons for this are, first, that physicists do almost nothing but pure experiments and have hardly any interest in theory and, second, that most mathematicians have turned to entirely abstract areas and do not concern themselves with applications. This reveals itself externally in that pure theoretical physics is taught from only two chairs (Berlin and Göttingen), and such an important chair as Munich has come entirely to an end. Theoretical physics currently finds no takers. Later it will all be different again, indeed, because otherwise physics would go completely to ruin; but I must make some allowance for trends and thoroughly busy myself with purely experimental researches as long as I still have to work for an external position.[41]

Wien was an extraordinary professor at Aachen waiting for his first call to a

39. Voigt to a colleague, 6 June 1899, STPK, Darmst. Coll. 1923.54.
40. Voigt to Sommerfeld, 3 Dec. 1899 and 24 Nov. 1902, Sommerfeld Correspondence, Ms. Coll., DM.
41. Wien to Sommerfeld, 11 June 1898, Sommerfeld Correspondence, Ms. Coll., DM.

university professorship, and positions in physics and how to get them were much on his mind. His remarks on the subject are revealing, for they show that a physicist in the 1890s could take it for granted that universities ought to have positions for theoretical physics. The point had been made that theoretical physics is a distinct and indispensable part of the physics curriculum.

It is a measure of the slow growth of theoretical physics as an independent teaching subject that, of the ordinary professors at the end of the nineteenth century, only Planck at Berlin, Voigt at Göttingen, and Volkmann at Königsberg had institutes for it, and that only two of these places, Berlin and Göttingen, as Wien knew, were worth mentioning. Moreover, theoretical physics might still lose ground by being reduced from an independent position to a secondary one, as had recently happened at Munich, a loss Wien felt keenly.

Below the level of chairs, many subordinate jobs in physics had been established in the universities from the 1870s, as we have seen. They assigned the teaching of theoretical physics to Privatdocenten and extraordinary professors, whose duty it was to support the teaching of the increasingly burdened, main physics professors. They met, that is, a practical need. They did not represent a response by physicists and government officials, certainly not at the beginning, to the recognized needs and claims of a new, full-fledged discipline, theoretical physics. Gradually the awareness of such needs did come, and that awareness was in large part responsible for Voigt's and Wien's sense of an imbalance in German physics and their feeling of dissatisfaction. The existence of the many subordinate jobs in theoretical physics was no longer seen to be sufficient.

At the end of the nineteenth century, most physicists were trained by physics professors whose work belonged to experimental physics. By this time there had come to be a good deal of opportunity for experimental research, since university physics institutes were increasingly well equipped for the purpose. The normal route to a career remained research in experimental physics, for which the best jobs, the directorships of the big physics institutes, now seemed to be reserved. With the recent advances in instruments and experimental methods, on the one hand, and in mathematical methods, on the other, physicists such as Kirchhoff who possessed a comparable mastery of experiment and theory inevitably appeared on the scene less often. What was needed to answer Voigt's and Wien's complaints was more good physicists who cultivated theory and, equally important, who directed institutes for training more theoretical researchers like themselves. Not long after Voigt's and Wien's strictures on German physics, in the early years of the twentieth century this development in theoretical physics was to be realized.

We conclude our discussion with two tables: table 1 summarizes the budgets of German physics institutes, table 2 the physics courses taught throughout Germany. They show that when Boltzmann was teaching theoretical physics at Munich, the subject was accepted everywhere as a necessary part of physics instruction. They also show that it was still taught largely by subordinate teachers. Boltzmann's presence at Munich was conspicuous, and his subsequent absence there was equally conspicuous.

Tables 1 and 2

TABLE 1. **Summary of the Budgets of Physics Institutes at German Universities and Technical Institutes in 1891**

	Aachen T.	Berlin U.	Berlin T.	Bonn U.	Braunschweig T.	Breslau U.	Darmstadt T.	Dresden T.	Erlangen U.	Freiburg U.	Giessen U.	Göttingen U.
Budget:	3000		3000	4400	2710	3762	1430	2700	2450	2900		4910
Is it sufficient?		no	yes	yes	no	no	no	no				no
Are you expected to cover these costs?												
Small building repairs:												
Changes in gas and water lines:										yes		
Building appliances, furniture:				yes		yes			yes	yes		yes
Gas, water, heat, cleaning:					yes 2000							
Assistant and servant salaries:				460	1500	144						
Number of assistants:	1		1	1		2	1	2	1	1		2
Salary of assistants:	2300		1800	1200	1200 1200	1000			1250	1200		1200 1200
Number of servants:	2		1	2	1	2	2	1½	1	1		2
Salary of servants:	1800 1350		1380	1200 960	1500	1500 900	1100 900		750	545+ f.l.q.		1100 600
Central heating:	yes		yes		yes	yes		yes		yes		
Separate building:							i.p.			yes		yes
Professor for electrotechnology:	yes		yes		yes			yes	yes			
Professor for theoretical physics:	e				e		e	e				yes
Professor for physical chemistry:												
Attendance:[1]	797	4611	1640	1386	273	1342	316	403	1678	1138	562	831
Adjusted attendance:	797	2305	1640	693	273	671	316	403	539	569	281	415
Adjusted budget:[2]	3000		3000	1740	1210	3418	1430	2700	2250	2500		4710
Budget per 100 students:	1500		183	250	445	500	455	675	415	437		1110
Adjusted number of assistants:[3]	3		2	2	1	3	2	4	1	1		3
Adjusted number of servants:[4]	3		2	2	2	3	2	2	1½	2		2

Greifswald U.	Halle U.	Hannover T.	Heidelberg U.	Jena U.	Karlsruhe T.	Kiel U.	Königsberg U.	Leipzig U.	Marburg U.	Munich U.	Munich T.	Münster A.	Rostock U.	Strassburg U.	Stuttgart T.	Tübingen U.	Würzburg U.
	1325+	1450	4000	4500	1100	3200	5270		4280	2143 (4000)	3100	1620		6000	2000	6600	
	no	yes	no	yes	no	no	no		no	no	yes?			yes	no	no	
			yes		yes				yes	yes							
			yes	yes	yes					yes	yes			yes			
			yes	yes	yes	yes	yes			yes		yes		yes		yes	
			yes	yes	clean.	yes	1260		1500					water		2000	
		1800	1700				1950									yes	
	1	1	1	1	1	2	1	2	2	1	2	1		3	1	1	
	1200	1350	1000	800+ f.l.q.	1200	1080+ f.l.q. 480	1200		1200	1200		1200		1425+ f.l.q. 900+ f.l.q. 400	1820	1280	
	1	1	1	1	2	1	2	2	1	2		1		3	1	1	
600			800+ f.l.q.	900+ f.l.q.	1600 800	1280+ f.l.q.	750+ f.l.q.		1000	1398 1398		1200		1800+ f.l.q. 1350+ f.l.q. 1050	1930	1560	
	yes	yes													yes	yes	
	yes		yes		yes				yes	i.p.	yes	i.p.		yes		yes	
		yes									yes				yes		
yes			yes	yes	yes				e	yes+ e				e		e	
								yes									
834	1483	580	1089	645	585	605	717	3242	952	3551	882	377	368	917	486	1393	1422
417	741	580	544	322	585	302	358	1621	476	1775	882	188	184	458	486	696	711
	1325+	1450	−400	2400	400	1000	1860		2580	3400	2900	1420		5500	2000	1760	
	178+	250	−74	745	67	330	515		545	191	325	755		1120	470	253	
	2+	2	1	2	1	3	2	3	3	3	3	1		4	2	2	
	2+	2	1	1	2	1	2	2	1	2	1	1		3	2	2	

NOTES FOR TABLE 1 (*preceding pages*)

The attendance at the time of the survey. 2. Budget minus operating expenses of the institute. 3. Assistants plus number of remaining physics chairs. 4. Servants plus 1 if there is central heating. Under salaries of assistants and servants, "f.l.q." stands for "free living quarters in the institute"; under professor for theoretical physics, "e" stands for "extraordinary professor"; and under separate building "i.p." stands for "in prospect." Under Erlangen University's central heating, there is an entry that reads, evidently, "Hausm.," which we have omitted from the table. The figures are in marks. We have not altered them in any way, nor have we corrected Lehmann's arithmetic.

In 1891 Otto Lehmann gathered this information by direct inquiry. Having recently moved to the Karlsruhe Technical Institute, he wanted to improve the arrangements for physics there; his survey showed, for example, that Karlsruhe ranked below the other schools in expenditure per student. The information is clearly incomplete: for Berlin, Giessen, and Leipzig Universities, Lehmann noted that the data were "obtainable only from the government"; for Greifswald University, "the institute is completely newly organized, request not yet approved"; for Rostock University, "the head of the institute is very ill and can't reply"; for Würzburg University, "no answer received." Lehmann also noted that for Halle University, "there exist two physics institutes, forgot to ask one of them"; and that for Dresden, where Lehmann had taught before moving to Karlsruhe, "data according to my recollection." The staff and budget details that are summarized in the table contain matters we discuss throughout this study, institute by institute, period by period. We reproduce Lehmann's summary here to give an impression, such as a German physics professor could gain by going to some trouble, of physics institutes throughout Germany near the end of the nineteenth century. The time, 1891, corresponds to our last survey of research in the *Annalen der Physik*. The table, dated by Lehmann 26 June 1891, is in Bad. GLA, 235/4168.

We will make several observations about the line in the table for theoretical physics, our subject. Two of the universities, Berlin and Giessen, for which the information of the table is incomplete, had extraordinary professors for theoretical physics in 1891. Königsberg also had one; its omission here, we suspect, owes to Lehmann's neglect to write to the second physics institute at Königsberg. The responses from Jena and Kiel of "yes" instead of "e" might be misunderstood; both universities had only extraordinary professors for theoretical physics at this time. Tübingen's response of "e" was premature; that year the Tübingen senate proposed that their Privatdocent for theoretical physics be promoted to extraordinary professor, but they were refused, and the promotion did not go through until 1895. To sum up the survey as far as it concerns theoretical physics at universities: in 1891 twelve of Germany's twenty universities had either an extraordinary or an ordinary professor for theoretical physics or, in the case of Munich, both.

NOTE FOR TABLE 2 (*opposite page*)

Announced for the summer semester 1892 and the winter semester 1892–93. The numbers beside the teachers' names give the hours per week required by the course. The letter (d) indicates that the teacher is a Privatdocent, the letter (e) that he is an extraordinary professor, and no letter that he is an ordinary professor. Free courses are indicated by (gr). Reprinted from Lexis, ed., *Die deutschen Universitäten* 2:164–65.

TABLE 2. **Summary of Physics Courses at German Universities in 1892–93**

Universitäten	Experimental-Physik	Theoretische Physik	Theorie der Elektricität und des Magnetismus	Wärmetheorie	Praktische Uebungen im Laboratorium: I. für Anfänger II. f. Geübtere
Berlin	Kundt 5, König (e) 4.	v. Helmholtz 4, Planck 4, Glan (d) 2, Rubens (d) 2.	Glan (d) 4.	Weinstein (d) 3.	Kundt I. 7, II. 39. Planck (Institut für theoretische Physik) 2.
Bonn	Hertz 4.	Lorberg (e) 4.	[Lorberg (e) 4.]	Lorberg (e) 2 gr.	Hertz I. 8, II. 54.
Breslau	Meyer 6.	Dieterici (e) 5.	—	[Dieterici (e) 4.]	Meyer 3 u. 6. Dieterici (e) 3 u. 6.
Erlangen	Wiedemann 5.	Ebert (d) 2.	[Ebert (d) 2.]	[Knoblauch (d) 2].	Wiedemann II. 40.
Freiburg	Warburg 5.	Warburg 2, Meyer (d) 2.	—	—	Warburg II.
Giessen	Himstedt 5.	[Fromme (e) 3.]	—	Fromme (e) 3.	Himstedt I. 12, II. täglich.
Göttingen	Riecke 4.	Voigt 5, Drude (d) 2.	[Drude (d) 2.]	—	Riecke II. 48 u. 4.
Greifswald	Holtz (e) 4.	[Oberbeck 2.]	Oberbeck 4.	—	Oberbeck I. 6, II. täglich.
Halle	Knoblauch 4.	[Dorn 2.]	Dorn 4.	[Dorn 4.]	Dorn 6.
Heidelberg	Quincke 5.	Quincke 3, Eisenlohr (e) 4.	—	—	Quincke II. täglich.
Jena	Winkelmann 5, Schaeffer (h) 4.	[Auerbach (e) 1.]	Auerbach (e) 3.	—	Winkelmann II. 48.
Kiel	Karsten 6.	[Weber (e) 2.]	—	Weber (e) 3.	Karsten Weber } 20.
Königsberg	Pape 5.	—	Volkmann (e) 4.	[Wichert (d) 1.]	Pape. Volkmann (e).
Leipzig	Wiedemann 6.	[Des Coudres (d) 2.]	—	—	Wiedemann 39.
Marburg	Melde 5.	[Feussner (e) 4.]	s. theor. Physik.	Elsass (e) 2.	Melde 12. Feussner (e) 12.
München	Lommel 5.	Grätz (d) 4.	Boltzmann 4.	[Grätz (d) 4.]	Lommel Narr } 15.
Münster	Ketteler 4.	Hittorf 3 (gr.)	Ketteler 2 (gr.)	[Hittorf 3].	Ketteler 9.
Rostock	Matthiessen 5.	—	—	Mönnich (d) 2.	Matthiessen 24.
Strassburg	Kohlrausch 5.	[Cohn (e) 3.]	Cohn (e) 3.	[Hallwachs (d) 2].	Kohlrausch I. 12, II. 39.
Tübingen	Braun 5.	[Waitz (e) 3.]	Waitz (e) 3.	[Waitz (e) 3.]	Braun I. 4, II. täglich.
Würzburg	Röntgen 5.	Heydweiller (d) 2.	Selling 4, Heydweiller (d) 2.	Geigel (d) 2.	Röntgen I. 10, II. täglich.

Bemerkungen

Ferner war angekündigt:

Die elektromagnetische Theorie des Lichts: Berlin Wien (d) 2. — Kinetische Gastheorie: Berlin Pringsheim (d) 2; Freiburg Zehnder (d) 1; Giessen Fromme (e) 3. — Ueb. elektr. u. magnet. Messmethoden: Berlin Arons (d) 2; Freiburg Zehnder (d). — Hydrodynamik: Erlangen Knoblauch (d) 2. — Krystalloptik in mathemat. Behandlung: Göttingen Pockels (d) 2. — Verwendung d. Elektricität in der Technik und Medicin: Halle Schmidt (d) 2. — Induct. u. Dynamomasch.: Kiel Hagen (d) 2. — Spektralanalyse u. ihre Verwendung: Halle Schmidt (d) 1. — Physikal.-chemische Theorien: Heidelberg Horstmann (h) 2; Leipzig Le Blanc (d) 2. — Diffusion des Lichts: Jena Abbe (h) 3. — Photometrie: Kiel Weber (e) 17. — Interferenz u. Doppelbrechung des Lichts: Leipzig des Coudres (d) 2; München Donle (d) 2. — Theorie des Mikroskops und seine Anwendung: Leipzig Ambronn (e) 2. — Photogrammetrie: Heidelberg Wolf. — Grundzüge der Elektrostatik: Strassburg Hallwachs (d) 2. — Potentialtheorie: Heidelberg Eisenlohr; Jena Auerbach 3 (s. Tab.); Kiel [Weber (e) 2]; Königsberg [Volkmann (e) 4]; München Boltzmann 3; Strassburg [Reye 4]; Würzburg Selling 4 (s. Tab.).

Die physikalisch-mathematischen Seminare s. u. — Ausserdem waren noch Uebungen, Repetitionen etc. angekündigt in Berlin (u. a. Praktischer Kursus für Mediciner: Kundt 3), Bonn, Göttingen, Halle, Erlangen, Giessen, Heidelberg, Jena, Leipzig, Strassburg, Tübingen, Würzburg.

Oeffentliche Vorlesungen in Berlin (5), Greifswald (1), Halle (2), Jena (2), Kiel (2), Königsberg (3), Marburg (3), Münster (2), Strassburg (2).

TABLE 2. **Summary of Physics Courses at German Universities in 1892–93**

22

Theoretical Physics at Leipzig

The first major institutional development in theoretical physics in Germany in the twentieth century was the establishment of an institute at Leipzig University. To show how this development came about and what it meant, we first discuss the earlier establishment at Leipzig of an extraordinary professorship for theoretical physics and the teaching and research of the man who held it. We then discuss the founding of the institute and the work of its first directors.

Drude's Early Research and Teaching

To help with the training of scientists and teachers, Leipzig University asked for and received an extraordinary professorship for theoretical physics in 1894. The first appointment to the new position went to Hermann Ebert, who left for a better position after only a few months. To continue without a break the lectures and exercises Ebert had introduced, with "good success," Leipzig urged the government to appoint a successor promptly. Passing over their Privatdocent Theodor Des Coudres as too young, they recommended Paul Drude, who was already well known to them since they had just compared him (unfavorably) with Ebert. Their initial objection to Drude had been what they regarded as an undue scientific dependence on his teacher Voigt and a hurried, incomplete appearance of some of his publications. In the meantime, however, their opinion of Drude had sharply risen with the appearance of his book *Physik des Aethers* (*Physics of the Ether*), which persuaded them that Drude now stood on his own two feet and had, moreover, a talent for clear and elegant exposition. Drude was offered the job, which he accepted.[1]

In Drude, Leipzig acquired a physicist well schooled in theoretical methods.

1. Leipzig U. Philosophical Faculty to the Saxon Ministry of Culture and Public Education, 1 Aug. 1894; Saxon Ministry to the Philosophical Faculty, 27 Aug. 1894; Drude Personalakte, Leipzig UA, PA 422. Drude to Heinrich Kayser, 10 Aug. 1894, STPK, Darmst. Coll. F 1 c 1897.

Originally intending to graduate in mathematics, Drude had turned to physics under Voigt's influence, writing a thesis under his direction in 1887. That same year he became Voigt's assistant in the Göttingen mathematical physics institute, and in 1890 he became Privatdocent as well.[2] In the years after his graduation, in addition to performing his duties for Voigt, he did a good deal of research while he waited, with growing impatience, for recognition and a call.[3]

Drude's purely theoretical dissertation, which he developed from Voigt's general theory of the motion of light in absorbing crystals, concerned the boundary conditions for reflection and refraction of light at the interface of absorbing crystals. Since Voigt's institute was equipped with measuring apparatus for testing theories of this type, in his next research Drude experimentally tested his theoretical conclusions using one of Voigt's crystals. This combination of theoretical study and laboratory measurement to which Voigt devoted his institute was to be "characteristic of all of Drude's scientific activity," earning him the reputation of being one of the few physicists who still "mastered equally theory and experiment."[4]

Drude graduated the year Hertz began publishing his experimental researches on Maxwell's theory, and in general Drude's formative years as a researcher coincided with the introduction of Maxwell's theory into Germany.[5] Drude did not immediately embrace Maxwell's theory; in his early optical work, he viewed the ether as a continuous, elastic solid and light as the mechanical oscillations of this body. Since the mechanical theory accounted for most optical phenomena in a consistent way, Drude argued, there was no urgent, inherent reason for its abandonment. With its special hypotheses, Maxwell's theory was to be approached gradually and critically, and Drude examined it together with the mechanical theory to decide what was essential to them both.[6]

In response to the seemingly endless and irresolvable controversies in optics, Drude published a major study in 1892 to determine the degree to which the contending theories of light "satisfy the requirements of practical physics." Physicists still debated the merits of optical theories from the early nineteenth century as well as their sequels by Kirchhoff, Voigt, and others, with the result that theoretical optics had come to seem mainly mathematical and philosophical speculation, bringing discredit to the whole subject.[7]

To this inconclusive "mathematical-philosophical" direction in theoretical optics, Drude opposed what he called the "practical-physical," which requires of an optical theory only that it describe the phenomena qualitatively in the most "eco-

2. Max Planck, "Paul Drude," Verh. phys. Ges. 8 (1906): 599–630, reprinted in Planck, Phys. Abh. 3: 289–320, on 291. Dean of the Göttingen Philosophical Faculty to Curator Ernst von Meier, 18 Jan. 1890, Drude Personalakte, Göttingen UA, 4/V c/205.

3. Woldemar Voigt, "Paul Drude," Phys. Zs. 7 (1906): 481–82, on 481.

4. Franz Kiebitz, "Paul Drude," Naturwiss. Rundschau 21 (1906): 413–15, on 413.

5. Walter König's talk, published together with Franz Richarz's talk, in Zur Erinnerung an Paul Drude (Giessen: A. Töpelmann, 1906), 18–19. Planck, "Drude," 298. Max Laue, "Paul Drude," Math.-Naturwiss. Blätter 3 (1906): 174–75, on 175.

6. Planck, "Drude," 291–92.

7. Paul Drude, "In wie weit genügen die bisherigen Lichttheorieen den Anforderungen der praktischen Physik?" Gött. Nachr., 1892, 366–91, 393–412, on 366.

nomical" way and quantitatively through mathematical aids. The partial differential equations and associated boundary conditions that accomplish this task Drude termed an "explanatory system." His "method" was similar to that of Hertz, who two years before, in 1890, had achieved "control of a discipline" through a systematic ordering of electrodynamic phenomena by means of a system of equations.[8]

The burden of Drude's 1892 study was to show that a correct explanatory system exists to which the various optical theories could be reduced. To construct it, he assumed certain equations from the mechanical theories, dispensing with the mechanical interpretations of the terms in the equations, since these went beyond the facts and only gave rise to familiar controversies in optics: the equations contain the light vector u, v, w, a second vector ξ, η, ζ derived from the first by the vector operation of curl, and an optical constant a.[9] To these, Drude added two sets of boundary conditions, which relate the two vectors at the interface of optical media. One set of boundary conditions corresponds to Neumann's optical theory in which the density of the ether is the same in all media and the vector of a linearly polarized light wave falls in the plane of polarization; the other set corresponds to Fresnel's theory in which the elasticity of the ether is the same in all media and the vector of linearly polarized light is normal to Neumann's vector. Fresnel's boundary conditions can be derived from Neumann's, Drude showed, if the light vector is associated with ξ, η, ζ instead of with u, v, w. Still another set of boundary conditions corresponds to Cauchy's theory. Drude regarded the common partial differential equations of these theories together with their boundary conditions as experimentally confirmed and therefore as constituting an explanatory system.[10]

From the side of the electromagnetic theory of light, Drude assumed the system of formulas Hertz had presented in 1890.[11] Electromagnetic constants appear in

8. Drude, "In wie weit genügen die bisherigen Lichttheorieen," 367–68.
9. For transparent isotropic media, Drude wrote:

$$\frac{\partial^2 u}{\partial t^2} = a\Delta u = a\left(\frac{\partial \eta}{\partial z} - \frac{\partial \zeta}{\partial y}\right),$$

with corresponding equations for the v and w components. In this set of equations, the symbol Δ stands for the operation $\partial^2/\partial x^2 + \partial^2/\partial y^2 + \partial^2/\partial z^2$. Drude, "In wie weit genügen die bisherigen Lichttheorieen," 369.

10. Drude, "In wie weit genügen die bisherigen Lichttheorieen," 369–71.

11. Following Hertz's formulation, Drude wrote the "system of formulas" relating the magnetic force L, M, N to the electric force X, Y, Z:

$$A\mu\frac{\partial L}{\partial t} = \frac{\partial Z}{\partial y} - \frac{\partial Y}{\partial z}, \qquad A\epsilon\frac{\partial X}{\partial t} = \frac{\partial M}{\partial z} - \frac{\partial N}{\partial y},$$

with corresponding equations for the M and N components and for the Y and Z components. From these Drude derived the following equations:

$$A\mu\frac{\partial^2 L}{\partial t^2} = \frac{1}{A\epsilon}\Delta L, \qquad A\epsilon\frac{\partial^2 X}{\partial t^2} = \frac{1}{A\mu}\Delta X, \text{ etc.}$$

these formulas in a certain combination, $1/A^2\mu\epsilon$ (where A is the reciprocal of the velocity of light, and μ and ϵ are the magnetic permeability and the dielectric constant), which Drude identified with the optical constant a. Then, by interpreting the magnetic force in these formulas as the light vector, Drude showed that the electromagnetic theory yields the partial differential equations and boundary conditions of Neumann's mechanical theory; conversely, by interpreting the electric force as the light vector, he showed that it yields Fresnel's.[12]

In this way, Drude compared the formulas of the various optical theories for a range of phenomena. Whenever he found it necessary to bring the equations into formal agreement, he freely added small extra terms to them without discussing their physical significance. He concluded that with minor departures many mechanical theories and the electromagnetic theory of light have explanatory systems that are identical for a large number of phenomena, so they may all be regarded as correct theories. To go from one to the other, it is only necessary to exchange the mechanical for the electromagnetic quantities entering the equations: the density, elasticity, and velocity for the magnetic permeability, dielectric constant, and field intensity.[13]

Drude acknowledged that special theories are useful in pointing the way to better explanatory systems, and in this regard the electromagnetic theory of light had recently proven more useful than the mechanical theories.[14] From the point of view of "practical physics," however, all that matters is the equations, so that the assumptions of the special theories in their derivations of the equations can eventually be discarded as unnecessary scaffolding.

The reflection of light from metal surfaces belonged to the phenomena Drude discussed in connection with explanatory systems. Hertz was interested in these same phenomena; guided by the electromagnetic theory of light, soon after his experiments on electric waves in air, Hertz had done experiments on metal reflection in a vain effort to arrive at the formulas of Voigt's metal optical theory.[15] In 1892 and 1893, in correspondence with Drude, Hertz explained that he lacked the time and patience to prepare himself properly for research on this topic. He would need six months or a year to study Cauchy's, MacCullagh's, Voigt's, Ketteler's, and all the other major optical theories, but since Drude already knew these theories, in "enviable measure," Hertz strongly urged him to work out a "useful, exact electrodynamic theory of metal reflection." He assured Drude that it would be important for "our views on the nature of electrical resistance, current, etc.": through an electrical approach to metal optics, physicists, with Drude's help, might enlarge their electrical understanding from the side of optics, just as they had already enlarged their optical

12. By setting $1/A^2\mu\epsilon = a$ and interpreting the magnetic force L, M, N as the light vector u, v, w, the equations in note 11 for the magnetic force are the same as those for the light vector in note 9 above. The same holds if the electric force is interpreted as the light vector. Drude, "In wie weit genügen die bisherigen Lichttheorieen," 377–78.

13. Planck, "Drude," 299–300.

14. Drude, "In wie weit genügen die bisherigen Lichttheorieen," 411.

15. Hertz to Voigt, 6 Feb. 1890, Voigt Papers, Göttingen UB, Ms. Dept.

understanding from the side of electromagnetism. Drude's electrical study of metal optics extended the analogy between optics and electromagnetic theory that had guided Hertz in his own recent experiments.[16]

From his researches in optics, Drude developed a course of lectures at Göttingen in 1892 on the electromagnetic theory of light, deepening, as he did, his grasp of Maxwell's work. Two years later he published the lectures as his first book, *Physik des Aethers*.[17] This "first German text" to treat electricity and optics from the standpoint of Maxwell's theory belonged to the earliest comprehensive books on the subject of Maxwell's theory. (Föppl published his lectures on Maxwell's theory at the technical institute in Munich the same year, 1894; J. J. Thomson had just published a treatise on the theory, which Drude did not see until he had finished his; and Henri Poincaré had just published his Sorbonne lectures on the same subject.) Drude called his text an introduction to Boltzmann's recently published Munich lectures and even to Maxwell's own writings.[18]

Drude presented his text as a work of theoretical physics in close touch with experiment, which had the purpose of introducing Maxwell's theory in an easily understood manner. Like Hertz, Drude presented Maxwell's theory as a mathematical description of the facts, though for didactic reasons he did not follow Hertz's form of presentation. Drude interpreted Maxwell's theory to say that action at a distance is excluded in electromagnetism: Hertz's experiments had decided this point, and the "entire progress that has been made in researches on the physics of the ether in recent times rests essentially on the consistent realization of the idea of contact forces."[19] The *Fortschritte der Physik* recommended the text for showing the advantages of the "modern, unified theory of electric phenomena."[20] Since Drude's text was widely studied by students and teachers, it was influential in introducing Maxwell's theory into the German universities.[21]

Drude as Theoretical Physicist at Leipzig

As extraordinary professor at Leipzig, Drude lectured and directed student research on topics in the electromagnetic theory. He did research himself on Hertzian electric waves, making precision measurements of electric wavelengths, and he also did research in optics, only now from the electromagnetic standpoint. His years at Leipzig

16. Hertz thought that if one assumed the optical equations for a strongly absorbing, anomalously dispersing medium, one could interpret the constants in the equations electrically in such a way that in the limit of very slow oscillations, the constants would coincide with the normal electrical constants for metals. Hertz cited recent work by Rubens and du Bois that seemed to show that with extreme ultrared rays, one could pass from optical to electrical phenomena. Hertz to Drude, 30 Apr., 15 May, 4 Aug. 1892, and 6 Feb. 1893, Ms. Coll., DM, 3208–10 and 3212.

17. Paul Drude, *Physik des Aethers auf elektromagnetischer Grundlage* (Stuttgart, 1894).

18. König, in König and Richarz, *Drude*, 22. Drude, "Vorwort," *Physik des Aethers*, v–viii.

19. Drude, *Physik des Aethers*, v–11, 342, 345.

20. Carl Brodmann's review of Drude's *Physik des Aethers* in the *Fortschritte . . . im Jahre 1894*, 1895, pt. 2, 407–10, on 410.

21. König, in König and Richarz, *Drude*, 22. Planck, "Drude," 301.

were highly productive; in 1897 alone, he published eleven substantial works, among them an invited report for a joint session for physics and related sciences at the German Association meeting that year.[22] In this report, which can be seen as a sequel to Hertz's report at the German Association eight years before, Drude asserted the goal of the reduction of all distance action in physics to contact action. The great obstacle to the realization of this goal was gravitation; referring to astronomical studies and to nearly fifty publications bearing on the problem of reinterpreting gravitation as contact action, Drude concluded that physicists did not yet have grounds for associating gravitation with electromagnetism. The simplification and unification of physical theory that contact action promised had not yet been fully achieved. But Drude thought that the promise of contact action, mediated by the ether, remained greater than that of the equally admirable goal of bringing uniformity to physical theory by interpreting all actions as distance ones, and upon this understanding he urged physicists to reformulate their account of the physical world. He proposed an "absolute" system of units on the basis of the "universal" properties of the ether; in the new system, the mean free path of ether atoms would serve as the unit of length and the velocity of light would serve as the unit of time, freeing the fundamental units from the accidental properties of substances, which had qualified the older system of "absolute" units.[23] For Drude, the language of measuring physics should conform to the dominant conception of physics, which was now Maxwell's, not Gauss and Weber's.

Toward the end of his time at Leipzig, Drude was invited by a publisher to write a textbook on optics, his specialty; because a modern electromagnetic treatment of optics did not exist and because writing a textbook would give him "deeper insight" into his subject, Drude accepted. He brought out his *Lehrbuch der Optik* (*Theory of Optics*) in 1900. In the presentation of the physical part of optics, he introduced the "electromagnetic point of view," presenting Maxwell's equations in Hertz's form and notation.[24] He had little to say about the alternative mechanical theories beyond enumerating the advantages the electromagnetic theory has over them.[25] The electromagnetic theory gives the "simplest and most consistent treatment of optical relations," Drude said, and it marks an "epoch-making advance in natural science" by bringing together electricity and optics, two formerly distinct subjects, into "relations which can be made the subject of quantitative measurements."[26]

To the presentation of optics in his textbook, Drude brought a large body of recent theoretical and experimental research on the properties of the ether, justify-

22. Wiener to Boltzmann, 6 May 1900, Wiener Papers, Leipzig UB, Ms. Dept. "Lebenslauf von Paul Drude," 12 Feb. 1899, Personalakte Drude, Giessen UA. Paul Drude, "Über Fernewirkungen," report at the 1897 German Association meeting, *Ann.* 62 (1897): i–xlix, supplement.

23. Drude, "Über Fernewirkungen," xlviii–xlix.

24. Paul Drude, *Lehrbuch der Optik* (Leipzig: S. Hirzel, 1900). Translated as *The Theory of Optics* by C. R. Mann and R. A. Millikan (New York: Longmans, Green, 1902).

25. The advantages Drude enumerated were the commonly cited ones: Maxwell's equations yield only transverse waves; their boundary conditions do not require special assumptions about the vibration of light; and the velocity of the waves can be determined from purely electromagnetic measurements. Drude, *Theory of Optics*, 261.

26. Drude, *Theory of Optics*, vi. 261.

ing his claim that "optics is not an old and worn-out branch of Physics, but that in it also there pulses new life." He showed that optics had acquired a larger domain of facts through the special theories that connected them, above all Maxwell's theory and, with regard to optical dispersion and the optics of moving bodies, Lorentz's "ion" theory, an extension of Maxwell's. In the last chapters of his text Drude treated the thermodynamics of radiation, including there a full discussion of blackbody radiation with Planck's derivation in 1899 and 1900 of the radiation law from "electromagnetic theory." Following Planck, Drude advanced another, improved "truly absolute system of units" based on the "universal" law of total blackbody radiation and the "universal" properties of the ether as expressed in the laws of gravitation and the velocity of light.[27]

Drude had been moved by one of the great conceptions of physics in the late nineteenth century, the "physics of the ether," which encompassed the domains of electricity, magnetism, light, and radiant heat and which pointed to the possibility of a total theory of physics. The treatise *Physik des Aethers*, with which Drude launched his academic career, had the mission of familiarizing German readers with the ideas of Faraday and Maxwell. When Walter König brought out a second edition of the treatise seventeen years later, after Drude had died, he could say that Drude's mission was accomplished, that now Maxwell's theory was a "secure possession," and for this reason König eliminated Drude's careful comparison of Maxwell's with the older mechanical theory.[28] As several more German Maxwellian texts had subsequently appeared, notably Cohn's and Max Abraham's, Drude's had now settled into a niche as just another introductory text.

With his research and teaching, Drude impressed his Leipzig colleagues, who encouraged the minister to accommodate his wishes when Heidelberg offered him a position in 1898. The minister complied with Drude's wish to be allowed to substitute for, when necessary, the ordinary professor of physics, the ailing Gustav Wiedemann, in lectures, exercises, and examinations, and he listened sympathetically to Drude's more ambitious wish, which was to be made a second ordinary professor of physics alongside Wiedemann. Receiving assurances about his future advancement and about funds for an assistant, subject to the institute director's approval, Drude refused Heidelberg's offer. With the "greater independence" he would enjoy at Leipzig, with his "special students" there, with his respectable income from salary and lecture honorariums, and with his freedom from duties in the elementary laboratory course, he was nearly the equal of an institute director. Drude's personal position at Leipzig was strengthened, and with it so was the position of theoretical physics.[29]

The question of the future of Drude and of theoretical physics at Leipzig soon came up in another context. Late in 1898 Wiedemann stepped down, which meant

27. Drude, *Theory of Optics*, ix. 527. We discuss Planck's work on blackbody radiation in chap. 25.
28. Paul Drude, *Physik des Aethers auf elektromagnetischer Grundlage*, 2d ed., ed. Walter König (Stuttgart: F. Enke, 1912), vii.
29. Dean of the Leipzig U. Philosophical Faculty to Saxon Ministry of Culture and Public Education, 13 July 1898; Minister to Philosophical Faculty, 30 July 1898; Drude to Philosophical Faculty, 30 July 1898, Drude Personalakte, Leipzig UA, PA 422. Drude to "Geheimrath," 16 and 29 July 1898, Bad. GLA 235/3135.

that Drude would no longer represent him as director of the institute. The ambitious Leipzig faculty recommended the "predominantly" experimental physicist Röntgen as Wiedemann's replacement, and they recommended only Röntgen, since in their eyes he stood head and shoulders above anyone else. When Röntgen turned them down, they reviewed several other physicists for Wiedemann's chair, including Drude. If someone other than Drude succeeded Wiedemann, some proposed Drude's elevation to ordinary professor for theoretical physics simultaneously with the new appointment. That proposal was not decided on, despite Friedrich Kohlrausch's advice to divide the responsibility for physics into two ordinary professorships and to give Drude one of them. To replace Wiedemann, the committee drew up the customary list of three favored candidates—they were Braun, Wiener, and Drude—and proposed only the first two, in that order.[30] Braun and Wiener had recently overseen the construction of new physics institutes, which gave them an advantage with the Leipzig faculty, who wanted their new professor to improve their institute. Braun declined the job and Wiener accepted it, which settled the question of the chair for "experimental physics" and freed the faculty to consider the question of a second physics chair, the "ordinary professorship for theoretical physics," which Drude had recently asked for. In the meantime, they did not want the conditions under which Drude served as extraordinary professor to be changed to his disadvantage with the appearance on the scene of a new institute director.[31]

Neither did Drude want to lose the independence he had gained during Wiedemann's illness. He had taken over the direction of the work of some of Wiedemann's doctoral students and had graduated some, and they were considered Drude's students even though they had signed up for Wiedemann's laboratory course (Drude did not offer such a course, so Wiedemann turned over to Drude the fees the students paid to him). Worried that Wiener, upon arriving in Leipzig, might claim the space in the "old building" that his students now occupied, Drude informed Wiener that he wanted to continue to direct the work of advanced students (he had five of his own at the moment), to announce that he was doing so in the lecture list (by offering a *Praktikum*), and to have an assistant of his own. But scarcely had he put these wishes and concerns to Wiener than he wrote to him again to take back his request to publicly announce his work with advanced students. He was anxious that he might have given Wiener the impression that he claimed "independent rights in the institute," which touched on the always delicate subject of the director's authority. Wiener was equally careful to delineate the director's rights: before receiving Drude's second letter, he had already replied to the first, advising Drude that he

30. Leipzig U. Philosophical Faculty to Saxon Ministry of Culture and Public Education, 5 Nov. 1898; Ministry to Philosophical Faculty, 8 Dec. 1898; minutes of the meetings of the commission for the reassignment of the professorship of physics, 9 and 18 Dec. 1898; draft of a letter, probably from the Philosophical Faculty to the Ministry, 17 Dec. 1898; Wiener Personalakte, Leipzig UA, PA 1064. Röntgen to Ludwig Zehnder, 18 Nov. and 8 Dec. 1898, in W. C. Röntgen, *W. C. Röntgen. Briefe an L. Zehnder*, ed. Ludwig Zehnder (Zurich, Leipzig, and Stuttgart: Rascher, 1935), 70–71.

31. Draft of letter to Saxon Ministry of Culture and Public Education, 17 Dec. 1898; Ministry to Leipzig U. Philosophical Faculty, 31 Jan. 1899; Wiener Personalakte, Leipzig UA, PA 1064. Röntgen to Zehnder, 19 Jan. 1899, in *Briefe an L. Zehnder*, 73.

could announce a laboratory course for independent researchers only if he had a "special institute" of his own, which he did not have. Until Drude got his own institute, they would avoid appearing "outwardly as rivals" if they jointly announced a laboratory course for advanced students and settled between them the question of space and fees. Drude was at first inclined to reject Wiener's proposal for a joint announcement since he had put "pressure" on Wiener, but he decided to accept it because not to would seem "ungrateful." Drude and Wiener assured one another that each looked forward to a "lively scientific life together" in Leipzig.[32]

In their effort to reach an understanding at the outset and avoid the familiar "collisions" between ordinary professor and extraordinary professor, Drude and Wiener sent one another telegrams between their almost daily letters. Their understanding had to be mutual, not enforced; for although Drude was only one year younger than Wiener, he was still an extraordinary professor while Wiener had been an ordinary professor for several years. Moreover, Wiener had to his credit, as the Leipzig faculty put it, "one work of *first rank*," the experimental proof of standing light waves, though otherwise he had few publications.[33] Drude had vastly more publications, only perhaps still lacking one of indisputably first rank, and he had long felt that his worth was inadequately recognized. He and Wiener discussed their future working relationship with tact.

Even as Drude and Wiener discussed their arrangements in Leipzig, the prospect of Drude's leaving Leipzig came up. Wiener told Drude that he was suggested for Giessen, where Wiener had been teaching. When the Giessen job went to someone else, to Wilhelm Wien, apparently with Wiener's approval, Drude reassured Wiener that the "sad affair" would not throw a shadow over their future collaboration in Leipzig. He did not blame Wiener for valuing Wien's scientific accomplishments above his own, and he was grateful for Wiener's encouragement of his hope to become ordinary professor for theoretical physics at Leipzig. But he felt his chances at Leipzig were not good because of past misunderstandings.[34]

Creation of an Ordinary Professorship for Theoretical Physics: Boltzmann's Appointment

Boltzmann received a confidential letter from the physical chemist Wilhelm Ostwald about a prospective *theoretical* physics position in Leipzig on the same day that Ostwald met with a commission to draw up a list of candidates for the *experimental* physics position. Since "in principle" the Saxon government was ready to establish the theoretical position, Ostwald wondered if Boltzmann would accept it if he were offered it. Boltzmann said he would if Leipzig met his needs, and he rejoiced that an ordinary professorship was about to be established for his "special science, theo-

32. Drude to Wiener, 19, 21, and 23 Jan. and 24 Feb. 1899; Wiener to Drude, 22 Jan. 1899; Wiener Papers, Leipzig UB, Ms. Dept.

33. Draft of letter to Saxon Ministry of Culture and Public Education, 17 Dec. 1898.

34. Drude to Wiener, 12 and 24 Feb. and 12 and 15 Mar. 1899, Wiener Papers, Leipzig UB, Ms. Dept.

retical physics," which was represented at that level at "so few universities in Germany." He made no secret of his discontent in Austria, where there were "far fewer students ready for scientific work" than in Germany, and where there were few scientific meetings and societies and no scientific stimulation. His activity there, his wife wrote to a friend, had the "character of a schoolmaster's drilling of candidates for secondary school education," which did not do justice to his talent or his aspiration.[35]

Several months after Ostwald had informed Boltzmann about the position at Leipzig, Ostwald wrote to him of a problem: Drude, the extraordinary professor for theoretical physics, whose presence at Leipzig University precluded the appointment of an ordinary professor for the same subject. Ostwald explained that Drude would like to become ordinary professor himself and for this promotion he had influential support. It was obvious that the necessary condition for Boltzmann's call *to* Leipzig was Drude's call *away*, which had not yet happened. Ostwald could only wish it would. Boltzmann regretted the disappearance of his hopes so soon after they had been raised.[36]

To make way for Boltzmann, Ostwald counted on Drude's rising reputation, which promised outside offers soon. Just that year, for instance, Boltzmann had placed Drude ahead of all other candidates for an extraordinary professorship for theoretical physics at Heidelberg. The confidence that the leaders of German physics placed in Drude was expressed by their intention to appoint him editor of the *Annalen der Physik*. Drude's, and with it Boltzmann's, opportunity came when Wien decided to stay only one year in Giessen, making his position available to his runner-up the year before, Drude. In the early spring of 1900, Ostwald wrote again to Boltzmann to say that his prospects were developing more favorably than he had dared hope. Drude was leaving, and the Leipzig faculty wanted Boltzmann and would press their choice on the ministry. When Ostwald told Boltzmann that his success was *"very probable,"* Boltzmann came out of his *"depressed mood."*[37]

During Wiener's first year as the experimental physics professor at Leipzig, he

35. Ostwald to Boltzmann, 9 Dec. 1898; Boltzmann to Ostwald, 13 Dec. 1898; Henriette Boltzmann to Ostwald, 29 Apr. 1899; in Wilhelm Ostwald, *Aus dem wissenschaftlichen Briefwechsel Wilhelm Ostwalds*, vol. 1, *Briefwechsel mit Ludwig Boltzmann, Max Planck, Georg Helm und Josiah Willard Gibbs*, ed. Hans-Günther Körber (Berlin: Akademie-Verlag, 1961), 22–30. Boltzmann regretted his move to Vienna from Munich. His wife wrote to friends about his melancholy. He wrote to his former Munich colleagues and to his future Leipzig ones about the inferior students in Vienna to whom he could not teach higher theoretical physics. He told Ostwald he was also dissatisfied with political conditions in Austria. Insertion in the minutes of the meeting of Section II of the Munich U. Philosophical Faculty, 30 Apr. 1896, Munich UA, OCI 22. Wiener for the Leipzig U. Philosophical Faculty to the Saxon Ministry of Culture and Public Education, 12 Mar. 1900; Wilhelm Ostwald, "Beibrief an den Minister in Sachen Boltzmann," 12 Mar. 1900; Boltzmann Personalakte, Leipzig UA, PA 326. Henriette Boltzmann to Leo Koenigsberger's wife, 13 Jan. 1895, STPK, Darmst. Coll. 1922.93.

36. Ostwald to Boltzmann, 5 May 1899; Boltzmann to Ostwald, 6 May 1899, in *Briefwechsel . . . Ostwalds*, 26–27.

37. Boltzmann to Leo Koenigsberger, 3 June 1899, STPK, Darmst. Coll. 1922.93. Drude to Wiener, 30 Dec. 1899, Wiener Papers, Leipzig UB, Ms. Dept. Saxon Ministry of Culture and Public Education to Leipzig Philosophical Faculty, 15 Mar. 1900, Drude Personalakte, Leipzig UA, PA 422. Hessen Ministry of the Interior to Giessen U., 24 Mar. 1900, Drude Personalakte, Giessen UA. Ostwald to Boltzmann, 10 Mar. 1900; Boltzmann to Ostwald, 13 Mar. 1900; in *Briefwechsel . . . Ostwalds*, 28.

made room for Drude in his institute and approved an assistant for him, for he believed that a theorist should do experimental work of his own and direct students in "partly experimental work." Since Drude did not have a special collection of apparatus, Wiener paid from the funds for his own institute for whatever Drude and his students needed. As Wiener repeatedly told Drude, he looked forward to a second institute for theoretical physics, which would give Drude independence. When Drude left for Giessen, Wiener applied the same reasoning to the needs of his replacement, Boltzmann.[38]

Wiener explained to the Leipzig faculty that Boltzmann's work belonged to "theoretical physics" rather than "mathematical physics," since it stressed "physical content" and the "connection with experimental physics." Boltzmann did not treat mathematics as an end in itself, which was done by others, notably by Carl Neumann, at Leipzig. Wiener and the faculty characterized Boltzmann to the ministry as penetrating, powerful, original, rich in ideas, as one of the last of Germany's recent great theoretical physicists, joining Helmholtz, Kirchhoff, and Clausius. They argued that at fifty-six Boltzmann was still young enough to make Leipzig a brilliant center for physics and the foremost university in the discipline of theoretical physics. With Drude's imminent departure, the time had come for the ministry to transform the extraordinary professorship into an ordinary professorship for theoretical physics, which it had expressed its willingness to do almost two years before. Then Leipzig would be in line with other universities, the faculty claimed, pointing to positions for theoretical physics alongside positions for experimental physics at Berlin and Vienna and even at intermediate and small universities such as Göttingen and Königsberg. The faculty recommended only Boltzmann for their position, since he was the "most important physicist in Germany and beyond." Moreover, he wanted to come to Leipzig, as Ostwald confirmed. Boltzmann was offered the job and, as expected, he accepted it.[39]

Ostwald had asked the government to act soon so that Boltzmann could have a say in the arrangements for theoretical physics in the new physics institute building now being planned. He explained that because of Boltzmann's research interests, these arrangements would not be elaborate, involving essentially seminar rooms with books and models. Wiener was more ambitious for Boltzmann, wanting in addition ample space for him, the theorist, and for his students to do experimental research. All along, Wiener had argued that Leipzig had to have an "institute for the theoretical physicist," for otherwise it could not attract anyone of significance. Hertz had not gone as theorist to Berlin, he pointed out to the ministry, mainly because he would have lacked an institute there. As another case in point, he cited Wilhelm Wien, who had said that he would never take a job without such an institute.

38. Wiener to Boltzmann, 25 Apr. and 6 May 1900, Wiener Papers, Leipzig UB, Ms. Dept.

39. Wiener to Dean Eduard Sievers of the Leipzig U. Philosophical Faculty, 10 Mar. 1900; letter to the Saxon Ministry of Culture and Public Education, drafted by Wiener and signed as well by Ostwald, Wilhelm Wundt, Heinrich Bruns, Otto Hölder, Carl Neumann, and Sievers, 12 Mar. 1900; Ostwald, "Beibrief," 12 Mar. 1900; Ministry to Philosophical Faculty, 4 Aug. 1900; Boltzmann Personalakte, Leipzig UA, PA 326.

Wiener's floor plans for the new building showed a theoretical physics institute, which occupied parts of three stories. It was a grand institute of its kind. The upper story contained the theorist's lecture hall, which could seat two hundred students, and beside it was a preparation room for his lectures. From the lecture hall, by elevator, the theorist descended to his collection of apparatus, alongside of which were his study and laboratory. From his laboratory he descended again, this time by spiral staircase, to more rooms for experimental work. The theorist also had access to the "big collection for experimental physics" as long as his needs did not conflict with those of the experimentalist. Since the plans had been drawn up with Drude in mind, Wiener asked Boltzmann what he thought of them.[40]

"Nice" and "suitable," Boltzmann replied. But unfortunately, he explained, he had little experience with institutes for theoretical physics. In Munich he had had few people working with him, since the more capable young men had worked with the experimentalist Lommel in his more complete institute. In Vienna, where he was now, the porters and assistants caused him "infinite annoyance" and, in general, he had been unlucky in his experience with the people working with him. Because of his bad eyesight, he could not do sustained experimental work and was "clumsy" in setting up apparatus. He would be willing to have advanced students work according to his ideas as long as they did not need to be shown everything. If Wiener thought that there were enough students sufficiently prepared and if the ministry was ready to approve what he needed (he asked if Drude had had a separate collection and separate funds), he supposed he could take charge of the projected institute. But he really doubted that he was the right man for it, especially as his old neurasthenia had returned.[41]

Boltzmann never taught in the institute that Wiener's floor plans envisioned, for he left Leipzig before the new building was completed. In the two years he was there, he did his usual teaching. He gave a comprehensive lecture course on theoretical physics, managing to get through most of the branches of his subject, beginning with his favorite, analytical mechanics. (Drude had not lectured on mechanics because the Leipzig mathematicians taught it, but Wiener had assured Boltzmann that no one could prevent him from teaching it if he wanted to.) He had capable students, even if there were not all that many of them, and he had good relations with Wiener. Yet he was unhappy in Leipzig, and he asked the Saxon government to release him for "reasons of health." To Wiener, later, he found it hard to say just why he felt unhappy. Perhaps it was the marshy climate or the North-German Protestant customs or yet some other cause. In any event, he felt better after he had returned to Vienna, where he again took up a position as theoretical physicist in the university.[42]

40. Ostwald, "Beibrief," 12 Mar. 1900. Wiener to Boltzmann, 25 Apr. and 6 May 1900.
41. Boltzmann to Wiener, 30 Apr. 1900, Wiener Papers, Leipzig UB, Ms. Dept.
42. Wiener to Boltzmann, 6 May 1900; Boltzmann to Wiener, 3 Jan. and 7 Feb. 1903; Wiener Papers, Leipzig UB, Ms. Dept. Saxon Ministry of Culture and Public Education to Leipzig U. Philosophical Faculty, 4 June 1902, Boltzmann Personalakte, Leipzig UA, PA 326.

The Leipzig Theoretical Physics Institute after Boltzmann: Des Coudres's Appointment

After Boltzmann left, Leipzig continued to have high expectations for its theorist, who was soon to have an institute of his own. The faculty looked outside German-speaking countries for its next candidate, the one European theoretical physicist to rank with Boltzmann, H. A. Lorentz. They learned from Lorentz privately that he wanted to stay where he was, in Leiden. His refusal forced the Leipzig faculty to consider younger, less eminent physicists, in particular, Wien and Drude, in that (now familiar) order. They limited the list of candidates to these two because they could not come up with a third who had either Wien's scientific renown or Drude's capacity for work.[43]

As they had with Boltzmann, the faculty commended Wien and Drude not only for their theoretical researches but also for their experimental ones and, perfunctorily, for their teaching. Wien and Drude were theorists who were also well-rounded physicists, and as such they were suited for the Leipzig position, as Wiener defined it. The faculty associated the accomplishments of the two men with those of eminent Austrian and German physicists from the recent past and the present, Boltzmann, Stefan, Planck, Hertz, Helmholtz, Riecke, and Voigt. They praised Wien as a theorist for taking on the "greatest tasks of physics," such as the determination of the interaction of ponderable matter with the ether and the connection of mechanics with electromagnetic phenomena. The faculty regarded Drude as less original than Wien, as someone who preferred to work on whatever problems were currently receiving most attention because of work on them by other physicists. They revived their old objection: as the faculty saw it, Drude had been handicapped by his training under Voigt, which had placed him within a "phenomenological," a "more formal mathematical," direction of physics. The direction was sometimes useful "if nothing better was available," but if it were used exclusively it would "kill every physical thought." Fortunately, Drude had increasingly freed himself from this one-sidedness, as was shown by his recent theoretical investigations that stressed "physical ideas." This justified his present recommendation.[44]

Because both Wien and Drude inclined toward theoretical research, the Leipzig faculty thought that either of them would accept an appointment as theoretical physicist. But the faculty also recognized that they both currently directed physics institutes that had superior facilities to those of the theoretical section of the physics institute planned for Leipzig. As they feared, the negotiations with Wien soon stumbled over the problem of facilities in the institute and of living quarters in or near the institute. Despite special pleading by Wiener, Ostwald, and the astronomer Heinrich Bruns for the best theoretical physicist (the training of their students was

43. Minutes of the meetings of the commission for the reassignment of the professorship for theoretical physics, 18 June, 3 and 8 July 1902; Wiener to Dean of the Leipzig U. Philosophical Faculty, 1 July 1902; Philosophical Faculty to Saxon Ministry of Culture and Public Education, 11 July 1902; Des Coudres Personalakte, Leipzig UA, PA 410.
44. Leipzig U. Philosophical Faculty to Saxon Ministry of Culture and Public Education (including earlier draft of this letter), 11 July 1902.

"influenced in a decisive way by the representative of theoretical physics"), the ministry failed to offer Wien enough to persuade him to accept. The Leipzig job would have given Wien more time for his research, which appealed to him. But he did not want to do only theoretical physics, and he doubted that his talents were suited for it.[45] Drude was then offered the job. The ministry promised him not only a good salary but also 60,000 marks for equipping his section of the institute and, on top of that, living quarters within the institute. But in the end, Drude, too, declined. He preferred to stay in Giessen, fearing that if he accepted the Leipzig job he would foreclose an experimental career. Although Wiener did not agree with him on this point, he conceded that Drude's further advancement would be easier from an experimental physics institute than from a theoretical one.[46]

The Leipzig faculty was proved right in their concern that if Wien and Drude turned them down, the choice would "no longer be easy." As the building schedule for the physics institute made the appointment an urgent matter, the faculty commission once more came up with names. The length of their list fluctuated this time, as it gradually acquired an ordering—and as it provoked a controversy over the desirable qualifications of a theoretical physicist. When the commission decided to recommend, in order, Des Coudres, Wiechert, Carl Runge, and Sommerfeld, Wiener wrote a separate, dissenting report, a copy of which he sent to Carl Neumann, who presumably was sponsoring Sommerfeld, a candidate Wiener objected to. Sommerfeld's candidacy contradicted the principles that had guided the establishment of the theoretical physics professorship in the first place, Wiener argued. Sommerfeld would not be capable of directing an institute and the laboratory work of students, in which case the newly created theoretical physics institute would be superfluous and worthless. If Sommerfeld's name were not eliminated, Wiener would inform the ministry that the commission had deleted two better suited men from the list of candidates against his wishes, Cohn because he was a Jew and Bjerknes because he was a foreigner.[47]

Wiener and Neumann's disagreement over Sommerfeld was based on their conflicting understanding of the nature of theoretical physics. In urging Sommerfeld's candidacy, Neumann argued that any advance in physics would be slow in coming and should depend on skilled mathematical work. His and Wiener's views were "very different," Neumann wrote to Sommerfeld: for him, Neumann, the "careful sifting and fashioning of what already exists," which only a mathematician could do, was necessary and important, whereas for Wiener the aim was a "certain *theoretical* [met-

45. Wilhelm Wien, "Ein Rückblick," in *Aus dem Leben und Wirken eines Physikers*, ed. K. Wien (Leipzig: J. A. Barth, 1930), 1–76, on 24. Wiener, Bruns, and Ostwald to Dean Wilhelm Kirchner of the Leipzig U. Philosophical Faculty, 12 Sept. 1902; Saxon Ministry of Culture and Public Education to Philosophical Faculty, 29 Sept. 1902; Des Coudres Personalakte, Leipzig UA, PA 410.

46. Drude to Wiener, 22 Sept., 11 Oct., and 4 Nov. 1902; copy of letter by Wiener to Drude, 1 Nov. 1902; Wiener Papers, Leipzig UB, Ms. Dept. Drude did advance rapidly: his outstanding record at Giessen led to his appointment in 1905 to the prestigious chair of physics at Berlin.

47. Leipzig U. Philosophical Faculty to Saxon Ministry of Culture and Public Education, 11 July 1902; note by Wiener, 30 Oct. 1902; minutes of the meetings of the commission for the reassignment of the professorship for theoretical physics, 6, 20, and 29 Nov. and 3 Dec. 1902; Wiener to Dean of the Philosophical Faculty and Wiener's accompanying "Separatbericht," 30 Nov. 1902; Des Coudres Personalakte, Leipzig UA, PA 410.

ronomic] time-keeping [Tact], which leads to new experimental investigations." Wiener
expected physics to advance quickly through new ideas, not gradually through mathematical refinement. Since Neumann thought that Wiener should have the final say
on a second physics position, Wiener won the argument, in a sense.[48]

In the end the commission kept their disagreement from becoming the concern
of the whole philosophical faculty by voting to remove Sommerfeld's name from the
list, and they removed Runge's name for similar reasons: from neither man did the
faculty expect the "introduction of fundamentally new physics ideas into theory."
Runge had never published a "theoretical-physical work of fundamental importance," though he had worked with other physicists on some experimental investigations. Sommerfeld, whose strength lay in the "forceful application of mathematics
to completed theoretical problems," had never done an experimental investigation.
What Leipzig wanted was a physicist whose strength lay in physics, not in mathematics, and the faculty believed that they knew of a man who fit their criteria: Des
Coudres, who would bring to Leipzig the "most physical stimulation and scientific
life." The faculty were relatively untroubled by the experimental nature of Des
Coudres's researches once they had been assured by Wien that Des Coudres could
give satisfactory lectures on theoretical physics.[49]

Unlike Wien and Drude, Des Coudres was not then a director of a physics
institute, but only the extraordinary professor for theoretical physics at Würzburg.
That made Leipzig's new job attractive by comparison; Des Coudres was only too
glad to move to a physics institute as head of its "theoretical physics department."
After a "rather big struggle," he got everything he wanted from the Saxon ministry,
which was largely what the ministry had promised Drude, though without living
quarters in the physics institute building; he said he did not care about living quarters (though when the physics building was finished he made an official residence
for himself from rooms not intended for that purpose). He got 60,000 marks for his
department and an annual budget of 3,000 marks, excluding the common budget
with Wiener for heat, light, water, and maintenance. Wiener rejoiced that Des
Coudres's institute was going to be so "richly equipped with apparatus." To make it
work, Des Coudres got permission to request an assistant immediately and a promise of a mechanic to begin in 1904. That year, at Des Coudres and Wiener's
request, the ministry elevated the department to an "independently funded teaching
institute" with the title "theoretical-physical institute" under Des Coudres's direction.[50]

48. Carl Neumann to Wiener, 29 Nov. 1902, quoted in Hans Salié, "Carl Neumann," in *Bedeu-
tende Gelehrte in Leipzig*, vol. 2, ed. G. Harig (Leipzig: Karl-Marx-Universität, 1965), 13–23, on 14–15.
Neumann to Sommerfeld, 22 May 1903, Sommerfeld Papers, Ms. Coll., DM.
    49. Leipzig U. Philosophical Faculty to Saxon Ministry of Culture and Public Education (draft), 6
Dec. 1902, Des Coudres Personalakte, Leipzig UA, PA 410.
    50. Des Coudres to Wiener, 13 and 17 Dec. 1902; Wiener to Des Coudres, 11 and 26 Dec. 1902;
Wiener/Des Coudres Correspondence, Leipzig UB, Ms. Dept. Des Coudres to Wiener, 24 Dec. 1902,
Wiener Papers, Leipzig UB, Ms. Dept. Saxon Ministry of Culture and Public Education to Leipzig U.
Philosophical Faculty, 20 Feb. 1903; Minister von Seydewitz to Leipzig U. Senate, 7 June 1904; Des
Coudres Personalakte, Leipzig UA, PA 410.

In his address at the dedication of the new institute building in 1905, Wiener recalled the days when a theoretical physicist had no rooms of his own and depended on the congeniality of the experimental physicist. But since today's theoretical physicist was no longer a man only of "sponge and chalk" but also of "drill and telescope," he needed a "special theoretical institute." Except for the theoretical emphasis in the lectures, the Leipzig theoretical physics institute was not much different from the experimental physics institute. Its budget, of course, was only a fraction of the latter's.[51]

Des Coudres's institute was located in the same new building as Wiener's, because the available space did not allow for separate buildings. It occupied rooms on three floors in the north wing, which favored "exact measurements." Wiener had been advised by a colleague to put up a "fire wall" to keep harmony between the two institutes, but he did not need one between himself and Des Coudres, and he hoped that future directors of the two institutes would have the same friendly understanding that they had. Four years later Des Coudres wrote an account of his institute in which he, too, stressed the advantage of a shared building. Through the door of the glass wall separating his collection from the bigger experimental physics collection, a regular exchange of apparatus took place, which was good, Des Coudres noted, for the state's exchequer as well as for science. Moreover, the joint physics colloquium was usually held in the theoretical lecture hall, where experimental demonstrations could not always be performed if the big collection were a kilometer away instead of next door. The new institute building embodied architecturally (if, in part, accidentally) a view of the relationship between the experimental physicist and the theoretical physicist. With their two institutes standing side by side, their apparatus and personnel moving freely between them, Wiener and Des Coudres's relationship was smooth. There was good reason why it should be, for the two physicists were not different in kind: Des Coudres, the theorist, did more "measuring," Wiener, the experimentalist, did more "experimenting," but both did experimental work, and that was the point.[52]

Boltzmann had drawn up the first list of apparatus for the theoretical institute, which was fairly modest and corresponded, Des Coudres explained, to Boltzmann's dispiritedness at the time. Although less modest than Boltzmann's, Des Coudres's list followed the same fundamental principle, which was to build the collection in certain directions and not try to cover all parts of physics equally. His collection soon included a number of costly pieces such as a powerful compression pump and a high-frequency dynamo that corresponded to his own research interests. With the help of a mechanic and an assistant, he was soon doing research and supervising the experimental work of five to seven advanced students in his institute.[53]

51. Otto Wiener, "Das neue physikalische Institut der Universität Leipzig und Geschichtliches," *Phys. Zs.* 7 (1906): 1–14, on 6.

52. Wiener, "Das neue physikalische Institut," 9. Theodor Des Coudres, "Das theoretisch-physikalische Institut," in *Festschrift zur Feier des 500jährigen Bestehens der Universität Leipzig*, vol. 4, *Die Institute und Seminare der Philosophischen Fakultät*, pt. 2, *Die mathematisch-naturwissenschaftliche Sektion* (Leipzig: S. Hirzel, 1909), 60–69, on 62–64.

53. Des Coudres, "Das theoretisch-physikalische Institut," 65–66, 68.

Des Coudres was primarily an experimentalist in his work. He had done an experimental dissertation under Helmholtz, after which he went to Leipzig as Wiedemann's assistant and Privatdocent only to leave in disappointment when another young physicist—Ebert briefly and then Drude—was brought in from outside as the new extraordinary professor for theoretical physics. He moved to Göttingen as Privatdocent, advancing to extraordinary professor for applied electricity there, in which capacity he also lectured on theoretical physics. In 1901 he moved to Würzburg as extraordinary professor for theoretical physics, where he remained until he was invited to return to Leipzig as Boltzmann's successor. Up to this time, he published steadily, two or three papers each year, a few on general subjects and on electrotechnology, a few on theoretical physics, and all the rest on experimental physics, the most important of which was his measurement of cathode rays undergoing magnetic deflection.[54]

Praised for his originality when he was recommended for Leipzig, Des Coudres proved unequal to Wiener's expectations. As ordinary professor for theoretical physics in Leipzig, he did not publish the new ideas that Wiener—correctly, as it turned out—believed would advance physics. In fact, he scarcely published any research at all; in nearly twenty-five years in Leipzig, he published only three papers, two of which belonged to high-pressure experimental physics. What Wiener came to value in his colleague was his contribution to the colloquium and his encouragement of students to work on good problems.[55] That in the second physics professor Wiener did not also acquire a productive researcher had in large measure to do with the position Wiener had designed for him. Wiener wanted a first-class man who could direct an institute filled with experimental researchers, but he could offer him only a second-class institute.

Wiener had studied under Kundt, who did not believe that physicists needed higher mathematics. Wiener carried over that attitude into his selection of the second physics professor at Leipzig. His colleague was to be a well-rounded physicist capable of doing both theoretical and experimental physics, as he was himself. For even as Wiener was planning the Leipzig institute building with space for a theorist, he was beginning a long series of theoretical works on the physical optics of various substances, which he would set his doctoral students to test experimentally. Later, in the 1920s, he worked intensively to construct a world picture based on a kinetic ether theory and a "fundamental law of nature," locating the uniformity of all physical occurrences in pure motions and eliminating all distance forces.[56] Although this theory bore little relationship to the theoretical needs of physics at the time, it,

54. Documents on Des Coudres's "Habilitation" in 1891 and withdrawal from Leipzig on 27 Sept. 1894; Leipzig U. Philosophical Faculty to Ministry of Culture and Public Education (draft), 6 Dec. 1902; Des Coudres Personalakte, Leipzig UA, PA 410. Des Coudres to a Leipzig professor, 23 Oct. 1894, STPK, Darmst. Coll. 1919.237. Otto Wiener, "Nachruf auf Theodor Des Coudres," Verh. sächs. Ges. Wiss. 78 (1926): 358–70; Wilhelm Wien, "Theodor Des Coudres," Phys. Zs. 28 (1927): 129–35.

55. Wiener, "Des Coudres," 364; Wien, "Des Coudres," 133.

56. Ludwig Weickmann, "Nachruf auf Otto Wiener," Verh. sächs. Ges. Wiss. 79 (1927): 107–18; Karl Lichtenecker, "Otto Wiener," Phys. Zs. 29 (1928): 73–78.

together with his earlier theoretical work, revealed Wiener as an experimentalist who could be his own theorist.

Having inspected the largest of Germany's university physics institute buildings, Leipzig's, on a "Studienreise" in Europe, the Japanese physicist Hantaro Nagaoka wrote to Ernest Rutherford in 1911 that size and money do not make a laboratory work, but only the people in it.[57] If the people at Leipzig did not always impress others as much as the building they occupied did, the responsibility belonged in part to Wiener, who had had the major say in the choice of the second ordinary professor for physics there. If he had been able to hold Boltzmann or attract Lorentz, or if he had followed Neumann's urging and gone for Sommerfeld instead, Leipzig would have had a first-class theorist for whom the theoretical institute would have seemed relatively commodious (as theoretical institutes were rare then, and where they existed they were cramped). Three years later, Röntgen brought Sommerfeld to Munich for a position and institute similar to Leipzig's, which assured that physics in Munich flourished as Wiener had hoped it would in Leipzig. By hiring Sommerfeld, Munich got a productive theorist who did not do experiments himself but who brought into his institute good experimenters to use its facilities and test his theories. By hiring Des Coudres, Wiener's final choice, Leipzig was without a strong representative of the theoretical side of physics, and in twentieth-century physics that was to prove a severe limitation. The year that Des Coudres moved into his new institute, 1905, was the year of relativity theory, which initiated a strong mathematical development within theoretical physics. That same year the light quantum entered physical theory, which with associated developments was to introduce a new armory of mathematical methods along with a theoretical physics of discontinuous atomic processes. In time, at Leipzig, Des Coudres was replaced by Werner Heisenberg, a theorist whose work belonged to the modern development. But that was not until the 1920s, which was late.

Boltzmann foresaw the likely difficulty and the missed opportunity for theoretical physics at Leipzig. He had welcomed the establishment there of a new and still rare ordinary professorship for theoretical physics, for the specialty and not just for himself. When he heard who his successor at Leipzig was to be, he wrote (in a consoling tone) to Wiener that Des Coudres was a promising young man, if possibly not the best. The problem was that Des Coudres lacked a "specifically mathematical head." Boltzmann hoped that Des Coudres would make himself more familiar with "theory, which one now after all understands to be mainly mathematics."[58]

57. Hantaro Nagaoka to Ernest Rutherford, 22 Feb. 1911, Rutherford Papers, Cambridge University Library.

58. Boltzmann to Wiener, 3 Jan. and 7 Feb. 1903, Wiener Papers, Leipzig UB, Ms. Dept.

23

Vienna Institute for Theoretical Physics

In 1890 the Vienna Privatdocent James Moser wrote to Hertz to give him a "picture of physics in Vienna." There were three ordinary professors for physics at Vienna University, Moser said: Victor von Lang, who lectured to medical students; Josef Loschmidt, who lectured to pharmaceutical students; and Josef Stefan, who lectured to teaching candidates. Lang had the medical students all to himself, an enviable burden, financially speaking; he tested four or five hundred of them a year, but, according to Moser, they avoided his lectures, where attendance often sank below ten. Loschmidt was a "chemist at heart," who taught physical chemistry as well as physics for pharmacists. Both Lang and Loschmidt had theoretical interests: Lang's were known, for example, through his text, *Einleitung in die theoretische Physik (Introduction to Theoretical Physics)* and Loschmidt's through his "great mathematical gift." The principal theorist at Vienna was, however, Stefan, who "in his lectures does *not* separate mathematical physics from experimental physics." Stefan was "unexcelled" as a lecturer, but because of the small number of serious students for the subject, he had only twenty-five to thirty in his class.[1]

This "picture" was soon to change. Loschmidt was seriously ill and his lectures had been temporarily taken over by Moser, who was then about to become his assistant; in 1891, Loschmidt was succeeded by Franz Exner. Stefan died in 1893; he was succeeded in the following year by Boltzmann, who, in effect, inaugurated the Vienna theoretical physics institute, though it then lacked the name. Stefan was referred to as professor of physics, but when Boltzmann was discussed as his replacement it was as professor for "theoretical physics." Titles were loosely used then; what was definite was Boltzmann's teaching assignment, which was for theoretical physics, and his authority over the physics institute. In 1900, as we have seen, Boltzmann left Vienna for Leipzig; in 1902 he returned to Vienna and was given the same teaching responsibility but a smaller institute carved out of the former one. This

1. James Moser to Hertz, 21 Mar. 1890, Ms. Coll., DM, 2982.

184

institute corresponded to his actual teaching and to custom. We discuss below this maneuver together with an associated reorganization of physics instruction at Vienna University.

Reorganization of Physics at Vienna and Boltzmann's Return

In 1901 a faculty commission at Vienna University reported on the question of Boltzmann's replacement. They listed the needs of the university in physics: one was that the lectures in theoretical physics be up to date, another that more practical exercises in physics be offered. The report elaborated on the second point. When Boltzmann came to Vienna in 1894 as director of the physics institute, he normally would have had responsibility for directing laboratory courses. But Boltzmann asked to be excused from this duty, which led to its being transferred to Exner. Because of the limited space and assistance at his disposal, Exner could not handle more than about fifty students, so that places in his laboratory were reserved for the next several years. Since laboratory work provided the "best school" for students of "all natural scientific disciplines," it was essential to expand it. At the Munich technical institute, the report pointed out, there were about two hundred students in the laboratory course, and at the Czech university in Prague there were so many students in the course that they spilled over into the German university. The lack of opportunity at Vienna University was a special "calamity" for teaching candidates, throwing a "shadow" over their entire future professional activity. Inspectors in the field had found that secondary teachers were incompetent at experiment, even though the newer rules placed great weight on it.[2]

The special concession to Boltzmann had resulted in a "curiosity," not to say a chaotic circumstance. Exner's title as professor of "physical chemistry" became "entirely illusory," as it was impossible for him to conduct the physics laboratory course and at the same time do justice to the subject he was originally assigned to teach. So Vienna University had a physics institute for laboratory instruction in physics in which no such instruction was given, while at the same time it had a physical chemistry laboratory in which physical chemistry was not taught. To impose a measure of reason on all of this, the faculty commission proposed that Exner be given the physics institute and his title be changed accordingly. In addition, it proposed that an ordinary professorship for theoretical physics be offered to Boltzmann (who, of course, had just left what was, in effect, that position at Vienna). Some rooms vacated by Exner would be offered to Boltzmann if he wanted them and other rooms offered to a new professor for physical chemistry, now to be appointed. The apparatus of the physics institute, the salaries, and the assistants would be divided between the professors for physics, theoretical physics, and physical chemistry.[3]

Boltzmann realized that his return to Austria was not going to be altogether

2. Report by the Vienna U. Philosophical Faculty's Commission on Boltzmann's replacement, written by Victor von Lang, 14 June 1901, Boltzmann file, Öster. STA, 4 Phil.
3. Report on Boltzmann's replacement, 14 June 1901.

straightforward. To prepare the way, he informed not only individual members of the Vienna faculty but also the Austrian minister of culture and education that Leipzig had not met his expectations and that he wanted to return. He was willing to give the minister what he demanded, his word of honor that if he came back he would never leave Austria again. On this point, Boltzmann was "morally bound," the minister assured the Austrian Emperor in his letter urging Boltzmann's reappointment. (While this assurance made practical sense to Austria, it did not make medical sense, for Boltzmann's main release from severe depression in the past had been to move to another job; like the doctors they consulted, the ministry underestimated Boltzmann's illness.)[4] Since Boltzmann commanded a large salary, it was all the more important for the ministry to determine that he was a good Austrian. Earlier, when Boltzmann had asked to be released from his Vienna job to go to Leipzig, the minister had told the Emperor that Boltzmann was not acting from any "unpatriotic motive." (In leaving Vienna, as the minister understood it, Boltzmann was acting from his dislike of "big city life" and his wish to be where he was better appreciated and had abler students who would spread his fame.)[5] Boltzmann's "love of the fatherland" was confirmed by his frequent references to his "Austrian origins" when abroad and by his longing to return. To win back this Austrian, a "scientific coryphaeus of the first rank," would be in the national interest, especially as Austria had recently lost some good scholars to German universities. The minister added that in Boltzmann's case any further threat from Germany was unlikely, since German ministries could hardly be interested in someone who had already turned down Berlin and had left Munich and Leipzig for Vienna.[6]

The assurances worked, and in 1902 Boltzmann was hired back. At first Boltzmann wanted his Vienna position restored exactly as he had left it, which would have made him director of the physics institute. But the faculty had already decided on Exner for that job—Exner was provisional director even then—and on a department for Boltzmann, who was to be limited strictly to theoretical physics and to a few institute and library rooms for that purpose. He would have no responsibility for conducting laboratory courses and for directing a "larger physics institute," a time-consuming business that he had little inclination for anyway. He accepted this reduced activity, especially since earlier he had not accepted all of his predecessor Stefan's responsibilities in the physics institute.[7] He was even willing to give up his old flat in the physics institute and some library space to make room for a teaching laboratory.[8]

The Vienna faculty argued that more than anything else its physicists needed a

4. "Vortrag" by the Austrian Minister of Culture and Education Wilhelm von Hartel, 20 May 1902, Boltzmann file, Öster. STA, 4 Phil.

5. "Vortrag" by the Austrian Minister of Culture and Education Hartel, 4 July 1900, Boltzmann file, Öster. STA, 4 Phil.

6. "Vortrag" by Hartel, 20 May 1902.

7. Entry for 26 Oct. 1900, Öster. STA, Phil, Physik, 4 G 867. "Vortrag" by Hartel, 20 May 1902; Boltzmann to Dean of Vienna U. Philosophical Faculty, n.d. [early 1902], Boltzmann file, Öster. STA, 4 Phil.

8. Boltzmann to an official, 5 June 1902, Öster. STA, 4 G Philosophie, physikal. chem. Institut.

new physics institute building; only then would satisfactory physics teaching be possible.[9] But for now they had to make do with a reorganization of what they had—along with a renaming of it. Lang's "physical cabinet" was now called the "first physics institute," Exner's "physical chemistry institute" the "second physics institute," and Boltzmann's "so-called 'physics institute' " the "institute for theoretical physics." Just as Lang and Exner had their staff, Boltzmann had his, an assistant together with funds for hiring either a mechanic or a helper. His teaching assignment was the usual one for a specialist in his subject, a comprehensive lecture course on theoretical physics, a seminar, and an occasional public lecture series.[10]

No sooner had Boltzmann received news of his reappointment at Vienna than a misunderstanding arose over the control and use of the physics apparatus. Boltzmann always set great store by apparatus in teaching, and he believed that Exner had agreed that Boltzmann would again be in charge of it all. Exner believed differently, and the ministry had to enter to help them reach an agreement, which was that Exner would select those pieces of apparatus from Boltzmann's old institute collection that should go back to Boltzmann for his new theoretical physics institute; the rest of the apparatus would go to Exner's institute. Boltzmann and Exner were to be allowed to use one another's apparatus, and the ministry assured Boltzmann that for what he had lost and still needed he would be given money to buy. Upon this understanding Exner declared himself ready to give his laboratory course, Boltzmann expressed guarded satisfaction, and the ministry felt momentary relief that they had averted "tension that could perhaps spoil collegial cooperation."[11] The faculty commission for the physics arrangements decided to deal with the question of the apparatus after Boltzmann's arrival in Vienna in the fall.

Boltzmann did not let the ministry forget his sacrifice; as soon as he began teaching, he reminded them that his institute had had to give up the "greatest part of its apparatus." The possibility of borrowing apparatus from Exner's institute did not entirely make up for the absence of it in his own institute, especially if the apparatus in question was in constant use. For this reason, additions to the inventory of his institute were "urgently needed." He asked for money to build up his collection and for student stipends to be diverted to a fund for buying apparatus. The ministry granted him what he wanted, so that he managed to avoid a quarrel with his colleagues over the collection that might have led to the faculty's involvement and the "most unfortunate ill-humor."[12]

9. Minutes of the Vienna U. Philosophical Faculty's Commission on Physics Instruction, 18 June 1902, Öster. STA, 4 G Philosophie, physikal. chem. Institut.

10. Austrian Ministry of Culture and Education to Boltzmann, 1 June 1902, Boltzmann file, Öster. STA. 4 Phil.

11. Boltzmann to "Sectionsrath," 5 June 1902; agreement signed by Franz Exner and two officials of the Ministry of Culture and Education, 23 June 1902; ministry official to Boltzmann, n.d. [June or July 1902]; Boltzmann to "Sectionsrath," 5 July 1902; ministry official to Boltzmann, 5 July 1902, Öster. STA, 4 G Philosophie, physikal. chem. Institut. Exner to Ministry, 22 Oct. 1902, Öster. STA, 4 G Philosophie, theoretisch-physikal. Institut.

12. Boltzmann to the Austrian Ministry of Culture and Education, 8 Nov. 1902, Öster. STA, 4 G Philosophie, theoretisch-physikal. Institut. Boltzmann got 1000 kronen for 1902 and again for 1903, and the ministry planned to give him more money for instruments the following year if he asked for it; he did and got it (18 and 28 Nov. 1903).

In his second year at Vienna, Boltzmann was given another teaching assignment in addition to his principal one for theoretical physics. While waiting for the faculty to suggest candidates for Mach's replacement, the ministry asked Boltzmann to lecture each semester on the "philosophy of nature and methodology of the natural sciences." It was not a big task, only two hours a week, and it brought him additional salary and a popular forum for presenting his philosophical views.[13]

Boltzmann's Teaching and Writing

Boltzmann held his theoretical physics lectures in the auditorium of the institute; it was not an overly large auditorium, and his lectures regularly filled it. As a lecturer, Boltzmann was forceful, humorous, and sometimes cutting. He offered his students his entire self, and they, the best of them, went away feeling that they had been shown a "whole new and wonderful world."[14] That was more important to them in the long run than the orderly equations (from which his whole argument could be reconstructed) that he wrote on the big blackboards behind the podium.[15] As a seminar director, Boltzmann drew students into his circle of interests. For example, his seminar on the topic of Hertz's mechanics prompted a gifted student, Paul Ehrenfest, to take up a suggestion he made about applying Hertz's approach to fluids; Ehrenfest soon turned the suggestion into a dissertation under Boltzmann's direction.[16]

Boltzmann extended his teaching beyond the lecture hall and seminar room by making his writing increasingly a part of it. His published research fell off through his succession of theoretical physics positions, beginning in Munich in 1889 and proceeding through Vienna, Leipzig, and back to Vienna again. In the 1890s he still published a good deal of research, most of it having to do with gas theory, but after 1900 he published practically nothing of this kind. Instead he published his popular lectures and his lectures on theoretical physics, which he hoped would reach a wide audience of open-minded colleagues and able students. The published lectures served Boltzmann's desire—the minister in Vienna called it his "almost morbid ambition"[17]—for recognition of his views, for scientific leadership.

It was as a theoretical physicist that Boltzmann wrote and spoke increasingly about his general views, placing himself at the center of a lively debate over the methods and future direction of theoretical physics. The whole recent advance of physical understanding, he believed, was due to the methods of research, which made it imperative for a theoretical physicist to reflect not only about nature but also about his methods of reflection about nature. In the face of what he saw as a

13. Austrian Ministry of Culture and Education to Dean of the Philosophical Faculty, 5 May 1903, Boltzmann file, Öster. STA, 4 Phil.

14. Ludwig Boltzmann, "Antrittsvorlesung, gehalten in Wien im Oktober 1902," in "Zwei Antrittsreden," Phys. Zs. 4 (1902–3): 274–77, on 277.

15. From recollections of Boltzmann's Vienna lectures on theoretical physics by Lise Meitner and Franz Skaupy, quoted in Broda, Boltzmann, 9–11.

16. Klein, Ehrenfest 1:66.

17. "Vortrag" by Hartel, 4 July 1900.

growing intolerance among his colleagues, Boltzmann pleaded for a plurality of methods, but also for a discriminating attitude toward them, as they were not all equally useful.[18]

Experimental physicists, Boltzmann explained, also had their methods of research, but these were simpler than the theorists', simpler because of the "continuously progressive" nature of experimental work. Theorists, unlike experimentalists, never seemed to settle their disputes, especially as their methods developed discontinuously, similarly to styles in art and literature. They viewed their preferred methods "very subjectively," through their "own spectacles."[19] But theorists, in their disputes, had a recourse that experimentalists did not have, or did not need; it was to publish books, which were generally based upon lectures, that covered the whole of their science, illuminating it from the perspective of their preferred methods. Boltzmann's published lectures on theoretical physics—covering his favorite parts of it, Maxwell's electromagnetic theory, gas theory, and analytical mechanics—were not syntheses of authoritative writings in the field but *his* version of theoretical physics.

If Boltzmann was the last of the line of great German theoretical physicists from the nineteenth century—that was how he was praised by faculties to ministries—he was unwilling to relegate the physical ideas on which he had built his reputation to the nineteenth century, to the past of his science. The methods of constructing theories from the ideas of mechanics and atomism had, in his view, led Maxwell, Clausius, Helmholtz, and other leading physicists of the recent past to their high accomplishments, as they had led him to his understanding of heat theory, and he argued for their continuing usefulness. He styled himself as a "reactionary" and the physics he belonged to as "classical," the perfecting of which he defined as the task of his life.[20]

In argument Boltzmann liked to magnify the numbers and strength of his adversaries, the anti-mechanists and anti-atomists. This was partly for effect, but from the mid-1890s he seemed truly pessimistic about the fate of his methods, at least in the short run: he published his lectures on gas theory in the hope that when the subject was "again revived, not too much will have to be rediscovered." He interrupted the publication of these lectures to publish his lectures on analytical mechanics, since his gas theory depended on that subject and the published lectures on gas theory could not include it all without becoming too bulky.[21] Boltzmann's published lectures on gas theory and analytical mechanics, each appearing in two parts and all appearing within the ten years from 1895 to 1904, gave his extended audience an ample introduction to his view of the nature of theoretical work in physics.

In his *Vorlesungen über Gastheorie (Lectures on Gas Theory)*, Boltzmann presented

18. Boltzmann's talk at the 1899 German Association meeting: "Über die Entwicklung," 198–227.

19. Boltzmann, "Über die Entwicklung," 201, 205; Ludwig Boltzmann, "The Relations of Applied Mathematics," in *International Congress of Arts and Science: Universal Exposition, St. Louis, 1904*, vol. 1, *Philosophy and Mathematics* (Boston and New York: Houghton, Mifflin, 1905), 591–603, on 591–93.

20. Boltzmann, "Über die Entwicklung," 205.

21. Ludwig Boltzmann, *Vorlesungen über Gastheorie*, 2 vols. (Leipzig, 1896–98), translated as *Lectures on Gas Theory* by Stephen G. Brush (Berkeley: University of California Press, 1964), 215–16. Ludwig Boltzmann, *Vorlesungen über die Prinzipe der Mechanik*, 2 vols. (Leipzig, 1897–1904), 1: v.

the subject for which he along with Clausius and Maxwell had done much of the developmental work. In these lectures, he introduced the concept of molecular disorder together with the calculus of probabilities, and he gave elementary derivations of the velocity distribution law, the H-theorem, the transport equation, the equipartition theorem, and other particular results of gas theory, all of which demonstrated the usefulness of his methods of theoretical physics. He took pains to make clear to his readers what kind of theory gas theory is: it is a mechanical picture or, in Maxwell's words, a "mechanical *analogy*," the kind of theory responsible for almost all "new discoveries." For Boltzmann, the key to the success of this picture is its use of atomism, which to date, he believed, had been the only satisfactory mechanical explanation of nature.[22]

In his complementary *Vorlesungen über die Prinzipe der Mechanik* (*Lectures on the Principles of Mechanics*), Boltzmann was concerned to show the mechanical picture underlying gas theory. He constructed *his* mechanical picture, which he regarded as classical, traditional, the "most exact and clearest" yet devised. It was an "unambiguous" picture, too, by which he meant what Hertz did by his requirement that the "picture must agree with the laws of thought" as well as with the facts.[23] Thought rather than experience was the proper starting point for the construction of a picture of nature; the present unclarity in the principles of mechanics, Boltzmann argued, had arisen from attempts by misguided phenomenologists to grasp nature directly instead of forming pictures of it in the mind.[24]

In this "deductive" approach to mechanics, Boltzmann proceeded from assumptions about the mutual accelerations of a system of material points rather than from our knowledge of the motion of real bodies. He did not have to address questions about the nature of matter, mass, force, and space by this approach, which suited his goal of constructing a useful and simple picture of the world instead of trying to fathom its greatest riddles. It was not until he was a hundred pages into his lectures that he came "at least in one respect into contact with reality," and even here he made only a brushing contact through the definition of a unit of mass in terms of the properties of one *real* substance, water. Further along, he took another "step nearer reality" by introducing a body in which the separations of material points are constant, an approximation to a rigid body. Of course, Boltzmann understood that the consequences of the picture were to be compared in detail with experience, since he was not playing a mere "game with thought pictures."[25]

Throughout his published lectures, Boltzmann treated, implicitly or explicitly, the topic of the methods of theoretical physics. He told his readers which parts of

22. Boltzmann, *Gas Theory*, 26–27.

23. Boltzmann, *Prinzipe der Mechanik* 1: 37–38.

24. Boltzmann, *Prinzipe der Mechanik* 1: 2–3.

25. Boltzmann, *Prinzipe der Mechanik* 1: 6, 99, 115. Boltzmann's views on analytical mechanics, as presented in his published lectures, are discussed in Broda, *Boltzmann*, 43–51, and in Dugas, *La théorie physique*, 61–67. Between the appearance of the two volumes of his mechanics lectures, Boltzmann delivered a series of lectures at Clark University in 1899 in which he developed the principles of mechanics first by the "deductive" method and then by the "inductive." The first lecture of this series is translated by J. J. Kockelmans, "On the Fundamental Principles and Basic Equations of Mechanics," in *Philosophy of Science*, ed. J. J. Kockelmans (New York: Free Press, 1968), 246–60.

Mach's teaching he accepted (for example, Mach's definition of mass), which parts of Kirchhoff's (for example, his characterization of a theory as a description), which parts of Hertz's (for example, his characterization of theoretical pictures), and so on. He told them as well what he did not accept; for example, he rejected the argument that a mechanics of volume elements instead of a mechanics of material points escapes a commitment to atomism.[26] (In his mechanics lectures, he kept his atomistic assumptions constantly before the reader's eyes by regularly writing the mathematical symbol sigma for a discrete sum instead of the integral sign for a continuous sum, even though he used the calculus throughout.)

Boltzmann presented mechanics as the foundation of all theoretical physics. This role for mechanics had both logical and historical justifications, though recently the whole program of constructing mechanical analogies throughout physics had been called into question. In the lectures on mechanics that Boltzmann delivered in 1897, he acknowledged that in future centuries or millenia a picture of nature might emerge that would be clearer and more comprehensive than the mechanical one; by the time he came to publish these lectures, only seven years later, Boltzmann could note that what he had expected to take centuries had already occurred: a cogent, nonmechanical picture of nature had emerged. By this he did not mean the old antiatomistic picture of the energeticists or phenomenologists, which he regarded as being as flawed and immature as ever, but a new atomistic picture, the "modern electron theory."[27] Now that a worthy successor to the mechanical world picture looked likely, Boltzmann was receptive: his understanding of theories as mental pictures rather than as claims for what the world is really made of allowed him that equanimity.

Boltzmann's life otherwise offered him little peace of mind. Talented lecturer that he was, he dreaded that his memory would fail him in front of his class. That was only one of many self-doubts that plagued him, adding to his burden of physical illness and recurring depressions. On a summer vacation from his teaching in Vienna, he committed suicide. It was 1906; he was sixty-two.[28]

Boltzmann's assistant Stefan Meyer temporarily took over the direction of the theoretical physics institute.[29] In the spring of 1907, the Vienna faculty proposed candidates for Boltzmann's job. First on their list was Planck, who visited Vienna and was impressed by its "new beautiful institute." It was only after much soul-searching that Planck turned the job down; he decided to stay where he was because of the "unexpected interest" the Berlin faculty took in the whole matter.[30] Following Planck on the list was Wien, but it would have taken more resources than Austria had to bring him to Vienna. The faculty's third choice was a much younger man, a native Viennese, Friedrich Hasenöhrl, who was at the time extraordinary professor

26. Boltzmann, *Prinzipe der Mechanik* 1: 4, 39–40.
27. Boltzmann, *Prinzipe der Mechanik* 2: 137–39, 335. In chapter 24 we discuss the directions in physics mentioned here: phenomenology, energetics, and the electron theory.
28. Broda, *Boltzmann*, 26; Klein, *Ehrenfest*, 76.
29. Entry for 7 Nov. 1906, Öster. STA, Phil, Physik, 4 G 867.
30. Planck to Wien, 19 June 1907, Wien Papers, STPK, 1973.110.

for general and technical physics at the technical institute in Vienna. What especially inclined the faculty to Hasenöhrl was his relationship to Boltzmann: he had been a favorite student of Boltzmann and he had already done teaching and research in Boltzmann's "new direction," assuring that he would continue the "tradition." His major work so far belonged to the electron theory, which Boltzmann had foreseen as the replacement of mechanics as the fundamental branch of physics.[31]

Hasenöhrl was killed in World War I, but not before he had formed what his colleague Stefan Meyer termed a "school" through his teaching at Vienna. Erwin Schrödinger, who entered Vienna University the year Boltzmann died, soon became Hasenöhrl's student. Boltzmann's "line of thought" was Schrödinger's "first love in science," and no other, he said, "has ever thus enraptured me or will ever do so again."[32] Schrödinger was invited to Vienna as Hasenöhrl's successor, but working conditions for Austrian professors had so deteriorated that he turned the job down.[33] His work in theoretical physics that compared in significance with Boltzmann's he did elsewhere.

Boltzmann's stature and mobility made him repeatedly the first choice of faculties and a participant in an unusual number of institutional developments in theoretical physics. In Vienna, he was the first occupant of an institute for theoretical physics, as he was in Munich and in Leipzig (and would have been in Berlin, too, if he had accepted the offer that eventually went to Planck). He had done his most important research before assuming any of these posts, but theorists such as Hasenöhrl, Sommerfeld, and Heisenberg who later occupied these institutes did some of their most important work in them.

31. Austrian Ministry of Culture and Education to the "Statthalter" of Lower Austria, 13 July 1907, Öster. STA, 4 Phil., theoretische Physik.

32. From Schrödinger's inaugural talk at the Prussian Academy of Sciences in 1929, in Erwin Schrödinger, *Science and the Human Temperament*, trans. J. Murphy and W. H. Johnston (New York: W. W. Norton, 1935), xiv.

33. Armin Hermann, "Schrödinger, Erwin," *DSB* 12: 217–23, on 218.

Physicists with Colleagues

At a variety of gatherings, German physicists, often together with colleagues from other sciences, had their pictures drawn or taken. The physicists were at times scientific collaborators, colleagues working in the same institution, members of the audience for general talks at the German Association, members of the specialized Berlin—later German—Physical Society, and, in the twentieth century, participants in highly specialized, invitational conferences, such as the Solvay Congress, devoted to specific problems of physics.

1. German Association Meeting in Jena, 1836. This drawing was made with such care that individual faces could be recognized. Reprinted from *Bericht über die Versammlung deutscher Naturforscher und Ärzte zu Jena 1836* (Weimar, 1837).

2

3

2. Rudolph Clausius with Colleagues. Left to right: standing August Kundt and Friedrich Kohlrausch; seated Georg Quincke and Clausius. Reprinted from Warburg, "Zur Geschichte der Physikalischen Gesellschaft," 36.

3. Max Born with Colleagues. Left to right: Wilhelm Oseen, Niels Bohr, James Franck, Oskar Klein, and, seated, Born. The occasion was the Bohr Festival at Göttingen in June 1922, at which Bohr gave invited lectures. Courtesy of Lehrstuhl für Geschichte der Naturwissenschaften und Technik, Universität Stuttgart.

4. Gustav Kirchhoff with Colleagues. Left to right: Kirchhoff, Robert Bunsen, and Henry Roscoe. The photograph was taken while Kirchhoff and Bunsen were visiting Roscoe in England in 1862. Reprinted from Roscoe, *The Life and Experiences,* 73.

4

5. Göttingen Physicists and Related Specialists in the Summer of 1907. In the front row from left to right are Ludwig Prandtl and Carl Runge, directors of the Institute for Applied Mathematics and Mechanics; Woldemar Voigt and Eduard Riecke, directors of the main Physics Institute; Hermann T. Simon, director of the Institute for Applied Electricity; and Max Abraham, Privatdocent for theoretical physics. Courtesy of Professor Dr. Friedrich Hund and Frau Maria Elisabeth Voigt.

5

6

6. Göttingen Physics Institute and Institute for Applied Electricity, 1906. The new main physics institute on the right housed the two departments for experimental and theoretical physics. Although relatively small, the new applied electricity institute on the left represented a part of the increasingly important third branch of physics, technical physics. The number of colleagues expanded accordingly, as can be seen from the photograph at the top of the page. Reprinted from *Die physikalischen Institute der Universität Göttingen*, 49.

7. Founders of the Berlin Physical Society, 1845. Left to right: standing Gustav Karsten, Wilhelm Heintz, and Hermann Knoblauch; seated Ernst Brücke, Emil du Bois-Reymond, and Wilhelm Beetz. Reprinted from the frontispiece to *Verh. phys. Ges. zu Berlin* 15 (1896).

8. Solvay Congress, 1911. Left to right: seated Walther Nernst, Marcel Brillouin, Ernest Solvay, H. A. Lorentz, Emil Warburg, Jean Perrin, Wilhelm Wien (behind), Marie Curie, and Henri Poincaré; standing R. Goldschmidt, Max Planck, Heinrich Rubens, Arnold Sommerfeld, F. A. Lindemann, Maurice de Broglie, Martin Knudsen, Fritz Hasenöhrl, G. Hosteler, E. Herzen, James Jeans, Ernest Rutherford, H. Kamerlingh Onnes, Albert Einstein, and Paul Langevin. Courtesy of Instituts internationaux de physique et de chimie.

7

Careers in Physics

Work in physics is shared; careers in physics are individual, even if repetitive, and they leave their record in copious official forms and documents. For our physicists, these varied from the diploma with its decorative profusion of typefaces to the utilitarian bureaucratic report covered with jottings and signatures of officials. The hard work in the physics institute that attended successful careers in physics was directed to teaching and research. Inevitably it included a good deal of time-consuming paperwork. The latter was partly generated by, and was partly in response to, paperwork of the government ministry to which it was responsible.

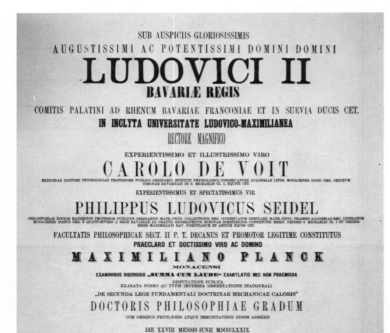

9. Max Planck's Announcement of Lectures at Munich University, 1880. Planck was teaching at Munich as Privatdocent. Courtesy of Archiv der Ludwig-Maximilians-Universität München.

10. Max Planck's Diploma from Munich University, 1879. Courtesy of Archiv der Ludwig-Maximilians-Universität München.

11. Prussian Ministry of Culture, Unter den Linden No. 4, Berlin. From the time Planck left Munich, six years after his graduation there, he pursued his career in conjunction with decisions made in this Berlin ministry. Reprinted from the frontispiece to Guttstadt, ed., *Die naturwissenschaftlichen und medicinischen Staatsanstalten Berlins.*

11

12. Max Planck while Privatdocent at Munich University. Courtesy of Dr. Dr. med. Hans Roos.

13

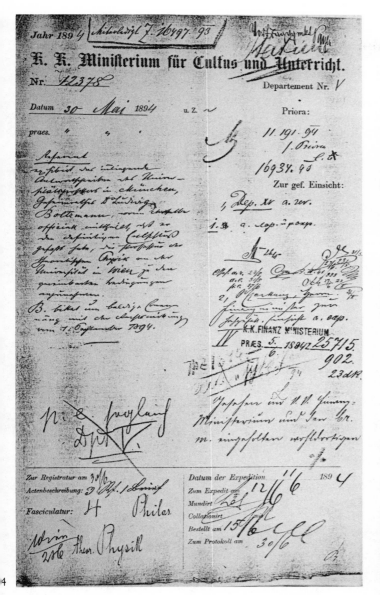

15

16

13. Berlin Physics Institute. Engraving of the new institute building appearing in the *Illustrierte Zeitung,* 29 Dec. 1877, p. 532, shortly before the institute moved from its old quarters in the university building to this location. The new building had been in use for only about ten years when Planck became the theoretical physicist there.

14. Document concerning Ludwig Boltzmann's Appointment at Vienna University, 1894. First page of a report on Boltzmann's willingness to return to Vienna as professor for theoretical physics. Its appearance, typical for this type of document, records the many official hands it passed through. Courtesy of Öster. STA.

15. Ludwig Boltzmann. Reprinted from the frontispiece to Stefan Meyer, ed., *Festschrift Ludwig Boltzmann.*

16. Austrian State Archive Today. This interior view shows the final home of the government's end of the myriad of documents relating to physicists' careers in Austria. Photograph by the authors.

Places of Physics

During the nineteenth century, the official workplace of physics expanded from a few rooms in a multi-purpose university building to an extensive suite of specialized rooms, preferably located within a building or wing of their own. The expansion was desirable, as laboratory instruction was offered to an increasing number of beginning students and as research facilities were provided for an increasing number of advanced students, assistants, and teachers. The floor space for laboratories came to be a comparative measure of physics institutes. Physics institutes built or remodeled for the purpose incorporated architectural features specific to the performance and organization of the work within them. It became common for institutes to specialize their facilities for research, as Bonn's new institute specialized its facilities for spectroscopy, corresponding to the director's interests and the local tradition; the additional financial argument for specialization can be appreciated by comparing Bonn's room-sized instruments for spectroscopy with Kirchhoff's spectroscope from the middle of the previous century. The principles of electricity and magnetism, which at the time of Weber and Gauss were the subject of physics research, by the twentieth century had come to be applied in alternating-current generators and other machinery, which supplied the many, complex electrical needs of the new physics institutes.

17. Leipzig University, Middle Paulinum. This part of the old monastery contained the physics collection until 1835. Reprinted from *Die Universität Leipzig, 1409–1909,* 17.

18

19

18. "Old University" of Freiburg. This building housed the mathematical-physical cabinet, which remained under the combined direction of the two professors for mathematics and physics until 1876. Courtesy of Freiburg SA.

19. Bavarian Academy of Sciences. This building was also the first home of the university when it moved to Munich in 1826. Courtesy of Munich SM.

20. Göttingen Magnetic Observatory, 1836. The south wall is removed to show the instruments: a clock, a theodolite and scale, and, suspended from the ceiling, a magnetometer. Reprinted from Wilhelm Weber, "Bemerkungen über die Einrichtung magnetischer Observatorien und Beschreibung der darin aufzustellenden Instrumente," in Weber's *Werke* 2:3–19, plate I.

21. Wilhelm Weber and C. F. Gauss's Magnetometer. The case containing the magnetic needle is shown from three sides, and the mounting for the mirror at the end of the case and other details are also shown. Reprinted from Weber, "Bemerkungen," plate III.

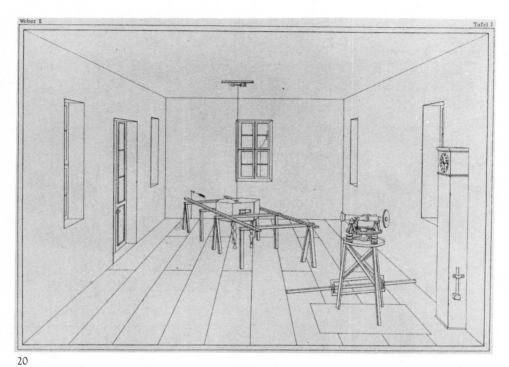

20

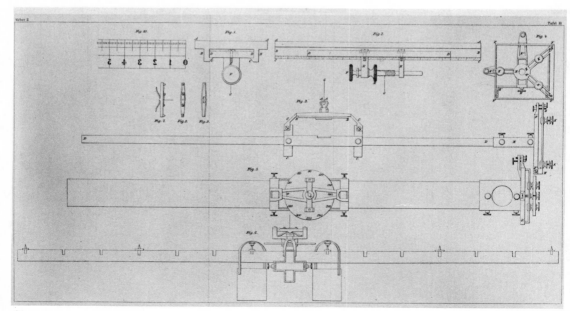

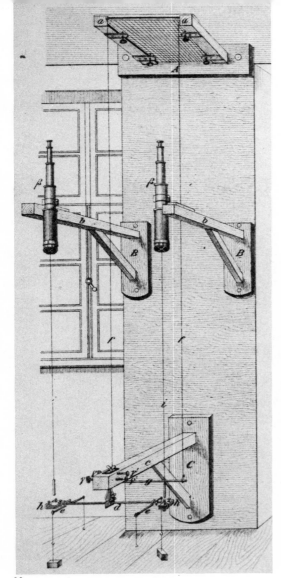

22. Gustav Kirchhoff's Apparatus for Testing Elasticity Theories, 1859. With this apparatus, Kirchhoff measured the ratio of cross-sectional contraction to longitudinal dilation of elastic steel wires, for which different theories yielded different predictions. Reprinted from Gustav Kirchhoff, "Ueber das Verhältniss der Quercontraction zur Längendilation bei Stäben von federhartem Stahl," in Kirchhoff's *Ges. Abh.*, 316–39, plate I, fig. 1, at end of volume.

23. Gustav Kirchhoff's Spectral Apparatus. Built for Kirchhoff by K. A. Steinheil in Munich, this apparatus passes a beam of light through telescope A, through the four prisms, and finally through telescope B, where it is observed. Reprinted from Hermann von Helmholtz, *Populäre wissenschaftliche Vorträge.* Vol. 3 (Braunschweig, 1876).

22

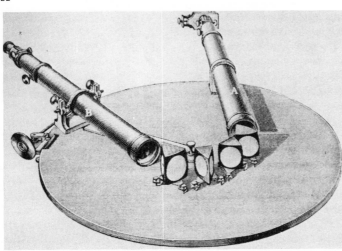

23

24. Rowland Concave Grating in the Bonn Physics Institute, 1913. The optically well-equipped new institute building contained two gratings like the one shown, of 6.5 meter radius of curvature, and several smaller gratings in addition. In this arrangement, the grating and camera were fixed while the slit and light source were moved in a circle. Reprinted from Kayser and Eversheim, "Das physikalische Institut der Universität Bonn," 1004.

25. Beginners' Laboratory for Heat and Mechanics at Bonn. Reprinted from Kayser and Eversheim, 1005.

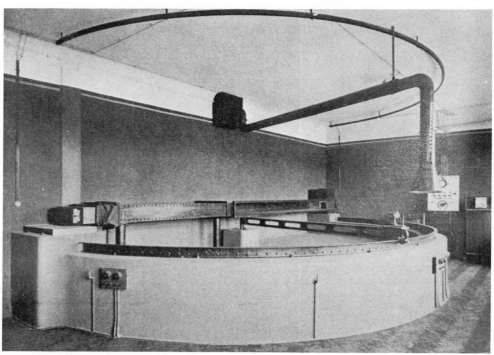

24

25

26. Workshop with Switchboard at Bonn. From here, the "heart of the works," electricity of various kinds was distributed throughout the institute. Reprinted from Kayser and Eversheim, 1006.

27. Machine Room at Bonn. The importance of electricity for physics institutes can be appreciated from this photograph. On the left, for example, is the alternating-current transformer. Reprinted from Kayser and Eversheim, 1007.

26

27

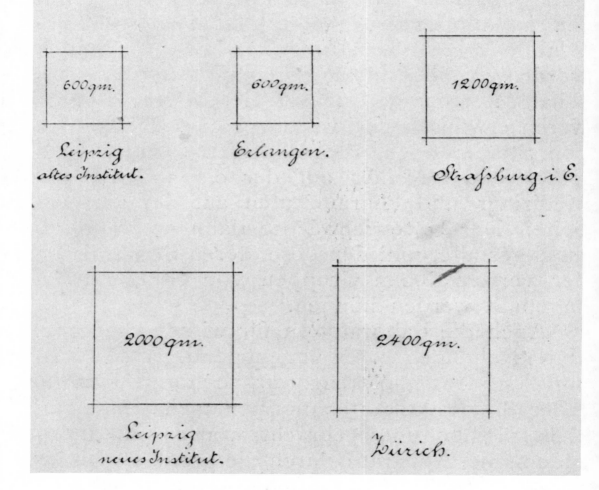

28. Laboratory Floor Space of Various Physics Institutes, 1906. The comparison shows that with respect to laboratory floor space, these institutes vary by a factor of four. The calculated areas included student and research laboratories but not preparation rooms for lectures, workshops, and the like. The calculation for the new Leipzig institute included the rooms for the theoretical physics institute. Reprinted from Wiener, "Das neue physikalische Institut der Universität Leipzig," 6.

29. Erlangen Physics Institute. Reprinted from Kolde, *Erlangen*, 470.

30. Leipzig Physics Institute. Reprinted from Wiener, "Das neue physikalische Institut der Universität Leipzig," 7.

29

0

31. Strassburg Physics Institute. Reprinted from *Handbuch der Architektur*, 234.

32. Zurich Polytechnic Physics Institute. Reprinted from *Festschrift zur Feier des fünfzigjährigen Bestehens des Eidg. Polytechnikums*, pt. 2, 336.

31

Theoretical Physicists Presenting Their Work

To students and colleagues, theoretical physicists presented their work, imparting and evaluating it and submitting it for criticism. The permanent form of presentation was by textbook, treatise, handbook, and article. Among the less permanent forms was the traditional lecture aided by chalk and blackboard.

33. Hermann von Helmholtz.
Courtesy of Wissenschaftliche
Verlagsgesellschaft.

210

34. Arnold Sommerfeld.
Courtesy of Lehrstuhl für
Geschichte der
Naturwissenschaften und
Technik, Universität Stuttgart.

35. Wolfgang Pauli. Courtesy of
Frau Franca Pauli.

36. Albert Einstein. Courtesy of
the Hebrew University of
Jerusalem.

34

35

36

24

New Foundations for Theoretical Physics at the Turn of the Twentieth Century

The close of the nineteenth century witnessed several remarkable experimental discoveries in physics. From his physics laboratory at Würzburg in 1895, W. C. Röntgen reported on a "new kind of rays," the penetrating X rays, or Röntgen rays, the nature of which was problematic at the time. It was a true discovery, something Röntgen had not anticipated as he experimented with electrical discharges in evacuated tubes. At this same time, by experiments with similar kinds of apparatus, the long-disputed nature of "cathode rays" was decided. The rays were not disturbances propagated through the ether as Hertz and other German physicists had thought, but streams of corpuscles as British physicists had argued and as Jean Perrin in France persuasively demonstrated in 1895. The most impressive confirmation of their corpuscular nature was reported from the physics laboratory at Cambridge University, where in 1897 J. J. Thomson measured the velocity and the ratio of charge to mass of the negatively charged corpuscles or "electrons," the name that had been proposed some years before and which came to be adopted. The several discoveries of the mid-1890s had some connections with one another: Thomson was led to study the nature of cathode rays by his work with Röntgen rays in the ionization of gases, and it was through an investigation of the relationship of Röntgen rays to his subject, fluorescence, that in 1896 the French physicist Henri Becquerel discovered certain penetrating radiations emitted by a uranium salt. He had discovered radioactivity, a phenomenon that drew the attention of his compatriots Pierre and Marie Curie, who soon discovered new radioactive elements, and of the New Zealander Ernest Rutherford who was then working with Röntgen rays in Thomson's laboratory at Cambridge. Becquerel, the Curies, Rutherford, his co-workers such as Frederick Soddy, and many other experimenters gradually brought clarity to the complex of radiations and transmutations characteristic of radioactivity. This extensive experimental work on Röntgen rays, electrons, and radioactivity posed fundamental problems for theoretical physics. Electrons, including high-speed electrons given off in radioactive decay, provided the means for testing the electron theory, one of the

211

contending comprehensive theories for physics at the turn of the century. Later experiments with radioactive sources led to the elucidation of the structure of the atom, which stimulated the invention of an exact atomic theory based on a new mechanics. Theoretical physics was also stimulated by—as it in turn stimulated—experimental work of another kind, not discoveries of new things existing in the world but precise measurements of, for example, spectral lines and their splittings, blackbody radiation, photoelectric emissions, and the motion, if any, of the earth through the ether.

In addition to the stimulus from experimental physics, theoretical physics had to respond to serious problems internal to itself. These arose in large part from the great developments of the nineteenth century, electrodynamics, energetics, and thermodynamics with its associated kinetic and statistical theories. The relationship between these theories and mechanics was a subject of widespread debate at the turn of the century. It was a time of intense questioning of the foundations of physics, and one of the liveliest questions then was the possibility—and the desirability—of the extension of mechanical modes of explanation throughout physics. At their meetings and in their publications, physicists asserted and debated the merits of the mechanical foundations of physics and their alternatives. Their reexamination of the foundations together with experimental advances were to lead them to the theories that marked the close of the "classical" period of theoretical physics and the advent of the "modern."

Mechanical Foundations

In 1896 the Leipzig physical chemist Ostwald remarked that recent textbooks on physics and chemistry confirmed his impression that the prevailing scientific world picture was formed of mechanical hypotheses.[1] In one way or another, if not always in the way Ostwald had in mind, physics textbooks in the late nineteenth century commonly ascribed to mechanics a leading role in physical theory. It was so with such standbys as Wüllner's text, which noted that physicists strove increasingly to found all phenomena on "one and the same fundamental cause, on motion," the science of which was mechanics, and Müller-Pouillet's text, which noted that much of heat theory was united with mechanics and that mechanics had been applied to light and electricity, too. Leopold Pfaundler, the present editor of the Müller-Pouillet text, concluded that "mechanics is really not a part of physics, but conversely the individual disciplines are parts or applications of mechanics." The newer texts by Kundt and by Warburg made a similar point: Kundt's published lectures on experimental physics introduced mechanics as the "foundation of the whole of physics and all natural sciences," and Warburg's characterized "mechanics or the theory of motion as building the foundation of physics"; the reason, Warburg pointed out, was the importance of motion: physicists perceived many phenomena directly as motion, and they successfully applied the laws of motion to many other phenomena that

1. Wilhelm Ostwald, "Zur Energetik," *Ann.* 58 (1896): 154–67, on 160.

they did not perceive directly as motion, such as those of light, electricity, and heat.[2]

In this connection we will briefly recall some already familiar texts on theoretical physics. They began, as Franz Neumann's published lectures did, with the "mechanical part of physics, since it builds the foundation for all remaining branches and contains the principles that find their application there." In his introductory text on theoretical physics, Volkmann regarded mechanics as the "fundamental discipline within the physical system." Voigt devoted roughly the first half of his text on theoretical physics to the mechanics of rigid and nonrigid bodies, in which he developed "mechanical theories" for the most important phenomena of all parts of physics. Kirchhoff, in his published lectures on theoretical physics, said that although in heat theory not all concepts had been successfully interpreted in mechanical terms, the reduction of all of physics to mechanics was still the goal to be pursued in the "fullest measure." Helmholtz concluded his published lectures on the motion of mass points by generalizing Hamilton's dynamical principle to apply to electrodynamics and thermodynamics, and Hertz's treatise on mechanics, taking its starting point in related work of Helmholtz's, affirmed the mechanical world picture. Even an ancillary text such as Heinrich Weber's edition of Riemann's lectures on the mathematical methods of physics might assert that "mechanics lies at the foundation of our entire theoretical natural science."[3] Boltzmann, who expressed his understanding of mechanics as the foundation of all natural science in his published lectures on mechanics, drew an analogy with politics: "If a nation achieves great success relative to its neighbors, it acquires a certain hegemony over them, indeed not infrequently subjugates and uses them. It is exactly the same with scientific disciplines. Mechanics soon acquired hegemony over all physics."[4]

Thermodynamic Questioning of the Molecular-Mechanical Foundations of Physics

In his inaugural lecture at the Prussian Academy of Sciences in 1894, Planck discussed the current status of mechanics in physics. He observed that the task of the theoretical physicist seeking a comprehensive view of nature was much harder than it had been a generation back; then it had been to reduce all natural phenomena to

2. Adolph Wüllner, *Lehrbuch der Experimentalphysik*, vol. 1, *Allgemeine Physik und Akustik* (Leipzig, 1882), 8. Leopold Pfaundler, ed., *Müller-Pouillet's Lehrbuch der Physik und Meteorologie*, 9th rev. ed., vol. 1 (Braunschweig, 1886), 15. August Kundt, *Vorlesungen über Experimentalphysik*, ed. K. Scheel (Braunschweig: F. Vieweg, 1903), xii. Emil Warburg, *Lehrbuch der Experimentalphysik für Studirende* (Freiburg i. B. and Leipzig, 1893), 1.

3. Franz Neumann, *Vorlesungen über mathematische Physik, gehalten an der Universität Königsberg*, ed. his students, *Einleitung in die theoretische Physik*, ed. Carl Pape (Leipzig, 1883), 1. Volkmann, *Einführung*, 41–43, 349. Voigt, *Kompendium der theoretischen Physik*. Kirchhoff, *Theorie der Wärme*, 2. Helmholtz, *Dynamik discreter Massenpunkte*. Hertz, *Principles of Mechanics*. Heinrich Weber, *Die partiellen Differential-Gleichungen der mathematischen Physik. Nach Riemann's Vorlesungen*, 4th rev. ed., 2 vols. (Braunschweig: F. Vieweg, 1900–1901), 1: 283.

4. Boltzmann, *Prinzipe der Mechanik* 1: 1; "Antritts-Vorlesung. Gehalten in Leipzig im November 1900," in "Zwei Antrittsreden," *Phys. Zs.* 4 (1902–3): 247–56, on 248.

mechanics, which was understood to be the way to acquire a comprehensive view of
nature, but "today this direct striving toward the highest goal has come to a halt, a
certain sobering has occurred." The reason is not that mechanics lacks sufficient
concepts, Planck explained, but that it allows too many mechanical explanations
for any process, each complicated and none clearly superior to the others. Planck
believed that only new ideas such as the communication of forces through a medium
might decide between the mechanical choices. The example of the whole recent
development of thermodynamics, which needs only its two fundamental principles,
might seem to indicate that physics was moving away from the "mechanical view of
nature" altogether. But this was to forget that the great connection of all the forces
of nature—the establishment of which, for Planck, was a timeless goal of physics—
is in its innermost nature an "identity" and that it can never be arrived at in physics
better than through mechanics. So Planck guardedly reaffirmed the continuing value
of the mechanical direction in theoretical physics in 1894, aware of the contemporary questioning of that direction, especially that based on the success of his own
preferred branch of physics, thermodynamics.[5]

There was another important reason for questioning the mechanical direction
in physics from the side of thermodynamics. Loschmidt had stated it in one form in
the 1870s in his objection to Boltzmann's interpretation of the second law, his argument turning on the reversibility with respect to time of the laws of the mechanics
of molecular motions. In another form, it was stated again in 1896 by Planck's
assistant at Berlin, Ernst Zermelo. Drawing on a theorem of 1890 by Poincaré,
which states that a mechanical system moving under conservative forces must in
time return arbitrarily close to its original state, Zermelo argued that the science of
irreversible processes, thermodynamics, could not be reduced to mechanics, at least
not to mechanics in its present form. For the second law to have general validity on
Boltzmann's terms, it would be necessary to assume highly improbable initial conditions, and to assume them would contradict the requirement for causality and, in
general, the "spirit of the mechanical view of nature itself." So unless there were to
be a new mechanics, for example, Hertz's, a "mechanical derivation of the second
law" must be assumed impossible.[6]

Zermelo's questioning of the mechanical direction was answered by Boltzmann,
who once again emphasized the probabilistic character of the second law of thermodynamics. From the standpoint of molecular theory the second law is a "mere law
of probability," not a theorem of ordinary mechanics. Zermelo had misunderstood
him, Boltzmann said: the second law does not require us to give up the use of
mechanical pictures; the recurrences predicted by Poincaré's mechanical theorem do
occur, but they are exceedingly rare and do not invalidate the probabilistic interpretation of the second law.[7]

Planck gave his opinion on the questions that divided Zermelo and Boltzmann

5. Max Planck, "Antrittsrede zur Aufnahme in die Akademie vom 28. Juni 1894," *Sitzungsber.
preuss. Akad.*, 1894, 641–44, reprinted in *Phys. Abh.* 3: 1–5, quotations on 1–2, 4.
6. Ernst Zermelo, "Ueber einen Satz der Dynamik und die mechanische Wärmetheorie," *Ann.* 57
(1896): 485–94, quotations on 492–94.
7. Ludwig Boltzmann, "Entgegnung auf die wärmetheoretischen Betrachtungen des Hrn. E. Zermelo," *Ann.* 57 (1896): 773–84, reprinted in *Wiss. Abh.* 3: 567–78, on 567. There was another round

in a letter to Graetz in 1897. Planck thought that they were the "most important questions that presently concern theoretical physics," and he was gratified to hear that in Graetz's Munich colloquium they had recently been debated. The probability calculus applies to the most probable state, Planck observed, but only mechanics can determine the state succeeding any given improbable state, and there is no reason to assume that physical changes always take place in the direction of greater probability. Planck believed that it was preferable to regard the second law as a natural law that is strictly valid than to try to save the kinetic theory of gases by assuming particular intitial conditions of the world, which would then have no further use. In all of this, Planck was in agreement with Zermelo. But he thought that Zermelo had gone too far in claiming that the "second law, as a law of nature, is altogether incompatible with any mechanical conception of nature. For it is quite a different matter if one passes from discrete mass points (like molecules in the gas theory) to continuous matter. I believe and even hope that in this way we will be able to find a rigorously mechanical interpretation of the second law; but this matter is obviously very difficult and requires time."[8]

Planck's questioning attitude entered his lectures at this time. The first lectures he published after coming to Berlin were on his chosen theme, the "entire field of Thermodynamics," which he presented from a "uniform point of view" and in which he included much of his own recent research. The year was 1897, soon after his argument with the energeticists, and in the text he again corrected their fundamental error of subsuming the second law of thermodynamics under the first. His own method was to introduce the first law and the "essentially" different second law as facts of experience, or as laws simply induced from the facts, and then to deduce from them the laws of physics and chemistry. He said that no one any longer disputed the first law, which is the principle of energy conservation applied to processes involving heat changes. But that was not the case with the second law, the law of increase of entropy, the unique "universal measure" of the fundamental irreversibility of all finite physical and chemical processes, and the more complex of the two laws. Planck allowed that the second law might have only limited validity, but if so, the limits are in nature and not in us; the reasoning behind the law makes use of "ideal" processes, which have nothing to do with our skill in experiment.[9]

In his lectures, Planck was concerned to place thermodynamics on the firmest possible foundations, on laws rooted in our universal experience of nature. He sharply separated its foundations in experience from its foundations within the "mechanical view of nature." He chose not to present the subject as a branch of mechanical theory, either the kinetic theory, which penetrates "deepest" but meets

of responses between Zermelo and Boltzmann; the whole exchange is discussed in Brush, *Motion We Call Heat* 2: 632–37.

8. Planck to Graetz, 23 May 1897, Ms. Coll., DM, 1933 9/30. This letter is discussed and quoted in part in Kuhn, *Black-Body Theory*, 27–28.

9. Max Planck, *Vorlesungen über Thermodynamik* (Leipzig, 1897), translated as *Treatise on Thermodynamics*, by A. Ogg (London, New York and Bombay: Longmans, Green, 1903), vii, ix, 38, 77–79, 86, 103.

with "essential difficulties," or Helmholtz's more general mechanical theory, which views heat as motion but leaves unspecified the type of motion. Planck believed that the mechanical theories had proven less fruitful than his own, which makes no assumption about the nature of heat. Planck did not consider his thermodynamic theory as final; one day, he said, it might yield to another theory that would correspond to "our aspiration for a uniform theory of nature, on a mechanical basis or otherwise," in which case the two thermodynamic laws "would not be introduced as independent, but would be deduced from other more general propositions."[10] This was not a casual speculation; for as we have seen, Planck elaborated on the unified mechanical theory of nature to come in his letter to Graetz at the time, in which he also thanked Graetz for his approving words about his published thermodynamic lectures.

Mach, who did not have Planck's belief of this time in the mechanical approach and certainly not his hopes for it, wrote in his treatise *The Science of Mechanics* that the "view that makes mechanics the basis of the remaining branches of physics, and explains all physical phenomena by mechanical ideas, is in our judgment a prejudice. . . . We have no means of knowing, as yet, which of the physical phenomena go *deepest*, whether the mechanical phenomena are perhaps not the most superficial of all, or whether all do not go *equally deep*." The "mechanical theory of nature" is historically intelligible, but it is nonetheless an "artificial conception" and destined, sooner or later, to be supplanted.[11] The theory does not correspond to the goal of physical science, which is the "*simplest* and *most economical* abstract expression of facts." The desired economy of scientific thought cannot be attained by mechanical hypotheses but only by concise, mathematical "descriptions," or "natural laws." Mach allowed that mechanical ideas might be usefully applied to illuminate the remaining branches of physics, heat, electricity, and so on, but it was to be understood that they were applied in the spirit of an analogy only. For Mach, physics is comparative, not mechanical.[12]

The "majority of modern scientists," Mach said, subscribed to Laplace's ideal of a total atomistic-mechanical determinism, and the atomistic as well as the mechanical met with Mach's principled skepticism. He had come to regard atoms as a convenience only, not to be taken as "realities behind phenomena." They are allowable as long as they are useful in research and are to be discarded once physical science is more advanced. The true elements of the world are not the atoms of the mechanical theory but colors, tones, and other sensations, the "real object of physical re-

10. Planck, *Thermodynamics*, ix–x.

11. Ernst Mach, *Die Mechanik in ihrer Entwickelung. Historisch-kritisch dargestellt* (Leipzig, 1883), translated as *The Science of Mechanics. A Critical and Historical Exposition of Its Principles*, from the 2d rev. German edition of 1889 by T. J. McCormack (Chicago, 1893), 495–96.

12. Ernst Mach, "Economical Nature of Physical Inquiry," talk at the Academy of Sciences in Vienna in 1882, in *Popular Scientific Lectures*, trans. T. J. McCormack (Chicago, 1895), 186–213, on 193, 207; *Mechanics*, 498; "On the Principle of Comparison in Physics," invited talk for the general session at the 1894 German Association meeting in Vienna, in *Popular Scientific Lectures*, 236–58, on 249–50.

search." This phenomenological understanding, Mach believed, would rid physical science of its unwanted metaphysics.[13]

"The world is our sensation," Mach said, and he anticipated that the results of the union of physiological psychology with natural science would "far outstrip those of the modern mechanical physics." Phenomena that appear purely mechanical must be recognized as being also physiological, which means being also chemical, electrical, thermal, and the like. From this recognition, it follows that—these are the concluding words of Mach's *Science of Mechanics*—"mechanics does not comprise the foundations, no, nor even a part of the world, but only an *aspect* of it."[14]

Mach carried over these ideas in 1896 to his treatise on the principles of the science of heat, treated historically and critically, which, like his treatise on mechanics, was based largely on his lectures. Thermodynamics provided him with concrete material for the further questioning of mechanical explanations. As before, he acknowledged that in their research on heat, physicists might find mechanical analogies useful, but he reminded his readers that analogies are not identities, and so they have no place in the physicists' presentations of the results of their research. In particular, Boltzmann's proof that the second law of thermodynamics corresponds to the principle of least action is not a proof of the mechanical nature of the processes but only a further development of the analogy between heat and living force. Equally unsatisfactory is the mechanical view of the second law based on the parallel between the increase of entropy and the increase of disordered motions among the molecules. "If one reflects," Mach said, "that an actual analogy of the *increase of entropy* in a purely mechanical system consisting of absolutely elastic atoms does not exist, one can hardly resist the idea that a violation of the second law—even without the help of [Maxwell's] demons—must be possible if such a mechanical system were the *real* foundation of heat processes." The entropy law went deeper. It and the energy law were the foundation of thermodynamics, not molecular mechanics.[15]

Energetic Foundations of Physics

The principle of conservation of energy provided Mach with an object lesson in the way physical science advances. Theoretical ideas such as the idea of heat as motion played a part in the genesis of the principle, but once established, it described an extensive set of facts directly and economically, without the need for any theoretical

13. Mach, "Economical Nature of Physical Inquiry," 188, 206–9. Mach's early endorsement of the atomic theory yielded to skepticism; in his talk on the history of the energy principle in 1872, he rejected atomism. Brush, *Motion We Call Heat* 1: 285–87. Erwin N. Hiebert, "The Genesis of Mach's Early Views on Atomism," in *Ernst Mach. Physicist and Philosopher*, vol. 6 of Boston Studies in the Philosophy of Science, ed. R. S. Cohen and R. J. Seeger (Dordrecht-Holland: D. Reidel, 1970), 79–106.

14. Mach, "Economical Nature of Physical Inquiry," 209, 212; *Mechanics*, 507. What mechanics needed, in Mach's view, was not extension but criticism; by the requirements of simplicity and economy of thought, Mach, in *Mechanics* and elsewhere, criticized Newton's absolute space and time, Newton's definitions and axioms of motion, and certain doctrines such as causality.

15. Ernst Mach, *Die Principien der Wärmelehre. Historisch-kritisch entwickelt* (Leipzig, 1896), 363–64, quotation on 364.

ideas.[16] Mach regarded the principle as more fundamental than any mechanical theorem of motion, and in this view he had the support of some other scientists.

In the second edition of his *Science of Mechanics* in 1889, Mach acknowledged Georg Helm's treatise on the energy principle, *Die Lehre von der Energie*, which had appeared since the first edition. Helm's goal was to "enunciate a general science of energetics," which was in accord with Mach's own researches. Mach said that he had "seldom read anything that . . . appealed in an equal degree to my mind."[17]

Helm was only then beginning his career: in 1888, the year following the appearance of his treatise on energy, he became extraordinary professor and four years later ordinary professor at the technical institute in Dresden. While he lectured on mechanics and mathematical physics in his new position, he advocated the program of energetics. In 1890, he carried out a task that he regarded as "essential" to the formal development of energetics: it was to derive from the energy principle the differential equations of mechanics, which included the equations of motion of a point moving freely or under constraints and Lagrange's form of the equations of motion. Helm concluded that since the energy principle leads to the equations of motion, it can be viewed analytically as a mechanical principle. He concluded, too, that the energy principle has advantages over other mechanical principles in that it extends beyond the purely mechanical to "all physical phenomena." Just as the energy principle has obvious advantages outside mechanics, it is now seen to have them—so Helm believed he had shown—inside mechanics, providing that science with its general founding principle. In this way, mechanics might be seen as being replaced by energetics as the source of irreducible principles for developing the laws of physics.[18]

Helm's derivation of mechanical laws embodied the goal of establishing physics on one fundamental principle: "*with each possible change the energy remains constant.*" In the past, the equations of motion had been derived from mechanical principles, and a unity of physical theory had been attained by extending those principles to the other parts of physics. Helm achieved a new unity by subsuming mechanics

16. Mach, "On the Principle of Comparison," 247–48.

17. The appendix to the 2d German edition of 1889 appears in Mach, *Mechanics*, 517. With the words quoted here, Mach commended not only Georg Helm's book, *Die Lehre von der Energie* (Leipzig, 1887), but also Josef Popper's *Die physikalischen Grundsätze der elektrischen Kraftübertragung. Eine Einleitung in das Studium der Electrotechnik* (Vienna, 1884). Mach approved of the energetic rejection of atomism and mechanism, but he was critical of energetics' own metaphysical tendencies. John T. Blackmore, *Ernst Mach. His Work, Life, and Influence* (Berkeley, Los Angeles, and London: University of California Press, 1972), 118–19.

18. The energy principle that Helm invoked in his derivation of the laws of mechanics was not the usual integral law, which is valid only for certain systems designated as conservative. From $T = \frac{1}{2}m(x'^2 + y'^2 + z'^2)$, he wrote the differential of the kinetic energy of a mass point: $dT = mx''dx + my''dy + mz''dz$; and he wrote the (in general not complete) differential of the work associated with the force X, Y, Z acting on the mass point: $dA = Xdx + Ydy + Zdz$. Then, from the "energy principle," the differential law $dT = dA$, he wrote the equation $(mx'' - X)dx + (my'' - Y)dy + (mz'' - Z)dz = 0$. For it to hold for all dx, dy, dz, Helm reasoned, the quantities in the brackets must all vanish: $mx'' = X$, $my'' = Y$, $mz'' = Z$, which are the equations of motion of a freely moving mass point. Georg Helm, "Ueber die analytische Verwendung des Energieprincips in der Mechanik," *Zs. f. Math. u. Phys.* 35 (1890): 307–20.

directly, and the extensions of mechanics indirectly, under the energetic foundations of physics.

In his efforts to establish the energetic viewpoint, Helm had an influential ally in Ostwald. Helm sent Ostwald his paper of 1890, remarking that a "unified construction of natural science from energetic ideas must, above all, be able to bring the most secure knowledge, mechanics, under this viewpoint."[19] Helm's and Ostwald's subsequent correspondence reveals that they did not always see eye to eye, but they agreed on the importance of an energetic derivation of mechanics and, in general, on the need to advance the understanding that energetic ideas are central to physics and chemistry.

While teaching chemistry at the polytechnic school in Riga and doing research on the catalytic action of monobasic acids, Ostwald had convinced himself that certain chemical processes could only be understood in energetic rather than atomistic terms. By his own testimony, his reading of Gibbs's papers on thermodynamics influenced this direction of his thought. In his inaugural lecture at Leipzig in 1887, he outlined a program of energetics, advancing it as an alternative to the program of atomism. In the second edition of his textbook on physical chemistry in 1892, he stressed energetic ideas and avoided all atomistic ones. At about the same time, in a series of papers he argued against the reduction of physics and chemistry to mechanics and for the reduction of the mechanical concept of mass, or matter, to that of energy. In connection with the replacement of mass by energy as a fundamental concept, he proposed changing the measures of exact natural science from the absolute system of mechanics to an energetic system in which the basic units are energy, length, and time.[20]

Following closely the writings on energetics, Boltzmann wrote to Ostwald in 1892 asking about his and Helm's ideas on the derivation of the fundamental equations of mechanics from the energy principle. He was then preparing for publication the second part of his lectures on Maxwell's electromagnetic theory of light, and he thought that the fundamental equations of this theory might be similarly derived, especially since he was not quite satisfied with their derivation from "purely mechanical principles."[21]

Now as later, Boltzmann showed his appreciation for what he regarded as promising about the energetics program. But his and Ostwald's preferred methods were largely opposed, even as they discussed them in their correspondence in a spirit of mutual respect. Concerning their presumed principal difference, Boltzmann explained his position to Ostwald: "To the dogma that nature can only be explained

19. Helm to Ostwald, 20 Jan. 1891, in *Briefwechsel . . . Ostwalds*, 73. Niles R. Holt, "A Note on Wilhelm Ostwald's Energism," *Isis* 61 (1970): 386–89, on 386–87.

20. Holt, "A Note." Wilhelm Ostwald, *Lehrbuch der allgemeinen Chemie*, vol. 1, *Stöchiometrie*, 2d ed. (Leipzig, 1891); "Studien zur Energetik," *Verh. sächs. Ges. Wiss.* 43 (1891): 271–88, and 44 (1892): 211–37. Erwin N. Hiebert, "The Energetics Controversy and the New Thermodynamics," in *Perspectives in the History of Science and Technology*, ed. D. H. D. Roller (Norman: University of Oklahoma Press, 1971), 67–86, on 75.

21. Boltzmann to Ostwald, 14 Apr. 1892, in *Briefwechsel . . . Ostwalds*, 7–8.

mechanically (by atomic motions), *I* would not like to oppose the other, namely, that it cannot and may not be explained by them. This is how I would express it according to my conviction; I do not argue with you if you express it according to yours."[22]

Boltzmann was not fully satisfied with this private discussion of his and the opposing energetic directions in physical science. He wanted to "provoke" a public discussion about the kinetic theory of gases, such as the one he attended in 1894 at the British Association meeting, and he regarded energetics as an appropriate theme for the coming meeting of the German Association in Lübeck in 1895. He encouraged Ostwald, as one of the "main representatives" of energetics, to participate. For Ostwald, the meeting provided a welcome occasion to bring before a wider audience the arguments for the superiority of energetics over the mechanics of atoms.[23]

In an address to a general audience at the Lübeck meeting, Ostwald claimed that almost all scientists subscribed to "scientific materialism," the view that matter and motion are the fundamental concepts and that every phenomenon is produced mechanically from atoms and the forces acting on them. He acknowledged that particular phenomena of heat and chemistry can be explained mechanically by the use of mechanical analogies, but he denied that the totality of the phenomena can be explained this way. It is evident, he said, from the reversibility of the equations of motion that mechanics cannot explain the irreversible events in our experience. The mechanical world picture cannot be defended scientifically, Ostwald concluded, it belongs to metaphysics.[24]

To overcome this scientific materialism—a desirable objective according to Ostwald—it must be recognized that the object of science is not to construct mechanical pictures but to connect measurable quantities with one another. The quantities of energy, space, and time are all that need enter the equations with which we describe natural phenomena. Energy replaces force and matter, the concepts of mechanics; energy is all we experience directly; energy is the most general invariant known to science. The abstract concept of energy offers a unified world view as clear as the mechanical one and without the latter's difficulties.[25]

In the physics section at the Lübeck meeting, Helm was invited to give a report on the current state of energetics, and following the example of the British Association, Boltzmann asked Helm to prepare it for publication before the meeting.[26] In his report, Helm extended the energy principle to the "fundamental formula of en-

22. Boltzmann to Ostwald, 11 June 1892, *Briefwechsel . . . Ostwalds*, 10–11.

23. Boltzmann to Ostwald, 1 June 1895, *Briefwechsel . . . Ostwalds*, 21–22.

24. Wilhelm Ostwald, "Die Ueberwindung des wissenschaftlichen Materialismus," *Verh. Ges. deutsch. Naturf. u. Ärzte*, vol. 67, pt. 1, 1st half (1895): 155–68.

25. Ostwald, "Die Ueberwindung."

26. At the meeting of the German Association two years before, in 1893, the physics section had decided to invite a report on a critical topic each year. In Vienna, in 1894, a report on radiation was given, and there a commission consisting of Boltzmann, Lang, Quincke, and Eilhard Wiedemann decided on the topic for the following year's meeting in Lübeck. *Verh. Ges. deutsch. Naturf. u. Ärzte*, vol. 67, pt. 2, 1st half (1895): 28. *Ann.* 55 (1895): i–ii. Helm to Ostwald, 27 Apr. 1895, in *Briefwechsel . . . Ostwalds*, 79–80.

ergetics."[27] Historically this extension had come about through two "essentially different" directions in physics, the mechanical and the thermodynamic. The mechanical direction Helm was concerned with was not the older mechanical world view
but the principles of perpetuum mobile and of analogy, which served to "transfer the
dynamical equations to all natural phenomena"; it, like the thermodynamic direction, shares the goal of energetics: to describe natural phenomena by a quantitative
method based on the energy principle. The "realism" of energetics consists in treating directly the various and equivalent forms of energy of our experience instead of
absorbing them in a mechanical picture.[28]

Following Helm's report at the meeting there was, in his words, a "stiff
fight." Apparently his only supporter in the debate was Ostwald, while his critics
included Boltzmann, the Göttingen mathematician Felix Klein, Nernst, Ostwald's
Leipzig colleague Arthur von Oettingen, Sommerfeld, and others. Boltzmann led
off the criticism, accusing energetics of being nothing more than a classification,
a mere "natural history" of various energies: for each of the branches of physics,
heat, electricity, magnetism, and light, energetics proposes a distinct form of
energy, each requiring its own laws. The mechanics of atoms, which energetics
renounces, is a route to "unified pictures" for these laws. From the fact that the
"mechanical conception of nature is not yet complete, it may not be concluded
that energetics is to be preferred, since the latter is much further still from completion."[29]

The well-attended debate on energetics was sufficiently lively that it was continued after a two-day interruption. At the second meeting, Helm expressed his disappointment over the unsympathetic reception of energetics in the previous debate; he
had assumed that he had been invited there "out of acknowledgment of at least the
most essential aspects of his scientific view," but instead Boltzmann had presented
his investigations as "erroneous" and his success as "vanishingly small." Helm hoped
that energetics would "survive rather well the execution intended for it," but he did
not wish to discuss it any further. Boltzmann in turn objected to the energeticists'
attack on mechanics: "The adherents of the old theory never objected to the attempt to continue to develop energetics alongside the old theoretical physics to see
if at some time it might achieve something approximately similar. Energetics, on
the other hand, had called the old theoretical physics an outmoded view and declared war on it. . . . It is therefore the old theoretical physics that one must wish

27. The fundamental formula is: $dE \leq \Sigma I \cdot dM$, which says that the total change in energy E is
equal to, or less than, the sum of products of intensities I and changes in corresponding capacities dM.
The inequality sign in the formula is required for irreversible processes. Helm's report, "Ueber den derzeitigen Zustand der Energetik," is briefly summarized in Verh. Ges. deutsch. Naturf. u. Ärzte, vol. 67, pt.
2, 1st half (1895): 29–30; a fuller version is "Ueber den derzeitigen Zustand der Energetik," Ann. 55
(1895): iii–xviii.

28. Helm, "Energetik," Verh., 29, and Ann., iv.

29. Helm to his wife, 17 Sept. 1895, in Briefwechsel . . . Ostwalds, 118–19. Hiebert, "Energetics
Controversy," 68–69. The published debate following Helm's report mentions as participants Helm, Ost-
wald, Boltzmann, Nernst, Oettingen, Ebert, Klein, and Hans Lorenz, among others. Verh. Ges. deutsch.
Naturf. u. Ärzte, vol. 67, pt. 2, 1st half (1895): 30–33. Boltzmann quoted in published debate, 30–31.

will be the victor in this war that was forced on it."[30] This concluding debate closed with an ovation for Helm, and he went away believing that he and Ostwald would look back on the day with satisfaction. He anticipated that the issue of energetics would be relegated to "literary discussion" from then on.[31]

The parties to the energetics debate of 1895 continued their arguments in the pages of the *Annalen der Physik* and elsewhere. Concerned about the energetic treatment of the "most cultivated discipline of theoretical physics, mechanics," Boltzmann pointed out an elementary error in the mathematics of Helm's derivation of the laws of mechanics.[32] He objected to Ostwald's claim that "today everyone thinks of atoms and forces as the final realities." He said that the "view that no other explanation can exist except that of the motion of material points, the laws of which are determined by central forces, had generally been abandoned long before Mr. Ostwald's remarks" at the Lübeck meeting. Mechanics provided only a picture, Boltzmann went on; it was the only rigorously developed picture to date, and it might be capable of being brought to perfection in the future. But Boltzmann allowed that energetics, upon further development, might "still be of the greatest use for science."[33] In an address in 1896, he emphasized his openness to energetic viewpoints, characterizing himself as a "passionate energeticist" because of his interest in the analogies between the various forms of energy. Only his enthusiasm did not extend to Helm and Ostwald's belief that the energy laws are the "fundamental laws of theoretical physics."[34] Ostwald published a rebuttal to Boltzmann in the *Annalen* in which he denied that the mechanical world view was a dead issue. He said that Boltzmann confused his own understanding with that of the majority of scientists, who not only adhered to the mechanical hypothesis but often regarded it as an "obvious truth" about nature.[35]

30. Published debate, 32–33.

31. Helm to his wife, 19 Sept. 1895, in *Briefwechsel . . . Ostwalds*, 119–20.

32. In the equation, derived from the energy principle, $(mx'' - X)dx + (my'' - Y)dy + (mz'' - Z) dz = 0$, Helm regarded dx, dy, dz as independently variable, as we showed in note 18 above. By setting dy and dz, but not dx, equal to zero, he obtained $mx'' - X = 0$, but by setting $dy = dz = 0$, he also assumed $y' = z' = 0$, since $dy = y'dt$, and $dz = z'dt$. So Helm proved that for the special case in which the material point has a nonvanishing velocity in the x-direction only, the equation $mx'' - X = 0$ holds. He did not prove that the three equations for the three coordinate directions all hold. In short, as Boltzmann pointed out, Helm confused the differential dx with the variation δx. Ludwig Boltzmann, "Ein Wort der Mathematik an die Energetik," *Ann.* 57 (1896): 39–71, on 40–41. That Boltzmann's criticism was effective is indicated, for example, by Voss's dismissal of Helm's derivation of 1890; Voss pointed out that with Helm's failure, the possibility of an energetic foundation of the laws of mechanics was not disproved in general; Pierre Duhem's derivation, for example, did not make the same error, and Voss withheld judgment on its future significance for theoretical mechanics. Aurel Voss, "Die Prinzipien der rationellen Mechanik," 1901, in *Encykl. d. math. Wiss.*, vol. 4, *Mechanik*, ed. Felix Klein and C. Müller, pt. 1 (Leipzig: B. G. Teubner, 1901–8), 3–121, on 115–16.

33. Boltzmann, "Ein Wort," 64, 71.

34. Ludwig Boltzmann, "Ein Vortrag über die Energetik," 11 Feb. 1896, *Vierteljahresber. d. Wiener Ver. z. Förderung des phys. u. chem. Unterrichts* 2 (1896): 38, reprinted in Boltzmann, *Wiss. Abh.* 3: 558–63, on 558.

35. Wilhelm Ostwald, "Zur Energetik," *Ann.* 58 (1896): 154–67, on 160.

For his well-known preference for the methods of thermodynamics over those of the kinetic theory of gases, Planck might have been expected to side with Helm and Ostwald and oppose Boltzmann.[36] But that was not where he came down on the issue. Planck had not taken part in the debate in Lübeck, but from the early 1890s he had taken a serious interest in energetics, especially in connection with the second law of thermodynamics, which Ostwald did not regard as a fundamental law of physics. Planck corresponded with Ostwald on the subject, and in 1896 he told Ostwald that he felt it was "high time" that he made his objections public.[37] He published his criticism of the "new energetics" in the *Annalen der Physik*; there he did not defend the mechanical world picture, which, he said, would require a deep and difficult study; rather, he demonstrated the inadequacy of energetics, a simpler task. The value of energetics for mechanics was much less than its supporters believed, Planck said; his more serious objection was that energetics failed to recognize a fundamental distinction between reversible and irreversible processes in nature. In general, Planck regarded energetics as lacking sound foundations and methods, as confusing its proofs with disguised definitions, and as being occupied with metaphysics instead of with science. He protested against any further development of energetics in its recent direction, since it had produced nothing of scientific value.[38] Its only success had been to incite younger scientists to "dilettantish speculations instead of a thorough absorption in the study of the present masterworks and thereby to lay fallow for years a broad and fruitful area of theoretical physics."[39]

From the side of energetics, Helm published a second treatise, *Die Energetik nach ihrer geschichtlichen Entwickelung*, conceived in the "strife" of the great debate at Lübeck. In this work, Helm presented energetics as a "unified development of thought"; he was concerned that energetics "be understood as a whole," not piecemeal as its opponents tended to understand it. He insisted that thermodynamics was the more complete and fruitful direction and supposed that anyone who did not see it this way still silently yearned for a real mechanical world behind the mechanical picture. He discussed the thermodynamic and mechanical directions of energetics

36. Max Planck, "Allgemeines zur neueren Entwicklung der Wärmetheorie," *Zs. f. phys. Chemie* 8 (1891): 647–56, reprinted in *Phys. Abh.* 1: 372–81, on 372–73. Hiebert, "Energetics Controversy," 73.

37. Planck to Ostwald, 27 Dec. 1895, in *Briefwechsel . . . Ostwalds*, 61.

38. Max Planck, "Gegen die neuere Energetik," *Ann.* 57 (1896): 72–78, reprinted in *Phys. Abh.* 1: 459–65. Hiebert, "Energetics Controversy," 76. The other parties to the debate continued to argue their positions. Helm responded to Boltzmann's and Planck's criticisms in the *Annalen der Physik* in 1896, and Boltzmann responded in the same place to all three. "I along with all who were on my side in Lübeck am no less convinced of the fundamental importance of the energy and entropy principles than Mr. Helm," Boltzmann said in a talk on energetics at the Düsseldorf meeting of the German Association in 1898, but he then went on, as he did in other talks and publications, to argue against what he saw as excessive claims made by the energeticists and to argue for the further development of theoretical physics through mechanical pictures. Georg Helm, "Zur Energetik," *Ann.* 57 (1896): 646–59. Ludwig Boltzmann, "Zur Energetik," *Ann* 58 (1896): 595–98; "Zur Energetik," *Verh. Ges. deutsch. Naturf. u. Ärzte* 70 (1898): 65–68, 74, reprinted in *Wiss. Abh.* 3: 638–41, quotation on 638.

39. Planck had the same concern as Boltzmann, who observed that young scientists "who do not have the mathematical criticism necessary for successful work in the area of theoretical physics" had begun to look to energetics for its promise of quick and easy rewards. Boltzmann, "Ein Wort," 64. Planck, "Gegen die neuere Energetik," 465.

and other matters he had taken up in his 1895 report. In his account of the mechanical direction, for example, he discussed the analogy principle, which had originated in Maxwell's use of Lagrange's equations of motion in electromagnetic theory,
and which Helmholtz had extended to thermodynamics and other parts of physics;
these Lagrangian studies, Helm claimed, embodied a thoroughly energetic spirit.[40]

Feelings ran high during the energetics controversy, and the participants repeatedly assured one another that the criticisms were not meant personally. Boltzmann,
as we have seen, was soon to join Ostwald at Leipzig, in large part through Ostwald's
efforts. In this connection Ostwald wrote to Boltzmann: "You see that I do not value
our scientific differences of opinion so highly as to lose hope for a prosperous and
useful cooperation. On the contrary, I hope for a great success for myself and for
science." Boltzmann wrote back: "I see that you are not personally angry with me,
but on the contrary your friendship, in which I have always prided myself, has remained unchanged."[41]

The draft report from the Leipzig faculty to the Saxon ministry recommending
Boltzmann's appointment contained a reference to Boltzmann's and Ostwald's recent
controversy. Boltzmann was the "main representative of the kinetic theory of physics
and chemistry," and in Ostwald, Leipzig already had the "main representative of the
energetics direction." That opened up an "extremely interesting perspective for a
scientifically useful exchange of ideas from both directions." Having second
thoughts, the faculty crossed out this whole passage, perhaps to spare the ministry
from contemplating what could—but did not—turn out to be a collision.[42]

If thermodynamics pointed the way to energetics in Germany, electromagnetism
at least pointed the way to an enhanced role for energy considerations. In this,
German physicists took their starting point in the work of British physicists in the
mid-1880s: J. H. Poynting concluded from Maxwell's theory that electromagnetic
energy moves continuously through space in trajectories normal to the electric and
magnetic oscillations, and Oliver Lodge extended Poynting's conclusion from electromagnetic energy to energy in general and in this way extended the principle of
conservation of energy.[43] German physicists responded to the notion of the flow,
identity, and localization of energy with sharply differing opinions.

One who responded positively was Wilhelm Wien, then a Berlin University
graduate who was about to habilitate there. Wien wrote to Hertz in 1890 to ask his
opinion of research he had undertaken into the question of the localization of en-

40. Georg Helm, *Die Energetik nach ihrer geschichtlichen Entwickelung* (Leipzig, 1898), quotations from
"Vorwort," v–vi. The Lagrangian methods of Helmholtz and others are discussed in the eighth and last
part of this treatise.

41. Ostwald to Boltzmann, 9 Dec. 1898; Boltzmann to Ostwald, 13 Dec. 1898, in *Briefwechsel . . .
Ostwalds*, 22–24.

42. Report of the Leipzig U. Philosophical Faculty to the Saxon Ministry of Culture and Public
Education, 12 Mar. 1900, Boltzmann Personalakte, Leipzig UA, PA 326.

43. J. H. Poynting, "On the Transfer of Energy in the Electromagnetic Field," *Phil. Trans.* 175
(1884): 343–61. Oliver Lodge, "On the Identity of Energy: In Connection with Mr. Poynting's Paper on
the Transfer of Energy in an Electromagnetic Field; and on the Two Fundamental Forms of Energy," *Phil.
Mag.* 19 (1885): 482–87.

ergy. He told Hertz that, like Poynting and Lodge, he wanted to endow energy with properties analogous to those of matter. His guiding idea was that the individual parts of energy have a traceable motion; his teacher Helmholtz did not like the idea, and Wien now tried it out on Hertz. Hertz did not like it much either, for reasons he included in his first paper on the equations of Maxwell's electrodynamics: given what we know of energy, he said, it is questionable if it "makes any sense at all to localize energy and to trace it from point to point."[44]

In the physics section of the German Association meeting that year, 1890, Wien reported on the current state of the theory of energy. Through the conservation law, he said, energy had acquired an objective significance comparable to that of matter: energy is immanent in all things and, along with matter, is responsible for the phenomena that act on our senses. Moreover, energy, through its conservation law, can be given an interpretation similar to that of matter: matter that disappears at one place, by the conservation law of matter, must reappear at another and do so by a continuous motion from the one place to the other. If the word "matter" is replaced by the word "energy," Wien argued, the statement still holds. We are allowed, then, to speak of a "current" of energy and, in cases in which only kinetic energy is involved, of the "velocity" of the energy. The equation of continuity applies to energy as well as to matter, and, in general, the laws of energy currents follow those of fluid mechanics. Wien explained that this new understanding of energy was made possible by the recent replacement of the physics of distance forces by the physics of continuous fields. In the report, Wien dealt chiefly with electromagnetic radiation. For this subject, the idea of energy current has most value; it has less for mechanics, but some there, too. Wien thought that any misgivings about the idea had to be epistemological in origin, not scientific. He concluded that this new way of regarding energy promised, as no other did, a unified view of all natural phenomena.[45]

Subsequent discussions brought out the difficulties of treating energy as analogous to, and of equal status with, matter. To speak of energy "particles," or "atoms," for example, could seem to be going too far in the direction of the "materialization of energy." Writers on the subject noted that energy particles could not be individualized as material particles are; moreover, energy particles are not impenetrable and do not have a natural volume, and no one yet had gone so far as to speak of indivisible atoms of energy or of the "inertia" of moving energy. In general, writers exercised caution in drawing physical conclusions from the idea of energy currents in electromagnetic theory.[46]

44. Wien to Hertz, 18 Mar. and 20 June 1890, Ms. Coll., DM, 3059, 3061. Heinrich Hertz, "Ueber die Grundgleichungen der Elektrodynamik für ruhende Körper," Ann. 40 (1890): 577–624, reprinted in his Ges. Werke 2: 208–255, on 234, and in the "Nachträgliche Anmerkungen" to this paper, Ges. Werke 2: 293–94. Hertz also pointed out that according to Poynting's theory, closed energy currents must occur in a static field produced by a magnet and an electrified body; to Hertz this seemed physically improbable (p. 234).

45. Wilhelm Wien, "Die gegenwärtige Lage der Energielehre," Verh. Ges. deutsch. Naturf. u. Ärzte, vol. 63, pt. 1 (1890): 45–49. Wien extended the theory of energy currents to fluids and elastic solids in "Ueber den Begriff der Localisirung der Energie," Ann. 45 (1892): 685–728.

46. In his text on Maxwell's theory, Föppl assumed the possibility of fixing the identity of a quantity

Although the specific program of energetics did not command a large following among physicists,[47] the importance of the concept of energy for physics was disputed by none. Around the turn of this century, certain energetic notions entered lectures and texts on physics, making it clear to the student that the energy concept, through the conservation principle and the transformability of the different forms of energy, connects the various departments of physics. Lommel's experimental physics text, which appeared in new editions at about yearly intervals, followed the historical order of presentation, beginning with mechanics. The only reason it did so was because beginners did not yet understand energy; ideally, Lommel said, a text "should necessarily begin with an empirical table of the forms of energy," which are simply "different manifestations of one and the same essence." Another successful text on experimental physics was Riecke's, which also began with mechanics rather than with the energy principle. Riecke believed in the pedagogical importance of the historical approach, but he left his readers in no doubt that energy is the most important concept in physics; for no other "penetrated and united in the same measure the various areas of phenomena." Mechanical pictures of molecules and their forces contain some arbitrariness, and Riecke preferred to treat molecular phenomena on the basis of their energy forms: molecular energy is bound to directly measurable surface and volume energies of bodies, an understanding Riecke attributed to his reading of Ostwald's studies on energetics. Ebert's text on experimental physics proceeded from the view that the law of conservation of energy yields the special laws relating to "all individual forms of energy relatively simply and directly." Ebert's text was exceptional, according to one reviewer: it could claim "for once to have taken seriously a scientifically unobjectionable, energetic presentation."[48]

Energetics did not gain the general recognition among physicists that Ostwald and Helm had hoped it would, but their promotion of it stimulated a general discussion of the foundations of physics. What advantages energetics enjoyed as an alternative to mechanical physics derived from thermodynamics, to which energetics as-

---

of energy, but he cautioned against the view, to which physicists seemed "more and more inclined," of seeing in the motion of localized, individualized quantities of energy the essence of all natural phenomena. August Föppl, *Einführung in die Maxwellsche Theorie der Elektricität* (Leipzig, 1894), 293–96. In his energetics treatise, Helm said that objections such as Hertz's to the idea of the motion of energy are valid only if the motion is regarded as mechanical and if the energy is regarded as a real substance; if the motion of energy is regarded as nothing more than an analogy, the objections fall away. Helm, *Energetik*, 349–50. Gustav Mie, in his Habilitationsschrift for mathematical physics at the technical institute in Karlsruhe, developed the energy current as a necessary consequence of the idea, arising from the influence of Maxwell's electromagnetic theory, that all natural phenomena are produced by contiguous rather than by distance actions; in extending the idea of an energy current to other parts of physics, Mie accepted the localization of energy and calculated with "energy particles" while at the same time rejecting any "individualization" of the particles. Gustav Mie, *Entwurf einer allgemeinen Theorie der Energieübertragung* (Vienna, 1898).

47. Einstein, for example, said that Planck's criticism of energetics "undoubtedly exercised a significant influence on his colleagues" because it showed that "energetics is worthless as a heuristic method." Einstein, "Planck," 1077.

48. Eugen Lommel, *Experimental Physics*, translated from the 3d German ed. of 1896 by G. W. Myers (London, 1899), vii, 29. Eduard Riecke, *Lehrbuch der Experimental-Physik zu eigenem Studium und zum Gebrauch bei Vorlesungen*, 2 vols. (Leipzig, 1896), 1: 3, 197. Hermann Ebert, *Lehrbuch der Physik, nach Vorlesungen an der Technischen Hochschule zu München*, vol. 1, *Mechanik, Wärmelehre* (Leipzig and

signed a reforming role for all of physical science. Energetics had little to say about electromagnetism, which was beginning to be seen by many physicists as an alternative to mechanics as the theoretical basis of their science. The electromagnetic questioning of the mechanical foundations of physics was to prove more productive scientifically than the energetic questioning had been.

Electromagnetic Foundations of Physics

Early German writers on Maxwell's electromagnetic theory presented its relationship to mechanical principles with varying emphases. As we have seen, Boltzmann derived the equations of the theory from mechanical ideas in his textbook of 1891 and 1893.[49] Korn wrote at this time that it was the duty of every theorist working in electricity to try to derive Maxwell's equations, preferably from mechanical hypotheses, as Boltzmann had just done.[50] Ebert wrote that the great advantage of Maxwell's over other electrical theories was that its equations could be derived mechanically. Helm, Voigt, Richard Reiff, Sommerfeld, and others published mechanical studies of Maxwell's theory in the early 1890s.[51]

By contrast, in his Maxwellian textbook of 1894, Föppl based the electromagnetic equations on experimental facts; he provided them with a mechanical derivation, too, but at the end of his text, not at the beginning. Föppl valued Boltzmann's approach, but he did not think it was the easiest to follow. Like Föppl, Drude derived the equations of electromagnetism from experimental facts in his Maxwellian textbook of the same year, and like Föppl, again, he regarded his text as an introduction to Boltzmann's. Drude did not believe that beginning students needed to

Berlin: B. G. Teubner, 1912), vi. Ebert's text was reviewed by Clemens Schaefer in *Phys. Zs.* 15 (1914): 813, who noted that its energetic presentation gave experimental physics an "entirely different face." Felix Auerbach was typical in his cautious appraisal of energetics: he defined physics as the study of "energy phenomena, especially of the changes in place, modality, and quality that energy experiences without changing its total quantity"; but he did not commit himself to an energetic presentation, pointing out that energetics had yielded few results so far and that most physicists still looked to mechanics as the foundation of physics. Felix Auerbach, *Kanon der Physik. Die Begriffe, Principien, Sätze, Formeln, Dimensionsformeln und Konstanten der Physik nach dem neuesten Stande der Wissenschaft systematisch dargestellt* (Leipzig, 1899), 1–2, 177.

49. Boltzmann, *Maxwells Theorie*; also Ludwig Boltzmann, "Über ein Medium, dessen mechanische Eigenschaften auf die von Maxwell für den Electromagnetismus aufgestellten Gleichungen führen," *Ann.* 48 (1893): 78–99.

50. Arthur Korn, *Eine Theorie der Gravitation und der elektrischen Erscheinungen auf Grundlage der Hydrodynamik*, 2 vols. (Berlin, 1892, 1894), 2: 1. This work was the basis of Korn's Habilitationsschrift at Munich University, which the physics professor Lommel commended as belonging to the "ever increasing efforts to derive the phenomena mentioned, especially the electric and magnetic ones, from the principles of mechanics." Dean to Munich Philosophical Faculty, 1 July 1895, Munich UA, OCI 21.

51. Hermann Ebert, "Versuch einer Erweiterung der Maxwell'schen Theorie," *Ann.* 48 (1893): 1–24, on 4. Georg Helm, "Die Fortpflanzung der Energie durch den Aether," *Ann.* 47 (1892): 743–51. Woldemar Voigt, "Ueber Medien ohne innere Kräfte und über eine durch sie gelieferte mechanische Deutung der Maxwell-Hertz'schen Gleichungen," *Ann.* 52 (1894): 665–72. Richard Reiff, "Die Fortpflanzung des Lichtes in bewegten Medien nach der electrischen Lichttheorie," *Ann.* 50 (1893): 361–67. Arnold Sommerfeld, "Mechanische Darstellung der elektromagnetischen Erscheinungen in ruhenden Körpern," *Ann.* 46 (1892): 139–51.

study the physics of the ether from mechanical principles; moreover, he regarded it as an open question whether the equations of the ether would be referred back to the equations of mechanics or whether the reverse would prove more useful.[52]

In the early 1890s, physicists could be found who disparaged mechanical representations of electromagnetism altogether. Wien, for example, in 1892 criticized the attempts by Maxwell and his successors to found the electromagnetic system of equations on Newtonian mechanics. To Wien's way of thinking, these attempts were too complicated or hypothetical to correspond to the canons of good physical theory. He urged physicists to follow the examples of Hertz and Heaviside by regarding Maxwell's system of equations and concepts as closed: in that way, they would see that Maxwell's system is completely analogous to that of pure mechanics and that their only connection is through the concept of energy.[53] Over the next years Wien did not change his opinion about the undesirability of mechanical approaches to electromagnetism, but he changed his opinion about the connection of the two subjects. With many other physicists, as we will see, he came to view electromagnetism as the foundation of mechanics.

In his published lectures on thermodynamics in 1897, Planck observed that thermodynamics might eventually be derived from an electromagnetic theory if not from a mechanical one. Two years later Planck concluded a talk on Maxwell's electromagnetic theory by noting its relationship to thermodynamics: "Perhaps one day we could succeed to an electromagnetic theory of heat—not by means of special new hypotheses, but simply by the further development of Maxwell's ideas on the connection between light and electricity." For some time this prospect guided Planck in a new series of researches on the second law of thermodynamics.[54]

These new researches built upon the application of thermodynamics to radiation processes, a subject already well established, especially through work Planck was familiar with, Kirchhoff's around 1860 and Boltzmann's and Wien's more recent work. As early as his inaugural lecture at the Prussian Academy in 1894, Planck mentioned his interest in, and approach toward, the thermodynamics of radiation: "there is hope that we will also be able to achieve a more detailed understanding of those electrodynamic processes that are directly conditioned by temperature—as in

52. Föppl, Einführung, v–xi, 266–73. Drude, Physik des Aethers, vi.

53. Wilhelm Wien, "Ueber die Bewegung der Kraftlinien im electromagnetischen Felde," Ann. 47 (1892): 327–44, on 328. Another example from this time of the questioning of the mechanical direction in electrical theory is a treatise on mathematical physics by Carl Neumann, who thought that mechanical explanations in physics were incomplete, contradictory, and overly complex. Having studied the analogies between electrodynamics and mechanics, specifically hydrodynamics, which Helmholtz, Kirchhoff, Boltzmann, and others had developed, Neumann convinced himself that the analogies lacked "deep foundations." Moreover, the explanation of heat phenomena requires *specifically* thermal principles, and since heat is intimately related to electricity, Neumann did not expect "*merely* mechanical principles" to succeed with electricity either. Carl Neumann, Beiträge zu einzelnen Theilen der mathematischen Physik, insbesondere zur Elektrodynamik und Hydrodynamik, Elektrostatik und magnetischen Induction (Leipzig, 1893), iii–iv, 205–6.

54. Planck, Thermodynamics, ix; "Die Maxwell'sche Theorie der Elektricität von der mathematischen Seite betrachtet," Jahresber. d. Deutsch. Math.-Vereinigung 7 (1899): 77–89, reprinted in Phys. Abh. 1: 601–13, on 613.

heat radiation in particular—without having to make the laborious detour through the mechanical explanation of electricity."[55] The following year, 1895, Planck submitted his first paper on the subject to the Prussian Academy. Building upon Hertz's treatment of electric oscillations using Maxwell's theory, Planck analyzed the absorption and emission of electromagnetic waves by electric resonators of dimensions small relative to the wavelength; he regarded this process as a way of understanding thermal equilibrium. In 1896, he followed up this first paper with another to the Prussian Academy, in which he introduced the notion of the damping of oscillations through radiation. By far the most important property of the damping is that it is conservative, which suggested to Planck the "possibility of a general explanation of irreversible processes through conservative actions—a problem that faces theoretical-physical research with greater urgency every day."[56]

Planck incorporated his results of 1895 and 1896 in his next work, which he entitled "On Irreversible Radiation Processes"; it came out in a series of papers in the proceedings of the Prussian Academy between 1897 and 1901.[57] Planck made clear his objectives from the start: the principle of conservation of energy and the principle of the increase of entropy should be placed on the same foundation, and since the energy principle requires that all natural processes be governed by conservative forces, the "fundamental problem" of theoretical physics is to refer the entropy principle to conservative forces too. Citing Zermelo's publications, Planck noted that the kinetic theory of gases assumes conservative forces; but because of the reversibility of the motions of molecules, every state of a system must eventually recur, so that this theory cannot completely solve the problem. In place of this theory, Planck proposed his analysis of a resonator in interaction with electromagnetic waves: the resonator is excited by absorbing waves and is damped by emitting them, and the incoming and the outgoing waves differ in form. In particular, the absorbed plane wave is emitted as a spherical wave, an irreversible change brought about by the conservative action of radiation damping of the resonator.[58] To facilitate the mathematical analysis of this irreversible influence on the waves, Planck imagined a geometrically simple arrangement: electromagnetic spherical waves are contained within a hollow reflecting sphere, at the center of which is an infinitesimal, linear resonator with a long, fixed wavelength and small damping.

Boltzmann was naturally interested in Planck's new approach to the second law

55. Planck, "Antrittsrede," 3.

56. Max Planck, "Absorption und Emission electrischer Wellen durch Resonanz," first published in the Prussian Academy's proceedings in 1895, then in *Ann.* 57 (1896): 1–14, and reprinted in *Phys. Abh.* 1: 445–58; "Über elektrische Schwingungen, welche durch Resonanz erregt und durch Strahlung gedämpft werden," first published in the Prussian Academy's proceedings in 1896, then in *Ann.* 60 (1897): 577–99, and reprinted in *Phys. Abh.* 1: 466–88, on 469–70.

57. This series of papers together with the preliminary papers in 1895 and 1896 is analyzed by Martin J. Klein in several publications, which include "Max Planck and the Beginnings of the Quantum Theory," *Arch. Hist. Ex. Sci.* 1 (1962): 459–79, on 460–64; "Thermodynamics and Quanta in Planck's Work," *Phys. Today* 19 (1966): 23–32, on 25–26; Ehrenfest, 218–24: by Hans Kangro, *Vorgeschichte des Planckschen Strahlungsgesetzes* (Wiesbaden: Franz Steiner, 1970), 125–48; and by Kuhn, *Black-Body Theory,* 34–37, 72–91. In our discussion of Planck here and, later on, of Planck and the quantum theory, we are largely indebted to the fundamental studies of Klein, Kangro, and Kuhn.

58. Max Planck, "Über irreversible Strahlungsvorgänge. Erste Mittheilung," *Sitzungsber. preuss. Akad.,* 1897, 57–68, reprinted in *Phys. Abh.* 1: 493–504, on 493–95.

of thermodynamics, which differed fundamentally from his own. He thought that Planck's might prove useful, but he was critical of an error in the reasoning: the equations of electrodynamics, like those of mechanics, could not do what Planck wanted of them. For these equations did not prohibit a reversal of the course of absorbed and emitted waves, so that if Planck wanted to explain irreversibility by electromagnetic waves, he had to assume a particular arrangement for their initial states.[59] Boltzmann's criticism resulted in Planck's introduction, in the fourth paper of the series, of an electromagnetic analog of molecular disorder, the concept Boltzmann had invoked in his derivation of the H-theorem; with the concept of "natural radiation," Planck excluded from consideration, as not occurring in nature, all radiation processes that lack the property of irreversibility; he did this by eliminating correlations between the phases of the electromagnetic waves and averaging the amplitudes of the field. Planck could now show that a certain function, which he called the combined "entropy" of the radiation and the resonator, can only increase, thereby proving the irreversibility of the radiation process under consideration. Planck thought that the extension of his theory, now incorporating the concept of natural radiation, to other cases would "succeed finally to a purely electromagnetic definition of entropy and with it also of temperature."[60]

In the fifth paper of the series, Planck generalized his approach. Instead of treating concentric waves about a resonator at the center of a reflecting sphere, he treated electromagnetic waves moving in all directions within a hollow reflecting cavity of any form and containing any number of resonators. The existence of an entropy of radiant heat is required by the second law of thermodynamics, so that if a body loses heat by radiation, its entropy decreases, which means that the radiation must acquire entropy for the law of entropy to be obeyed. Planck defined an expression for the "electromagnetic entropy," which, he showed, can only increase and which attains its maximum value only when the resonators and the electromagnetic field are in equilibrium. He regarded this definition as unique, a necessary consequence of the application of the principle of the increase of entropy to electromagnetic radiation. He also derived the law, which Wien had derived by other reasoning, for the distribution of energy over the wavelengths in the normal spectrum of blackbody radiation; Planck regarded this law, too, as a necessary consequence of the second law. Finally, because entropy determines temperature at equilibrium, Planck gave an electromagnetic definition of temperature.[61] At the close of the fifth paper, Planck introduced a "natural system of units." The formula for the radiation entropy contains two "universal" constants, a and b, the values of which Planck calculated from recent blackbody radiation measurements. Upon joining to these

59. Ludwig Boltzmann, "Über irreversible Strahlungsvorgänge I," Sitzungsber. preuss. Akad., 1897, 660–62; "Über irreversible Strahlungsvorgänge II," Sitzungsber. preuss. Akad., 1897, 1016–18; "Über vermeintlich irreversible Strahlungsvorgänge," Sitzungsber. preuss. Akad., 1898, 182–87, reprinted in Wiss. Abh. 3: 615–28.

60. Max Planck, "Über irreversible Strahlungsvorgänge. Vierte Mittheilung," Sitzungsber. preuss. Akad., 1898, 449–76, reprinted in Phys. Abh. 1: 532–59, on 533, 536, 556.

61. Max Planck, "Über irreversible Strahlungsvorgänge. Fünfte Mittheilung," Sitzungsber. preuss. Akad., 1899, 440–80, reprinted in Phys. Abh. 3: 560–600, on 592–97.

two constants the constant velocity of light and the gravitational constant, Planck deduced a complete system of measures for physics to replace the old, arbitrary, absolute system of centimeter, gram, and second.[62]

In late 1899, Planck submitted yet another long paper on irreversible radiation processes, this one to the *Annalen der Physik*. Planck repeated his point that Maxwell's equations alone are insufficient for the task of explaining the second law in its application to radiant heat. To these equations must be added the hypothesis of natural radiation, which describes radiation in which the energy is distributed "completely *irregularly*" over its constituent individual partial oscillations. This hypothesis "leads, in a purely electromagnetic way, necessarily to the validity of a law analogous to the second law of thermodynamics," just as in the kinetic theory of gases, the hypothesis of molecular disorder leads, in a purely mechanical way, to such a law.[63]

In his researches in the 1890s on the foundations of irreversible radiation processes, Planck worked with the equations of electromagnetism instead of those of mechanics. He did this while still holding out the prospect of an eventual derivation of Maxwell's equations from the laws of continuum mechanics. He left open, that is, the question of the foundations of physics. His results in the thermodynamics of radiation were unaffected by it, though as a theorist his interest in it was always there and would intensify in the future.[64]

In his work in the 1890s Planck did not yet make use of the electron concept, though he would in his later work. It was that concept, when joined to the electromagnetic theory, that would provide physicists with a well-defined program for placing physics on electromagnetic foundations. From the 1890s, German physicists increasingly studied, developed, and modified the electron theory, especially in the form given to it by the Dutch theoretical physicist H. A. Lorentz.

Primarily as a result of his electron theory, Lorentz acquired a reputation in Germany comparable to Boltzmann's. German faculties, as we have noted, looked repeatedly to Lorentz, as they did to Boltzmann, to fill ordinary professorships in theoretical physics. Although Lorentz declined the offers from Germany, he maintained close working relationships with German physicists, and through their researches, the near contemporaries Lorentz and Boltzmann urged German physicists in common directions: Maxwell's electromagnetic theory and molecular mechanics.

62. Planck, "Über irreversible Strahlungsvorgänge," 599–600. In the new system of measures, the natural units are chosen so that each of the four constants—*a*, *b*, and so on—has the value 1. The resulting new units are for length 4.13×10^{-33} cm, for mass 5.56×10^{-5} gr, for time 1.38×10^{-43} sec, and for temperature 3.50×10^{32} C°. The advantage of the new system is clearly not its convenience but its universality. Planck prized results of universal significance in physics, to which his work on the thermodynamics of radiation was directed. Klein discusses the significance of natural units for Planck in, among other places, "Thermodynamics and Quanta in Planck's Work," 26–27.

63. These observations Planck had made in earlier publications; the *Annalen* paper he intended mainly as a summary. But in a note he added in proof, he remarked that recent experiments showed departures from Wien's distribution law for blackbody radiation. That result had serious implications for Planck's theory, which he could not take up in the present paper. His long work on irreversible radiation processes had not, after all, come to a conclusion, but only to a brief pause. Planck, "Über irreversible Strahlungsvorgänge," *Ann.* 1 (1900): 69–122, reprinted in *Phys. Abh.* 1: 614–67, on 614–21, 662.

64. Kuhn, *Black-Body Theory*, 31.

From the beginning of his career, Lorentz was strongly drawn to the theoretical side of physics. The subject of his theoretical dissertation at the University of Leiden in 1875 was physical optics, which he treated systematically—the first to do so—from the standpoint of the new electromagnetic theory of light. His starting point was Helmholtz's action-at-a-distance theory of electromagnetism, since he regarded it as less dependent on unconfirmed hypotheses than Maxwell's contiguous-action theory. The quality of his dissertation and his scientific promise in general were early recognized. In 1877 he was appointed to the new chair of theoretical physics at the University of Leiden, the first chair of its kind in Holland. Lorentz's main interest in the 1880s was not electromagnetism and optics but the molecular-kinetic theory of heat; he published, for example, a correction of the proof of the *H*-theorem, which Boltzmann accepted. In the 1890s he returned to his earlier interests. Following Hertz, Lorentz now approached electromagnetism through contiguous rather than distance forces; but he had two objections to Hertz's treatment of the electrodynamics of moving bodies, and these prompted his first publication on the electron theory in 1892. First, because of conflicting optical evidence, Lorentz objected to Hertz's assumption that moving ponderable bodies carry with them the ether they contain; Lorentz preferred to regard the ether as Fresnel did, as immovable; for Lorentz the ether is stationary and completely transparent to bodies moving through it. Second, he objected to Hertz's postulation of the bare field equations. He referred to Maxwell's application of Lagrangian mechanics for deriving the electromagnetic equations without assuming a detailed mechanism as "one of the most beautiful" chapters in Maxwell's *Treatise on Electricity and Magnetism*. Remarking that "one has always tried to return to mechanical explanations," Lorentz continued to approach the subject in Maxwell's, rather than Hertz's, spirit. He noted that Boltzmann had been guided by the "same fundamental idea"; after completing his own paper, Lorentz said, Boltzmann's Munich lectures on Maxwell's theory came into his hands, which have as their "principal object the mechanical explanation begun by Maxwell."[65]

65. Tetu Hirosige, "Origins of Lorentz' Theory of Electrons and the Concept of the Electromagnetic Field," *HSPS* 1 (1969): 151–209, especially the discussion on 186–96. Lorentz's first publication on the electron theory is "La théorie électromagnétique de Maxwell et son application aux corps mouvants," *Arch. néerl.* 25 (1892): 363, reprinted in *Collected Papers* 2: 164–343, quotations on 168–69. Studies of Lorentz's electron theory include Gerald Holton, "On the Origins of the Special Theory of Relativity," *Am. J. Phys.* 28 (1960): 627–36; Stanley Goldberg, "The Lorentz Theory of Electrons and Einstein's Theory of Relativity," *Am. J. Phys.* 37 (1969): 982–94; K. F. Schaffner, "The Lorentz Electron Theory [and] Relativity," *Am. J. Phys.* 37 (1969): 498–513; Arthur I. Miller, *Albert Einstein's Special Theory of Relativity* (Reading, Mass.: Addison-Wesley, 1981), 25–40. Among recent studies of other electron theorists are Robert Lewis Pyenson, "Physics in the Shadow of Mathematics: The Göttingen Electron-Theory Seminar of 1905," *Arch. Hist. Ex. Sci.* 21 (1979): 55–89; Stanley Goldberg, "The Abraham Theory of the Electron: The Symbiosis of Experiment and Theory," *Arch. Hist. Ex. Sci.* 7 (1970): 7–25; Arthur I. Miller, "A Study of Henri Poincaré's 'Sur la Dynamique de l'Electron,'" *Arch. Hist. Ex. Sci.* 10 (1973): 207–328. In the following discussion of the electron theory, we draw frequently on Russell McCormmach, "H. A. Lorentz and the Electromagnetic View of Nature," *Isis* 61 (1970): 459–97; "Einstein, Lorentz, and the Electron Theory," *HSPS* 2 (1970): 41–87; "Lorentz, Hendrik Antoon," *DSB* 8 (1973): 487–500. We thank *Isis* and Charles Scribner's Sons, publisher of the *Dictionary of Scientific Biography*, copyright American Council of Learned Societies, for permission to excerpt from two of these articles.

Lorentz ascribed identical properties to the ether everywhere instead of the varying densities and elasticities of Fresnel and Neumann. Lorentz's ether has no mechanical connection with matter; the interaction of the two occurs solely through small, ponderable, rigid bodies carrying positive or negative charges, which he assumed to be contained in all molecules of ordinary bodies. He referred to them in 1892 as "charged particles," in 1895 as "ions," and only after 1899 as "electrons," from which his theory derived its permanent name. From a set of hypotheses about the stationary ether and electrons and from d'Alembert's principle, Lorentz mechanically deduced the equations of the electromagnetic field and the equations of motion of an electron in the field.

In contrast to Maxwell and Hertz, Lorentz provided a clear, simple interpretation of electric charge and current and of their relation to the electromagnetic field. He explained that a body carries a charge if it has an excess of one or the other kind of electrons, and an electric current in a conductor is a flow of electrons; correspondingly, a dielectric displacement in a nonconductor is a displacement of electrons from their equilibrium positions. The electrons of Lorentz's theory create the electromagnetic field, the seat of which is the ether; the field in turn acts ponderomotively on ordinary matter through the electrons embedded in material molecules.

Because Lorentz separated ether and matter, he needed only one pair of directed magnitudes—one electric and one magnetic—to define the field at a point, and this regardless of whether or not matter is present at the point. In this way he answered Hertz's formal objection to a stationary ether; namely, that it requires two sets of directed magnitudes to define the field at a point, one for matter and one for ether. Lorentz also answered Hertz's objection that particulate electric fluids belong to action-at-a-distance, not contiguous-action, electrodynamics. By means of his concepts of a stationary ether and of electrons transparent to it, Lorentz constructed a consistent electrodynamics in which, at once, he rejected action at a distance and retained particulate electric fluids. For this reason, Lorentz characterized his theory as a fusion of Continental and British, or Maxwellian, electrodynamics. He retained the clear understanding of electricity of the theories of Weber and Clausius, and at the same time he accepted the crux of Maxwell's theory, the propagation of electric action at the speed of light.[66]

In his next major presentation of the electron theory, in 1895, Lorentz no longer derived the equations of the theory from mechanical principles but postulated them instead. This time he systematically went over the problem of the effects of the earth's motion through the ether, which he had only touched on in 1892. Since the ether of the electron theory is not dragged by a body moving through it, the earth has an absolute velocity relative to it. The question arises whether or not the earth's absolute velocity is detectable through optical or electromagnetic effects of the accompanying ether "drift" or "wind." The magnitude of the effects of the wind

66. Lorentz, "La théorie électromagnétique," 229.

is measured theoretically by the ratio of the speed of the earth's motion v to the speed of light c. The ratio is small for the earth, but not so small as to be beyond the reach of observation.[67]

The effects of this wind, however, were not observed, and for his theory to be credible, Lorentz had to explain their absence. He showed that, according to the theory, an unexpected compensation of actions eliminates all effects of the ether wind to first-order approximation (neglecting terms involving the very much smaller second and higher powers of v/c). He analyzed the absence of first-order effects of the ether wind in phenomena such as reflection, refraction, and interference with the aid of a formal "theorem of corresponding states." The theorem asserts that to first-order accuracy, no experiments using terrestrial light sources can reveal the earth's motion through the ether. By introducing transformations for the field magnitudes and spatial coordinates and a "local time," Lorentz showed that to first-order approximation the equations describing a system in a moving reference frame are identical with those describing the corresponding system in a frame at rest in the ether, for which Maxwell's equations hold exactly.

In more technical terms, Lorentz's reasoning is as follows: To explain the absence of optical effects of the earth's motion through the ether, Lorentz transformed Maxwell's equations to axes attached to a moving dielectric. The equations he used for this purpose were the familiar Galilean equations from mechanics for the transformation of the spatial coordinates supplemented by the transformation, his own, of the absolute time to the "local" time t' of the dielectric moving with velocity v. To the first order in v/c, he showed that the equations that describe the passage of light in a moving dielectric are, in electromagnetic units:[68]

$$\text{div}' \, \mathbf{D}' = 0,$$
$$\text{div}' \, \mathbf{H}' = 0,$$
$$\text{rot}' \, \mathbf{H}' = 4\pi \, \frac{\partial D'}{\partial t'},$$
$$\text{rot}' \, \mathbf{E} = - \frac{\partial H'}{\partial t'},$$

These four equations are Maxwell's equations written in vector notation and, as the primes indicate, for axes moving with the dielectric. They have the same form as Maxwell's equations for a dielectric at rest, a correspondence with far-reaching con-

67. H. A. Lorentz, *Versuch einer Theorie der electrischen und optischen Erscheinungen in bewegten Körpern* (Leiden, 1895), reprinted in his *Collected Papers* 5: 1–137.

68. In these equations, $\mathbf{D}'$ is the dielectric displacement, $\mathbf{H}'$ the magnetic force, and $\mathbf{E}$ the electric force, which is proportional to $\mathbf{D}'$ through factors that depend on the nature of the dielectric and the frequency of the light. The shorthand terms "div'" and "rot'" stand for the vector operations of divergence and rotation, or curl, referred to the spatial coordinates and local time of the moving dielectric. Lorentz's definition of local time, $t' = t - (v_x/c^2)x - (v_y/c^2)y - (v_z/c^2)z$, was a mathematical convenience; his transformations for the field variables were guided by the physics of moving systems: $\mathbf{D}' = \mathbf{D} + 1/4\pi c^2 \cdot \mathbf{v} \times \mathbf{H}$ and $\mathbf{H}' = \mathbf{H} - 1/c^2 \cdot \mathbf{v} \times \mathbf{E}$. Lorentz used slightly different notation in his 1895 paper; namely, $\mathbf{p}$ for $\mathbf{v}$ and V for c.

sequences. If anywhere in the system at rest there is darkness, $\mathbf{D} = \mathbf{E} = \mathbf{H} = 0$, it follows that there will be darkness, $\mathbf{D}' = \mathbf{E} = \mathbf{H}' = 0$, at the corresponding place in the moving system. Since a ray of light is determined by the absence of light at its boundaries, the same laws for the reflection and refraction of light rays must hold in the moving as in the stationary system, and for the same reason the alternating patterns of dark and light resulting from the interference of light must be the same in both systems.

Lorentz's first-order approximation accounted for nearly all null experimental results in the optics and electrodynamics of moving bodies. Second-order experiments had been performed, however, in which theoretically no compensating actions should occur. The most important second-order experiments were A. A. Michelson's interferometer experiment in 1881 and his and E. W. Morley's more accurate repetition of it in 1887. Their experiments failed to produce evidence of second-order ether wind effects, which were expected on the basis of the stationary ether assumed by the electron theory. The only solution Lorentz could think of was the contraction hypothesis, which FitzGerald proposed independently at about the same time. Lorentz had published an approximate form of the hypothesis in 1892; in 1895 he published an exact form, which states that the arms of the interferometer contract by a factor of $\sqrt{1 - v^2/c^2}$ in the direction of the earth's motion through the ether. Lorentz regarded the hypothesis as dynamic, requiring the molecular forces determining the shape of the interferometer arms to propagate through the ether analogously to electric forces.[69]

Lorentz came into personal association with physicists in Germany for the first time when he accepted Boltzmann's invitation to address the physics section at the Düsseldorf meeting of the German Association in 1898. The subject, the state of motion of the ether, was central to the electron theory and to optics and electrodynamics generally. Paired with, and preceding, Lorentz's report was one by Wien, and both discussed the main experiments bearing on the decision between a moving and a stationary ether. Lorentz rejected Stokes's "dragged" ether as incompatible with stellar aberration, which provided the main support for Fresnel's and his own view of the ether as stationary. Wien discussed an implication of Lorentz's preferred stationary ether for the laws of mechanics. Whereas a moving ether presupposes a similarity between its properties and those of ordinary matter, a stationary ether has properties all its own. The stationary ether can, for example, exert forces but no forces can act on it, in violation of Newton's third law of motion, the law of action and reaction, as Lorentz had pointed out in 1895. Although Wien doubted that

---

69. Lorentz refined his theorem of corresponding states in later presentations of the electron theory, culminating in his paper of 1904 in which he gave a general proof, using the "Lorentz transformations," of the undetectability of the earth's absolute motion through the ether, replacing the separate explanations for the absence of first-order and of second-order effects. "Electromagnetic Phenomena in a System Moving with Any Velocity Less Than That of Light," *Proc. R. Acad. Amsterdam* 6 (1904): 809, reprinted in his *Collected Papers* 5: 172–97. The Michelson and Morley interferometer experiment and its relationship to the electron theory and the contraction hypothesis are discussed in Loyd S. Swenson, Jr., *The Ethereal Ether: A History of the Michelson-Morley-Miller Aether-Drift Experiments, 1880–1930* (Austin and London: University of Texas Press, 1972).

mechanics could account for all ethereal processes, he urged further theoretical study of this point. Lorentz responded to the "repeatedly expressed objection" cited by Wien by conceding that it is a deficiency of the electron theory that in it Newton's third law appears "insignificant and accidental," and he hoped that a clarification of concepts would explain how action and reaction between observable bodies is obeyed. There is a "certain satisfaction" in seeing the third law hold for elementary actions, he said, but there is no absolute need that it do so: the third law does not necessarily have universal validity, and so neither does Newtonian mechanics.[70]

Independently of Lorentz, the electron theory was introduced and developed by other physicists, notably by Wiechert, who was just beginning his teaching career at Göttingen, and by Joseph Larmor at Cambridge. In whatever formulation, the electron theory received a strong impetus from the experimental demonstration of electric particles. In 1897, the year of J. J. Thomson's experiments at Cambridge, several German physicists—especially Wien, Wiechert, and Walter Kaufmann, an assistant at the Berlin physics institute—experimentally established to their own satisfaction the particulate interpretation of cathode rays and then went on to develop it. These particles were to be identified with the hypothetical particles of the electron theory: universal atoms of electricity, the same whatever their source and however produced.

Even before his demonstration of electric particles, Thomson calculated a result that had an important bearing on their interpretation. From Maxwell's theory, he showed that the self-induction of a moving charged sphere results in an effective mass that varies with the velocity of the sphere. In 1898 at Göttingen, Des Coudres, referring to Thomson's and Wiechert's measurements on cathode rays, raised the possibility that the mass of the streaming negatively charged particles arises entirely from self-induction and is therefore only "apparent." That same year Wiechert repeated an earlier suggestion of his that all matter consists of positive and negative particles and, as a result, that not only the mass of cathode-ray particles but all mass may be electromagnetic in nature and not an original property of matter.[71] Here then was the prospect of a new mechanics based on an electromagnetic interpretation of mass and, beyond that, of an electromagnetic foundation of all of physics.

Lorentz gave close consideration to this prospect. In an address at Leiden in

70. H. A. Lorentz, "Die Fragen, welche die translatorische Bewegung des Lichtäthers betreffen"; Wilhelm Wien, "Ueber die Fragen, welche die translatorische Bewegung des Lichtäthers betreffen"; *Verh. Ges. deutsch. Naturf. u. Ärzte*, vol. 70, pt. 2, 1st half (1898): 56–65 and 49–56, respectively.

71. Theodor Des Coudres, "Ein neuer Versuch mit Lenard'schen Strahlen," *Verh. phys. Ges.* 17 (1898): 17–20. Emil Wiechert, "Bedeutung des Weltäthers," *Physikal.-ökonom. Ges. zu Königsberg* 35 (1894): 4–11, followed Helmholtz's proposal of 1881 of an atomistic constitution of electricity; Wiechert speculated that the "electric atom" is a local modification of the ether and that the mass of matter is partly or wholly electromagnetic in origin. In 1896, Wiechert formulated an electron theory independent of, but similar to, Lorentz's, in which he showed that upon the assumption that matter consists of charged particles, a complete electrodynamic theory of the interaction of matter and field can be expressed: "Über die Grundlagen der Elektrodynamik," *Ann.* 59 (1896): 283–323. In 1898, Wiechert continued to speak of the possibility that all inertia is electromagnetic, but he believed that it was still premature to assert that atoms of ordinary matter are nothing but aggregates of simpler electrical atoms: "Hypothesen für eine Theorie der elektrischen und magnetischen Erscheinungen," *Gött. Nachr.*, 1898, 87–106.

1900, he discussed the extent to which the phenomena of the various parts of physics can be regarded as consequences of the electron theory. He had explained theoretically the simpler of the magnetic splittings of atomic spectral lines, which in 1896 Pieter Zeeman had discovered with the aid of a diffraction grating of high resolving power. By assuming that electrons move within atoms, emitting light as they do, Lorentz and Zeeman could determine from the magnetic effect on the motion and the corresponding frequencies of the emitted light the ratio of the charge to the mass of the electron and the negative sign of the charge. It was an impressive achievement combining precision measurement and electron-theoretical calculation to probe the interior of the atom. Lorentz now looked upon atomic structure as an ultimate goal of the theory; he thought that chemical forces have their origin in electrons, and although he thought it was still premature to identify molecular forces with electric ones, he was confident that in his analysis of Michelson's experiment, he had proved that these forces propagate similarly through the ether. Of the great forces in nature, only gravitation seemed to remain outside the reach of the electron theory; but because electric charges are inseparable from ponderable matter, Lorentz was confident that gravity could not be unrelated to electromagnetism. The difficulty was that astronomical facts seemed to demand a velocity for gravitational actions much greater than that for electromagnetic ones. In the same year, 1900, Lorentz published a possible solution to the velocity question from the viewpoint of the electron theory, demonstrating at least one way of annexing gravitation to an electromagnetic physics. Lorentz's approach was inspired by Mossotti's gravitational theory of 1836, recently revived by Weber and Zöllner, according to which a ponderable particle is a composite of two opposite electric atoms, and the attraction between two such ponderable particles is greater than their repulsion. Substituting electrons for the older electric atoms and states of the ether for distance forces, Lorentz derived a gravitational law that is identical with Newton's when there is no motion between the two attracting bodies; but if there is motion, the new law of attraction involves velocity-dependent terms as well, and these, Lorentz noted, are analogous to terms appearing in the laws of Weber, Riemann, and Clausius. Applying his law to the secular motion of Mercury, he found that it does not fit nearly as well as Weber's. But he was not especially troubled, since his chief purpose had been to show that gravitation can be understood as an action that propagates through the ether with a speed no greater than that of light. Convinced that the properties of the ether, as revealed in electromagnetic investigations, set conditions on all theories, he supposed that the ether does not allow the communication of any action—gravitational, molecular, or electromagnetic—at speeds greater than that of light.[72]

In 1900 Lorentz gave another talk in the physics section of the German Association, and again he entered into an exchange with Wien. His subject was "apparent mass," and he and Wien discussed assumptions about the structure of the electron on which the formula for the apparent mass depends. The connection between

72. H. A. Lorentz, "Elektromagnetische Theorien physikalischer Erscheinungen," *Phys. Zs.* 1 (1899–1900): 498–501, 514–19, reprinted in *Collected Papers* 8: 333–52; "Considérations sur la pesanteur," *Versl. Kon. Akad. Wet. Amst.* 8 (1900): 603, reprinted in *Collected Papers* 5: 198–215.

this question and the larger question of the electromagnetic foundations for physics is clear from Wien's opening remarks:

> I have tried to go beyond Lorentz's point of view in the sense that I asked myself if we could not do just with the apparent mass and leave out the inertial mass and replace it by the electromagnetically defined apparent mass to give a uniform representation of the mechanical and electromagnetic phenomena. Up to now, after all, the magnetic and mechanical phenomena have been linked only by the energy principle. I have tried to pose the question of whether by starting from Maxwell's theory we could not attempt to encompass mechanics, too. This would provide the opportunity of founding mechanics on electromagnetism now that Lorentz has developed a conception of the law of gravitation according to which gravitation is said to be closely related to electrostatics. We would then have to assume that matter is composed of nothing but very small positive and negative charges which are at a certain distance from each other. Under these conditions, ponderable mass is not constant but a function of velocity; that is, we would also get terms that are functions of the even powers of the ratio of velocity and velocity of light. The numerical factor with which the second term is multiplied is a function of the curvature of the trajectory and also of the shape of the electric charge. As we choose slightly different forms of electric molecules, we obtain different numerical factors. They cancel out in the case of ordinary terrestrial motions since the velocity is very small. However, in the case of planetary motions we may perhaps come up with something, for in that case we get velocities for which second-order terms must be considered. If we assume a certain form of charge that leads to the simplest electromagnetic field, these terms enter [the calculation] in this way: the accelerations of two bodies due to gravitation are the same, except for a small numerical difference, as they would be if bodies of constant mass attracted each other according to Weber's law. The electromagnetically defined mass enters the calculation as if not Newton's but Weber's law held.[73]

Lorentz said that he agreed with Wien on the essential points.

Lorentz's work on gravitation encouraged Wien to publish in 1900 the similar ideas he had sketched in the discussion at the German Association meeting. In this paper, Wien assumed that ponderable matter is constituted of positive and negative charges, a point he believed would be "granted by all physicists today." He assumed further that the mass of these charges is entirely electromagnetic in origin. He said it was too soon to say if molecular forces were electrostatic, but Lorentz's explanation of Michelson's experiment made it likely. For the explanation of gravitation, he followed Lorentz. Applying Maxwell's principles, he derived an approximate motion of an ellipsoidal mass around a second, fixed mass. Except for a negligible discrep-

73. H. A. Lorentz, "Über die scheinbare Masse der Ionen," paper given at the German Association meeting in 1900, printed in *Phys. Zs.* 2 (1901): 78–80; his and Wien's discussion is on 79–80.

ancy in a coefficient, his result for the anomalous motion of the planet Mercury was identical with Weber's. Wien's main interest was, however, more fundamental than the law of gravitation: it was to bring together the "now completely isolated areas of mechanical and electromagnetic phenomena and to deduce each valid differential equation from a common foundation." This was one of the first tasks of physics, and the natural way to carry it out, Wien conceded, was to derive the electromagnetic laws from mechanical ones; after all, Maxwell himself had shown that a mechanical derivation of the electromagnetic equations is possible; others such as Kelvin and Boltzmann had approached the problem in similar ways, and Hertz had openly declared that his intention in his reformulation of mechanics was to supply proper principles for describing electromagnetic as well as mechanical phenomena. But Wien's intention in 1900 was the reverse of Hertz's. He thought it was more promising to regard the electromagnetic equations as the general, exact equations and to derive from them the equations of mechanics as special cases. From Maxwell's theory and from the assumption of the electrical constitution of matter, Wien deduced Newton's first law of motion as the law of conservation of electromagnetic energy. He showed that Newton's second law follows from the identification of the work done by a force with a change in electromagnetic energy. Finally, he showed that Newton's third law applies to forces acting between charges at rest in the ether but fails for moving charges. Newton's laws of motion, then, together with Newton's law of gravitation, are approximately correct, and the correct laws are those of the electron theory.[74]

The electron theory was the electrodynamics of greatest authority by 1900, and its implications for the foundations of physics in general were widely recognized. Boltzmann, for example, in his inaugural lecture at Leipzig in 1900, pointed out that mechanical explanations had extended their dominion throughout all natural science only to lose favor at home, in theoretical physics; following Hertz's electrical researches, he said, electromagnetism had become so important that some physicists challenged the "mechanical hegemony in theoretical physics" and sought to replace it by an electromagnetic one by deriving, "conversely, the laws of mechanics from the theory of electromagnetism." Boltzmann elaborated on the electromagnetic challenge in his inaugural lecture at Vienna two years later: until recently, he said, the physical disciplines "appeared increasingly to transform themselves gradually into special chapters of mechanics," but doubts arose about the reduction of electromagnetism to mechanics and with them doubts about the adequacy of the mechanical world picture. Previously, physicists regarded the law of inertia as the first fundamental law of nature, one that explained everything while remaining itself unexplained; but now they thought differently, persuaded by Maxwell's equations that a

74. Wilhelm Wien, "Ueber die Möglichkeit einer elektromagnetischen Begründung der Mechanik," *Arch. néerl.* 5 (1900): 96–104, reprinted in *Ann.* 5 (1901): 501–13, quotations on 501, 504. Wien's derivation was widely noted. Voss, for example, in his survey of rational mechanics in 1901, referred to recent efforts to reduce all phenomena to states of the ether as the "tendency of an electrical world view, which presently has spread to many physicists"; citing Wien, Voss noted that if this tendency were to continue, the foundations of mechanics would have an "entirely different character." Voss, "Die Prinzipien der rationellen Mechanik," 40.

massless electric particle, an electron, moves as though it has an inertial mass owing to the action of the ether; from this result and from the hypothesis that all matter is made up of massless electrons, physicists were inclined to believe that all mass is apparent and that the laws of mechanics are special cases of electromagnetic laws. In short, Boltzmann said, physicists want "no longer to explain everything mechanically," but instead they look for a mechanism, the ether, for the explanation of all mechanisms. He added that the ether itself is still "completely obscure."[75]

Experimental means for directly investigating electrons made it possible to study their dynamical laws, in particular, the theoretical dependence of their mass on velocity. For this dependence to become observable, since it is of the second order in v/c, physicists needed to obtain electrons moving with a speed close to that of light. The opportune discovery of radioactivity at the end of the nineteenth century provided them with the means. In 1901 Kaufmann began to report results from his experiments with Becquerel rays emitted by radium salts, and in an invited talk at the German Association meeting that year he related them to efforts then underway to establish an electromagnetic physics, which he saw as the continuation of a thirty-year historical development going back to Weber. He identified a number of problems remaining to be solved in electromagnetic physics: the proof that all mass is apparent, which would eliminate the "unfruitful" mechanical explanations of electrical phenomena; the experimental confirmation of the electron theory of gravitation; the proof that matter is composed solely of electrons; and the explanation of chemical periodicities by stable dynamical arrangements of assemblies of electrons.[76] The next year, 1902, reporting his experimental results in the physics section of the German Association meeting, Kaufmann drew the conclusion: *"The mass of electrons in Becquerel rays depends on the velocity; the dependence is exactly represented by Abraham's formula. The mass of electrons is accordingly of purely electromagnetic nature."*[77]

Kaufmann's reference was to the first publication on the electron theory by his then Göttingen colleague, the Privatdocent for theoretical physics Max Abraham. From experimental results, Abraham concluded that the inertia of electrons is probably entirely electromagnetic, the precondition of a "purely electromagnetic foundation of mechanics," as Wien had proposed. To analyze properly Kaufmann's deflection experiments, Abraham needed to know the "transverse" mass of an electron, which is the inertia opposing any acceleration normal to the direction of motion of the electron. The electromagnetic energy of the electron determines its "longitudinal" mass, or the inertia opposing any acceleration in the direction of motion; it does not, however, determine the transverse mass, since an acceleration normal to

75. Boltzmann, "Zwei Antrittsreden," 255, 276.

76. Walter Kaufmann, "Die Entwicklung des Elektronenbegriffs," talk given in a common session of the two main groups at the 1901 German Association meeting, published in *Phys. Zs.* 3 (1902): 9–15, on 14–15.

77. Walter Kaufmann, "Die elektromagnetische Masse des Elektrons," paper read in the physics section at the 1902 meeting of the German Association, published in *Phys. Zs.* 4 (1903): 54–56, quotation on 56.

the motion does no work. To calculate the transverse mass, Abraham drew on the concept of an electromagnetic momentum in the ether, which Poincaré had recently introduced to preserve the principle of action and reaction in Lorentz's theory. The values Abraham found for the two masses are different, and as a result the force on the electron and its acceleration are not, in general, in the same direction, as they are in ordinary mechanics. Similarly, the electromagnetic momentum is not proportional to the velocity, as it is in ordinary mechanics. At high enough velocities, therefore, electron dynamics departs in several respects from the dynamics of ponderable bodies.[78]

In 1903, Abraham corrected what he considered to be a critical defect in Lorentz's theory: he calculated that the deformable electron requires a nonelectromagnetic inner elastic potential energy to maintain equilibrium, which conflicts with the program of founding the electron theory on strictly electromagnetic principles. Abraham proposed in place of this electron an electron that does not undergo deformation when in motion, for which purpose he made the kinematic description of a rigid body a basic equation of the theory. He introduced the idea of rigid connections, as they are discussed in Hertz's mechanics, since these connections do no work in maintaining the shape of an electron, preserving electromagnetic energy; no nonelectromagnetic energy is then required. At the same time that he borrowed from Hertz's mechanics, Abraham pointed out that his electromagnetic direction was diametrically opposed to Hertz's mechanical direction.[79]

In 1904, in a revision he undertook of Föppl's 1894 text on Maxwell's electromagnetic theory, Abraham remarked that Föppl's derivation of Maxwell's law of induction from Lagrange's equations was not necessarily favorable to the mechanical world picture. In 1905, Abraham brought out a companion volume to Föppl's; this, Abraham's own, presented the electron theory. By placing electron dynamics on an electromagnetic basis, he hoped to "shake the foundations of the mechanical view of nature": after perfecting electron dynamics, the next steps in establishing an "electromagnetic world picture," he said, were to put the forces between electrons and atoms on an electromagnetic basis and to explain gravitational and molecular forces by atoms conceived of as assemblies of electrons. Abraham noted that the "electromagnetic world picture is so far only a program."[80]

78. Max Abraham, "Dynamik des Electrons," *Gött. Nachr.*, 1902, pt. 1, pp. 20–41, quotation on 21.

79. Max Abraham, "Prinzipien der Dynamik des Elektrons," *Ann.* 10 (1903): 105–79. Like Abraham, Bucherer also replaced Lorentz's hypothesis, according to which the electron contracts in the direction of its motion: Bucherer proposed in 1904 that the electron not only contracts but also dilates in the direction normal to its motion in such a way that the volume of the electron remains constant. He argued for his hypothesis again in 1905, observing that he, Abraham, and Lorentz himself regarded the Lorentz theory defective because the inner nonelectromagnetic energy of deformation demanded by the contractile electron precludes a purely electromagnetic theory; the volume-preserving, deformable electron diminishes, though not entirely removes, the objection. Alfred Bucherer, *Mathematische Einführung in die Elektronentheorie* (Leipzig: B. G. Teubner, 1904); "Das deformierte Elektron und die Theorie des Elektromagnetismus," *Phys. Zs.* 6 (1905): 833–34.

80. Max Abraham, *Theorie der Elektrizität*, vol. 2, *Elektromagnetische Theorie der Strahlung* (Leipzig: B. G. Teubner, 1905), especially 143–47.

In the laws of the older mechanics, the velocity of light does not enter as it does in their electromagnetic counterparts; but, as we will see, it enters in the laws of the new mechanics of the early twentieth century.[81] Owing to certain formulas in the electrodynamics of moving bodies, the expectation arose that electrons cannot travel faster than light. Nor, by implication, should ordinary matter be able to travel faster, since it contains electrons. The notion of a limiting velocity posed an "interesting problem," Abraham said in 1902, and he referred to Des Coudres who had taken up the problem experimentally two years earlier in the Göttingen physics institute. Using vacuum tubes operating with high potentials, Des Coudres, whose interest in the problem was related to the question of electron mass, tried in vain to accelerate electrons to velocities exceeding the velocity of light. On theoretical grounds, the prospect of succeeding had seemed unlikely to him, since such velocities would require infinite energies.[82]

For velocities greater than the velocity of light, Maxwell's theory predicts strange phenomena. These phenomena were analyzed by, among others, Wiechert, who in 1904 pointed out that for an electric body moving uniformly with a velocity exceeding that of light, electrodynamics requires a resisting mechanical force of electrodynamic origin; an outside mechanical force is necessary to keep the body moving uniformly, which contradicts Newton's first law of motion.[83] For the resisting force, Wiechert used a formula from a recent comprehensive paper on the electron theory by Sommerfeld. In 1905, Sommerfeld took up the problem again; failing to find a theoretical unforced motion for electrons moving faster than light, he concluded that this might be the only example of a "sensibly posed physical problem" without a solution, and he suspected that such a motion is impossible.[84] Wien, in an invited talk on electrons at the German Association meeting in 1905, referred to Sommerfeld's work to support his own similar conclusion that the assumption of electron velocities greater than the velocity of light leads to results of little physical probability.[85]

The German Association meeting in 1903 devoted a common session to the timely theme of the "present state of mechanics." The session contained three in-

81. Léon Rosenfeld, "The Velocity of Light and the Evolution of Electrodynamics," *Nuovo Cimento,* supplement to vol. 4 (1957): 1630–69.

82. Abraham, "Dynamik des Electrons," 23. Theodor Des Coudres, "Zur Theorie des Kraftfeldes elektrischer Ladungen, die sich mit Ueberlichtgeschwindigkeit bewegen," *Arch. néerl.* 5 (1900): 652–64. Des Coudres referred to Heaviside's early discussion, in 1888, of velocities greater than the velocity of light and to J. J. Thomson's theoretical argument in 1893 against their possibility and also to G. F. C. Searle's in 1897.

83. Emil Wiechert, "Bemerkungen zur Bewegung der Elektronen bei Ueberlichtgeschwindigkeit," *Gött. Nachr.,* 1905, 75–82.

84. Arnold Sommerfeld, "Zur Elektronentheorie. II. Grundlagen für eine allgemeine Dynamik des Elektrons," *Gött. Nachr.,* 1904, 363–439, on 384–402; "Zur Elektronentheorie. III. Ueber Lichtgeschwindigkeits- und Ueberlichtgeschwindigkeits-Elektronen," *Gött. Nachr.,* 1905, 201–35, on 201–4.

85. Wien thought the assumption of deformable electrons preferable to that of rigid electrons, since it forbids velocities greater than the velocity of light. Wilhelm Wien, "Über Elektronen," *Verh. Ges. deutsch. Naturf. u. Ärzte* 77 (1905): 23–38; published separately as *Über Elektronen. Vortrag gehalten auf der 77. Versammlung deutscher Naturforscher und Ärzte in Meran,* 2d rev. ed. (Leipzig and Berlin: B. G. Teubner, 1909), 27–28.

vited reports: one by the Göttingen astronomy professor Karl Schwarzschild on celestial mechanics, one by Sommerfeld on technical mechanics, and one by the Leipzig professor Otto Fischer on physiological mechanics. The subsequent discussion referred mainly to Schwarzschild's talk, and it centered on Lorentz's electron theory, although that had not been Schwarzschild's subject.

Ostwald introduced into the discussion his doubts about the reliability of Newton's mechanics, specifically the conservation of the motion of the center of gravity, which the pressure of light on ponderable matter seems to violate. Schwarzschild responded that astronomers were already familiar with the failure of this law, to which Drude added that physicists were too: the law holds only for ponderable masses and not for the ether, the reasons for which Lorentz had given. The failure of Newton's third law was particularly disturbing to Oettingen, an honorary ordinary professor at Leipzig who taught elementary physics along with other sciences: ". . . I would leave the auditorium feeling very uneasy if Newton's 3rd law of motion were really overturned if the fact of light pressure violated the center of gravity law." Wien told Oettingen that the question of the law of the center of gravity relates to the foundations of electrodynamics: if Lorentz's theory is accepted, then the law does not generally hold, a point which Poincaré had studied carefully. Boltzmann added "for Mr. von Oettingen's peace of mind" that "if one considers the ether to be nothing, the law of the center of gravity definitely does not hold," but if a very small mass is attributed to the ether, the law can be assumed to hold for the ether and matter taken together. Oettingen pointed out that the same question arises in the case of gravitation: "Depending on whether the motion of my hand is noticed by Sirius at once or only after some time, we get conservation of the center of gravity or its brief displacement." Schwarzschild answered that Lorentz had shown that gravity can be assumed to propagate with the velocity of light, but that its confirmation lies beyond the accuracy of astronomical observations. Wien elaborated on the untestability of Lorentz's theory of gravitation, according to which the law of the center of gravity holds only for bodies at rest; if the bodies are moving, the law holds no more for gravitation than for light pressure; but the terms departing from the law contain the square of the ratio of the velocity of the bodies to the velocity of light, a quantity "so small that we can hardly hope for a decision from experiments."[86]

At an earlier time but no longer in 1903, a meeting on the state of mechanics might have been an occasion for affirming Boltzmann's mechanical "hegemony." The emphasis on electromagnetism in a discussion ostensibly on mechanics points up the widespread interest in the implications of the electron theory for a new gravitational theory and for a "purely electromagnetic foundation for mechanics."[87] In a

86. The "Referate über den gegenwärtigen Stand der Mechanik" at the 1903 German Association meeting consisted of: Karl Schwarzschild, "Über Himmelsmechanik"; Arnold Sommerfeld, "Die naturwissenschaftlichen Ergebnisse und die Ziele der modernen technischen Mechanik"; and Otto Fischer, "Physiologische Mechanik"; published in *Phys. Zs.* 4 (1903): 765–73, 773–82, and 782–89. The discussion following the three talks is on 789–93.

87. The quotation is from the Göttingen physicist Paul Hertz at about this time: the "electron theory owes its significance to, among many other things, the circumstance that it opens a way to a purely electromagnetic foundation of mechanics"; in "Die Bewegung eines Elektrons unter dem Einflusse einer longitudinal wirkenden Kraft," *Gött. Nachr.*, 1906, 229–68, on 229.

talk at the German Association meeting in 1905, Richard Gans, a Privatdocent teaching theoretical physics at Tübingen, observed that Lorentz's electron theory comprehended electricity, magnetism, optics, and radiant heat in a fully satisfactory way, but not yet gravitation: since "without question our world picture will be simpler if we place it on unified foundations," Gans sketched an electromagnetic understanding of gravitation, building on Lorentz's and Wien's theories of 1900.[88] It was important to Gans, as it was to Wien, the Greifswald physicist Gustav Mie, and others, that through an electromagnetic theory of gravitation, the two measures of mass—the inertial and the gravitational—are necessarily proportional, not fortuitously so, as they are in the older physics in which gravitation and electromagnetism are separate.[89]

The possibility of a mechanical interpretation of electromagnetism continued to interest German physicists well into the twentieth century. Planck's student Hans Witte, later professor at the technical institute in Braunschweig, gave a report on the state of the problem in the physics section at the 1906 meeting of the German Association. His starting point was the recognition that for a long time physicists had been trying to found all of the branches of their science on a "unified system of concepts." Through the energy principle, they had attained a measure of unity, so that now there were only two separate branches, mechanics and electrodynamics; thermodynamics appeared to be completely fitted under the other two. That left as possible routes to the desired unity the reduction of electrodynamics to mechanics, the reverse reduction, or the derivation of both from a common foundation, or *Urprinzip*. With Planck's encouragement, Witte had tried the first route, deciding that what was needed was a systematic, general study of all conceivable mechanical theories. He identified nine types, and from his analysis of them he concluded that a mechanical explanation of electromagnetism is impossible on the assumption of a continuous ether and unacceptable on the assumption of a discontinuous ether. In arriving at this conclusion, he eliminated from consideration all types of mechanical explanations that do not depend on the assumption of a world ether, since he was confident that physics could not do without it. Witte's preference for a theoretical physics free from mechanical reduction is clear from his concluding remarks:

> A unified description or, if one likes, explanation of all physical phenomena . . . [is] in no way excluded. In this case, either the attempt initiated by W. Wien at an electromagnetic foundation of mechanics or the attempt at founding mechanics as well as electrodynamics on a common base, which could be designated neither as mechanical nor as electrodynamic, would replace the mechanical foundation of all of physics. Admittedly, often enough an opposite view has been expressed. One has maintained, with all determination, the a priori wisdom that the "nature" of electricity must lie absolutely in the states of hidden motions

88. Richard Gans, "Gravitation und Elektromagnetismus," *Phys. Zs.* 6 (1905): 803–5, on 803.
89. Gustav Mie, *Moleküle, Atome, Weltäther* (Leipzig: B. G. Teubner, 1904,), 131–32. Wien, "Ueber die Möglichkeit," 508; "Über Elektronen," 37.

and tensions, that all of physics must admit of being explained completely mechanically, that at each place in the entire world there must be a "thing in itself," in the true sense of the word, as an exhaustive mechanical principle of explanation. Possibly it is so; but no one has been able to give an a priori compelling proof for such assertions.[90]

Abraham, who heard the report, did not think that Witte had laid to rest the debate over the mechanical foundations of physics, since he believed that the "new conceptions of Lorentz" could only be carried out by endowing the electromagnetic field with masses and hidden motions. Mie was not as "optimistic" as Abraham about the possibility that mechanical theories would become important for the further development of electrodynamics, and in his own research he had not intended a mechanical theory, as Witte seemed to think. Neither did Mie think that a "purely electromagnetic explanation of nature" was possible. The direction that the unified foundations of physics would take was still very much up in the air.[91]

It was in part because the electron theory came so close to being a universal electromagnetic physics that it was modified and promulgated as an electromagnetic view of nature. But the questions that were posed in the search for electromagnetic foundations for physics could not all be adequately answered. Meanwhile, the rapid development of experimental atomic physics and of radiation theory in the early years of this century brought into question the viability of any comprehensive theory founded on laws and concepts of contemporary physics. At the same time, as we will see in the next section, the theory of relativity came accompanied by a call for the abandonment of the ether, a concept integral to the electromagnetic world picture. Moreover, the central question of the electromagnetic origin of mass gradually lost interest with the acceptance of the theory of relativity, which gave another reason for the dependence of mass on velocity. The theory of relativity in its application to electrodynamics yielded all of the testable laws of motion of electrons without having to decide their shape, substance, and charge distribution. It offered physics foundations of a different kind, ones concerned with the measurement of physical events in space and time.

Relativistic Foundations of Physics

In 1905 the *Annalen der Physik* published a paper on electrodynamics by Albert Einstein, a recent graduate of the Zurich Polytechnic who was then working as a patent examiner in Switzerland. Einstein began his paper by calling attention to an asymmetry in "Maxwellian electrodynamics," by which he understood Lorentz's version of it.[92] The asymmetry had to do with two descriptions of a phenomenon for

90. Hans Witte, "Über den gegenwärtigen Stand der Frage nach einer mechanischen Erklärung der elektrischen Erscheinungen," paper given at the 1906 meeting of the German Association, published in *Phys. Zs.* 7 (1906): 779–86, quotation on 784.
91. Discussion following Witte's report, *Phys. Zs.* 7 (1906): 785–86.
92. Albert Einstein, "Zur Elektrodynamik bewegter Körper," *Ann.* 17 (1905): 891–921. Among the

which, Einstein held, there should be only one: the current induced in a conductor by a magnet depends only on their relative motion, yet by the Maxwell-Lorentz theory the description of what happens depends on whether the magnet or the conductor moves. Einstein next brought up the failure of experiments to detect the earth's motion through the ether. These two and other examples, he said, suggest the absence of phenomena that correspond to the idea of absolute rest "not only in mechanics but also in electrodynamics."[93] It had been shown to first approximation that the same coordinate systems for which the equations of mechanics hold are also valid for the equations of electrodynamics and optics, a result that Einstein raised to the level of a postulate, the "principle of relativity": the laws of nature are the same for all observers regardless of any uniform motion they may have with respect to one another. To this Einstein added as a second postulate the constancy of the velocity of light to all observers regardless of the motion of its source. This second postulate incorporated into relativity theory a cardinal feature of the stationary ether of the electron theory. But that was all; Einstein said that for his purposes the ether was "superfluous," since he had no need of an "absolutely stationary space." From his two postulates, Einstein analyzed the measurement of space and time, leading to the relativity of the simultaneity of physical events and of lengths and time intervals. He deduced the equations of transformation of the coordinates and the time of observers in uniform relative motion in the x-direction, which are identical with the transformation equations Lorentz had arrived at by different reasoning; they preserve the form of the "Maxwell-Hertz" equations of the electromagnetic field in empty space:

$$\tau = \frac{t - vx/V^2}{\sqrt{1 - (v/V)^2}}, \, \xi = \frac{x - vt}{\sqrt{1 - (v/V)^2}}, \, \eta = y, \, \zeta = z,$$

where x, y, z are the spatial coordinates and t the time in one reference system, and ξ, η, ζ and τ are the corresponding quantities in a second reference system moving at velocity v with respect to the first.[94] (In cases in which the velocity v is small compared to the velocity of light V, the Lorentz transformations approximate the so-called Galilean transformations under which Newton's laws of motion of mechanics are invariant: $\tau = t$, $\xi = x - vt$; these transformations do not, however, leave the Maxwell-Hertz electromagnetic equations invariant.) With his new "kinematics," Einstein gave a method for solving problems of the electrodynamics of moving bodies by transforming the electric and magnetic field quantities to coordinate sys-

---

best historical studies of Einstein's theory of relativity of 1905 are several papers by Gerald Holton, which are collected in his book *Thematic Origins of Scientific Thought: Kepler to Einstein* (Cambridge, Mass.: Harvard University Press, 1973), and subsequent papers, especially, "Einstein's Scientific Program: The Formative Years," in *Some Strangeness in the Proportion*, ed. Harry Woolf (Reading, Mass.: Addison-Wesley, 1980), 49–65; Goldberg, "Lorentz Theory"; Tetu Hirosige, "The Ether Problem, the Mechanistic Worldview, and the Origins of the Theory of Relativity," HSPS, 7 (1976): 3–82; Arthur I. Miller's several papers, which are largely incorporated in his book, *Albert Einstein's Special Theory of Relativity*.
    93. Einstein, "Zur Elektrodynamik," 891.
    94. Einstein, "Zur Elektrodynamik," 892, 902.

tems stationary with respect to the moving bodies. By adding convection currents to the Maxwell-Hertz equations, Einstein arrived at the equations of Lorentz's electron theory, which he showed were compatible with the principle of relativity; he concluded his paper with a set of testable laws of motion of the electron.

Einstein's postulates addressed a fundamental opposition in the foundations of physics, that between Newtonian mechanics and Maxwellian electrodynamics. They had implications that extended beyond the electrodynamic and optical problems his paper expressly treated. The dependence of mass on velocity was now seen to apply to all ponderable bodies, not just to electrons, and the same was true of the role of the velocity of light as the maximum possible velocity in nature.[95] Based on postulates, Einstein's relativity theory resembled thermodynamics with its first and second laws, which had guided him in his search for a solution to the difficulties of electrodynamics. As a theory consisting of statements about light signals, clocks, and rigid measuring rods, it required specific laws of physics to be invariant under the Lorentz transformations, and it connected specific laws that otherwise appeared independent. The laws might apply to any class of physical phenomena, thermal, electromagnetic, or whatever, so that relativity theory came to be recognized as belonging to the foundations of physics as a whole.[96]

The early responses in Germany to Einstein's theory of relativity varied from Planck's sympathetic reading and active support of it to the Munich physicist Korn's prediction to Sommerfeld that the theory would "soon disappear from the scene again." For his part, Sommerfeld felt as uncomfortable at first with Einstein's "deformed time" as with Lorentz's "deformed electron,"[97] and other German physicists had other difficulties with the theory; but they addressed it, and that made a difference in the end. Mathematicians also took it seriously, sometimes becoming, as Riecke observed, "hypnotized by the elegance of the rules of calculation."[98] At least one of them, Hermann Minkowski, did work on it of great importance for physics, reformulating it in 1907 as a theory of four-dimensional space-time and persuading physicists such as Sommerfeld of its correctness.[99] Many German theorists and a number of experimentalists published their investigations of relativity theory, clarifying it, testing it, defining problems within it, and in general establishing its kinematical relationships throughout the parts of physics. By 1910 or 1911 the theory

95. Einstein, "Zur Elektrodynamik," 919–20.

96. Martin J. Klein, "Thermodynamics in Einstein's Thought," *Science* 157 (1967): 509–16, on 515.

97. Korn to Sommerfeld, 31 Dec. 1907, Sommerfeld Correspondence, Ms. Coll., DM. Sommerfeld to Lorentz, 12 Dec. 1906, Lorentz Papers, AR. We discuss Planck's response to Einstein's theory below.

98. Riecke's description of the Göttingen mathematicians' interest in the theory. Letter to Johannes Stark, 13 Oct. 1911, Stark Papers, STPK.

99. Sommerfeld, "Autobiographische Skizze," AHQP. Sommerfeld to Lorentz, 9. Jan. [1909 or 1910?], Lorentz Papers, AR. Wien to Hilbert, 15 Apr. 1909, Hilbert Papers, Göttingen UB, Ms. Dept. The Göttingen mathematician Hermann Minkowski is discussed in, for example, Gerald Holton, "The Metaphor of Space-Time Events in Science," *Eranos Jahrbuch* 34 (1965): 33–78; Robert Lewis Pyenson, "Hermann Minkowski and Einstein's Special Theory of Relativity," *Arch. Hist. Ex. Sci.* 17 (1977): 71–95; Peter Louis Galison, "Minkowski's Space-Time: From Visual Thinking to the Absolute World," *HSPS* 10 (1979): 85–121.

was fairly widely accepted in Germany. It had also begun to be recognized as a universal theory in its own right rather than as an elaboration of Lorentz's electron theory, as its earlier familiar designation as the "Lorentz-Einstein" theory implied.[100]

Planck, one of the physicists who early recognized that the theory of relativity would become part of the foundations of physics, was influential in building confidence in it.[101] Not long after the event, Einstein wrote that the "determination and warmth" with which Planck stood up for relativity theory was responsible "in great part for the attention this theory so quickly received among colleagues."[102]

As the advisor on theoretical physics for the *Annalen der Physik*, in 1905 Planck was already familiar with Einstein's work. For five years, Einstein had regularly submitted papers to that journal, the most important of which treated thermodynamics and statistical mechanics, subjects of particular interest to Planck at the time. Einstein extended these studies to a related interest of Planck's, blackbody radiation, in 1905.[103] Einstein's relativity theory of the same year set Planck to work; it was the subject, Max Born observed, that "caught Planck's imagination more than anything else."[104]

The implications of Einstein's theory for the foundations of physics appealed to Planck, who based his own work on principles of general validity drawn from thermodynamics. Planck reported on relativity theory in the Berlin physics colloquium in the fall of 1905, according to Max Laue, who was there. In 1906 he presented papers on the theory at meetings of the German Physical Society and of the German Association; at the first of these meetings he applied the theory to mechanics, at the second he examined recent measurements of electrons for their agreement or disagreement with the theory. In 1907, in a paper presented to the Prussian Academy, he applied the theory to thermodynamics and, in particular, to the radiation of a blackbody. In 1908 he gave another talk on relativistic dynamics at the German Association. After that he did little more work on the theory. By then, a number of others in Germany had begun to develop formalisms useful to the theory and to treat a variety of problems posed by it.[105]

100. Tetu Hirosige, "Theory of Relativity and the Ether," *Jap. Stud. Hist. Sci.*, no. 7 (1968): 37–53, on 46–48.

101. Planck's response to relativity theory is discussed in Stanley Goldberg, "Max Planck's Philosophy of Nature and His Elaboration of the Special Theory of Relativity," *HSPS* 7 (1976): 125–60; Planck's and others' responses are discussed in Stanley Goldberg, "Early Response to Einstein's Theory of Relativity, 1905–1911: A Case Study in National Differences," (Ph.D. diss., Harvard University, 1969). Miller discusses Planck's and others' responses in *Albert Einstein's Special Theory of Relativity*, Planck's especially on 360–62, 365–67.

102. Einstein, "Planck," 1079.

103. Einstein's early thermodynamic, statistical mechanical, and quantum researches are analyzed in Martin J. Klein, "Einstein's First Paper on Quanta," *The Natural Philosopher*, no. 2 (1963): 59–86; "Einstein and the Wave-Particle Duality," *The Natural Philosopher*, no. 3 (1964): 3–49; "Einstein, Specific Heats, and the Early Quantum Theory," *Science* 148 (1965): 173–80; "Thermodynamics in Einstein's Thought"; "No Firm Foundation: Einstein and the Early Quantum Theory," in Woolf, ed., *Some Strangeness*, 161–85; in Thomas S. Kuhn, "Einstein's Critique of Planck," in Woolf, ed., *Some Strangeness*, 186–91, and *Black-Body Theory*, 170–87.

104. Born, "Planck," 173.

105. Max Laue, "Mein physikalischer Werdegang. Eine Selbstdarstellung," in *Schöpfer des neuen Weltbildes. Grosse Physiker unserer Zeit*, ed. H. Hartmann (Bonn: Athenäum, 1952), 178–210, on 192.

Einstein singled out three contributions by Planck to the theory of relativity. Planck, he said, developed the relativistic equations of motion for a material point; he showed that the principle of least action has the same fundamental significance in the new theory as in classical mechanics; and he developed the relativistic connection between energy and inertial mass. The first two of these contributions appeared in a paper Planck gave at the German Physical Society in the spring of 1906. Planck began it by stating the transformation equations of the "principle of relativity," which Lorentz had introduced in 1904 and Einstein, in "still more general wording," in 1905. The transformations relate two coordinate systems with uniform relative motion, and the principle says that each system has the same right to be used to express the "fundamental equations of mechanics and electrodynamics." To see if the principle leads to any absurdities, Planck examined its consequences for mechanics, determining the equations of motion to replace Newton's. Centering a primed coordinate system on a moving body of mass m for which Newton's equations of motion hold, $m\ddot{x}' = X'$, etc., he then transformed the equations to an unprimed coordinate system moving with respect to the first, resulting in the following equations of motion:

$$\frac{d}{dt}\left[\frac{m\,\dot{x}}{\sqrt{1 - q^2/c^2}}\right] = X, \text{ etc.}$$

Here q is the velocity of the mass point in the unprimed coordinate system, and the factor $1/\sqrt{1 - q^2/c^2}$ is a consequence of the space and time transformations of the relativity principle. If q is small compared with the velocity of light c, the equations of motion are indistinguishable from Newton's, so that the relativistic equations appear as a generalization of Newton's. Planck concluded by introducing the "kinetic potential" and showing that the new equations of motion can be presented in the form of Hamilton's principle of least action.[106]

In this first paper, Planck recommended relativity theory on the theoretical grounds of simplicity and generality. He also spoke of recent measurements of moving electrons that seemed to go against it, but as these measurements presupposed a complicated theory, Planck thought that it was still possible that further experiments would favor relativity. The meeting of the German Association in the fall of 1906 provided Planck with an early occasion to discuss the experimental evidence. Kaufmann's "subtle measurements" of electromagnetic deflections of β rays emitted by radium bore on the leading questions of the "mechanics of electrons," in particular, on the decision between the two most developed theories, the "Lorentz-Einstein" relativity theory and Max Abraham's spherical electron theory. From the side of the "spherical theory," the theoretical values for the electrical deflection of the rays were closest to the observed values; but they were not close enough for Planck to see in them a "definitive confirmation" of that theory and a refutation of the relativity theory. The theoretical meaning of the observed values needed further clarification

106. Einstein, "Planck," 1079. Max Planck, "Das Prinzip der Relativität und die Grundgleichungen der Mechanik," *Verh. Phys. Ges.* 8 (1906): 136–41, reprinted in *Phys. Abh.* 2: 115–20.

before they could provide a definitive decision between the two theories. Planck gave advice on how to make further measurements on electrons; as in his work on radiation theory, in his work on relativity theory he emphasized the importance of experimental confirmation, and he was inclined to think that relativity was confirmable.[107]

Planck's talk at the German Association meeting was followed by a long discussion, which was joined by a good number of the leading developers of electron dynamics. They introduced issues arising largely from electron theory, which had to be dealt with in assessing the claims and merits of relativity theory. In addition to its current apparent experimental advantage, Planck agreed with Abraham that "the important advantage of the sphere theory would be that it is a purely electric theory" and that if it could be worked out it would be "very beautiful." But Planck found "more sympathetic" of the two the "Lorentz-Einstein theory" with its postulate that absolute translatory motion cannot be shown to exist. He was the only one who did. His point of view was characterized as "pessimistic" by Sommerfeld, who regarded Abraham's "purely electromagnetic theory" as more innovative than relativity theory. "On the question of principles formulated by Mr. Planck," Sommerfeld elaborated, "I would suspect that the gentlemen under forty will prefer the electrodynamic postulate, those over forty the mechanical-relativistic postulate." Needless to say, Sommerfeld preferred the electrodynamic postulate as the way to the future physics. The alternative, the relativity postulate, was associated with the older and apparently outmoded mechanical point of view in physics. Lorentz's electron theory, which in this discussion was identified with the relativity postulate, was for that reason seen to stand in opposition to the electromagnetic viewpoint in physics that it had originally done much to inspire. Planck's theoretical preference for the relativity postulate was based in large part on the unity it brought to electrodynamics and mechanics through a "general mechanics" encompassing both. In response to the Bonn physicist Alfred Bucherer's complaint that Planck had not given enough attention to his own electron theory, Planck asked Bucherer if he could give his equations a Lagrangian form. Bucherer had not looked into that. "But that would be very important," Planck said, "since through the Lagrangian form the equations of motion of the electron are reduced to those of general mechanics." Planck made a related point about the unity of mechanics and electrodynamics in response to Kaufmann's objection that the "epistemological value" of the relativity postulate is not great because it applies only to uniform translatory motion, not to the more general case of accelerated motion. That restriction has no importance in this connection, Planck said; what is important is "that what is not demonstrable in mechanics is also not demonstrable in electrodynamics."[108]

107. Max Planck, "Die Kaufmannschen Messungen der Ablenkbarkeit der β-Strahlen in ihrer Bedeutung für die Dynamik der Elektronen," *Verh. Phys. Ges.* 8 (1906): 418–32, reprinted in *Phys. Abh.* 2: 121–35, on 130–31. Planck gave this paper in the physics section at the German Association meeting in Stuttgart, 19 Sept. 1906. Within a short time, by 1908, physicists had experiments that were more exact than Kaufmann's and that favored the Lorentz-Einstein theory. Miller, *Albert Einstein's Special Theory of Relativity*, 254.

108. This discussion following Planck's talk was published in *Phys. Zs.* 7 (1906): 759–61.

Planck's student Kurd von Mosengeil wrote his dissertation on the application of the relativity principle to the radiation of a moving blackbody; it was published in 1907, the same year Planck's former student and current assistant Max Laue began publishing on relativity.[109] Planck himself continued to work on the theory, increasingly confident of its prospects. After the 1906 German Association meeting, Planck wrote to Wien that "concerning the relativity principle, I really see no difficulty yet," and he wrote to Wien in the following spring that he was still "occupied with the Lorentz-Einstein relativity theory."[110]

In the summer of 1907, Planck presented a paper at a meeting of the German Physical Society in which he treated the dynamics of moving systems according to the relativity principle. Because blackbody radiation in an evacuated hollow cavity is "of all physical systems the only one whose thermodynamic, electrodynamic, and mechanical properties can be explained with absolute exactness independently of the conflict between special theories," Planck began his study with the relativistic dynamics of such radiation. From there he proceeded to the principle of least action, under which, he noted, Helmholtz had subsumed mechanics, electrodynamics, and the two main laws of thermodynamics in their application to reversible processes. By subsuming the laws of blackbody radiation under the least action principle, Planck extended the realm of validity of the principle beyond where Helmholtz had left it.[111]

From the combined principles of relativity and least action, Planck deduced consequences he hoped would lead to conclusive experimental tests of the relativity principle. He derived, as Einstein had earlier and in a different way, the law equating the mass of a body at rest with its energy content, and he discussed the seemingly remote possibility of testing the predicted, but extremely small, change in the mass of a body upon a change in its energy. "If only $(v/c)^2$ were not so absurdly small! It is a true misery," Planck complained to Lorentz.[112]

Having been criticized for asserting the compatibility of the relativity and least action principles, Planck was reassured when Einstein wrote to him that he did not accept the criticism. Planck replied to Einstein that "as long as the representatives of the relativity principle are still such a modest band, as is the case now, it is doubly important that they agree among themselves."[113] Planck regarded the invariance of the least action principle under the relativity principle as a major conclusion, since

109. Kurd von Mosengeil, "Theorie der stationären Strahlung in einem gleichförmig bewegten Hohlraum," *Ann.* 22 (1907): 867–904. Mosengeil died in late 1906, and Planck saw the dissertation through publication; it is included in Planck, *Phys. Abh.* 2: 138–75.

110. Planck to Wien, 15 Oct. 1906 and 24 May 1907, Wien Papers, STPK, 1973.110.

111. Max Planck, "Zur Dynamik bewegter Systeme," *Sitzungsber. preuss. Akad.*, 1907, 542–70, reprinted in *Phys. Abh.* 2: 176–209.

112. Planck, "Zur Dynamik bewegter Systeme," 202, 205. Planck to Lorentz, 19 Oct. 1907, Lorentz Papers, AR. Planck made this remark in connection with the problem of the experimental measurement of the influence of the earth's motion on the intensity of blackbody radiation. The enormous value of the square of the velocity of light, c^2, compared with the square of the velocity of any observed body, v^2, posed difficulties for all of the proposed experimental tests. This difficulty is pointed out in connection with the ether-drift problem in Swenson, *Ethereal Ether*, 66.

113. Planck to Einstein, 6 July 1907, EA. Bucherer was the critic in question.

the least action principle is the foundation of a "general dynamics," which encompasses not only mechanics but also electrodynamics and thermodynamics. In a talk at the German Association meeting in 1908, Planck developed further this general dynamics. He recalled the "stir" caused by the denial of Newton's third law—the principle of action and reaction, which contains the law of conservation of momentum—in Lorentz's electron theory. He recalled, too, that the difficulty was partially overcome, thanks especially to Abraham, by the introduction of an electromagnetic momentum in addition to the only known momentum to date, mechanical momentum. He now showed that Einstein's relativity theory permits a definition of momentum that contains both the mechanical and the electromagnetic forms; the corresponding generalized law of action and reaction Planck called the "inertial law of energy."[114]

As Planck came to occupy a public position in German physics similar to Helmholtz's before him, he called attention to the increasing unity of the physical world picture, to which his own researches were directed. In a talk at Leiden University in 1908, he began by characterizing as the highest goal of the study of nature the ordering of physical phenomena under a "unified system, where possible within a single formula." There are two methods for attaining this goal, he said: one is to place at the center of the world picture a single concept or principle, as Ostwald's energetics and Hertz's mechanics do; the other is to admit into the picture only what appears confirmed through direct experience, as Kirchhoff's definition of mechanics stipulates. Both methods are indispensable. Only Mach's prescriptions, and proscriptions, for a physics in which reality is identified with human sense perceptions do not offer a useful approach to a unified world picture; the history of the physical world picture, Planck said, shows a progressive liberation of physical ideas from the vagaries of time, place, and individual intellects. If Mach were right, there would be as many different physical world pictures as there are different intellects. But that is not the case; the physical world picture has achieved a large measure of unity. In this connection, Planck believed that the present great division of physics into mechanics and electrodynamics, into the physics of matter and the physics of the ether, was probably not final, that a suitably "generalized view of mechanics" was possible that would include electrodynamics within a "general dynamics."[115]

Planck developed the themes of his Leiden talk in the first of a series of lectures on theoretical physics at Columbia University in the following year, 1909. With the development of the physical world picture, there has been a "real obliteration of personality," Planck told his American audience, which is why the "same physics is made in the United States as in Germany." He lectured on the parts of theoretical

114. Max Planck, "Bemerkungen zum Prinzip der Aktion und Reaktion in der allgemeinen Dynamik," *Verh. Phys. Ges.* 10 (1908): 728–32, reprinted in *Phys. Abh.* 2: 215–19. Planck gave this talk at a physics session of the German Association in Cologne, 23 Sept. 1908.

115. Max Planck, "Die Einheit des physikalischen Weltbildes," *Phys. Zs.* 10 (1909): 62–75, reprinted in *Phys. Abh.* 3: 6–29, quotations on 6, 10. Planck gave this talk in Leiden, 9 Dec. 1908. Planck also speculated here that the division of physics in the future would be between reversible and irreversible processes rather than between mechanics and electrodynamics.

physics he had worked on himself, beginning with thermodynamics, proceeding through radiation theory, and concluding with general dynamics, which incorporates the least action and relativity principles. In the part dealing with general dynamics, he argued that the least action principle, in the interpretation Helmholtz gave to it, has the advantage over all of the other principles originating in mechanics because it applies to "processes other than mechanical" ones. Since for Planck all irreversible processes in nature can be considered as reversible elementary processes, and since all elementary processes can be considered as consequences of the least action principle, the latter is the basis of a "representation of a unified system of theoretical physics," which was the principal object of Planck's lectures.[116]

To general audiences, Planck spoke repeatedly on the theme of the replacement of the older "mechanical world picture" by a developing "physical world picture." In his Columbia University lectures, he cautioned that his derivation there of the Maxwell-Hertz electromagnetic equations from the least action principle does not imply a mechanical explanation; for that principle now belongs to general dynamics, no longer only to ordinary mechanics. In the same lectures, he noted that relativity theory excludes the definition of the velocity of a ponderable body with respect to the Lorentzian stationary ether: thus, as Einstein pointed out, the "ether drops out of the theory and with it the possibility of mechanical explanation of electrodynamic processes, i.e., of referring them to motions"; he added that this conclusion is not particularly interesting anymore, since his student Witte had already proved the impossibility of a mechanical explanation for a continuous ether.[117] When Planck addressed a general session at the 1910 meeting of the German Association, his subject was the "position of recent physics on the mechanical view of nature," and here again he stressed that the relativity principle was incompatible with any mechanical view of the ether. The golden age of the mechanical viewpoint was the nineteenth century, which culminated in Hertz's attempt to express all physical phenomena in terms of matter and motion; for Planck in 1910, the once highly useful mechanical world picture had only historical significance for physics. Physics had acquired a different unified foundation.[118]

116. Max Planck, *Eight Lectures on Theoretical Physics Delivered at Columbia University in 1909*, trans. A. P. Wills (New York: Columbia University Press, 1915), quotations on 6–7, 97–99.

117. Planck, *Eight Lectures*, 111, 118–19.

118. Max Planck, "Die Stellung der neueren Physik zur mechanischen Naturanschauung," *Phys. Zs.* 11 (1910): 922–32, reprinted in *Phys. Abh.* 3: 30–46. Planck gave this talk at the Königsberg meeting of the German Association, 23 Sept. 1910.

25

Institutional Developments, Research, and Teaching in Theoretical Physics after the Turn of the Century

In the early years of the twentieth century, theoretical physics was still taught at most German universities by an extraordinary professor. Of the few universities with separate institutes for theoretical physics directed by ordinary professors, those at Königsberg and Leipzig did not contribute strongly to advances in the field at this time. But Berlin's institute continued to be important, as did Göttingen's, and Munich's new institute in 1906 quickly joined them as one of the centers for theoretical physics. The field was strengthened when several outstanding young theoretical physicists working abroad were attracted to positions in Germany. On the eve of World War I, theoretical physics in Germany was a flourishing discipline.

Theoretical Physics at Berlin and Göttingen

Theoretical physics had been well established at Berlin University through the work of Kirchhoff and Helmholtz by the time Planck was brought there in 1889 to head its new institute for theoretical physics. After three years as extraordinary professor, Planck was promoted to ordinary professor, his predecessor Kirchhoff's rank and the customary one of an institute director. At Berlin Planck was in close contact with outstanding physicists, above all with Helmholtz. Scientifically and personally, Helmholtz made a deep impression on Planck: a word of praise from Helmholtz, Planck recalled, would make him happier than any worldly success.[1]

In 1894 Helmholtz died, and Planck was no longer the theoretical physicist "beside Helmholtz," as Kundt had described him when he first came to Berlin.[2] Planck was now *the* theoretical physicist in Berlin. At the same time, he acquired an official responsibility for theoretical physics for all of Germany, replacing Helmholtz as the designated advisor on matters of theoretical physics for the *Annalen der*

1. Planck, *Wiss. Selbstbiog.*, reprinted in *Phys. Abh.* 3: 382.
2. Kundt to Graetz, n.d. [Dec. 1888], Ms. Coll., DM, 1933 9/18.

Physik.[3] In 1894 Kundt also died, and he was replaced by Warburg, another experimental physicist, which meant that Planck's position as the university's principal theorist remained unchanged.

Planck's institute for theoretical physics was modest, its budget a miniscule fraction of that of the physics institute.[4] His institute's main activity was teaching, which meant lecturing, assigning and correcting written exercises, and directing some advanced work. Enrollments in Planck's general lecture course showed a gradual increase from an average of fifty-five in 1896–97 to an average of a hundred and thirty-five in 1903–4. The number of participants in exercises or independent work showed a comparable increase: from eighteen in 1890 to eighty-nine in 1900 to one hundred and forty-three in 1909. Planck's obligatory annual reports on the activities of the institute were terse. Besides enrollments, they stated the odd extra expense such as the repair and tuning of the institute's harmonium and the replacement from time to time of the assistant: Ernst Zermelo by Max Abraham, Abraham by Hermann Diesselhurst. That was about all. By contrast, reports on the Berlin physics institute by Warburg typically contained two full pages of published dissertations and papers as well as impressive attendance figures. Warburg lectured to audiences of three hundred, and sometimes more wanted to attend than the hall could hold, and his many teaching subordinates and assistants accommodated large groups of students working in the laboratories.[5]

In Warburg's institute the research students, along with Warburg, all but lived as one great "family,"[6] while in Planck's there was not much institute life to speak of. That suited Planck, who worked alone and expected his students to do the same; he did not encourage them to work closely with him under his supervision.[7] His reports for the years around the turn of the century mentioned no published dissertations, and by 1910 he counted only fifteen dissertations as having come out of his institute.[8] The reports do not mention the expected help he gave dissertation students who were not his own. Nor do they mention his own publications, which at this time included the theory of blackbody radiation, the groundwork for what was to become the leading theory of atomic physics.

Outside the institute Planck had other duties, which he appealed to when warding off new demands on his time. When he declined to work on a proposed new physics journal, he explained that any time left over after his teaching and research was taken up in presiding over the Physical Society and in editing the *Annalen der Physik*.[9] When he declined to contribute an article to an encyclopedia, he explained

3. In 1895, in volume 54 of the *Annalen der Physik*, Planck's name replaced Helmholtz's on the masthead as editorial collaborator; the editors at this time were Gustav and Eilhard Wiedemann.

4. In 1909 Planck's institute received 700 marks, while the physics institute received 26,174 marks. Lenz, *Berlin* 3: 446.

5. Lenz, *Berlin* 3: 446. This discussion is based largely on the annual reports by Planck on his institute in the series *Chronik der Königlichen Friedrich-Wilhelms-Universität zu Berlin;* the years considered here are 1896–97 through 1903–4.

6. James Franck, "Emil Warburg zum Gedächtnis," *Naturwiss.* 19 (1931): 993–97, on 995–96.

7. E. Lamla's recollection, in Alfred Bertholet, et al., "Erinnerungen an Max Planck," *Phys. Bl.* 4 (1948): 161–74, on 173.

8. Planck, "Das Institut für theoretische Physik," 277.

9. Planck to Otto Lummer, 8 Jan. 1898, Wrocław UB, Lummer Nr. 219.

that his years at Berlin University had taught him that after his lectures and "eternal meetings, examinations, and written reports," what time remained he needed for his own researches.[10]

Planck's researches at the turn of the century were to prove his most influential. As a part of his electromagnetic investigations in thermodynamics, he introduced into physics a radiation law and accompanying theory containing two new universal constants, the "Boltzmann" and "Planck" constants, k and h, which were to enter ubiquitously in atomic physics. Planck's path to the radiation law brought him into repeated contact with experimentalists, and in this connection his new location in Berlin turned out to be advantageous. Beginning in the 1890s, many experimentalists in Germany, Kundt's former students prominent among them, investigated the laws of blackbody radiation, and these experimentalists were largely concentrated in Berlin. Another of their locations was Hannover, which Planck was in touch with through correspondence.

In Berlin, researchers on blackbody radiation were affiliated with several institutions, with the university, the technical institute, and the Reichsanstalt. The latter entered because the optical group of the scientific section sought a useful primary luminous standard and a scientific foundation for the study of radiation effects; while radiation research represented only a minor part of the work of this institution as a whole, the optical group devoted a fair amount of effort to it in the 1890s and, from 1898, all of its resources.[11]

While Wien was employed in this optical group, he carried out his fundamental studies of radiation, the most important of which were done unofficially and were theoretical.[12] In 1893 he derived a form of what was later called the "displacement law," a new relation between blackbody radiation and the second law of thermodynamics. The law reads, in Wien's words: "In the normal emission spectrum of a blackbody, with a change in temperature each wavelength is displaced such that the product of the temperature and the wavelength remains constant." The starting point of Wien's derivation was Boltzmann's proofs of the existence of radiation pressure, using the second law of thermodynamics, and of Stefan's law of the total radiation of a blackbody, using Maxwell's electromagnetic theory.[13] Wien capped his

10. Planck to Sommerfeld, 11 Sept. 1899, Sommerfeld Papers, Ms. Coll., DM. The encyclopedia in question is the *Encyklopädie der mathematischen Wissenschaften.*

11. Cahan, "Physikalisch-Technische Reichsanstalt," 292–311.

12. But Wien also took an interest in the experimental side of the problem: in 1895 he and Lummer published an experimental method for investigating the laws of blackbody radiation. To improve on the earlier techniques, which had to make do with imperfectly "black" bodies, they proposed to bring a hollow body to uniform temperature and to observe the radiation through a small opening. To try their proposal, they acquired hollow spheres of porcelain and metal capable of reproducing the radiation of a theoretical blackbody to "any approximation." Their "blackbody" was not a cavity enclosing radiating bodies, as Kirchhoff originally described it; here the empty hollow body itself created the radiation. Wilhelm Wien, "Temperatur und Entropie der Strahlung," *Ann.* 52 (1894): 132–65; Wilhelm Wien and Otto Lummer, "Methode zur Prüfung des Strahlungsgesetzes absolut schwarzer Körper," *Ann.* 56 (1895): 451–56, on 453; Kangro, "Vorgeschichte," 106.

13. Wilhelm Wien, "Eine neue Beziehung der Strahlung schwarzer Körper zum zweiten Hauptsatz der Wärmetheorie," *Sitzungsber. preuss. Akad.*, 1893, 55–62, on 62. This and other radiation studies by Wien are analyzed by Kangro, *Vorgeschichte*, 45–48, 93–113, who points out, p. 47, that the name "displacement law" was introduced in 1899 and that only later was the law restricted to the wavelengths corresponding to the maximum energy of blackbody radiation for different temperatures.

thermodynamic studies with a law describing the distribution of energy in the emission spectrum of a blackbody, which he derived with the help of Maxwell's distribution law for molecular motions in thermal equilibrium. From the assumption that each molecule probably contains electric charges, capable of emitting electromagnetic waves with a wavelength and an intensity that depend on the velocity of the molecule, and from his results of 1893 and 1894 and from the Stefan-Boltzmann law, Wien derived the following distribution law: the intensity ϕ_λ of radiation in the interval λ to $\lambda + d\lambda$ is

$$\phi_\lambda = \frac{C}{\lambda^5} e^{-c/\lambda\theta},$$

where C and c are constants and θ is the temperature. Before the publication of his paper containing this law, Wien received some confirmation of his derivation from the Hannover experimental physicist Friedrich Paschen, who independently inferred from his empirical results a law equivalent to Wien's.[14] The same year that Wien announced his distribution law, he accepted an extraordinary professorship at the technical institute in Aachen, so that he was no longer in a position to undertake the experimental testing of his law. Others would do this, but since it required improved absolute measuring methods, more perfect blackbodies, and the like, it was to be some time before new measurements were available.[15]

When they were, they called into question the accuracy of Wien's law. Planck was especially concerned; for although Wien had derived the law from doubtful molecular assumptions, Planck believed that he himself had rederived it with the authority of the second law of thermodynamics. The confidence he placed in his derivation is evident from his remarks on it in the fifth paper of his series on irreversible radiation processes: "the limits of validity of this law, in case there are any at all, coincide with those of the second fundamental law of the theory of heat." Further experimental tests of the law gained "all the more fundamental interest" because they were at the same time tests of the second law; Planck urged that they be undertaken.[16]

Like Wien, Planck did not do the experimental tests himself. He always worked from the theoretical side, leaving the experiments entirely to others, a division of labor which proved to be no barrier to—and probably facilitated—progress in understanding blackbody radiation. Planck, the purely theoretical physicist, contributed to the rapidly developing experimental field by supplying it with an accurate radiation law; in turn, Planck was guided to the law in part by new experimental results.[17]

14. Wilhelm Wien, "Ueber die Energievertheilung im Emissionsspectrum eines schwarzen Körpers," *Ann.* 58 (1896): 662–69.

15. Hans Kangro, "Das Paschen-Wiensche Strahlungsgesetz und seine Abänderung durch Max Planck," *Phys. Bl.* 25 (1969): 216–20; *Vorgeschichte,* 149–79.

16. Planck, "Über irreversible Strahlungsvorgänge. Fünfte Mittheilung," 597.

17. For the history of blackbody research leading to the quantum theory, there exists a number of careful and detailed studies. Among the older studies, Léon Rosenfeld, "La première phase de l'évolution

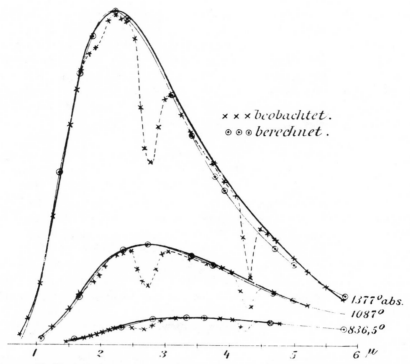

× × × *beobachtet*.
⊙ ⊙ ⊙ *berechnet*.

1377°abs.
1087°
836,5°

Fig. 37. Blackbody Radiation Curves, 1899. Reprinted from Otto Lummer and Ernst Pringsheim, "Die Vertheilung der Energie im Spectrum des schwarzen Körpers," *Verh. phys. Ges.* 1 (1899): 23–41, on 34.

Through Wien's, and even more through Planck's, derivation of the blackbody radiation law, the experimentalist Paschen at the end of 1899 said that the law "appears to be a rigorously valid law of nature" and that its "constants have general significance." The remaining problem for experimentalists, as Paschen saw it, was to measure the constants exactly and to "investigate within what boundaries the law and with it Planck's assumptions can be found to be valid."[18] Earlier that year the

de la Théorie des Quanta," *Osiris* 2 (1936): 149–96. Among the recent studies, Klein, "Max Planck and the Beginnings of the Quantum Theory,"; "Thermodynamics and Quanta in Planck's Work"; Ehrenfest, 217–30; and "Planck, Entropy, and Quanta, 1901–1906," *The Natural Philosopher*, no. 1 (1963): 83–108; Armin Hermann, *The Genesis of the Quantum Theory (1899–1913)*, trans. C. W. Nash (Cambridge, Mass.: MIT Press, 1971), 5–28; Kuhn, *Black-Body Theory*, 92–113; Kangro, *Vorgeschichte*, 149–223. Our purpose here is not to give another self-contained history of blackbody research leading to the quantum theory, but to sketch the development to illustrate features of the practice of theoretical physics around the turn of the century. In this we are guided by Kangro's observation that the route to Planck's law of blackbody radiation is an excellent example of the interaction between experiment and theory in modern physics, but our discussion here is greatly indebted to all of these sources.

18. Friedrich Paschen, "Über die Vertheilung der Energie im Spectrum des schwarzen Körpers bei höheren Temperaturen," *Sitzungsber. preuss. Akad.*, 1899, 959–76, on 959. Paper presented by Planck to the Prussian Academy of Sciences on 7 Dec. 1899.

experimentalists Otto Lummer and Ernst Pringsheim began to report on systematic departures of the measurements from Wien's (and Paschen's) law (see figure 37).[19] These departures, Lummer and Pringsheim said in early 1900, "gained in theoretical interest" as a result of Planck's improvement over Wien's derivation of the law.[20] The deviations of the "Wien-Planck" law from the measurements were seen to increase with the temperature of the blackbody and with the wavelength of the radiation observed. Using special prisms, improved bolometers, new forms of blackbodies, absolute measuring methods, and varying arrangements, experimentalists were able to extend their measurements to ever longer wavelengths and higher temperatures.[21] In addition, for a better fit, they proposed a variety of laws for describing the extended spectrum of this "black radiation."[22] At meetings and in technical publications, the experimentalists critically discussed the theoretical assumptions underlying the energy distribution law, and in turn the theorists critically discussed the measurements; experimentalists criticized one another's work and theorists did the same.[23] The appeal to extraneous causes to explain the deviations of the measurements from the theoretical curves became increasingly implausible. At the German Association meeting in late 1900, when Wien suggested that the recent experiments on blackbody radiation might be in error owing to the absorption of long waves by

19. Otto Lummer and Ernst Pringsheim, "Die Vertheilung der Energie im Spectrum des schwarzen Körpers," *Verh. Phys. Ges.* 1 (1899): 23–41, on 34. Paper given at a meeting of the German Physical Society on 3 Feb. 1899. The curves in figure 37 are Wien's and Paschen's blackbody distribution law plotted for three absolute temperatures and compared with the experimental measurements. The departures are evident here. The ordinate of the graph represents the energy of the radiation, the abscissa the wavelength.

20. Otto Lummer and Ernst Pringsheim, "Ueber die Strahlung des schwarzen Körpers für lange Wellen," *Verh. Phys. Ges.* 2 (1900): 163–80, on 166. Paper given by Pringsheim at a meeting of the German Physical Society on 2 Feb. 1900.

21. Kangro, "Das Paschen-Wiensche Strahlungsgesetz und seine Abänderung durch Max Planck."

22. The expression "black radiation" came into common use at this time. Max Thiesen, one of the blackbody experimenters, justified it: "The concept of the (complete) blackbody was used by Kirchhoff in 1862; at the same time, he showed how the radiation from such a body can be realized independently of an actual blackbody. Since then, one has become accustomed to regard radiations independently of the radiating body, and I therefore recommend that radiation with the properties of that emitted by a blackbody be designated by a special name, most simply *black radiation;* the paradox of the expression disappears upon closer consideration, since the trivial and the scientific concept of blackness are not coextensive." "Über das Gesetz der schwarzen Strahlung," *Verh. Phys. Ges.* 2 (1900): 65–70, on 65. Paper given at a meeting of the German Physical Society on 2 Feb. 1900.

23. The critical "theorists" in this connection were Planck and Wien. Planck wrote on 22 Mar. 1900 that "the questions pending between the observers are also an inducement for me to state clearly, and to undertake a sharp criticism of, the theoretical suppositions that lead to the expression for the radiation entropy." "Entropie und Temperatur strahlender Wärme," *Ann.* 1 (1900): 719–37, on 720. In this paper Planck retracted his earlier conclusion that Wien's law is the only law that satisfies the second law of thermodynamics. But he still thought that Wien's law is correct; the experimental evidence against its general validity was not yet compelling to him. Wien wrote on 12 Oct. 1900 that "lately the theoretical and experimental investigations of the radiation of blackbodies have been the subject of many discussions" and that he "would like to discuss this question critically in still greater detail." "Zur Theorie der Strahlung schwarzer Körper. Kritisches," *Ann.* 3 (1900): 530–39, on 530. In this paper Wien repeated criticisms he had recently made of Planck's derivation of his, Wien's, radiation law. Wien's derivation was in turn criticized by Eugen Jahnke, Otto Lummer, and Ernst Pringsheim, "Kritisches zur Herleitung der Wien'schen Spectralgleichung," *Ann.* 4 (1901): 225–30; submitted on 12 Dec. 1900.

the atmosphere, he was answered by Pringsheim: "This source of error I consider completely excluded by the type of experiment." In a similar vein, when Wiener asked, "Is it really certain that even with the longest wavelength the absorption of the bolometer is taken precisely into account?" Pringsheim answered, "One knows that very precisely. We have checked that very precisely. For long waves it is much easier to produce a blackbody. There soot is already 92% of a blackbody, porcelain 90%." When Wiener persisted, "But for very long waves soot becomes transparent," Pringsheim answered, "That may be, but it does not matter here," which was the final word on the matter.[24] Blackbody experimenters could place confidence in the precision of their measurements.

In October 1900, at a meeting of the German Physical Society, Planck proposed another "improvement," this time not of the derivation of Wien's law but of the form of the law itself. He remarked on the experimentalists Ferdinand Kurlbaum and Heinrich Rubens's measurements of long waves, reported on at this same meeting but already known to Planck who had been informed of them privately before the meeting; they confirmed Lummer and Pringsheim's earlier assertion that Wien's law does not have "general significance" but is only a "boundary law" good for short waves or low temperatures; for the other limit, that of long waves and high temperatures, Kurlbaum and Rubens had found a nearly linear relationship between energy and temperature. On the basis of these empirical findings, Planck proposed an interpolation formula to agree with both limits. This improvement over Wien's law depended on a new expression for the entropy of a linear resonator in equilibrium with the radiation field, since by Planck's electromagnetic theory the energy distribution law is determined once the entropy is expressed as a function of the energy of the resonator's oscillations. To obtain this improved law, Planck "constructed completely arbitrary expressions for the entropy, which, although more complicated than Wien's expression, still seem to satisfy all demands of thermodynamic and electromagnetic theory as completely as it did." Of these expressions, "one especially pleased" him for its simplicity. Using it together with the second law of thermodynamics and Wien's displacement law, Planck deduced the energy distribution law:

$$ E = \frac{C\lambda^{-5}}{e^{c/\lambda t} - 1} \quad , $$

where c and C are constants. It gave as good a fit with the observed values as the best of the other laws established to date, and Planck recommended it to the experimentalists for testing.[25] This was done: on the same night following the meeting,

24. Discussion following Wien's report, "Die Temperatur und Entropie der Strahlung," at the Aachen meeting of the German Association in September 1900, submitted on 20 Oct. 1900 to the *Phys. Zs.* 2 (1900): 111. At the same meeting, the talk by Ernst Pringsheim, "Über die Gesetze der schwarzen Strahlung nach gemeinschaftlich mit O. Lummer ausgeführten Versuchen," was followed by the discussion we give here; submitted on 14 Nov. 1900 to the *Phys. Zs.* 2 (1900): 154–55, quotations on 155.

25. Max Planck, "Ueber eine Verbesserung der Wien'schen Spectralgleichung," *Verh. Phys. Ges.* 2 (1900): 202–4; paper given at a meeting of the German Physical Society on 19 Oct. 1900. In place of the earlier connection between entropy S and energy U, from which Planck had derived Wien's law,

Rubens compared his measurements with Planck's formula, and the next morning he called on Planck to tell him of the satisfactory agreement he had found.[26] Planck had found something that looked important.

It was only a beginning for Planck, who did not regard the relative simplicity of the improved energy distribution law as the same thing as a theoretical understanding. It had "only a very limited value," he said in his Nobel prize lecture, since even if it proved accurate, it was at best "happily guessed": "Therefore, from the day of its formulation, I was occupied with the problem of obtaining for it a true physical meaning."[27] Planck was on his way to finding that "true physical meaning" when a few days after presenting his law to the Physical Society he received a letter from Lummer with his and Pringsheim's new law for the energy distribution of blackbody radiation. "I immediately started to work on deciphering the theoretical meaning of your new formula," he replied to Lummer, "but unfortunately I have absolutely no idea what to make of it." From Lummer and Pringsheim's formula, Planck determined the corresponding entropy, his own theoretical starting point, finding it "so colossally complicated that probably even the most skillful mathematician would not succeed in putting it into a useful form."

> If the prospect should exist at all of a theoretical derivation of the radiation law, which I naturally assume, then, in my opinion, this can only be the case if it is possible to derive the expression for the probability of a radiation state, and this, you see, is given by the entropy. Probability presumes disorder, and in the theory I have developed this disorder occurs in the irregularity with which the phase of the oscillation changes even in the most homogeneous light. A resonator, which corresponds to a monochromatic radiation, in resonant oscillation will likewise show irregular phase changes, and on this the concept and the magnitude of its entropy are based. According to my formula, the entropy of the resonator would come to:
>
> $$S = \alpha \log \frac{(\beta + U)^{\beta + U}}{U^U},$$
>
> and this form very much recalls expressions occurring in the probability calculus. After all, in the thermodynamics of gases, too, the entropy S is the log of a probability magnitude, and Boltzmann has already stressed

---

$d^2S/dU^2 = \text{const.}/U$, he now wrote $d^2S/dU^2 = \alpha/U(\beta + U)$, which led to the improved distribution law replacing Wien's. On 7 Oct. 1900, twelve days before the meeting at which Planck presented his new radiation formula and Ferdinand Kurlbaum and Heinrich Rubens presented their new measurements, Planck spoke with Rubens about his and Kurlbaum's experiments then in progress. Max Laue, "Rubens, Heinrich," *Deutsches biographisches Jahrbuch*, vol. 4, *Das Jahr 1922* (1929): 228–30, on 230. Kangro, *Vorgeschichte*, 200.

26. Planck, *Wiss. Selbstbiog.*, reprinted in *Phys. Abh.* 3: 394.

27. Max Planck, *Die Entstehung und bisherige Entwicklung der Quantentheorie* (Leipzig: J. A. Barth, 1920). This talk was given in Stockholm on 2 June 1920; reprinted in *Phys. Abh.* 3: 121–34, quotation on 125.

the close relationship of the function X^X, which enters the theory of combinatorials, with the thermodynamic entropy. I believe, therefore, that the prospect would certainly exist of arriving at my formula by a theoretical route, which would then also give us the physical significance of the constants C and c.[28]

As he said, Planck "naturally" assumed the existence of a theoretical derivation of his new radiation law and furthermore that the route to it lay somehow through Boltzmann's "theory of combinatorials." His letter to Lummer, written one week after the meeting of the Physical Society, reporting on his work on the blackbody problem, agrees with his recollection at the time of his Nobel prize: The question of a theoretical derivation of the law "of itself led me to a consideration of the connection between entropy and probability, thus to Boltzmann's train of ideas; after several weeks of the most strenuous work of my life, the darkness lifted, and a new, unexpected perspective began to dawn on me."[29]

By now Wien and Paschen had conceded that their original law failed for long waves, as Lummer and other experimentalists had previously reported. Only Wien still held on to part of his original derivation, believing that there might have to be two sets of theoretical hypotheses, one for his law, which was confirmed for short waves, and a second for another law for long waves.[30] Planck was not interested in this but in a single set of hypotheses for a single law, which he soon found and which he referred to in a letter to Wien: "I now also have a theory for [my new formula], which I will lecture on here at the Physical Society in 4 weeks."[31]

At this meeting of the Physical Society, on 14 December 1900, Planck opened his now famous paper: "Gentlemen! Several weeks ago I had the honor of directing your attention to a new equation that seemed suitable to me for expressing the law of the distribution of radiating energy over all areas of the normal spectrum."[32] He went on to give the theory for the entropy formula he had included in his letter to Lummer, in the direction he had indicated there, using Boltzmann's combinatorials.[33] To connect the entropy formula for a single resonator with probability, Planck analyzed a large number of resonators: he regarded the time-average energy of a single resonator—the quantity entering the entropy formula—as an average over the energies of a large number of independent, identical resonators at the same time,

28. Letter from Planck to Lummer, 26 Oct. 1900, Wrocław UB, Lummer Nr. 222.
29. Planck, *Die Entstehung und bisherige Entwicklung der Quantentheorie*, 125.
30. Wien, "Zur Theorie der Strahlung schwarzer Körper. Kritisches."
31. Planck to Wien, 13 Nov. 1900, Wien Papers, STPK, 1973.110.
32. Planck, "Zur Theorie des Gesetzes der Energieverteilung im Normalspectrum," *Verh. Phys. Ges.* 2 (1900): 237–45, on 237; paper given at a meeting of the German Physical Society on 14 Dec. 1900.
33. Rosenfeld inferred the existence of an entropy formula similar to the one Planck cited in his letter to Lummer early in Planck's search for a theoretical derivation, from which Planck worked backwards to the combinatorial formula: "La première phase de l'évolution de la Théorie des Quanta," 165–66. Klein has recognized this as Planck's starting point in "Max Planck and the Beginnings of the Quantum Theory," 469–70. It is clear from Planck's remarks in his letter to Lummer that the resemblance of the entropy formula to formulas of the probability calculus pointed him in the direction of Boltzmann's combinatorial approach to counting complexions in his probabilistic definition of entropy, as Kuhn has argued in *Black-Body Theory*, 100–101.

which allowed him to introduce an adaptation of Boltzmann's probabilistic method for calculating a "simultaneous energy distribution." To carry out the calculation, Planck had to consider the total energy available to the resonators oscillating at frequency v "as composed of a quite definite number of finite equal parts," or "energy elements," hv; here h entered as a "natural constant," the value of which, as calculated from blackbody radiation measurements, was 6.55×10^{-27} erg sec. From the average resonator energy, Planck calculated the entropy of a resonator and from it, his electromagnetic theory, and the second law of thermodynamics, he deduced the formula for the energy distribution of blackbody radiation at frequency v, which he designated by u_v:

$$u_v = \frac{8\pi h v^3}{c^3} \cdot \frac{1}{e^{hv/k\theta} - 1}.$$

It is easy to see that this formula exactly agrees with Planck's earlier "happily guessed" one, only now the constants c and C are replaced by h and k, universal constants on which he placed great significance.[34]

That same month, December 1900, Planck's formula received further confirmation from Rubens and Kurlbaum, who extended their measurements to "as large a temperature range as possible." They noted, correctly, that "for short waves and low temperatures, [Planck's] equation approximates Wien's, for long waves and high temperatures Lord Rayleigh's equation, and encompasses both as limiting cases"[35]

34. Planck, "Zur Theorie des Gesetzes der Energieverteilung im Normalspectrum," 238–40, 242. The constant c in the equation is the velocity of light. The first factor in the law comes from Maxwell's theory and entered Planck's earlier version of Wien's law; the resonator energy U_v is converted to field energy u_v through the relation $u_v = 8\pi v^2/c^3 \cdot U_v$. The constant k in the second factor relates the entropy to the logarithm of the probability in Boltzmann's formulation of the second law. Boltzmann had not actually introduced this constant, and Planck calculated it here for the first time from his radiation law. With a value for k, Planck calculated Avogadro's constant and the "Boltzmann-Drude constant," or the average kinetic energy of an atom at unit absolute temperature; with Avogadro's constant, he calculated Loschmidt's constant and the charge of the electron. The exactness of these calculated constants depended on that of k, which was sufficient for Planck to conclude that his values surpassed "by far all previous determinations of these magnitudes"; he regarded the experimental test of these values as an important and necessary "problem for further research" (p. 245). Klein, Ehrenfest, 226, 229.

Our brief discussion of Planck's theory is meant to give an idea of Planck's theoretical goals and the sources he drew on to realize them. To see how he combined the sources and how he interpreted his results requires a close, technical analysis of his reasoning. This has been given by Klein and Kuhn in the publications cited in note 17 above. They have arrived at different interpretations of Planck's thought and have located differently the introduction of energy discontinuity, the concept by which physicists have come to distinguish "modern" from "classical" physics. Klein sees it in Planck's work in 1900; Kuhn sees it later, in 1906, when Einstein and Paul Ehrenfest recognized that discontinuity is demanded by Planck's law.

35. Heinrich Rubens and Ferdinand Kurlbaum, "Über die Emission langwelliger Wärmestrahlen durch den schwarzen Körper bei verschiedenen Temperaturen," Sitzungsber. preuss. Akad., 1900, 929–41, on 931, 933. Paper presented by Kohlrausch to the Prussian Academy of Sciences on 25 Dec. 1900. Rayleigh's equation, which is compared here with Wien's and Planck's, reads: $E = C \cdot 1/\lambda^5 \cdot \lambda T \cdot e^{-c/\lambda T}$. A few months before, Lord Rayleigh had discussed Wien's law in the Philosophical Magazine; deciding it was improbable because of its behavior at infinitely high temperatures, Rayleigh proposed the above law in its place. Measurements of "Reststrahlen" were based on the behavior of some crystals, which selec-

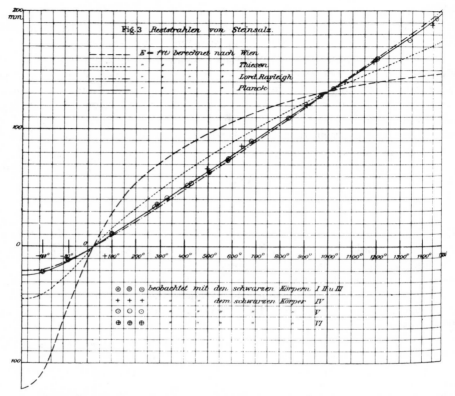

Fig. 38. Blackbody Radiation Curves, 1901. Reprinted from Heinrich Rubens and Ferdinand Kurl-baum, "Anwendung der Methode der Reststrahlen zur Prüfung des Strahlungsgesetzes," *Ann.* 4 (1901): 649–66, on 660.

(see figure 38). At the German Association meeting the next year, Pringsheim noted that "Planck has replaced his theoretical derivation of Wien's equation by another train of thought, which makes his new spectral equation . . . seem theoretically probable"; and although from the experimental side the question was not yet closed as to whether or not Planck's law is a "complete expression of black radiation," it

---

tively reflect waves of certain lengths much stronger than they do waves of other lengths. The selective wavelength for "Steinsalz," rock salt, is about 51.2 μ. The curves here display measurements, taken with several blackbodies, of the intensity of radiation for the wavelength of rock salt as a function of temperature. That is, one of the independent variables in the radiation law, wavelength, is held constant in "Reststrahlen" measurements while the other independent variable, temperature, is allowed to change. The observations are compared with several important theoretical curves, including the best, Planck's. The remaining law being compared here is Max Thiesen's, $E = C \cdot 1/\lambda^5 \cdot \sqrt{\lambda T} \cdot e^{-c/\lambda T}$. Rubens and Kurlbaum included the results from their Prussian Academy paper above in a paper early in 1901 in the *Annalen der Physik*, from which figure 38 is taken. In this figure Planck's law takes the place of Lummer and Jahnke's in the comparable figure in the Prussian Academy paper. "Anwendung der Methode der Reststrahlen zur Prüfung des Strahlungsgesetzes," *Ann.* 4 (1901): 649–66, figure on 660.

"deserves to be preferred to all other spectral equations established so far and in any case comes very close to the truth."[36] Future experimental tests confirmed Planck's law with ever greater completeness.

The solution to the blackbody problem, at the stage where we leave it, was achieved rapidly through a close interaction of theoretical and experimental work, providing an early example of what was to become an increasingly familiar pattern of research in atomic physics. The research we have sketched here involved several of the principal developments in our account of German theoretical physics around the turn of the century. They include the bringing together of thermodynamics, statistical mechanics, and the electromagnetic theory of radiation, fundamental theories of the second half of the nineteenth century. They include the establishment of specialized positions in physics: Planck, the first German theorist by preference to work wholly in positions for theorists, carried out protracted mathematical-theoretical studies on radiation that went beyond the competence, or at least fell outside the concerns, of the experimental specialists in blackbody radiation. They include ever more refined precision measurements of physical laws and constants. Research on the blackbody problem points up the ways theoretical and experimental specialists came together to report on their work and to learn from, and to criticize, the work of others at meetings and through correspondence and conversation. It points up, too, the collaboration of specialists working in various institutions, one of which, the Reichsanstalt, embodied a related development we have discussed: the employment of physicists in institutes intended to further industrial technology. Finally, the developments connect, as we show below, with the teaching of theoretical physics, which returns us to the activities in Planck's Berlin institute for theoretical physics.

In 1901 Planck published a second theoretical derivation of the new blackbody radiation law in the *Annalen der Physik* and several more papers having to do with blackbody theory. After that, he published nothing more on the subject until 1906, when he brought out a new textbook on heat theory. The book was based on the lectures he gave in the winter semester of 1905–6, the place in the lecture cycle on theoretical physics where he covered the theory of heat. It was a kind of sequel to his first published lectures in Berlin in 1897; both contained much research of his own on thermodynamics. His purpose in 1906 was to present the "entire theory of radiant heat on unified thermodynamic foundations."[37] To accomplish this, in his *Vorlesungen über die Theorie der Wärmestrahlung* Planck brought together many of the principal theoretical developments of nineteenth-century physics. These included the mechanical theory of heat and Kirchhoff's extension of it to the phenomena of

---

36. Otto Lummer and Ernst Pringsheim, "Temperaturbestimmung mit Hilfe der Strahlungsgesetze," submitted on 9 Oct. 1901 to the *Phys. Zs.* 3 (1902): 97–100, on 97–98. Paper given by Pringsheim at the Hamburg meeting of the German Association in September 1901.

37. Max Planck, *Vorlesungen über die Theorie der Wärmestrahlung* (Leipzig: J. A. Barth, 1906). This text is discussed in Kuhn, *Black-Body Theory*, 114–34.

radiation in his radiation law, the Stefan-Boltzmann law for the total energy of radiation, Wien's displacement law relating the frequency of maximum radiation to the temperature, Boltzmann's relation between entropy and probability from the theory of gases, and Maxwell's electromagnetic laws, to which a theory of radiation now had to conform. Planck referred his audience to the theoretical writings by Kirchhoff, Boltzmann, Helmholtz, and others, including himself, as he also referred them to much experimental work on the thermodynamic properties of radiation.

Lecture by lecture, Planck incorporated these developments as he gradually built up the theory of radiant heat. He began with optics, which he specialized to Maxwellian optics. Then, as he needed them, he introduced thermodynamic considerations and then, from gas theory, the probabilistic interpretation of the second law. From this theoretical foundation, he derived the law, his law, describing the distribution of energy over the frequencies in blackbody radiation, and he concluded by deriving the H-theorem electromagnetically. He discussed the distinction between reversible and irreversible processes in nature, the reason for the complex foundation of the theory. Planck's lectures on radiant heat belonged to his lectures on all parts of theoretical physics, and he needed all parts to construct the theory. He carefully explained the contributions to it from mechanics, electrodynamics, and thermodynamics, pointing out the insufficiency of each by itself, a consequence of the distinction between reversibility and irreversibility. He showed that the calculus of probabilities coupled with the hypothesis of an elementary disorder bridges the exact microscopic descriptions of mechanics and electrodynamics with the average macroscopic descriptions of thermodynamics.

Planck took pains to make clear to his students the need for introducing probability into the foundations of the theory. At first sight, he said, probability would seem incompatible with electrodynamics, since Maxwell's equations determine unambiguously the future course of an electrodynamic process. The flaw in the objection is that the intensity of radiation absorbed by a resonator does not determine the oscillations of the resonator unambiguously. If physicists do not want to renounce their goal of explaining thermodynamic processes by electrodynamics, they must supplement the latter; it is the same in the theory of gases, where mechanics has to be supplemented by the hypothesis of the molecular disorder of the motions of heat. In radiation theory, the hypothesis of natural radiation, of an elementary disorder in electromagnetic waves, justifies the introduction of the "foreign element" of probability in the derivation of the entropy of a resonator and with it of the complete form of the law of blackbody radiation. The energy element $h\nu$ entering the probabilistic calculation does not disappear from the final result for the entropy of a resonator as does the analogous energy element in the derivation of the entropy of a gas; the "new universal constant" h expresses the "real difference" in the two cases.[38]

After deriving the energy element by calling on Wien's displacement law, Planck derived it again by dividing up the phase space of the electromagnetic state of a resonator in such a way that h defines an area of equal probability. The constant

38. Planck, *Theorie der Wärmestrahlung*, 129–40, 153.

now provides another basis for calculating the number of complexions in the analysis of probability, and it gains another physical significance, one which Planck regarded as more fundamental than that of determining the size of the original energy element. Since the dimensions of the constant are the same as those from which the "principle of least action derives its name," Planck proposed calling h the "quantum of action" or "element of action."[39]

With these lectures on the theory of radiant heat, Planck revealed to his students in Berlin and to his readers elsewhere his approach to theoretical physics in general. The resonator, which he analyzed at length, is the mediating agency that brings matter and radiation in a blackbody into thermal equilibrium; the resonator, like the blackbody, is an idealization, the simplest conceivable system for the purpose: two electric quantities of opposite signs capable of oscillating on a line between them. There was no reason to think that such a system really exists in nature, nor was there any reason for Planck to worry if none did. He had Kirchhoff's law, which is expressed in terms of a "universal" function that requires of the systems it governs only their conformity to the laws of physics, not their reality. Planck built his radiation theory from, and toward, such functions, constants, and laws that are independent of the differences between materials, that are universal. His starting point, the assumption of the "universal validity" of the two main laws of thermodynamics, justified his extension of them from matter to radiation. The laws of probability are "universal," too, and so, Planck reasoned, they should have a close connection with thermodynamics. The law of the normal spectrum of a blackbody is a "universal" law. Entropy and probability are related through the "universal constant of integration," k, which applies to "terrestrial" and to "cosmic" systems alike, and to it is joined the "new universal constant" h. Since the formulas containing k and h have "absolute validity," these two constants together with the speed of light and the gravitational constant provide the basis for a "universal" system of "natural units" for length, time, mass, and temperature.[40]

Planck's lectures on the theory of radiant heat were a fitting culmination of the entire lecture cycle. In the semester preceding these lectures, Planck had introduced his students to optics, in the semester before that to electricity and magnetism, and in the two semesters before that to mechanics. The sequence was necessary, as we have seen, since the theory of heat called upon all of the major theories the students had learned. Coming at the end of the series of lectures on the major divisions of physics, the theory of radiant heat illustrated a fundamental point: theoretical physics has a unity, and its various parts can be made to yield universally valid results when skillfully joined.

Planck did not leave his students with the idea that the theory of radiation was a closed book. It had not yet been brought to a "fully satisfactory conclusion," since the significance of the constant h was still undetermined. Planck recognized that this constant was probably no less significant than the elementary electrical quantum, the electron charge; "naturally," he said, h must receive a "direct electrodynamic

39. Planck, *Theorie der Wärmestrahlung*, 155.
40. Planck, *Theorie der Wärmestrahlung*, 60, 135, 137, 153, 163–65.

significance," by which he meant in terms of the electron theory. He was accurate in anticipating what would become the center of interest of his radiation theory, though in 1906 he could not have foreseen the far-reaching changes in the foundations of theoretical physics that the pursuit of h would entail.[41] Neither, of course, could the young physicists who heard Planck's lectures on radiant heat, James Franck, Robert Pohl, Walter Meissner, and Wilhelm Westphal, whose careers would be affected by the subsequent development of the quantum theory.[42] Einstein, for the past year a regular reviewer for the *Beiblätter* of the *Annalen der Physik,* praised Planck's lectures in the *Beiblätter* of 1906 for bringing together Kirchhoff's, Wien's, and Planck's own researches into a "wonderfully clear and unified whole."[43] That had been precisely Planck's intention.

Soon after his lectures on radiant heat were published, Planck was approached by a publisher to write a brief textbook covering the "entire theoretical physics." Planck tried to get Sommerfeld to write it with him. He was not a writer of textbooks, he told Sommerfeld; he was much more inclined to do individual research, going forward rather than doing the retrospective work of collecting older materials for composing textbooks. Sommerfeld declined, pointing out that the necessary unity of such a work was best realized if one physicist wrote it.[44] Only much later, after World War I, would Planck publish his lectures on all parts of theoretical physics, now the traditional expectation of chairholders in the field.

Göttingen acquired its long-awaited new physics building in 1905, and with it Voigt acquired a good number of bright, solid rooms for his institute. His and Riecke's institutes were not separated by any horizontal or vertical divisions, as might be expected; rather he and Riecke each had his own lecture hall with preparation and apparatus rooms, while they shared certain rooms, such as the library and machine room, and the remaining space was divided roughly in half, Riecke receiving the larger rooms and Voigt receiving more rooms. The principal difference between the two institutes was that the director of one, Riecke, gave the lecture course in experimental physics and the director of the other, Voigt, the lecture course in theoretical physics. More experimental work was done in Riecke's institute, but the "aims and means of both laboratories," Voigt said, were "essentially the same."[45]

41. Planck, *Theorie der Wärmestrahlung,* 153, 179, and 221. In a letter at the time, Planck elaborated on the electrodynamic significance he anticipated for h. The introduction of the "*finite* quantum of oscillation $\epsilon = h\nu$ signifies a new hypothesis foreign to the resonance theory," so that a "new element enters the theory that in no case can be deduced in a logical way." Progress might be made by assuming that resonance oscillations consist in the motion of electrons; for it seems that the "existence of an elementary quantum of electricity," the electron charge, provides a "bridge to the existence of an energetic elementary quantum h, especially as h is of the same dimensions and also of the same order of magnitude as e^2/c." Planck to Paul Ehrenfest, 6 July 1905. This letter is quoted in full, and in connection with Planck's 1906 lectures, in Kuhn, *Black-Body Theory,* 132, 288–89. In a long correspondence with Lorentz, Planck discussed the physical meaning of h and of the related need for the "electron theory to be enlarged by a new hypothesis." Planck to Lorentz, 1 Apr. 1908, Lorentz Papers, AR.

42. Wilhelm Westphal's recollections in Bertholet, et al., "Erinnerungen," 167.

43. Einstein's reviews in the *Beiblätter* in 1905–7 are discussed in Martin J. Klein and Allan Needell, "Some Unnoticed Publications by Einstein," *Isis* 68 (1977): 601–4.

44. Planck to Sommerfeld, 24 Feb. 1909, Sommerfeld Correspondence, Ms. Coll., DM.

45. Woldemar Voigt, "Rede," in *Die physikalischen Institute . . . Göttingen,* 37–43, on 38.

Voigt outfitted his institute according to his interests. He had a good deal of apparatus for optics and crystal physics, related subjects which he had worked on throughout his career and which dominated his students' dissertations. In the lowest cellar floor of the new building, for example, he kept apparatus for studying the gradual deformation of crystals under constant temperature. And because he was interested in the Zeeman effect, he had begged and borrowed spectroscopic apparatus and a powerful electromagnet to do research on it. The firm of Friedrich Krupp gave the institute the iron parts for the massive mounting of a concave grating, and at the dedication of the new building another industrialist promised a spectroscope. These were small gifts compared with what industrialists gave Göttingen's applied physics institutes, but they were important, for without them and other private contributions the institute would have been crippled, as Voigt saw it. The Prussian government and Voigt did not see eye to eye on the support for the research that Voigt believed his institute existed to carry out.[46]

Poor employment prospects made German students less preoccupied with examinations and more inclined to spend time doing experiments, Voigt told the Göttingen curator. Voigt's expanded quarters in the new building gave students their opportunity. From the start, over thirty students attended the laboratory, and the number of advanced students grew steadily. In the old institute, Voigt seldom had more than five advanced students working at a time, but in the first year of the new institute, 1906, he had thirteen, and in 1910 he had twenty-two.[47] With the increase in numbers of students, the cost of running the institute naturally increased, rising to two or two and a half times the official institute budget. To make ends meet, Voigt did not keep the fees the students paid him to work in the laboratory, money which institute directors usually pocketed; instead he used them to supplement the budget. While he tried, in vain, to get the budget raised, he made do with extra allowances from time to time.[48]

The Göttingen curator did not understand why Voigt spent so much money every year and why he always asked for more instead of living within his budget as he was supposed to. Voigt explained: he intended his institute in "overwhelming measure for physical *research*," and he regarded it as his "*first duty*" to admit as many students as he had room for, whatever the cost to himself. The official budget was sufficient for "*elementary*" instruction but not for "*higher*" instruction, meaning scientific research. Voigt reminded the curator that research had an "educational force of great significance" and was bound to help the student in his later professional life.[49]

In addition to the ever-present running costs, materials, laundry for all the students, and taxes on lights in the rooms where research was carried out, there always

46. Voigt, "Rede," 41–42.

47. Voigt to Göttingen U. Curator Osterrath, 18 Mar. 1907, Göttingen UA, XVI. IV. C. v. Voigt to Prussian Minister of Culture von Trott zu Solz, 30 Oct. 1910, Göttingen UA, 4/V h/35.

48. Voigt to Osterrath, 18 Mar. 1907, 18 and 20 Mar. 1908, 11 Jan. 1910, 31 Jan. 1911, and 21 Mar. 1912; Göttingen UA, XVI. IV. C. v. Voigt to Prussian Minister of Culture von Trott zu Solz, 11 Mar. 1911, Göttingen UA, 4/V h/35.

49. Voigt to Osterrath, 18 and 20 Mar. 1908 and 31 Jan. 1911. Osterrath to Voigt, 19 Mar. 1908, Göttingen UA, XVI. IV. C. v.

seemed to be an extra expense. One year it was for the mechanic and other skilled workers, the plumber, the locksmith, and the cabinetmaker. Another year it was for instruments, such as the small pieces of auxilliary apparatus needed by students in the course of their research. Ordinarily, the mechanic could make the small apparatus, saving the institute money, but he lacked sufficient time, as he had to serve Riecke's institute as well as Voigt's. When Voigt needed large apparatus he looked to outside donors or borrowed from the Prussian Academy of Sciences or elsewhere. On occasion he approached the ministry directly with a request; for example, for money to bring in electrical equipment for spectroscopic research. As justification, Voigt explained that it was necessary to keep students from losing heart when they compared their obsolescent spectroscopic equipment with the up-to-date equipment they read about in physics journals. He reminded the ministry that it was the "right and duty" of a provincial institute to be excellently equipped in "*one* area," as at Göttingen in spectroscopy.[50]

With the expanded activity of the new institute, Voigt found himself overburdened. He complained that it was "uncommonly great intellectual work" to oversee the twenty or so students who worked at scientific research, usually throughout the entire day. One assistant was no longer enough, and rather than restrict the number of students, Voigt hired a second assistant at his own expense in 1909. When in the following year he asked that the new assistant's salary be included in the budget, he explained that he was oppressed by his duties; he was after all over sixty, and he had "in great part spent his vigor in the service of the institute."[51] In 1911 he was again elected prorector of the university, which post he could accept, he told the curator, only if he could be relieved of some of the "almost crushing work load of the direction of the institute." He asked if he might hire two assistants at his own expense. He was told he might.[52] Because of too much work he often had to be excused from giving his public lectures. In 1912 the attendance in his optical laboratory was so great that he had to spend four afternoons in it instead of the regular two, which was for him and for his assistants yet another "very noticeable" increase in work.[53]

For fifty semesters, Voigt recollected, he had run the theoretical physics institute in Göttingen, and with satisfaction he had seen it grow in activity and importance. But through the years, the government had rejected his requests for increased support, which he came to interpret as unfriendliness, as disinterest in research. It was not even a question of supporting Voigt's own research, which he paid for out of his own pocket with help from the Göttingen Society of Sciences. It was a question of proper support for the institute, which Voigt suspected would be casually granted to his successor, while he had been denied it despite his years of devoted service. This state of affairs was "hardly dignified," he told the curator, and having to work under such conditions year after year was "tiring and embittering."[54]

50. Voigt to Osterrath, 20 Mar. 1908 and 21 Mar. 1912. Voigt to Prussian Minister of Culture von Trott zu Solz, 26 Nov. 1912, Göttingen UA, XVI. IV. C. v.
    51. Voigt to Trott zu Solz, 30 Oct. 1910 and 11 Mar. 1911.
    52. Voigt to Göttingen U. Curator, 25 July 1911, Voigt Personalakte, Göttingen UA, 4/V b/267c.
    53. Voigt to Göttingen U. Curator, 17 Apr. 1912, Voigt Personalakte, Göttingen UA, 4/V b/267c.
    54. Voigt to Osterrath 11 Jan. 1910 and 31 Jan. 1911.

Voigt experienced frustrations familiar to directors of physics institutes, but there was an additional source of frustration for him. As a theoretical researcher and teacher, he looked on while other theoretical physicists such as Planck and Lorentz moved, as he saw it, in the "pure ether of the most general questions." His situation limited him to digging "like a mole in the earth after small specialties." His "subordinated" work at Göttingen restricted him to working with students on "elementary" questions.[55] His duties as director of a large institute equipped for experimental work did not allow him time to reflect. Voigt's view of the desirable activity of a theoretical physics professor was becoming unworkable, and that was, at least in part, the source of Voigt's problems and recurring complaints.

After his *Kompendium* in 1895–96, Voigt did not write another text covering all of theoretical physics. Instead he went back to writing texts on the parts of his science; for example, in 1903 and 1904 he brought out a text on thermodynamics, which he hoped would do some good since there was "no comprehensive German thermodynamics," and since Planck's published lectures on the subject gave too much weight to thermochemistry;[56] he followed up this text with another specializing in thermochemical and thermoelectrical transformations. As with Planck, Volkmann, and other writers of physics texts, Voigt's choice of which lectures to publish depended on his research interests. In the late 1890s and early 1900s, Voigt did a good deal of research on the interactions of the natural forces, which tended to fall between the more standard topics; in large part, it belonged to crystal physics and concerned, for example, piezoelectricity, piezomagnetism, pyromagnetism, and magneto- and electro-optics. In 1908 Voigt brought out a text based on his lectures on electro-optics, for which the motivating discovery was the Zeeman effect and the appropriate theory was Lorentz's electron theory; one of the "most brilliant" results in all of physics, Voigt said, was the probable identification of the bound electrons responsible for the emission and absorption of light, as in the Zeeman effect, with the free electrons studied in cathode-ray experiments. To Voigt, a "main attraction" of magneto- and electro-optics was the "lively interactions between theory and observation": here experiment gave the first stimulus, then theory made sense of it, reworked it, and gave renewed stimulus to experiment, a repeating cycle.[57]

In 1910 Voigt published his last textbook, nearly a thousand pages long, which was again based on lectures he gave at Göttingen. The subject was crystal physics, and by dedicating the text to his teacher Franz Neumann, Voigt pointed to work spanning a career of thirty-six years. Now that his research "might be approaching its end," he took this way of showing that the scattered and seemingly unconnected investigations that he had devoted to this "large and wonderful" subject were guided by a striving for unity; he presented crystal physics as a "rigorously closed, unified whole."[58]

55. Voigt to Lorentz, 19 May 1911, Lorentz Papers, AR.
56. Woldemar Voigt, *Thermodynamik*, 2 vols. (Leipzig: G. J. Göschen, 1903–4). Voigt to Sommerfeld, 18 Oct. 1902, Sommerfeld Correspondence, Ms. Coll., DM.
57. Woldemar Voigt, *Magneto- und Elektrooptik* (Leipzig: B. G. Teubner, 1908), iv, 3, 73.
58. Woldemar Voigt, *Lehrbuch der Kristallphysik (mit Ausschluss der Kristalloptik)* (Leipzig and Berlin: B. G. Teubner, 1910), viii, 13.

In crystals, the properties of molecules appear in their purest and most complete form, Voigt said at the beginning of his text; he anticipated that crystal physics would be the path to the solution of the ultimate questions of physics, those concerning molecular processes.[59] He valued crystal theories based on assumptions about the forces of molecules, which he took to be electrical, corresponding to the prevalent electrical view of matter, but he disapproved of "structure theories" that began with assumptions about the molecular constitution of crystals, which he believed had proved fruitless in the past. In any case, it was not molecular ideas but the thermodynamic potential that lent unity to Voigt's presentation.[60]

In Voigt's treatment of crystals, the various parts of physics came together naturally, a consequence of his mathematical-geometrical ordering of the phenomena. He did not, for example, treat all of the mechanical phenomena of crystals first, then all of the heat phenomena, and so on; rather he grouped the phenomena according to the kind of directed quantities they required for their representation: interactions between two vectors, representing force and flow, and the more complex interactions between two tensors. For Voigt, the properties of mathematical quantities expressed the symmetries of the physical world and were more useful than the traditional divisions of physics for organizing the materials of crystal physics. The purpose of his text was not only to help researchers correctly pose problems in crystal physics but also to show teachers of physics the value of symmetry considerations. Lectures on symmetry should enter every "theoretical-physics course," Voigt urged.[61]

Nature appealed to Voigt for its order, not its disorder. Crystals exhibit a high degree of order because of the repeated angles of their edges, and to suggest their appeal, Voigt called on an image drawn from his other great intellectual love, music: if every player in an orchestra played the same piece, but not in unison, the result would have no esthetic appeal; molecules in gases, liquids, and amorphous solids produced "music" of just that sort. But if the players played in harmony, the result would be like the "music" of crystals. That was why certain phenomena occurring with "wonderful manifoldness and elegance" in crystals occurred only in "sad monotone average values" in other bodies. "According to my feeling," Voigt said, "the music of physical regularities is intoned in no other field in such full and rich chords as in crystal physics."[62]

Voigt had made the same comparison between music and crystal physics years before, at the opening ceremonies for the new physics building at Göttingen in 1905. He had spoken of this kind of musical physics, of crystal physics, as "old-fashioned physics in the strictest sense," by which he meant that it was pursued for pure understanding apart from any possible technical application. He contrasted the

59. Voigt, *Kristallphysik*, 5.
60. Voigt, *Kristallphysik*, 110–11, 120–21.
61. Voigt, *Kristallphysik*, vii, 133, 226, 305–6. By the "physical symmetry" of a crystal, Voigt understood the number and distribution of its physically equivalent directions (p. 17).
62. Voigt, *Kristallphysik*, 4.

machines of his colleague Ludwig Prandtl in the department for applied mechanics, which bent, drilled, and tore iron bars, with the delicate apparatus he used to measure the elasticity of crystals.[63] Crystal physics—with the exception of crystal optics, now a "closed" subject—attracted a few, very few, physicists, and to his colleagues, Voigt sometimes spoke plaintively of his work. It was "*hermit's* work, in areas that otherwise did not excite interest," he told Sommerfeld.[64] When he told Kayser that he would send him a copy of his text on crystal physics, he added that it would appear to him "as a guest from a rather strange world."[65]

In 1915, in honor of the tenth anniversary of the theory of relativity, the editors of the *Physikalische Zeitschrift* reprinted a paper of Voigt's from 1887 on the Doppler principle. This "very early precursor" of relativity theory contained what became known as the Lorentz transformations, which Voigt had derived from a study of the elastic light-ether. Physicists liked to recall this independent discovery, a curiosity that pointed up Voigt's remove from the developments that "modern" physics came to be identified with.[66]

Voigt had reservations about Einstein's relativity principle as a natural law, and although he thought that the quantum theory had "far greater practical significance" than relativity theory, he did not contribute directly to it either.[67] He did embrace Lorentz's electron theory, spoke of it as a "revolution" in principles, problems, and methods, and appreciated it as a triumph of the molecular viewpoint in physics. His and his students' work on magneto-optics, in fact, proved valuable for the development of atomic physics, but he never accepted the atomistic treatment of physical phenomena as a general way of doing physics, believing it was not yet justified by the facts.[68] While he could appreciate the heuristic value of atomistic and other provisional hypotheses, he preferred the cautious method of phenomenology, which was to draw principles from experience and to build from them a body of mathematical consequences.[69]

"Day and night," Voigt's work would not leave him alone; it is "my misfortune," he told a colleague, "that I am so very passionate about my work."[70] If he did not develop theories of the consequence of some of his colleagues', he was a great "calculator" and a "master of precise experimental art," and he published more than any contemporary German theorist.[71]

63. Voigt, "Rede," 39–40.

64. Voigt to Sommerfeld, 25 Nov. 1909, Sommerfeld Collection, Ms. Coll., DM.

65. Voigt to Kayser, 20 July 1910, STPK, Darmst. Coll. 1924.22.

66. Woldemar Voigt, "Über das Dopplersche Prinzip," reprinted, with editorial comments, from the *Gött. Nachr.*, 1887, in *Phys. Zs.* 16 (1915): 381–86. Arnold Sommerfeld, "Woldemar Voigt," *Jahrbuch bay. Akad.,* 1919 (1920): 83–84, on 84. Försterling, "Voigt," 221.

67. Woldemar Voigt, "Phänomenologische und atomistische Betrachtungsweise," in Warburg, ed., *Physik,* 714–31, on 730. Försterling, "Voigt," 221.

68. Voigt, "Betrachtungsweise," 722–23, 729. Sommerfeld, "Voigt," 83–84.

69. Woldemar Voigt, "Ueber Arbeitshypothesen," *Gött. Nachr.,* 1905, 102.

70. Voigt to Runge, 1 Mar. 1902, Ms. Coll., DM, 1948/53.

71. Voigt published more studies in theoretical physics than anyone else in Germany around the time of his call to Göttingen; in the two years following his call, he produced nearly a quarter of all German publications in theoretical physics. And he kept it up, publishing over a hundred articles in the *Annalen der Physik* alone. Försterling, "Voigt," 217.

Voigt was a superb teacher and practitioner of methods for solving problems in theoretical physics. The symmetry considerations and transformation properties he stressed were increasingly recognized as central to the methods of theoretical physics. In the fields he worked in—optics, elasticity theory, and crystal physics, for example—vectors with their single directionality proved inadequate and tensors with their two directionality indispensable; he did much to introduce tensors into the teaching and practice of theoretical physics, repeatedly calling attention to the name "tensors," his name, which he wanted physicists to adopt for these quantities.[72] Despite his constant reminder to readers of his texts that he was stressing the physical over the mathematical in his presentations, he came to adopt a "more mathematical-formal than physical-visual" way of proceeding in his research, as his recommenders for a post as corresponding member of the Prussian Academy noted.[73] But this was also the direction that much of modern theoretical physics was to take.

New Institute for Theoretical Physics at Munich

The Berlin and Göttingen institutes for theoretical physics, as we have seen, differed greatly in their facilities and in the kind of research their directors and students did. Munich's institute for theoretical physics, the next such institute after Leipzig's to be established in Germany, was no less distinct in its arrangements and scientific personalities. The events that led to the establishment of the theoretical physics institute at Munich and to the appointment of Arnold Sommerfeld as its first director were set in train by W. C. Röntgen, the recently appointed director of the physics institute at Munich. Röntgen had come to Munich with a clear idea of a proper physics establishment for a major university, which included a prominent place for theoretical physics. Because of his stature in physics, Röntgen was able to realize his idea, though it took time.

Würzburg, where Röntgen was professor for experimental physics before coming to Munich, was one of the several German universities that still did not have a separate professor for teaching theoretical physics at the end of the nineteenth century. When Röntgen received a call to Freiburg University in 1895, the Würzburg philosophical faculty drew up a long report on the conditions under which he would remain at Würzburg; a major condition was the establishment of an extraordinary professorship for theoretical physics. The ministry responded favorably to the requests, though with respect to the theoretical physics position only with vague prom-

72. In many writings, Voigt advocated the use of tensors, which had earlier found limited use in physical science; for example, in his first book on crystal physics, *Die fundamentalen physikalischen Eigenschaften der Krystalle* (Leipzig, 1898); in the revision in 1901 of his *Elementare Mechanik*, 10 ff., and in papers such as "Der gegenwärtige Stand unserer Kenntnisse der Krystallelasticität," *Gött. Nachr.*, 1900, 117–76; "Ueber die Parameter der Krystallphysik und über gerichtete Grössen höherer Ordnung," *Ann.* 5 (1901): 241–75; and "Etwas über Tensoranalysis," *Gött. Nachr.*, 1904, 495–513.

73. Max Planck, "Wahlvorschlag für Woldemar Voigt (1850–1919) zum KM," in *Physiker über Physiker*, 154–55.

ises. Röntgen agreed to stay at Würzburg—neither the experimental facilities nor his income at Freiburg would have measured up to those at Würzburg—and for this he was awarded a decoration, but the requests were not met for financial reasons. In 1898 Röntgen received a call to Leipzig University, and again he negotiated for an extraordinary professorship for theoretical physics at Würzburg, threatening to leave if his request was not met. The Würzburg academic senate supported Röntgen's demand, noting that the extraordinary professorship had been repeatedly requested, and the Bavarian government declared itself willing "to postulate in the next state budget the means for . . . the establishment of an extraordinary professorship for theoretical physics." The ministry of the interior now appealed to the Bavarian prince regent, pointing out that "Dr. Röntgen's leaving would mean an all the more painful loss for Würzburg University since there is a lack of outstanding younger physicists, and hence a replacement [by someone his equal] . . . would encounter great difficulites." Despite these appeals, Röntgen did not achieve his goal; Würzburg's extraordinary professorship for theoretical physics was established only in 1901, and by then Röntgen had already left.[74]

Röntgen had not wanted another experimental physicist for the anticipated theoretical physics position at Würzburg, since he himself worked almost exclusively in the experimental direction. Rather he wanted someone who was "well up on the new theoretical way of presentation" and who, if possible, was also a "productive" researcher.[75] Röntgen wanted a physicist of the same description to work beside him at Munich University, where in 1899 he was appointed ordinary professor for experimental physics.

In the summer of 1904, the Bavarian ministry of the interior notified Röntgen that the Imperial ministry of the interior wanted him to come to Berlin to succeed Kohlrausch as president of the Reichsanstalt. Röntgen needed time to think it over. The Reichsanstalt stood high in the opinion of the scientific world, and it offered Röntgen unparalleled resources for his own researches, but he was unsure how well suited he was for a nonacademic job like that. He asked the Bavarian minister to give him good reasons for declining the Berlin offer, which amounted to providing

---

74. Rector of Würzburg U. to the Bavarian Ministry of the Interior, 15 Feb. 1895; Würzburg U. Senate to the Ministry, 25 Feb. 1895; note on Röntgen's refusal of Freiburg's offer, 1 Mar. 1895; Senate to the Ministry, 27 Nov. 1898; the Ministry's response, 30 Nov. 1898; Ministry of the Interior to Prince Luitpold, 12 Dec. 1898, Röntgen Personalakte, Bay. HSTA, MK 17921. Röntgen wrote to Zehnder, whom he had in mind for a salaried extraordinary professorship for theoretical physics at Würzburg, the position he had been trying to establish for a "series of years," that he had just gone to Munich to see about the professorship and learned that there was no money for it; this meant that it would be buried for the next two-year budget period, if not for longer. Röntgen to Zehnder, 10 July 1895, in Röntgen, *Briefe an L. Zehnder*, 36–37.

75. In 1898 Röntgen reported to Zehnder that he now expected approval of the Würzburg extraordinary professorship for theoretical physics. Although earlier he had told Zehnder he wanted him for the position, he now said he did not. Zehnder, who was returning to Würzburg to be Röntgen's assistant, was an experimentalist; Röntgen regarded it as his "duty" to see that the man appointed to the professorship represented "*theoretical* physics." Röntgen to Zehnder, 8 Dec. 1898, in Röntgen, *Briefe an L. Zehnder* 71.

conditions at Munich for teaching and research corresponding to the "state and significance of physics." In its provisions for physics, Röntgen thought, Munich had fallen behind other German universities.[76]

Röntgen pointed to some deficiencies that needed remedying: owing to the high cost of instruments, he needed more money for running his institute; he needed a third assistant because the professor of pharmacy asked him to give a new laboratory course; and he wanted to see an extraordinary professorship for applied physics created. Most important, he needed an ordinary professor for theoretical physics. He had long maintained that instruction in physics is "only half provided for so long as the ordinary professorship for theoretical physics has not been filled." The Munich faculty understood that his "main wish" was to acquire a man of the "first rank" for theoretical physics, even if that meant paying out a "very high salary." Röntgen told the minister that it hardly needs to be said that the second ordinary professor of physics at Munich must have had success in *"theoretical* physics." By this criterion, Röntgen intended specifically to exclude from consideration the local theoretical physicist Leo Graetz.[77]

The minister gave Röntgen quieting assurances that he had no desire to give the vacant position to Graetz, but he could not make a binding promise since appointments were the exclusive right of the crown. He agreed with Röntgen on the main point, the need for an ordinary professor for theoretical physics at Munich, though he did not know where the money would come from. The minister was hardly anxious about losing Röntgen, since his income in Munich was substantially higher than it would be in Berlin; besides, Röntgen himself did not think he was the "right man for the job" in Berlin. So he and the minister entered into negotiations to improve physics in Munich.[78]

In the ten years since Boltzmann left Munich, the faculty had repeatedly expressed its wish to restore the discipline of "higher theoretical physics," while recognizing that there were few "outstanding, first-class workers in this field." Röntgen got the ministry to assure him it would do all it could to raise the necessary money. But the ministry's promise to put the matter before the Landtag did not result in anything, nor did its appeal to the university to make available money from terminated salaries. Just when it seemed to Röntgen that this affair would come to nothing like others before it, the financial problem was solved, at a stroke. Upon his own initiative, the prince regent made available a "considerable sum" to add to the available funds to bring to Munich an outstanding theoretical physicist. Röntgen knew the man he wanted; drawn to the "burning questions of electrons," he wanted

76. Bavarian Ministry of the Interior to Röntgen, 3 Aug. 1904; Röntgen to Bavarian Minister of State Wehner, 15 Aug. 1904, Röntgen Personalakte, Bay. HSTA, MK 17921.

77. Röntgen to Wehner, 15 Aug. 1904. Munich U. Senate to Section II of the Philosophical Faculty, 5 Nov. 1904; draft report by the Philosophical Faculty to the Academic Senate, 17 Nov. 1904, Munich UA, OCI 31.

78. Bavarian Minister of the Interior to Röntgen, 3 and 29 Aug. 1904; apparently Munich U. Curator to Minister, 25 Sept. 1904, Röntgen Personalakte, Bay. HSTA, MK 17921. Röntgen to Zehnder, 11 Oct. 1904, in Röntgen, *Briefe an L. Zehnder*, 91–92.

the creator of the electron theory, H. A. Lorentz. With Lorentz, he believed he could establish physics "on an excellent basis in Munich," making it comparable with physics in Berlin, Göttingen, and Leipzig, where there were already institutes for theoretical physics and where the "best conditions existed in most respects."[79]

If Lorentz were to come, Europe's most famous theorist would join Europe's most famous experimentalist to make Munich a center of physics. Lorentz was strongly tempted, but in the end he declined after Leiden improved his position. Failing Lorentz, Röntgen wanted the "next best theoretical physicist"; he asked Lorentz for advice on a theorist who could lecture in German and who fit the following description: "We need no mathematician but a physicist, who is equipped and familiar with the whole armory of mathematics, but who knows exactly what physics needs and who does not lose himself in speculations that are unfruitful for physics. Further, we need a man about whom one may assume, on the grounds of his previous accomplishments, not only that he can teach the theoretical physics produced by others, but that he will accomplish something competent in this, his own field of work, which one may regard as an essential advance."[80]

To locate this theorist, the Munich faculty appointed a commission consisting of Röntgen and three others. The outcome of their deliberations was a list of twenty-one candidates, all from Germany with the exception of the Dutchman C. H. Wind and the Austrians Hasenöhrl and, once again, Boltzmann. Three local candidates were on it, the applied physicist Föppl and the two men who had given theoretical physics lectures there for many years and whom Röntgen considered below standard, Graetz and Korn. To Graetz, who had already requested a promotion to ordinary professor, the commission gave special attention; they asked eight outside physicists their opinion of Graetz, and their unanimous, "objective and authoritative judgment" was that he was not in the same league with the leading candidates, confirming the commission's judgment. Rarely had an appointment been gone into "in so thorough a manner," the commission said of its own work. They presented their recommendations to the faculty, which in turn would recommend them to the academic senate.[81]

The three candidates the commission recommended were Cohn and Wiechert in first place and Sommerfeld in second. About Sommerfeld, the youngest of the three, the commission was divided, but he had come highly recommended by Lorentz, Boltzmann, and others. The commission's report stressed the research in the-

79. Munich U. Curator to Bavarian Minister of the Interior, 25 Sept. 1904. Röntgen to T. Bovari, 31 Mar. 1905, quoted in W. Robert Nitske, *The Life of Wilhelm Conrad Röntgen: Discoverer of the X-Ray* (Tucson: University of Arizona Press, 1971), 216. Röntgen to Zehnder, 6 Jan. 1905, in Röntgen, *Briefe an L. Zehnder*, 93–94. Ulrich Benz, *Arnold Sommerfeld. Lehrer und Forscher an der Schwelle zum Atomzeitalter, 1868–1951* (Stuttgart: Wissenschaftliche Verlagsgesellschaft, 1975), 47. Röntgen's remarks on the "burning questions of electrons" are referred to in a letter from Sommerfeld to Lorentz, 12 Dec. 1906, Lorentz Papers, AR.

80. G. L. de Haas-Lorentz, ed., *H. A. Lorentz. Impressions of His Life and Work* (Amsterdam: North-Holland, 1957), 97. Minutes of the meeting of the Munich Philosophical Faculty, 6 Mar. 1905, Munich UA, OCI 31. Röntgen to Lorentz, 14 Mar. 1905, Lorentz Papers, AR.

81. Bavarian Ministry of the Interior to the Munich U. Senate, 6 July 1905, Munich UA, OCI 31. Commission report to Section II of the Munich U. Philosophical Faculty, 20 July 1905, Munich UA, E II-N Sommerfeld.

oretical physics by the three candidates and only mentioned their experience and competence in lecturing on theoretical physics. Like Lorentz, both Wiechert and Sommerfeld had done their most important research in electron theory, and although Cohn was not an "unconditional follower of the electron theory," he had done his most important research in electromagnetic field theory, which was closely related.[82]

The original premise of the commission's search seemed borne out: there were few theorists who were at the same time first class and likely to come to Munich, despite the ample salary and promising conditions there. Wiechert and Sommerfeld did not even hold physics positions: Wiechert's was in geophysics, though he was known to want to move to "pure physics" under the right conditions; Sommerfeld's was in mechanics, though he had recently worked within the "circle of interests of theoretical physics." Cohn was still only an extraordinary professor for theoretical physics after more than twenty years in that position, and he had not published much, though what he had was good. In its wisdom, the ministry offered the job to Wiechert, who had just been promoted to ordinary professor for his science at Göttingen. When Göttingen improved his position even further to keep him, the ministry turned to Sommerfeld. At last Röntgen could report to a colleague that the ministry was making the "necessary efforts (?!)" to "complete" physics at Munich, that after all kinds of struggles, the "ordinary professorship for mathematical physics has been filled, namely, as you know, by Sommerfeld." Sommerfeld was, in the language of the appointment, "ordinary professor for theoretical physics" and "conservator of the mathematical-physical collection of the state."[83]

Röntgen had said at the beginning of the search for a Munich theoretical physicist that "we need no mathematician." That criterion had worked against Sommerfeld's candidature at Leipzig only a few years before, but Sommerfeld had worked on the electron theory in the meantime, and he was respected by the senior theorists Röntgen would have liked to bring to Munich in the first place. Sommerfeld was not the proven theorist of first rank that Röntgen wanted, but Röntgen was encouraged by the response in Munich to Sommerfeld's recent talk on Maxwell's theory and the electron theory. Röntgen was satisfied that he had gained a "good colleague and collaborator," someone he could "talk with in a stimulating way about physical things."[84]

82. Minutes of the meeting of Section II of the Munich U. Philosophical Faculty, 2 June and 20 July 1905, Munich UA, OCI 31. Commission report, 20 July 1905.

83. Commission report, 20 July 1905. Sommerfeld's salary consisted of 5,400 marks from university funds and 2,000 marks from the fund for the general conservatorium of the scientific collection of the state. Information from various documents in 1906 in Sommerfeld's Personalakte, Munich UA, E II-N, Sommerfeld. Sommerfeld's salary was nowhere close to the 12,000 marks that Röntgen thought might be needed to attract a first-rate theorist to Munich, but it was in line with the salaries of other natural scientists at Munich and higher than a good many. The justification for the substantial salary was, in part, the small lecture honorariums the professor of theoretical physics in Munich could expect. Probably Munich U. Curator to Ministry, 25 Sept. 1904; "Übersicht über die Gehaltsverhältnisse von ordentlichen Professoren der K. Universität München," Bay. HSTA, MK 17921. Cohn never received a call, according to Friedrich Schur in a letter to Wien, 15 May 1910, Wien Papers, Ms. Coll., DM. Röntgen to Zehnder, 30 Dec. 1905 and 25 Dec. 1906, in Röntgen, *Briefe an L. Zehnder*, 109–11.

84. Röntgen to Zehnder, 27 Dec. 1906, in Röntgen, *Briefe an L. Zehnder*, 112. Sommerfeld to Lorentz, 12 Dec. 1906.

During the negotiations with Sommerfeld, Röntgen had received another call, this time to Berlin University after Drude's unexpected death there. He declined, giving as his reason the prince regent's personal interest in the matter of the Munich theoretical physicist, which made possible an appointment.[85] The appointment had cost him much time and effort, and he had had to trade on a prestigious offer from Berlin to help secure it; but in the end, he had got what he wanted, a *theoretical* physicist. Röntgen, it would seem, could now put theoretical physics out of his mind and turn his undivided attention to his own institute, which had never had so many independent laboratory workers as right then. But this was not to be.[86]

Röntgen's reorganization of physics instruction at Munich did not take place smoothly, nor could it have. For over twenty-five years Graetz, and for over ten years Korn, had lectured on theoretical physics at Munich by the time of Sommerfeld's appointment, and both were extraordinary professors (Korn by title and rank only and otherwise a Privatdocent). As the principal lecturers on theoretical physics, they felt that they had a claim on the new position and that it had been overlooked.

Korn and Graetz were both Jews, from the large Jewish community in Breslau; to have overcome that handicap in the competition for the Munich position, they would have needed scientific reputations they did not have. They both had decent records of publication, and they were interested in the same subject that Sommerfeld and the other leading candidates were; but their work was not regarded as outstanding, and their approach was seen to belong to a stage of theoretical physics that had all but passed. Graetz, for example, was dissatisfied with Lorentz's theory because it ascribes to electrons the properties they must possess without making them "mechanically understandable," and in 1901 he published a mechanical representation of the world ether to account for electric and magnetic phenomena; this approach he no doubt used in the course of lectures on the electron theory he announced for the winter semester of 1904–5, the time of the Munich search for a theorist. Korn, too, argued for a mechanically explicable world ether, which drew Sommerfeld's scorn in 1901; earlier Sommerfeld himself had worked within the mechanical direction but no longer, and his own work on the electron theory, as we have noted, proceeded from the more recent electromagnetic rather than the older mechanical viewpoint.[87]

Once Sommerfeld was brought in to teach theoretical physics, Korn and Graetz recognized that they had been displaced and that their teaching had been damaged. They tried to retain a teaching function within the university, which brought them into—unequal—conflict with Röntgen, the cause of their displacement.

During Sommerfeld's first semester at Munich, Korn asked to be released from his teaching duties because Sommerfeld had taken away his students. As compensa-

85. Letter from the Bavarian Minister of the Interior, 18 July 1906; Röntgen to the "Generaladjutant" of the Prince Regent, 3 Aug. 1906; Röntgen Personalakte, Bay. HSTA, MK 17921.

86. Röntgen to Zehnder, 30 Dec. 1905.

87. Leo Graetz, "Ueber eine mechanische Darstellung der elektrischen und magnetischen Erscheinungen in ruhenden Körpern," *Ann.* 5 (1901): 375–93. Sommerfeld to Lorentz, 21 Mar. 1901, Lorentz Papers, AR.

tion, he asked for a teaching assignment for analytic and applied mathematics and for a position as state examiner. He was offered instead an assignment for mathematics without salary, which he refused; he required a salaried ordinary professorship to stay in Munich. From the start, Röntgen was unsympathetic to Korn's plight, regarding Korn's talk of resignation as a ruse to get a promotion. He opposed Korn's request for an ordinary professorship of mathematics because, he said, Korn was not good enough. Eventually, in 1908, Korn left Munich, but not without a feud with Röntgen, which spilled over into the newspapers.[88]

With Graetz, Röntgen avoided a feud of this nature, but he did not avoid a protracted disagreement with him over rights within his physics institute. Sommerfeld's appointment only exacerbated the tension between the two that had existed almost from the moment Röntgen arrived in Munich. During the illness of Röntgen's predecessor, Lommel, Graetz had performed the duties of the director of the institute and had in this way acquired an unintended independence there, which brought him into conflict with Röntgen, once he had taken over. Their disagreements occupied the faculty and senate and led, finally, to Graetz's withdrawal as director of the student laboratory in Röntgen's institute, which meant that he forfeited one half of his teaching assignment. The other half of his assignment was to lecture regularly on theoretical physics, which became "superfluous" once Sommerfeld was appointed. Röntgen conceded that it must be an "unpleasant experience" for Graetz, "however one may think about him otherwise," to be shelved like that, especially at his age, which was fifty.[89]

Several years earlier, in 1903, the faculty had anticipated the hopelessness of Graetz's position once the ordinary professorship for theoretical physics was filled again. They had recommended that he be given an honorary professorship when the time came. Now that Sommerfeld had been hired, the faculty wanted to express its appreciation of Graetz's long service at Munich and even a measure of human sympathy by reviving their recommendation. While it was out of the question to allow Graetz the rights of a regular ordinary professor, since that would make him equal to Röntgen or Sommerfeld in their own institutes, it was reasonable to make him an honorary one. So as not to burden Sommerfeld with a second theoretical lecturer, the faculty wanted Graetz's assignment changed from "theoretical physics" to simply "physics." His elevation to ordinary professor implied that he could not direct the student laboratory in Röntgen's institute, since that was the task of a mere assistant, and, besides, Graetz had long since given up the laboratory. The faculty's resolution was acted upon, which would seem to have settled Graetz's position. The difficulty with it, however, was that it left Graetz no place within either of the physics institutes and assigned him nothing specific to teach. This led to conflict with Röntgen once again;[90] for now, as ordinary professor of "physics," Graetz felt more justified than ever to claim a share of the Munich physics facilities, now in Röntgen's insti-

88. Entries from 19 Dec. 1906 and after, in Korn's Personalakte, Munich UA, E II-N, Korn. Röntgen to Zehnder, 27 Dec. 1906. Sommerfeld to Stark, 6 Mar. 1908, Stark Papers, STPK.

89. Section II of Munich U. Philosophical Faculty to Academic Senate, 6 Dec. 1906, Munich UA, OCN 14. Röntgen to Zehnder, 27 Dec. 1906.

90. Röntgen to Minister von Wehner, 15 Aug. 1904. Section II of the Philosophical Faculty to Academic Senate, 6 Dec. 1906. Entry on 16 Apr. 1908, Munich UA, E II-N, Graetz.

tute.[91] Disregarding the large numbers of students attracted to Graetz's lectures, Röntgen rejected—and urged the faculty to reject—all of Graetz's requests.[92]

Although Graetz lacked an institute at Munich, he was a successful teacher within the means he had available to him.[93] He offered a complete program in physics at the university. Parallel to Sommerfeld's courses, he announced sequential lectures on all parts of theoretical physics. From the winter semester of 1908–9 on, he announced a two-semester experimental physics course, which Röntgen, of course, offered, too. Graetz's was in phase with Röntgen's: part I in the winter, part II in the summer. Graetz also regularly announced an "Introduction to Independent Researches," which both Röntgen and Sommerfeld also announced.

Graetz long outlasted Röntgen, who gave up teaching in 1912 and retired in 1920.[94] Graetz taught at Munich until the advent of the Third Reich, which was about the time Sommerfeld began to make plans for his own retirement. Graetz's researches of twenty years had not always belonged "to the best," as Planck put it gently, and after 1905 he published no more researches at all. His most valued writings were articles in Adolph Winkelmann's handbook of physics and his textbooks, which he continued to write and to revise after 1905; generations of students learned from his textbook *Electricity and Its Applications*, which they referred to as *The Great Graetz*, and which was in its tenth edition in 1903, its twenty-first in 1922.

During the unsettled affairs of Korn and Graetz, the new professor Sommerfeld lectured systematically on the branches of theoretical physics, as he was hired to do. From the start, the faculty anticipated that the new professor would want an institute; when they questioned that four rooms in the projected new main building of the university would satisfy his needs, Röntgen explained that these rooms were thought of only as workrooms, not as a complete institute. They would need to set aside more space for the new theoretical physicist. For his first several years at Munich, however, Sommerfeld had to make do with a few rooms in the old building where the Bavarian Academy of Sciences met. One of his first students, P. P. Ewald, described his introductory call on Sommerfeld there; he rang the bell and after much rattling of keys he was admitted by the mechanic and led down a long, dim hallway flanked by cabinets. At length, he came to a space illuminated by daylight, with four rooms off it. One of the rooms was a small lecture hall with benches, lectern, and a great blackboard. Sommerfeld sat in another room, the assistant in another, and in the last a student was doing an experiment on the dynamics of turbulent currents, a favorite theme of Sommerfeld's.[95]

91. Graetz to Röntgen, 6 and 24 May 1908; Röntgen to Graetz, 20 May and 5 June 1908, Munich UA, OCN 14.

92. Röntgen to Section II of the Munich Philosophical Faculty, 20 Jan. 1909, Munich UA, OCN 14.

93. K. Kuhn, "Erinnerungen an die Vorlesungen von W. C. Röntgen und L. Grätz," *Phys. Bl.* 18 (1962): 314–16. Section II of the Munich U. Philosophical Faculty to Academic Senate, 6 Dec. 1906.

94. Bavarian Ministry of the Interior to the Prussian Ministry of Culture, 14 Jan. 1912; Bavarian Ministry to Röntgen, 27 Jan. 1912, and related documents; Röntgen Personalakte, Bay. HSTA, MK 17921.

95. Dean Richard Hertwig of Section II of the Munich U. Philosophical Faculty to the members of Section II, 25 Nov. 1904, Munich UA, OCI 31. P. P. Ewald, "Erinnerungen an die Anfänge des Münchener Physikalischen Kolloquiums," *Phys. Bl.* 24 (1968): 538–42, on 539.

The assistant in the institute was Peter Debye, who had moved from Aachen with Sommerfeld to take up the position that Boltzmann had established. Sommerfeld's request for retaining the position evidently caused surprise among the faculty and ministry, who believed that a theoretical physicist needed only a desk and his paper, pencils, and books. It helped him in his request that he was "conservator of the mathematical-physical collection of the state" as well as ordinary professor for theoretical physics; in his capacity as conservator, he had research instruments at his command and funds to add to them. In 1910 the collection was transferred from the academy's building to the university's; the institute for theoretical physics, which was ready that year, was better equipped and larger, with a lecture room on the ground floor for about sixty students and four nearby rooms for Sommerfeld and his co-workers and another place for physical models. The basement contained an additional four rooms for experiments and also a workshop and darkroom. There was a small library for journals and texts, which was open to participants in the mathematical-physical seminar. In 1911 Sommerfeld acquired a second assistant. For a theoretical physics institute, Sommerfeld's was impressive, much like an ordinary physics institute, if on a somewhat smaller scale than most.[96]

To build up theoretical physics in Munich, Sommerfeld used the familiar combination of lecture, seminar, and colloquium. In his comprehensive lectures, his ability to evoke the harmony between mathematical thought and physical phenomena was remarkable, persuading at least one abstract mathematician to become a

96. Sommerfeld, "Das Institut für Theoretische Physik," 291. Benz, *Sommerfeld*, 50. Paul Forman and Armin Hermann, "Sommerfeld, Arnold (Johannes Wilhelm)," *DSB* 12 (1975): 525–32, on 527. From a letter by Sommerfeld two years later, we can get an idea of the aids, short of experimental apparatus, he believed useful in teaching theoretical physics. The letter was in response to a request for advice from Heinrich Konen, the extraordinary professor teaching theoretical physics at Münster University. In 1912 the Prussian ministry included several hundred marks for theoretical physics instruction in the Münster physics budget together with a one-time grant for initial equipment. The exact amount of the latter was to depend on Konen's needs, and it was to determine these that he appealed to Sommerfeld. Konen considered asking for six kinds of teaching aids: (1) models, insofar as they belonged to physics and not to mathematics; (2) plates, some 120 of them, to be prepared by draughtsmen from the local mechanics school, which would include drawings of important mathematical functions, graphs, and diagrams of lines of force, level surfaces, lines of curvature, diffraction, and other calculable examples; (3) transparencies, the basis of a large collection, to begin with some 200 pictures from the literature, for example, of tables, of theoretically important experimental arrangements for comparison, of diffraction figures, the Zeeman effect, hydrodynamic experiments, and other important researches, and, in general, of physics materials that were unsuited for the blackboard, plates, or demonstration; (4) aids for calculation, such as a calculating machine, planimeter, tables of important functions, perhaps a harmonic analyzer, and the slide-rule, which the students must also own themselves; (5) assistance; and (6) a circulating library, which the minister had often refused before. Konen intended Sommerfeld to be specific as to which makes of models and calculating aids he should have, but not with regard to plates and transparencies for which Konen had already drawn up a long list. In the margin of Konen's letter, Sommerfeld made notes to himself, presumably the basis of his reply. His notes are (with some abbreviations spelled out): alongside items (1) and (2), "Bessel's functions. Wave surfaces. Radiation surfaces. Gibbs's surfaces. Thermodynamic surfaces for water (Ritter). Entropy surfaces. For seminar purposes." On the top of the next page of Konen's letter, Sommerfeld wrote: "Auerbach. Maxwell," which would presumably refer to Auerbach's book on graphical presentations of physics, published that year, and some parallel writings by Maxwell. Alongside (3), Sommerfeld wrote: "Hydrodynamics. Stark, Kaufmann. Wilson photographic apparatus," and alongside (4)–(6): "Jahnke Emde," presumably referring to Jahnke and Emde's mathematical manual, Sommerfeld's recommendation of Gruner's second assistant at Zurich, and his note "cheap at . . . 120 marks," followed by the manufacturer's address.

theoretical physicist. His other, advanced lectures presented the latest developments in research. As one of the first converts to relativity theory, he gave special lectures on it in his institute in the winter of 1908–9; he later recalled these lectures with satisfaction, believing them to be the first given on that subject. In the same year, he prepared material on the quantum theory to include in his lectures. He was one of the first to develop Niels Bohr's quantum theory of the atom, the subject in which he was to make his major contributions, and by the winter of 1914–15 he was lecturing on it, too. As a director of the physical section of the mathematical-physical seminar, he introduced relativity theory in the seminar in 1911. The seminar allowed for more give-and-take than lectures; it was there, for example, that Gustav Hertz "acquired the feeling for what physics really is." Sommerfeld's colloquium, which was organized in 1908 on a student's suggestion, was originally intended as a forum in which advanced students would present their work to less advanced ones. Sommerfeld gave it his blessing but stayed away at first, leaving Debye to preside. Later he joined in, and physicists from Röntgen's institute and colleagues from other sciences also made a practice of coming to the colloquium, which was held in Sommerfeld's institute from the time it moved to the new university building; as in Sommerfeld's lectures and seminar teaching, in his colloquium the relationship of statistical, relativity, and quantum theories to the structure of classical theoretical physics was repeatedly discussed. From the start of his teaching at Munich, Sommerfeld encouraged the discussion of problems that were to make the greatest difference to the foundations of twentieth-century physics. Around him a "school" of theoretical physics came into being, a recognizable "style of research and of presentation of results."[97]

The working relationships within the institute for theoretical physics were unusual for the time. Sommerfeld did not insist on the social distance between the professor and his subordinates that some of his colleagues did; as one of his students recalled, he encouraged a "free exchange of ideas without any formality or restriction." He invited the collaboration of his students and assistants, who often influenced his own views on physics. He had them to his house, and he met with them in cafes before or after colloquiums. He took them along to the alpine ski hut he owned jointly with the institute mechanic, where physics discussions were as demanding as the sport. He had the "rare ability to have time to spare for his pupils, in spite of all his duties and his scientific work," Born pointed out. Near the end of his life, Sommerfeld summed up his approach as a teacher: "Personal instruction in the highest sense of the word is best based on intimate personal acquaintanceship."[98]

Essential to Sommerfeld's effectiveness as a teacher and as a researcher was his openness to new trains of thought. He broke with the classical physics in which he

97. Sommerfeld, "Autobiographische Skizze." Ewald, "Erinnerungen," 541–42. Gustav Hertz to Sommerfeld, 16 Jan. 1927, Sommerfeld Papers, Ms. Coll., DM. Benz, Sommerfeld, 71–72, 76–77.
98. Max Born, "Arnold Johannes Wilhelm Sommerfeld, 1868–1951," Obituary Notices of Fellows of the Royal Society 8 (1952): 275–96, on 286, reprinted in Ausgewählte Abhandlungen 2: 647–59; "Sommerfeld als Begründer einer Schule," Naturwiss. 16 (1928): 1035–36. Benz, Sommerfeld, 51–52, 66–68. Forman and Hermann, "Sommerfeld," 530. Arnold Sommerfeld, "Some Reminiscences of My Teaching Career," Am. J. Phys. 17 (1949): 315–16, on 315.

was trained to become a leader of modern theoretical physics, and he taught powerful mathematical methods to those who would break with it more sharply than he
did. He was, in Born's words, "one of the most distinguished representatives of the
transition period between classical and modern theoretical physics." His gift did not
lie so much in the "divination of new fundamental principles . . . or the daring
combination of two different fields of phenomena into a higher unit" but rather in
conceptual clarity and in the "logical and mathematical penetration of established
or problematic theories and the derivation of consequences which might lead to
their confirmation or rejection."[99]

When Wien complained to Sommerfeld in 1898 about German disinterest in
theoretical physics, he complained in particular about the failure of Munich to represent the subject by an ordinary professor. The grounds for his special complaint
were removed when Sommerfeld himself became that professor. In remarkably short
order, he created in Munich a "center of theoretical physics research," and he went
on to produce more doctorates than any other theorist over the next quarter century.
In outstanding measure, Born thought, Sommerfeld's abilities included the "discovery and development of talents." "What I especially admire about you," Einstein
told Sommerfeld, "is that you have, as it were, pounded out of the soil such a large
number of young talents."[100]

In 1912 the diffraction of Röntgen rays by crystals was demonstrated for the first
time; it was one of the great discoveries of the twentieth century, Sommerfeld
thought the greatest, and he was proud that it had been made in his institute. The
elements that entered into the discovery illustrate the relationship between theory
and experiment that was possible within the physics resources of Munich University,
which Röntgen had done much to establish. The original discoverer of Röntgen rays
was now rather inaccessible to younger physicists; he had grown overly careful in his
work, never finishing it, and he stayed away from the colloquium and was generally
skeptical of the new theories discussed in it. But in his institute, a good deal of
experimental research was being done by the younger men. Walter Friedrich, for
one, had just finished a dissertation on Röntgen rays under Röntgen, and Sommerfeld had hired him in his own institute as second assistant, the new position he had
just acquired. Friedrich was an *experimental* assistant, in keeping with Sommerfeld's
concern to have rapid experimental confirmation of theoretical conjectures. In
bringing Friedrich into his institute, Sommerfeld wanted experimental work done on
Röntgen rays in connection with his theory of Bremsstrahlung, which incorporated
his recent recognition of the significance of the quantum. In addition to Friedrich,
Sommerfeld's institute contained several young theorists, among them Laue, who
had come there from Planck's institute in 1909. In Munich, Laue found that there
was much interest in Röntgen rays. He also found that there was interest in crystals,

99. Born, "Sommerfeld," 647, 654.
100. Max Planck, "Arnold Sommerfeld zum siebzigsten Geburtstag," *Naturwiss.* 26 (1938): 777–79,
reprinted in *Phys. Abh.* 3: 368–71, on 370. Forman and Hermann, "Sommerfeld," 529–30. Einstein to
Sommerfeld, [14?] Jan. 1922, in *Albert Einstein/Arnold Sommerfeld Briefwechsel*, ed. Armin Hermann
(Basel and Stuttgart: Schwabe, 1968), 97–98, on 98. Born, "Sommerfeld als Begründer," 1035.

and when Sommerfeld's student Ewald asked Laue a question about the space lattices formed by the atoms of crystals, Laue thought of an experiment to try involving Röntgen rays and crystals. It was to look for interference phenomena with Röntgen rays by passing them through crystals, the regularly spaced atoms of which would act as a kind of diffraction grating for rays for which no ordinary grating had worked. His suggestion was discussed in the institute and in nearby cafes, and it met with a good deal of theoretical skepticism. Sommerfeld himself was skeptical at first and did not want his assistant Friedrich to waste precious time on it. Laue persisted; he got Friedrich to offer to try the experiment in his spare time, and to help him, Laue enlisted the doctoral student Paul Knipping, who was preparing Röntgen-ray experiments in Röntgen's institute. When Sommerfeld heard of the successful outcome of the experiment, he became enthusiastic and placed the resources of the institute at Friedrich and Knipping's disposal. The resulting first paper had a collaborative form: a "Theoretical Part" by Laue followed by an "Experimental Part" by Friedrich and Knipping, and it was presented by the director of the institute in which the work was done, Sommerfeld, to the Bavarian Academy of Sciences.[101]

The work on Röntgen-ray diffraction involved the collaboration of physicists from both of Munich University's physics institutes. The theoretical physics institute was itself a microcosm of physics with its experimental facilities and personnel, which expedited the testing of Laue's theoretical suggestion. The preparation of the experiment required the lively and—almost too—critical discussion by physicists working on problems related in some way to Sommerfeld's interests. For Laue, the appearance of a predicted orderly set of dark spots on a photographic plate behind an irradiated crystal demonstrated the value of theoretical anticipations; after all, Röntgen rays had been sent through crystals before, but without the anticipation of interference effects. He saw the discovery as a confirmation of his theoretical approach through the "great general principles" of physics, here those of the wave nature of Röntgen rays and the general lattice structure of crystals. The discovery confirmed, as well, he believed, the scientific stimulation of Sommerfeld's institute, which he had hardly found anywhere else; that was so, even though he did not fit in well personally with Sommerfeld's circle. To Debye, the discovery confirmed Sommerfeld's good judgment of Laue's ability and showed the effect of Sommerfeld's liberal direction of the institute. Röntgen was at first dubious about the interpretation of the experiment as a demonstration of Röntgen-ray interference, but he soon convinced himself of its correctness. In general, Röntgen had reason to feel pleased with the results of his efforts at reestablishing theoretical physics in Munich. Ten years after these efforts, he told Sommerfeld that the "Munich mathematical-physical school has indeed become one of the first and best in the world."[102]

101. Benz, *Sommerfeld*, 58–62. Laue, "Mein physikalischer Werdegang," 193–96. Paul Forman, "The Discovery of the Diffraction of X-Rays by Crystals: A Critique of the Myths," *Arch. Hist. Ex. Sci.* 6 (1969): 38–71. Walter Friedrich, Paul Knipping, and Max Laue, "Interferenz-Erscheinungen bei Röntgenstrahlen," *Sitzungsber. bay. Akad.*, 1912, 303–22.

102. Laue, "Mein physikalischer Werdegang," 197. Laue to Sommerfeld, 3 Aug. 1920; Peter Debye to Sommerfeld, 13 May 1912; Röntgen to Sommerfeld, 6 Jan. 1915; Sommerfeld Correspondence, Ms. Coll., DM. Otto Glaser, *Dr. W. C. Röntgen* (Springfield, Ill.: Charles C. Thomas, 1945), 126.

In this section, as elsewhere, we have encountered anti-Semitism, and because it affected the careers of many theoretical physicists, we should say something more about it. If a candidate for a position or a promotion in a German university were Jewish, even if he had converted to Christianity, the faculty and the ministry would likely regard it as a weighty consideration. Reports of faculty meetings, letters by faculty members, and jottings by ministry officials spoke clearly, if at times guardedly, on this point. For example, the report comparing the three candidates for the Leipzig chair of physics in the late 1890s did not mention the religion of Röntgen and Braun but did that of Wiener, who was said to stem from an "old Giessen evangelical family";[103] Wiener was not Jewish, but with his name he might have been. Only a few years before, in connection with Kundt's replacement at Berlin, the Prussian ministry had gone to lengths to determine if Wiener's ancestors were Jewish. Asked by the ministry about Wiener's origins, his colleagues accepted the request as proper and provided information favoring his candidacy, though it did not get Wiener the Berlin job. That job went to the faculty's first choice, Warburg, who was Jewish, a choice the Prussian ministry official in charge tried to avoid.

If a physicist were outstanding, the question of his Jewish origins might be played down. Riecke said of Theodor von Kármán, who was going to Aachen to take up a position there in 1913: "He is a Jew and possesses the properties of his race, but he is without any doubt a great talent."[104] Röntgen hoped that Warburg would overcome the anti-Semitism at Berlin, and he was pleased when Warburg was hired there.[105] But when Korn, whom Röntgen did not think much of as a physicist, threatened to resign his post in Munich, Röntgen resorted to stereotypes: "after all, only very rich people can get away with such means, and they use it, too, if they have the necessary Semitic impudence."[106] When Ehrenfest told Sommerfeld that he wanted to become a Privatdocent in his theoretical physics institute at Munich, Sommerfeld formally asked Röntgen's opinion. To the "confessional question" that Sommerfeld had raised, Röntgen replied that from what he had heard, Ehrenfest was "fiery, critical, dialectical," in short, a "Jewish type." Röntgen advised "caution," doubting that Sommerfeld could train Ehrenfest to become a *"physicist."*[107] Sommerfeld got more determined advice from his former assistant Debye, who described Ehrenfest as a Jew of the " 'high priest' type," who would stifle any fresh ideas and exert an "extremely noxious influence" in Sommerfeld's institute.[108] When at this time Lorentz asked Ehrenfest to become his successor at Leiden, Debye explained

---

103. Draft letter from the Leipzig Philosophical Faculty to the Saxon Ministry of Culture and Public Education, 17 Dec. 1898, Wiener Personalakte, Leipzig UA, PA 1064.
104. Riecke to Johannes Stark, 1 Feb. 1913, Stark Papers, STPK.
105. Zehnder's report to Röntgen on the opposition in Berlin to Warburg because he was a Jew, and Röntgen's reply to Zehnder, 19 Dec. 1894, in Röntgen, *Briefe an L. Zehnder*, 29–30. Warburg was only one of several nineteenth- and early-twentieth-century German physicists of Jewish background who were directors of institutes; they included Magnus and Rubens in Berlin and Hertz in Bonn.
106. Röntgen to Zehnder, 27 Dec. 1906.
107. Röntgen to Sommerfeld, 12 Apr. 1912, Sommerfeld Correspondence, Ms. Coll., DM.
108. Debye to Sommerfeld, 29 Mar. 1912, Sommerfeld Correspondence, Ms. Coll., DM.

that Ehrenfest owed his extraordinary success to Einstein, who did not regard the "race question" as lightly as Sommerfeld might have thought.[109]

Faculties looked down on physicists who showed "pushiness, cheek, pettiness [*Krämerhaftigkeit*]," unpleasant peculiarities they associated with Jews. If the Jew were Einstein, the faculty might be reassured about his "personal character,"[110] but if he were Cohn, Auerbach, or Abraham, for instance, he would not advance as rapidly or as far as his talents would warrant. Among academic fields in Germany, theoretical physics had an unusually high proportion of Jews, who found more opportunity there than in the more desirable field of experimental physics.

Extraordinary Professorships for Theoretical Physics

At the beginning of the twentieth century, as earlier, most extraordinary professorships and the occasional personal ordinary professorships were for theoretical physics. And as earlier, most men holding these positions were primarily experimentalists, who, if all went well, usually went on to become directors of experimental physics institutes.[111] Only a few of them were interested in theory as much as in experiment, for example, Mie at Greifswald and Rudolf Weber at Rostock, and only a few, such as Pockels at Heidelberg and Born at Berlin, were dedicated primarily to theory.

Since most universities still did not have separate institutes for theoretical physics, as before, the extraordinary professor for the subject usually had to work within the institute directed by the experimental professor. But now, when this arrangement caused problems between theorist and experimentalist, the theorist—even below the rank of ordinary professor—wanted his field to be treated as a proper academic discipline. Freiburg, Heidelberg, and Halle provide examples.

At Freiburg in 1903 a chair for an extraordinary professor for mathematical physics was established together with an institute of sorts, one which enjoyed a limited measure of independence. The man appointed to it was Johann Koenigsberger, a physicist who was not known to be a theorist. When asked about him, Voigt said that he did not know how to judge Koenigsberger's "theoretical power," but he regarded him as a "good observer."[112] Son of the mathematician Leo Koenigsberger, Johann studied mathematics and natural science first at his father's university, Heidelberg, then at Freiburg, and finally at Berlin, where he graduated in 1897 with a

109. Debye to Sommerfeld, 3 Nov. 1912, Sommerfeld Correspondence, Ms. Coll., DM.

110. Because the Zurich University physics professor Alfred Kleiner knew Einstein personally, he was able to quiet his faculty's doubts about admitting a Jew. Not wishing to inscribe on its banner anti-Semitism as a principle, the Zurich faculty recommended Einstein's appointment. Quotations from the letter by Dean C. Stoll of Section II of the Zurich U. Philosophical Faculty to the Director of Education for the Canton of Zurich, 4 Mar. 1909, STA K Zurich, U. 110b. 2, Einstein.

111. They were, among others, Theodor Des Coudres and Mathias Cantor at Würzburg, Walter Kaufmann at Bonn, Johann Koenigsberger at Freiburg, Heinrich Konen at Münster, Gerhardt Schmidt, Arthur Wehnelt, and Rudolf Reiger at Erlangen, August Becker at Heidelberg, Edgar Meyer at Tübingen, and Ernst Pringsheim at Breslau.

112. Voigt's response to an inquiry about candidates for a theoretical physics position in Bonn. Letter to Kayser, 24 Oct. 1902, STPK, Darmst. Coll. 1924–22.

thesis on crystal physics under Warburg's supervision. He then returned to Freiburg, where he became Franz Himstedt's assistant and a Privatdocent, and where he began to establish a reputation in experimental physics and, especially, in mineralogy.[113]

To provide space for instruction in Koenigsberger's "mathematical physics institute," the Baden ministry put up a small sum of money. The lecture hall was carved out of the main physics institute by removing a wall between two small rooms. A salary was provided for a servant, who was also on call in the main physics institute.[114] Lacking rudimentary facilities such as enough tables and chairs, a cabinet, and the like, he had to ask the ministry for extra allowances beyond the small budget for maintaining the institute. By 1907 he had three doctoral candidates, who could not be expected to buy their own apparatus. He offered to buy a spectrograph for the institute and to be gradually reimbursed from the budget; the ministry had no objection to this arrangement, which cost the government nothing.[115] For an assistant, Koenigsberger got by with a volunteer, since he lacked any prospect of a salary for a regular one. As the teaching picked up in his institute, he regularly spent money out of his own pocket to cover the costs.[116]

The impoverished condition of Koenigsberger's institute brought him into conflict with Himstedt, the director of the main physics institute. Koenigsberger believed that Himstedt, on whose favor he depended, treated him unfairly, and to document his grievances, he mailed out a questionnaire in 1913 asking theoretical physicists throughout Germany about their arrangements. He explained to the recipients of the questionnaire that Himstedt had taken back two rooms he had let Koenigsberger's doctoral students use for several years, which left Koenigsberger's institute with one room that had to serve at the same time as workroom, director's room, and space for research and teaching apparatus. That made it all but impossible for Koenigsberger and his students to test "theoretical considerations" experimentally. He planned to petition the Freiburg natural science and mathematics faculty.[117]

The conflict had to do mainly with the control of the work of doctoral students in Koenigsberger's institute. Koenigsberger claimed that as director of an institute he should have complete control, while Himstedt regarded Koenigsberger as only a guest, and he required Koenigsberger's students to come to him before they began working in the space on loan to Koenigsberger. Himstedt claimed that he had generously offered Koenigsberger the use of two rooms, which he badly needed himself; for the crowded conditions of the physics institute required Himstedt to limit the number of his own doctoral students, and it required the laboratory for teaching candidates to be set up in the windowless space under the podium of the lecture

113. Schroeter, "Koenigsberger," 236–37. *Aus der Geschichte der Naturwissenschaften an der Universität Freiburg i. Br.*, ed. E. Zentgraf (Freiburg i. Br.: Albert, 1957), 22.

114. Alexander von Dusch and Franz Böhm, officials in the Baden Ministry of Justice, Culture, and Education, on the "extraordinary professorship for theoretical physics," 25 Nov. 1904, Bad. GLA, 235/7769.

115. Koenigsberger to Freiburg U. Philosophical Faculty, 16 Dec. 1907, and to Baden Ministry of Justice, Culture, and Education, 22 June 1908, Bad. GLA, 235/7769.

116. Koenigsberger to Baden Ministry of Justice, Culture, and Education, 7 June 1912 and 18 Feb. 1914; Ministry to Koenigsberger, 24 Feb. 1914; Bad. GLA, 235/7769.

117. Koenigsberger to Auerbach, 8 Jan. 1913, Auerbach Correspondence, STPK.

hall. For years the conflict went unresolved until, in 1919, Koenigsberger appealed to the ministry. The ministry in turn asked two members of the Freiburg faculty to look into the matter and settle it. Their conclusion was that Koenigsberger's list of complaints was unpersuasive, but they noted correctly that the quarrel was "renewed proof" that the space of both the physics and the mathematical physics institutes was inadequate.[118]

Koenigsberger's counterpart in Baden's other university, Heidelberg, was Friedrich Pockels. Like his teacher Voigt, Pockels was a skillful observer who was best known for his "comprehensive theoretical developments," and Koenigsberger described him as "essentially a theorist."[119] Pockels published a text on the mathematical methods of physics, a work of over three hundred pages on a single partial differential equation; he published another on crystal optics, his specialty, which was described as the most rigorous mathematical presentation of the subject.[120] By 1900 he was regarded as one of the more promising young theoretical physicists in Germany.[121]

That year Pockels was invited to replace Lenard as the extraordinary professor for theoretical physics at Heidelberg. By accepting the position, Pockels unwittingly joined the struggle for independence for the subject of theoretical physics that Koenigsberger was engaged in at Freiburg. The new Heidelberg position had been created in 1896 following a request by the science faculty for a salaried extraordinary professor of physics who was to take on a "part" of the theoretical and practical instruction "under the direction" of the head of the physics institute, Quincke. When the position was offered to Lenard, he insisted that his appointment be for the "(independent) representation of theoretical (or mathematical) physics" together with the support of the experimental professor in the laboratory course, provided that that "support" be limited to teaching; in the choice of his theoretical lectures he insisted on "complete independence" and in the laboratory course on "no kind of assistant's duties." The Baden government appointed him to the extraordinary professorship only, omitting any reference to laboratory instruction, but the faculty and Lenard himself understood his assignment to be as he had defined it.[122] As

118. Koenigsberger to Baden Ministry of Culture and Education, 15 and 28 Feb. and 18 Oct. 1919; Himstedt to Freiburg U. Science and Mathematics Faculty, 21 July 1919; letter by the Freiburg physiologist Deecke, 29 July 1919; Bad. GLA, 235/7769.

119. Ernst Dorn to [Wilhelm Wien?], 25 Jan. 1903, Ms. Coll., DM. Koenigsberger, "Pockels," 19.

120. Friedrich Pockels, Über die partielle Differentialgleichung $\Delta u + k^2 u = 0$ und deren Auftreten in der mathematischen Physik (Leipzig, 1891); Lehrbuch der Kristalloptik (Leipzig and Berlin: B. G. Teubner, 1906). Koenigsberger, "Pockels," 19.

121. Voigt and Boltzmann, among others, thought so, though Boltzmann thought him "more of a pure mathematician." Voigt to Leo Koenigsberger, 6 June 1899, STPK, Darmst. Coll. 1923.54. Boltzmann to Leo Koenigsberger, 3 June 1899, STPK, Darmst. Coll. 1922.93.

122. Dean Adolph Stengel of the Heidelberg U. Science and Mathematics Faculty to Baden Ministry of Justice, Culture, and Education, 30 Apr. 1894; Science and Mathematics Faculty to Ministry, 15 June 1896; Ministry to Science and Mathematics Faculty, 3 June 1896; Minutes of the Science and Mathematics Faculty meeting on 10 June 1896; Lenard's acceptance of Heidelberg position, 4 Oct. 1896; and Heidelberg U. Senate to Heidelberg U. Faculty, 2 Nov. 1896; Bad. GLA, 235/3135. Riese, Hochschule, 146–47.

Walter König pointed out later, there was no separating the extraordinary professorship from assistance in laboratory instruction, since a large part of the extraordinary professor's income, beyond his modest salary, came from the fees of the laboratory students. When Lenard received an offer to direct an experimental institute elsewhere, the Heidelberg senate urged the ministry to keep Lenard and suggested that his extraordinary professorship for theoretical physics be changed into an ordinary one for the duration of his tenure there. The ministry on the contrary considered it ill advised to enhance the second physics position in any way.[123]

With Lenard gone, Heidelberg tried to attract first Drude and then Walter König. Both had an understanding of the job similar to Lenard's: they envisioned an independent position with means that amounted to those of a modest institute. From the start Lenard had requested money for instruments for his own research, which the budget of the physics institute could not accommodate, and he had received four thousand marks "for the acquisition of scientific aids (apparatus, instruments, and such objects)." This instrument collection would now go to his successor. Beyond that, Drude could expect from Heidelberg what his own university, Leipzig, had already assured him: an ordinary professorship for theoretical physics, freedom from any obligation to assist the professor of physics in the laboratory, a salary considerably larger than that which Lenard had received, and a decent number of students. The Heidelberg institute facilities and arrangements, which he went to see first-hand, led him to decline the Heidelberg position without further negotiations for the "single" reason that he enjoyed "greater independence" in his present position, including the opportunity to have research students of his own.[124] König next received the offer of the Heidelberg extraordinary professorship of theoretical physics, and he too immediately expressed his concern about its "hardly independent character." But since he did not have a university position then, he tried, in vain, as it turned out, to have the position made over into one he could accept. As extraordinary professor of theoretical physics he expected not only to do research of his own, for which the one room that Quincke could offer him might have sufficed, but also to have research and doctoral students; for their work and for his own, he needed three or four rooms entirely at his disposal, the means to equip and furnish them, a budget of two or three thousand marks, a servant or mechanic, and an assistant later. With regard to his theoretical physics lectures, he said that it would be out of keeping with the "character" of the field if they were given in "completely abstract mathematical form"; sometimes König would have to show the experimental bases of the theories, and so he required a small, appropriately equipped lecture hall as well. In addition to his lectures on theoretical physics and on any special subject of physics for which the main physicist at Heidelberg might have no time, König offered to take on courses in applied physics, especially electrotechnology, for teach-

123. Heidelberg U. Senate to Baden Ministry of Justice, Culture, and Education, 2 Mar. 1898; Walter König to "Geheimrat," 12 Feb. 1899; and Baden Ministry of Culture and Education to Heidelberg U. Science and Mathematics Faculty, 11 Jan. 1913; Bad. GLA, 235/3135.

124. Baden Ministry of Justice, Culture, and Education to Heidelberg U. Senate, 3 May 1898; Lenard's request for money for instruments, 6 Sept. 1896; and Drude to "Geheimrath," 16 and 29 July 1898; Bad. GLA, 235/3135. Grants to Lenard in 1897, Heidelberg UA, IV 3e Nr. 53, 1875–1929.

ers; this was to enhance his income, one may assume, and also, as he explained, to increase the usefulness of a "second physical institute at Heidelberg."[125]

Anxious that after nearly two years of negotiations, their theoretical physics position might be turned down by a third candidate after Drude and König, the Heidelberg science faculty tried to make it more appealing by freeing it of the obligation of teaching the laboratory course with Quincke and by securing a lecture hall for the new extraordinary professor. The ministry answered their proposal with "the opinion that a special institute, separate from the physical institute, . . . cannot be created" and that the new professor must in every respect work within the existing physics institute. Pockels, who was then teaching mathematical physics at the technical institute in Dresden, was given the assignment for theoretical physics and made "director of the theoretical-physical apparatus," Lenard's collection of instruments.[126]

Although his means fell far short of what physicists were beginning to take for granted for an extraordinary professorship for theoretical physics, Pockels had advantages at Heidelberg that physicists in theoretical positions did not always have. He had a small annual budget, which he could administer independently of the physics institute, an instrument collection of his own, and enough students to offer a lecture cycle and sometimes seminar exercises. He felt himself an independent specialist in his teaching, even though he was still working within the physics institute and, at least in his eyes, his position vis-à-vis the institute director had never been officially clarified.[127] With Quincke, his relations were friendly. But when Lenard returned to Heidelberg to replace Quincke, difficulties arose between him and Pockels over the autonomy of theoretical physics. Pockels found Lenard an unfriendly colleague, so much so that he made it known he wanted to leave Heidelberg.[128] In 1911, the Baden ministry issued an order that extraordinary professors representing a "special subject" should be voting members of the university faculty—ordinarily composed only of ordinary professors—in matters concerning their subject. Lenard opposed this order vigorously. He believed that he was a good enough theoretical physicist to direct and examine doctoral candidates in physical theory, and in principle he wanted to exclude purely theoretical doctoral dissertations. After a year of conflict over this issue, Pockels asked the faculty that theoretical physics be recognized as a "special subject" in the sense of the ministerial order. At a full faculty meeting,

125. Heidelberg U. Science and Mathematics Faculty to Baden Ministry of Justice, Culture, and Education, 28 Dec. 1898, and König to "Geheimrat," 12 Feb. 1899, Bad. GLA, 235/3135.

126. Dean Stengel of the Heidelberg U. Science and Mathematics Faculty to Baden Ministry of Justice, Culture, and Education, 16 Nov. 1899 and 23 Jan. 1900; and Ministry to Heidelberg U. Senate, 8 Dec. 1899; Bad. GLA, 235/3135. *Ruperto-Carola. Sonderband. Aus der Geschichte der Universität Heidelberg und ihrer Fakultäten*, ed. G. Hinz (Heidelberg: Brausdruck, 1961), 440. In the faculty debate about freeing Pockels of the obligation to assist Quincke in the laboratory, a move toward greater independence for the theoretical physicist that was suggested by Leo Koenigsberger, Quincke insisted on the institute director's unlimited rights in the institute; but in this instance his point of view did not prevail. Riese, *Hochschule*, 147.

127. Dean Stengel to Baden Ministry of Justice, Culture, and Education, 23 Jan. 1900. Pockels to Wilhelm Wien, 14 Mar. 1903, Wien Papers, Ms. Coll., DM. Pockels to Jakob Laub, 12 Mar. 1912, Ms. Coll., DM, 1961/22.

128. Voigt to Kayser, 8 Mar. 1908, STPK, Darmst. Coll. 1924.22. Riese, *Hochschule*, 147.

Lenard spoke at length against Pockels's request, but all the other members of the faculty voted for it, and so did Lenard in the end. Practically, Pockels's "victory" did not improve matters much, since he was still dependent on Lenard's good will for room in the physics institute. Lenard reluctantly ceded him three rooms in the new physics institute built in 1912, but he had Pockels officially prohibited from designating these rooms as an "institute" or "department for theoretical physics." With respect to doctorates in theoretical physics, Lenard could still find extraneous reasons to reject them. For its part, the Baden ministry of culture took the view that the position of the extraordinary professor should not be strengthened, this being the way to avoid conflict in the physics institute; it held that the position was not meant for life but as a transitional position for a younger man.[129]

Nevertheless, in a limited sense, Pockels had made the case for regarding theoretical physics as an independent subject. Even Lenard had been won over to the extent of considering the establishment of an ordinary professorship for theoretical physics at Heidelberg after Pockels's early death in 1913, but only if he could get a really good man such as Einstein. The description he sent to Sommerfeld, Planck, and Wien of the kind of theoretical physicist he wanted to replace Pockels sounded much like a description of Pockels: a physicist who possessed a good, thorough education in all of physics and a good mathematical talent.[130]

Apart from the personalities involved, Pockels's unhappiness at Heidelberg—if either Drude or König had taken the job, he would have felt the same unhappiness—originated in a general problem that theoretical physics professors encountered in their desire to have an independent teaching field. The Heidelberg faculty understood that the goal of the theorist had to be the training of doctoral candidates. For the extraordinary professor for theoretical physics, the opportunity to direct doctoral candidates in theory and in experiment related to theory was important for at least two reasons: he got more satisfaction from his teaching, and he was in a better position to advance to an ordinary professorship, which was most likely to be in experimental physics. For the latter reason theorists also still generally expected to do experimental work in connection with their theoretical work, and for that they required laboratories. The faculty regretted the limitations of space in their physics institute, which made it impossible for the extraordinary professor for theoretical physics to have a number of laboratory rooms at his disposal.[131]

As long as the theoretical physicist's activities were limited to the space in the physics institute controlled by the experimental professor, there could be no question of his complete independence. The degree of independence turned on his personal relations with the experimentalist, which were often satisfactory; but the nature of the arrangements in physics institutes worked against them, as our example of Halle will show.

129. Pockels to Laub, 12 Mar. 1912. Baden Ministry of Culture and Education to Heidelberg U. Science and Mathematics Faculty, 11 Jan. 1913. Riese, Hochschule, 147.
130. Lenard to Sommerfeld, 4 Sept. 1913, Sommerfeld Correspondence, Ms. Coll., DM.
131. Dean Stengel to Baden Ministry of Justice, Culture, and Education, 23 Jan. 1900.

In 1895 the Halle personal ordinary professor for theoretical physics Ernst Dorn succeeded Knoblauch as the Halle ordinary professor for experimental physics and director of the physics institute.[132] At the same time he, the faculty, and the curator called for an extraordinary professor for theoretical physics to take over his old duties. Pockels's name was brought up, but the position went to Karl Schmidt, who had been Privatdocent for "physics," with a preference for theoretical questions, since 1889. The minister described Schmidt's assignment in 1895 as theoretical physics in "agreement" with Dorn; the agreement was never reached in practice, as over the next years, the degree of Schmidt's independence with respect to Dorn, the relation of theoretical to experimental physics, and the distribution of space in the institute were repeatedly disputed. Since Dorn could hardly use his higher rank as ordinary professor to discourage the understandable ambitions of an extraordinary professor for a special field that Dorn had once represented as ordinary professor himself, he asserted the priority of experimental over theoretical physics through his authority and sole responsibility as director of the physics institute. In this veiled form, he all but denied any independence to theoretical physics at Halle. From Dorn's point of view, Schmidt did not recognize the correct relationships within the institute; Schmidt tried to assert rights for himself that amounted to, in effect, a co-directorship. Schmidt gave lectures on applied physics alongside those on theoretical physics, which threatened Dorn from another direction as well. Schmidt lectured to railway officials on electrotechnology in 1896–97, for example, and he wanted to make it a regular practice, which Dorn saw as an embryonic electrotechnology institute growing at the expense of his physics institute. In 1912 Schmidt was made personal ordinary professor, but it was not until 1915 with the establishment of a provisional laboratory for applied as well as pure physics that he acquired a working independence at Halle.[133]

The assertion of independence by representatives of theoretical physics in their claim to control doctoral students was accompanied by signs of increasing organizational separation between the two parts of physics. One sign was that extraordinary professors for theoretical physics now often had to stay in their positions permanently while other beginning theoretical physicists did not advance at all. They were not being called to experimental ordinary professorships, and there were few theoretical professorships to which they might be promoted. At Strassburg, Cohn remained extraordinary professor for thirty-five years before being given the title of ordinary professor in 1918, which was only shortly before his retirement from teaching. At Jena, Auerbach remained extraordinary professor for thirty-five years. For others, too, such as Koenigsberger, Pockels, and Mathias Cantor, who took up their positions in the years after 1900, their extraordinary professorships were their final

132. Communication from Dr. H. Schwabe, director of the Halle University Archive.

133. To look ahead: when Karl Schmidt retired in 1927, he was replaced by A. G. S. Smekal, who was personal ordinary professor and director of a new laboratory for theoretical physics. It was not until 1934 that his position was transformed into a regular ordinary professorship, which at last completed the development of theoretical physics at Halle to full independence. Communication from Dr. H. Schwabe.

positions.[134] Another sign of the separation—as encouraging as the other was discouraging to theorists—was the growing number of successful careers wholly within theoretical physics. Several of them began at Zurich University, our final example.

At Zurich University, at the level of extraordinary professor, theoretical physics was taught in succession by theorists of exemplary talent: Einstein, Laue, and Debye, all of whom were to move on to prominent positions in theoretical physics in Germany. Their successor was Schrödinger, Zurich's first ordinary professor for the subject. The eminence of Zurich University in theoretical physics was not necessarily to be expected, since it was not an established center for research in physics. The director of the Zurich physics institute Alfred Kleiner was not a "great physicist," Einstein remarked (but that made no difference to Einstein, who believed that "scientific repute and great personality do not always go together," and that Kleiner had the latter).[135] Kleiner had an understanding of the needs of physics teaching, and he was a good judge of talent for research, as his early support of Einstein showed.

At the beginning of the twentieth century, the University of Zurich was behind the times. At the other Swiss universities in Basel, Bern, Geneva, and Lausanne, as at German universities, there were at least two physics professors, one for experimental physics and one for theoretical physics. But there was only one physics professor at Zurich, and Kleiner presented the case for a second: in the nineteenth century, physics teaching had developed in two directions, Kleiner argued, both of which had become "very extended disciplines"; while earlier it was customary for one man to teach both directions, the rise of teaching laboratories made that impossible, and the solution adopted everywhere was for one man to teach the popular elementary course and another man to teach the "scientific part par excellence" of physics, which meant small courses. Kleiner himself had been giving both the experimental and the theoretical lectures at Zurich, but now, in 1909, he no longer wanted to carry the burden alone. Zurich had just acquired a physics laboratory, which justified a second physics professor, Kleiner argued, and the faculty agreed. The only question that remained was whom to hire.[136]

The Zurich faculty had to be informed about the new type of specialist they

134. Capable extraordinary professors for theoretical physics encountered obstacles to promotion in addition to those attending the separation of their science. That was the case, for example, if they made a practice of criticizing influential authorities, or if they were Jewish, as many of them were, a point we have discussed. The same obstacles could stand in the way of the earlier advancement from Privatdocent to extraordinary professor for theoretical physics. They did, for example, in the case of Max Abraham. Einstein to Kleiner, 3 Apr. 1912, ETHB. Commission to the Mathematics and Natural Sciences Section of the Philosophical Faculty, Zurich University, 6 Apr. 1912, STA K Zurich, U. 110b. 2, Laue. Arnold Sommerfeld, "Abraham, Max," *Neue deutsche Biographie* 1:23–24. Max Born and Max Laue, "Max Abraham," *Phys. Zs.* 24 (1923): 49–53, on 53.

135. Carl Seelig, *Albert Einstein: Eine dokumentarische Biographie* (Zurich: Europa-Verlag, 1952), 98. Translated by M. Savill, *Einstein: A Documentary Biography* (London: Staples, 1956). Where we use the English translation, we indicate it.

136. Dean C. Stoll of the Zurich U. Philosophical Faculty, Section II, to Director of Education of Canton Zurich, 4 Mar. 1909, STA K Zurich, U. 110b. 2, Einstein.

were asked to admit as extraordinary professor. They were told that there are two types of theoretical physicist. First, there are the "pure mathematicians," who stay away from experiment and only concern themselves with theory; they deliver their lectures and otherwise have no duties; they include Clausius and Kirchhoff from the last century and Planck and Sommerfeld in the present. The other type of theoretical physicist does experimental research, exemplifying the saying that the "theoretical physicist is no longer the man of sponge and chalk"; he is represented by Voigt and Des Coudres, who have institutes equipped for this purpose. In the event, Zurich tried both kinds of theorist, Debye among the latter and Einstein, Laue, and Schrödinger among the former, though at least one of them, Laue, did some experiments while at Zurich.[137]

Through successive appointments, the Zurich position developed in the direction of greater autonomy and importance, owing to the interests of the individual physicists holding it, to the university's teaching needs and its experience with the position, and to the development of theoretical physics as a teaching field. Einstein possibly and Debye definitely were assigned not only to lecture on theoretical physics but also to help Kleiner in the laboratory. By the time Laue was appointed, however, Zurich had recognized that theoretical physics is a large teaching field and that the theorist must concentrate on it and not share the experimental duties, as Debye had done. Laue was assigned theoretical teaching only, and he taught only the standard number of hours a week, not an excessive number as Einstein and Debye had done.[138]

Einstein and Laue remained at Zurich for two years, Debye for one year, and all three remained a far shorter time than the six-year term of their appointments. The only argument *against* appointing Debye, the faculty said, was his gifts, which were the main argument *for* him as well. The gifts would not remain long hidden, and the faculty anticipated that Debye would soon be called away, as he was. It did not help Zurich to require Debye to inform the government before entering into binding negotiations for a position abroad; he duly reported to them the calls that came and simply accepted the one he wanted, despite an increase in salary and the prospect of an ordinary professorship at Zurich. Debye and the other young theorists in their first positions at Zurich were viewed abroad as talents easily enticed across national frontiers. Einstein moved from Switzerland to Austria; Laue was invited to America but moved to Germany; and Debye was invited to Germany but moved to Holland. Zurich learned to expect that if it attracted talented researchers in theoretical physics to a subordinate position such as an extraordinary professorship, there would be a constant turnover, and the physics professor Kleiner and the faculty would have to go to a good deal of trouble to keep it occupied and its occupants happy. At

137. Dean of the Zurich U. Philosophical Faculty to Director of Education of Canton Zurich, 22 Mar. 1911, STA K Zurich, U. 110b. 2, Debye.

138. Director of Education of Canton Zurich to "Regierungsrat," 6 May 1909, STA K Zurich, U. 110b. 2, Einstein. Director to "Regierungsrat," 6 Apr. 1911, STA K Zurich, U. 110b. 2, Debye. Director to "Regierungsrat," 16 July 1912, STA K Zurich, U. 110b. 2, Laue.

times, it meant, too, that Kleiner together with various assistants had to pitch in to keep the theoretical lectures going.[139]

After Laue left Zurich in 1914, the position remained vacant through World War I and for several years after. Available qualified theorists were scarce and none of them could be attracted to Zurich. The position had to be redefined in any event, since theoretical physics had recently acquired a "high scientific significance" and had become an "extended teaching field"; only an ordinary professorship could properly serve it and attract an established physicist.[140] The new ordinary professorship for theoretical physics was given to Schrödinger, who had taught theoretical physics briefly at Stuttgart and Breslau, and who went on to justify Zurich's confidence in the promise of the discipline. Einstein, Debye, and Laue had all done research while teaching at Zurich, but it was Schrödinger who settled in at Zurich and who did his most significant work there: wave mechanics, his accomplishment, ranks among the most significant work in twentieth-century theoretical physics.

The theorists at Zurich were not hired for their reputations as teachers. Einstein had no such reputation yet, having lectured only one semester as a Privatdocent and having cancelled his second semester's lectures because only one student turned up. Kleiner said it was too soon to judge Einstein's teaching ability; all he could do was predict that it would prove satisfactory, as it did. Upon arriving in Zurich, Einstein found himself "*very* busy" with lectures and with less time left over than when he worked in the patent office in Bern, but he soon reported that "now keeping school agrees better with me and gives me pleasure." He recalled of this time: "I did not lecture brilliantly. Partly because I was never very well prepared and partly because the condition of 'what-was-to-be-discovered-in-the-future' rather got on my nerves."[141] That did not seem to bother the students, who valued his gifts and his accessibility. He even attracted some students from the neighboring Zurich Polytechnic. When he was invited to move to Prague, a number of students and assistants urged the government to keep him, pointing out that his lectures were a "great pleasure" and that he was certain to create great fame for "this newly created discipline at our university."[142] Unlike Einstein, Laue was already known as a lecturer when he came to Zurich; in fact, he was known as a bad lecturer, which Einstein thought was the reason he had not been given an appointment before. The Zurich faculty found fault with Laue's soft, rapid, unclear speech, but that did not weigh in

139. Dean of the Zurich U. Philosophical Faculty to Director of Education of Canton Zurich, 22 Mar. 1911; letters concerning Debye's appointment at Zurich, April 1911; Debye to "Regierungsrat," 16 Dec. 1911 and 4 Jan. 1912; and Director to "Regierungsrat," 10 Jan. 1912; STA K Zurich, U. 110b. 2, Debye.

140. Excerpt of the minutes of a meeting of the "Regierungsrat" on the "professorship for theoretical physics," 28 July 1921, STA K Zurich, U. 110b. 3, Schrödinger. Armin Hermann, "Schrödinger, Erwin," DSB 12:217–23.

141. Dean Stoll to Director of Education of Canton Zurich, 4 Mar. 1909. Einstein to Besso, 17 Nov. and 31 Dec. 1909, in *Albert Einstein-Michele Besso Correspondance 1903–1955*, ed. P. Speziali (Paris: Hermann, 1972), 16–18, on 16–17. Einstein to Laub, n.d., in Seelig, *Einstein*, English trans., p. 90.

142. Letter signed by fifteen students and assistants to Director of Education of Canton Zurich, 23 June 1910, and Director of Education to "Regierungsrat," 12 July 1910, STA K Zurich, U. 110b. 2, Einstein.

the balance with Laue's "epoch-making investigations," above all his recent discovery of X-ray diffraction.[143] Einstein, Laue, Debye, and Schrödinger all worked on relativity or quantum theory or both, and it was their work on these and other modern problems that impressed the Zurich faculty and officials.

As the early history of the Zurich position shows, around the beginning of this century, theoretical physics was understood to occupy a university teacher full time, and for a teaching assignment in this field there existed a number of young, capable theorists. Einstein came to Zurich as the author of the theory of relativity and other major theoretical studies and with the support of Kleiner, who had known him and had followed his work for several years. Einstein had not studied at a university and had not worked under a leading physicist, theoretical or otherwise; he had made his own way in physics. It was different with Debye, Laue, and Schrödinger, who had. all graduated from universities and had done their dissertations under well-known theoretical physics professors. By the time of Schrödinger's appointment at Zurich, his theoretical teacher Hasenöhrl had died, but Debye's and Laue's theoretical teachers were Sommerfeld and Planck, respectively, who strongly recommended their protégés for this and other positions. Planck's student Laue received a recommendation from Sommerfeld as well, for Laue had spent the last several years in Sommerfeld's institute. Laue's predecessors at Zurich, Einstein and Debye, also had a say in Laue's appointment there.[144]

Each time the Zurich position became available, a number of qualified candidates were considered, most of whom were German if not Swiss. From those proposed for the position at the time it was offered to Laue, Einstein concluded, correctly, that Zurich was making a "choice of a *theorist*"[145] By then there were many young German and German-speaking physicists who were recognized as "theorists," and like Debye and Laue, many of them had been trained by other theorists.

New Ordinary Positions for Theoretical Physics

From Zurich University, Einstein moved to the German university in Prague as ordinary professor, replacing Lippich, who retired. The Prague faculty committee to recommend candidates for the chair identified the "central problem of theoretical physics" today and for a long time to come as the relationship of mechanics to electrical theory. Whereas earlier, they said, mechanics was regarded as fundamental and the foundation of electricity, now electrical theory was regarded as the fundamental of the two. The faculty recommended Einstein in first place for his "epoch-making" work on the electrodynamics of moving bodies, his theory of relativity,

143. Einstein to Kleiner, 3 Apr. 1912, ETHB. Director of Education of Canton Zurich to "Regierungsrat," 16 July 1912.

144. Dean Stoll to Director of Education of Canton Zurich, 4 Mar. 1909. Dean of the Zurich U. Philosophical Faculty to "Regierungsrat," 30 Mar. 1911, STA K Zurich, U. 110b. 2, Debye. Report by the Commission to the Mathematics and Science Section of the Philosophical Faculty, 6 Apr. 1912, STA K Zurich, U. 110b. 2, Laue.

145. Einstein to Kleiner, 3 Apr. 1912.

which bore on this problem. The ministry, exercising its prerogatives, did not follow the committee's recommendation but offered the job to the second candidate, an Austrian, who declined it. The offer then went to Einstein, who was appointed in January 1911. At the same time, his chair and the institute associated with it were given modern names: the chair was now for "theoretical" rather than "mathematical" physics, and the "cabinet of mathematical physics" became the "institute for theoretical physics." The mathematical-physical seminar was divided, the latter half being attached to the new chair as the "seminar for theoretical physics."[146]

"My position and my institute here give me much pleasure," Einstein wrote upon arriving in Prague. Yet not everything in his "fine institute with a magnificent library" satisfied him; the Prague students were less industrious and capable than the Swiss, and there were never many of them, at the maximum thirteen in his lectures and six in his seminar. He was discouraged by the students' indifference to his field, and he felt alien in Prague and longed to return to Zurich.[147]

While he was teaching at Prague, Einstein received offers from Vienna, Utrecht, and Leiden, where Lorentz wanted Einstein to come as his replacement. Einstein told Lorentz that to "hold your chair would be something immensely oppressing to me," and in this connection he mentioned their colleague Hasenöhrl for whom Einstein always felt compassion because he "must sit in Boltzmann's chair." Before moving to Prague, Einstein had promised his alma mater, the Zurich Polytechnic, that he would tell them of any future offers, so that they might make him one, too. When they offered to restore a long-empty chair for him, he accepted.[148]

Since 1902, when Hermann Minkowski left, the Zurich Polytechnic's chair for "higher mathematics" had remained unfilled, which was a source of concern for the department responsible for training secondary school teachers of mathematics and physics. Einstein was deemed qualified for the chair by his publications and by recommendations from the "first authorities." Of the recommendations, Planck's was the most decisive. It was taken from Planck's 1909 Columbia University lectures on theoretical physics and liberally quoted in the official Swiss report on Einstein; Einstein's new concept of time, Planck said in these lectures, placed the highest demand on physicists' capacity for abstraction and power of imagination and surpassed in "boldness" all that had been done in speculative philosophy and epistemology, making non-Euclidean geometry seem like "child's play" by comparison; Planck compared the "revolution" in the physical world view that the relativity principle had brought about with the introduction of the Copernican world system. The Swiss report acknowledged that Einstein was not a spellbinding teacher, but what really

146. Information on Einstein's appointment from his Personalakte in the Central State Archive of the Czechoslovak Socialist Republic, as given in Jan Havránek, "Die Ernennung Albert Einsteins zum Professor in Prag." *Acta Universitatis Carolinae—Historia Universitatis Carolinae Pragensis* 17, pt. 2 (1977): 114–30, on 120, 125, 128. József Illy, "Albert Einstein in Prague," *Isis* 70 (1979): 76–84, on 76–78.

147. Einstein to Besso, 13 May 1911 and 4 Feb. 1912, in *Einstein-Besso Correspondance*, 19–20, 45–47, on 19 and 45. Einstein to Lucien Chavan, 5 July 1911, quoted in Seelig, *Einstein*, and in Ronald Clark, *Einstein: The Life and Times* (New York: World, 1971), 137.

148. Einstein to Lorentz, 18 Feb. 1912, Lorentz Papers, AR. Illy, "Einstein," 83–84.

mattered was his abundance of "creative thoughts," which promised a continuation of his uncommon success.[149]

At the Zurich Polytechnic, Einstein was not required to give general lectures to large audiences or to direct a student laboratory; his narrower assignment was to offer "theoretical physics" to teaching candidates in their final semesters. As at Prague, at Zurich he got a handsome salary, larger than that of any other teacher of the abstract sciences. He asked for this salary, since he would have few students and a correspondingly small share of student fees.[150]

Einstein had crossed Germany from Switzerland to Austria and back, and he had considered moving to Holland, too. But he had not yet worked in Germany, even though he had close connections with German physicists. He published regularly in German journals, above all in the *Annalen der Physik,* and Laue, Sommerfeld, and other German theoretical physicists had paid him visits in Switzerland. He had begun to attend the meetings of the German Association, first in Salzburg in 1909, recently in Karlsruhe in 1911, and he had discussed physics with German physicists for several days at the 1911 Solvay Congress in Brussels. The desirability of attracting Einstein to Germany was evident, and the only question was how to do it.

Einstein's academic advancement was typical of, if a little more rapid than, that of other talented young theoretical physicists. Between 1908 and 1912, he passed through the ranks from Privatdocent through extraordinary professor to ordinary professor for theoretical physics, a position he held at two places. During these years, no ordinary professorships for theoretical physics had become available in Germany; in 1913, however, there was a vacancy in the Prussian Academy of Sciences, and Planck and Nernst called on Einstein in Zurich, soon after his return there from Prague, to discuss it with him. Planck drafted a proposal to the Prussian ministry of culture, which was signed by him together with Nernst, Rubens, and Warburg, all of whom knew Einstein from the Solvay Congress, if not from elsewhere. In describing his contributions to "theoretical physics," they noted in each case the stimulus it provided to experimental physics. They spoke of his deepening of "classical theory" through the kinetic theory of matter; they extolled his quantum theory of specific heats, which provided the foundations for the new "kinetic atomism," and his relativity theory and his new gravitational theory, although it had not yet been confirmed. For completeness, they also mentioned his light-quantum hypothesis, which they thought had missed the mark. In November 1913 Einstein was duly appointed an ordinary member of the mathematical-physical class of the academy with a salary comparable to that of the highest paid university physics institute directors, though without the prospect of substantial student fees from large lectures. His main responsibility in Berlin was to his own research, but he was to have light administrative duties as director of the Kaiser Wilhelm physics institute, which was

149. President R. Gnehm of the Swiss Education Council to the Swiss Department of the Interior, 23 Jan. 1912, A Schweiz. Sch., Zurich, 1–442, 1912.
150. President Gnehm to Department of the Interior, 23 Jan. 1912.

still in the planning stage. He was also allowed to lecture at Berlin University as an ordinary professor there.[151]

"Not without a certain feeling of discomfort I see the Berlin adventure come closer," Einstein wrote to a friend.[152] Only the year before, he had explained his preference for Zurich over Leiden in part because he could do his "protracted sitting on a few scientific eggs at a less exposed and illuminated spot." Berlin was even more exposed than Leiden, yet he decided to move there in the end: "I couldn't resist the temptation," he told Lorentz, "of accepting a position in which I am freed of all duties, so that I can devote myself wholly to brooding."[153] To Ehrenfest he explained: "I accepted this peculiar sinecure because giving lectures grates so oddly on my nerves, and I don't have to lecture there at all."[154] In April 1914, at age thirty-five, he left Zurich for Berlin, where he remained in the same position for the next twenty years. "Theoretical physics here is looked after," Planck observed after Einstein's appointment at the Prussian Academy and in anticipation of Laue's appointment at Berlin University.[155] Einstein would deliver the "scientific eggs" that his German promoters were confident he could; with his move to Berlin, theoretical physics in Germany gained immeasurably in stature.

In the world of physics, Einstein's value had been increasingly recognized, which was reflected in the salaries he received and the duties he was assigned. His position at the Zurich Polytechnic had been made to correspond, as the Swiss school officials put it, in "especially great measure to the individuality of Einstein";[156] his position in Berlin corresponded to it, too. Einstein came to Berlin through a special arrangement with the Prussian Academy, as had his predecessor there in theoretical physics, Kirchhoff, some forty years before. But whereas Kirchhoff had come at the end of a typical career in physics, Einstein did so from a career entirely within theoretical physics.

Nearly Einstein's contemporary—he was five years younger—Debye had early opportunities comparable to Einstein's, and he progressed through the ranks with the same ease. Born in Holland, an engineering graduate from the technical institute in Aachen, Debye moved with Sommerfeld from Aachen to Munich to continue his studies. Graduating from Munich in 1908, he stayed on there until, as we have seen, he moved to Zurich University as extraordinary professor for theoretical physics in 1911. In 1912 he moved on to the university in Utrecht.[157]

---

151. Proposal of Einstein as ordinary member of the Prussian Academy of Sciences, 12 June 1913, signed by Planck, Nernst, Rubens, and Warburg; Secretary Roethe's letter of appointment to Einstein, 22 Nov. 1913, and Einstein's acceptance, 17 Dec. 1913; in *Albert Einstein in Berlin 1913–1933*, pt. 1, *Darstellung und Dokumente*, ed. Christa Kirsten and Hans-Günther Treder (Berlin: Akademie-Verlag, 1979), 95–96, 101, respectively.

152. Einstein to Besso, n.d. [end of 1913], in *Einstein-Besso Correspondance*, 50.

153. Einstein to Lorentz, 18 Feb. 1912 and 14 Aug. 1913, Lorentz Papers, AR.

154. Einstein to Ehrenfest, n.d. [winter 1913–14], quoted in Klein, *Ehrenfest*, 296.

155. Planck to Wilhelm Wien, 31 July 1913, Wien Papers, STPK, 1973.110.

156. President Gnehm to the Department of the Interior, 23 Jan. 1912.

157. Friedrich Hund, "Peter Debye," *Jahrbuch der Akademie der Wissenschaften in Göttingen*, 1966, 59–64.

Sommerfeld, Debye's former teacher, was unenthusiastic about Utrecht and urged Debye to wait for a call to Leiden to succeed Lorentz. That would have been "like a medal," Debye told him, but it offered him no more substantial prospects than Utrecht did, and he did not feel like waiting. At Utrecht, Debye had a purely theoretical job, which was one of its attractions. If he had stayed in Zurich, he would have spent years putting a laboratory into order and neglecting his own work in the process. Besides, as the theorist in Utrecht, Debye saw an opportunity to apply what he had learned from Sommerfeld; by crossing Dutch thoughtfulness with German boldness, he would create a new strain of physicist.[158]

Soon an opportunity arose for Debye to return to Germany to fill a vacancy that had long been anticipated. For some years Riecke had spoken of the downward "curve" of his activity in the Göttingen experimental institute,[159] and he had made plans to retire in 1914; in 1915 he died, at age seventy. Voigt, who was only five years younger than Riecke, planned to retire at about the same time, which suggested to the Göttingen curator and to others in "mathematical-physical circles" that it would be a good time to collapse the two physics institute directorships into one. They did not expect to find two physicists who could work together as "peacefully" in a shared institute as Riecke and Voigt had done.[160] Debye was to enter into the new plans for physics at Göttingen.

When in 1913 Debye gave a series of invited lectures at Göttingen, the local mathematician David Hilbert decided that Debye should be obtained for Göttingen.[161] Since Debye would have to have an institute, Voigt declared himself willing to turn over his sooner than he had intended; he would still share the institute and the teaching of theoretical physics, but the institute would be directed solely by Debye. Voigt told the Göttingen curator that despite the "great love and sacrifice" he had expended on the institute, he regarded Debye's call as too important to allow his feelings to stand in the way. In September 1914 Debye was appointed personal ordinary professor with the understanding that the "personal" was to be dropped in the course of reducing the two directorships to one.[162]

As Debye was known for both his experimental and his theoretical research, he was regularly considered for both experimental and theoretical positions.[163] Al-

158. Debye to Sommerfeld, 23 and 29 Mar. and 3 Nov. 1912, Sommerfeld Correspondence, Ms. Coll., DM.

159. Riecke to Stark, 6 Jan. 1907, Stark Papers, STPK. In his lecturing, Riecke was still lively, but to his colleague Voigt his lectures seemed out of date, too elementary for the kind of students then coming from the secondary schools. Voigt wanted Riecke's replacement to be a "true *experimenter*," who would completely reform the experimental lectures, lengthening them, supplementing them with a theoretical session, and replacing antiquated demonstrations with modern ones. Voigt to Stark, 23 and 27 June 1914, Stark Papers, STPK.

160. Göttingen U. Curator to Prussian Minister of Culture, 18 Apr. 1914, Riecke Personalakte, Göttingen UA, 4/V b/173.

161. Walther Gerlach, "Peter Debye," *Jahrbuch bay. Akad.*, 1966, 218–30, on 224.

162. Voigt to Göttingen U. Curator, 29 Mar. 1915, Voigt Personalakte, Göttingen UA, 4/V b/267c. Voigt held the positions of "department head" and dean of the philosophical faculty in 1917 and of "acting director" of the physics institute in 1918. He died in December 1919. Prussian Minister of Culture to Debye, 19 Sept. 1914, Debye Personalakte, Göttingen UA, 4/V b/278.

163. Despite his aptitude in both, Debye was more often identified with theoretical physics than

though he did research in both, as a practical matter he valued the division of experimental and theoretical responsibilities in teaching. He declined the offer Zurich University made to him in 1915; he was asked to be their experimental professor, with a salary comparable to Göttingen's, with the understanding that he would lecture on theoretical physics in addition to lecturing on experimental physics and directing laboratory work. He explained to the Swiss authorities that the additional theoretical lecturing would eat into his research time and, moreover, would greatly weaken the position of theoretical physics and with it that of physics in general in Zurich.[164] The duties he was assigned at Göttingen from 1916 on, when he received a regular ordinary professorship, were burdensome enough: as both director of the entire physics institute and director of the mathematical physics department within the institute, he was required to give the theoretical lectures and, at least every other year, an advanced experimental course.[165]

By the summer of 1916, after a series of intricate ministerial moves, physics in Göttingen was directed by an ordinary professor, Debye, and an extraordinary professor instead of by two ordinary professors as it had been since the time of Weber and Listing. The extraordinary professor was the former Privatdocent Robert Pohl, who was given rooms in the institute previously assigned to Riecke, and who directed the experimental department and gave the big experimental lectures.[166] The organization of work was much as before.[167] In 1920 Pohl would be elevated to ordinary professor.[168]

In 1914 Laue left Zurich to return to Germany, to Frankfurt am Main. There an old private society, the Physical Society, had maintained a set of institutes for physics and other natural sciences. The institutes with their separate directors and staff held classes and colloquiums, and advanced students did independent research in their laboratories. It was all much as in a university, which the society trans-

with experimental. For example, to the Bonn faculty, Debye was a "theorist" and so lacked the "penetrating knowledge of the countless experimental methods and aids" that were needed to direct one of the "largest Prussian physics institutes." Bonn U. Philosophical Faculty to the Ministry of Culture, Aug. 1919, Kayser Personalakte, Bonn UA.

164. Debye to Director of Education of Canton Zurich, 30 Mar. 1915, STA K Zurich, U. 110b. 3, Edgar Meyer.

165. Prussian Minister of Culture to Debye, 28 Feb. 1916, Debye Personalakte, Göttingen UA, 4/V b/278.

166. Robert Pohl belonged to the commission for examining physicians and Debye to the commission for examining teachers, which corresponded to the division of their lectures. Pohl as well as Debye had the right to give doctoral examinations. Minister of Culture to Debye, 19 Sept. 1914 and 26 Feb. 1916.

167. The negotiations and paperwork that went into this seemingly small rearrangement can be appreciated from the steps required to bring it about. Riecke was relieved of his duties in April 1914, and his professorship was abolished after his death in 1915. In the summer of 1915, Voigt formally retired as director of the department for mathematical physics and was replaced by Debye, who was already co-director of the mathematical physics seminar. For this step, an extraordinary professorship was created for Debye, which was to be abolished after Voigt stopped teaching. This position was given to Pohl in April 1916 at the time that Debye was made regular ordinary professor. The whole affair was so complicated that at some point before Voigt's retirement, an official tried to straighten it out, as is evidenced by notations all over the cover of Voigt's personnel file. Göttingen UA, 4/V b/203.

168. Runge to Göttingen U. Curator, 20 Oct. 1920, Pohl Personalakte, Göttingen UA.

formed itself into on the eve of World War I. Laue was called to Frankfurt University as the first ordinary professor for theoretical physics, joining Richard Wachsmuth, who had been the director of the society's physics institute for many years.[169]

So, as we have seen, shortly before World War I Germany acquired three ordinary professors who were outstanding theoretical physicists: Debye and Laue came to Germany to assume ordinary professorships for theoretical physics, and Einstein came to assume a position at the Prussian Academy of Sciences with the right, which he exercised, to lecture as a professor in the university there. For the discipline of theoretical physics, what is notable is that all three had early careers entirely within theoretical positions. Across Central Europe, from Zurich to Prague, there were positions for German-speaking theoretical physicists, and the "theoretical physicist" had become a recognized and sometimes much sought after specialist.

169. Richard Wachsmuth described the Physical Society to Lorentz in a letter informing him of his election as honorary member, 11 Feb. 1912, Lorentz Papers, AR. Ludwig Heilbrunn, *Die Gründung der Universität Frankfurt a. M.* (Frankfurt a. M.: Joseph Baer, 1915), 54–55, 232. *Jahresbericht des Physikalischen Vereins zu Frankfurt am Main, 1907–1908* (Frankfurt a. M., 1909), 74–75.

26

Theoretical Physics Unbound: Examples from the Theories of the Quantum, Relativity, the Atom, and the Universe

In the early years of the twentieth century, German physicists took up a wide range of theoretical problems connected with recent experimental discoveries, and they continued to probe the foundations of physics as they had in the late nineteenth century. They informed, stimulated, and criticized one another's work in the usual ways, by letters and visits, by journal papers, by talks and discussions at professional meetings, and, because of the importance of recent developments in the quantum theory, by an unusual international conference, the Solvay Congress of 1911. In this chapter we sketch several examples of individual and concerted work by German physicists on theoretical problems. Our selection is narrowly limited and traditional; the examples belong to, and by no means exhaust, the theoretical problems that the physicists of the time regarded as especially important, all of which have been closely examined by historians. To portray certain features of the work of theoretical physicists at the time our study ends, we take up again the special theory of relativity, the subsequent general theory of relativity, and the early quantum theory. Theoretical physics experienced some of its greatest advances, and German theoretical physicists played a significant and often leading part in them.

Quantum Theory at the Salzburg Meeting of the German Association

Einstein had introduced relativity theory in 1905 through electrodynamics, and one of the implications he drew from his relativistic electrodynamics was the localization of radiant energy: he reasoned that without the ether, which his theory dispensed with, a continuously distributed energy in space was an "absurdity." This conclusion agreed with another of his 1905 works, one which dealt specifically with this point; guided by the statistical understanding of the second law of thermodynamics, Einstein argued that the experimentally confirmed Wien distribution law for the high-frequency limit of blackbody radiation implies that the electromagnetic field has a particulate structure. He proposed that high-frequency radiation behaves thermody-

304

namically as though it, analogously to the particles of an ideal gas, consists of a "finite number of [independent] energy quanta which are localized at points in space, which move without dividing, and which can only be produced and absorbed as complete units."[1]

Particulate reasoning was associated with certain work on radiation phenomena but not with recent work on light. The reasoning behind Einstein's proposal of light quanta in 1905 did not convince his colleagues at the time: Planck, Laue, Wien, Sommerfeld, and other early supporters of Einstein's relativity theory all rejected his hypothesis of light quanta.[2] Their principal argument was that interference phenomena—and also diffraction, refraction, and other phenomena of physical optics—demand a wave interpretation of light. Planck, for instance, assumed that processes in the ether are "represented *exactly* by Maxwell's equations," the equations of the wave theory, which maintain the clear "opposition between ether and matter."[3] With Lorentz, who was also unpersuaded of light quanta, Planck exchanged technical letters on the problem of the interaction of the ether with resonators. What was needed, Planck told Lorentz, was to add to the electron theory a "new hypothesis," which would make the constant h a property of resonators and not of the ether. In this way, the failure of the old theory would be restricted to Hamilton's equations of motion, which apply to the resonators, and would not extend to the theory of the pure ether;[4] the difficulties, that is, would be conveniently concentrated in the molecules of matter.[5] German physicists did not see the need for so "radical" a step, as Planck put it, of light quanta, but they acknowledged the subtlety of Einstein's theoretical arguments. That was shown by their persistent efforts to answer them.

Only one other leading German physicist advocated light quanta at this time, the experimentalist Johannes Stark. He had already applied the basic quantum relationship between energy and $h\nu$ to the determination of wavelengths of band spectra and to other phenomena of interest to experimentalists when, in 1909, he came out in support of Einstein's hypothesis. His advocacy soon brought him into open conflict with Sommerfeld, and it brought him disapproval, sometimes behind the

---

1. Einstein, "Antwort auf Plancks Manuskript," [1908?], quoted in Seelig, *Einstein*, 122; "Über einen die Erzeugung und Verwandlung des Lichtes betreffenden heuristischen Gesichtspunkt," *Ann.* 17 (1905): 132–48; translated by A. B. Arons and M. B. Peppard as "Einstein's Proposal of the Photon Concept—a Translation of the *Annalen der Physik* paper of 1905," *Am. J. Phys.* 33 (1965): 367–74.

2. Laue to Einstein, 2 June 1906, Ms. Coll., DM, 1973–6; Wien to Stark, 4 Oct. 1909, Sommerfeld Correspondence, Ms. Coll., DM. Sommerfeld to Lorentz, 9 Jan. [1909 or 1910], Lorentz Papers, AR. Einstein's light quantum hypothesis and responses to it are discussed by Klein in, among other places, "Einstein's First Paper on Quanta" and "Einstein and the Wave-Particle Duality," *The Natural Philosopher*, no. 3 (1964): 3–49; by Russell McCormmach in "J. J. Thomson and the Structure of Light," *Brit. Journ. Hist. Sci.* 3 (1967): 362–87, on 370–72; by Hermann in *Genesis of the Quantum Theory*, 50–71; and by Kuhn in *Black-Body Theory*, 176–82.

3. Planck to Einstein, 6 July 1907, EA.

4. On the relationship between the ether and matter, the main difference of opinion between Lorentz and Planck at the time of their correspondence was that Lorentz thought that the ether could only absorb energy in portions of $h\nu$, whereas Planck thought that the electron could only give up energy in these portions. Planck to Lorentz, 1 Apr. and 7 Oct. 1908, 24 Apr., 16 June, and 10 July 1909, Lorentz Papers, AR.

5. Planck to Wien, 27 Feb. 1909, Wien Papers, STPK, 1973.110.

scenes, by other theorists such as Planck and Wien. Sommerfeld, who had been in recent correspondence with Wien about the implications of Planck's energy element for Maxwell's theory, charged Stark with being "completely wrong" about what he called the "ether-wave hypothesis," the antithesis of Stark's light-quantum hypothesis; Stark entered into something "very risky" with this theoretical direction, he warned. Conceding that Stark knew experimental physics better than he did, Sommerfeld reminded Stark that he himself was much the superior theorist, and he advised Stark to read certain chapters in a standard electromagnetic text. Offended by Sommerfeld's "school-masterly" way of speaking to him, Stark felt that his good reputation as an experimentalist was threatened. The theory itself was hardly the central issue, since what Stark liked about it was the "visual conception" it offered for experimental purposes.[6] As the "constitution of light is above all an experimental problem," Stark said, he regarded the theory of light quanta as having "only heuristic value," and before long he abandoned it.[7]

Einstein, looking forward to meeting Stark at the German Association meeting in Salzburg in 1909, wrote to him: "You can hardly imagine how hard I have tried to think of a satisfactory mathematical formulation of the quantum theory. But so far I have not been successful with it."[8] In Salzburg Einstein talked on problems of radiation, which gave him an opportunity to discuss both the relativity theory and the light-quantum hypothesis. The latter topic, as the quantum problem in general, was still not widely known, so that Einstein's talk had considerable importance for German physics. The Salzburg meeting was also the first occasion for German physicists—with the exception of a few who had gone to Bern for the purpose—to see Einstein in person.

Einstein set out to persuade his audience of the need for a new understanding of radiation. What he asked them to do was go back to the old Newtonian corpuscular, or "emission," theory of light, and he asked them to do this without giving up the wave theory of light, the alternative theory that physicists had preferred since the early nineteenth century. Since he knew that they would not find it easy to agree to this new understanding, he brought forward a combination of arguments to support his opinion "that the next phase of the development of theoretical physics will bring us a theory of light that can be understood as a kind of fusion of the wave and emission theories of light." Einstein emphasized the importance of Lorentz's electron theory for the development of the viewpoint he was presenting, even

6. Wien to Sommerfeld, 18 May and 15 June 1908, and to Stark, 4 Oct. 1909, Sommerfeld Correspondence, Ms. Coll., DM. Wien, for example, criticized Stark's latest view of the "energy element" as being incompatible with the facts of interference. Sommerfeld to Stark, 4, 10, and 16 Dec. 1909; Stark to Sommerfeld, 6 and [12 or 13?] Dec. 1909, and 24 Apr. 1910, Stark Papers, STPK. Stark's light-quantum ideas are discussed in Hermann, *Genesis of the Quantum Theory*, 72–86, and in Kuhn, *Black-Body Theory*, 222–25.

7. Stark to Lorentz, 4 Aug. 1910, Lorentz Papers, AR. Kuhn, *Black-Body Theory*, 224–25.

8. Einstein to Stark, 31 July 1909, quoted in Armin Hermann, "Albert Einstein und Johannes Stark. Briefwechsel und Verhältnis der beiden Nobelpreisträger," *Sudhoffs Archiv* 50 (1966): 267–85, on 279.

though he rejected the theory's stationary ether. Lorentz's theory decisively separated the electromagnetic field from matter, interpreting each entity as something existing independently. As support for his view of light, Einstein called on his relativity theory: according to the equivalence of mass and energy, light transfers mass from the emitting to the absorbing body, and in the space between, it behaves as an independent entity, not as a state of a hypothetical ether. It was not, however, from the theory of relativity that Einstein drew his main arguments, but from the second law of thermodynamics and its statistical-mechanical interpretation and from Planck's law of blackbody radiation. From these sources, Einstein derived expressions for the fluctuations of the energy and momentum of light, which contain two parts, one corresponding to what one would expect from the wave theory of light and the other from the corpuscular theory. As both parts were indispensable, so were both theories of light—through a connection yet to be determined.[9]

Members of the audience at Einstein's Salzburg talk recalled that among the physicists joining the discussion, only Stark favored light quanta. Stark appealed for "greater good will" toward the hypothesis, while the chairman of the session, Planck, along with most of the other physicists there opposed it. The outstanding question to decide was "where we should look for these quanta," Planck said. If, as Einstein proposed, physicists were to assume that light has an atomistic constitution, Maxwell's equations would have to be given up, a step that Planck regarded as unnecessary. Maxwell's equations could be preserved if the quantum were located not in the pure radiation field but in its interaction with matter, in the processes of absorption and emission. The problem was difficult; for the electron theory was inadequate to explain emission, and if the resonators of the quantum theory could explain it, the explanation would evidently be at the expense of the laws of mechanics, and it would lie outside the laws of present-day electrodynamics. To arrive at the necessary detailed accounting of the behavior of resonators for short time intervals and rapid accelerations, Planck suggested this:

> Perhaps we may assume that an oscillating resonator does not have continuously varying energy but that its energy is a simple multiple of an elementary quantum. I believe that if we use this hypothesis we can arrive at a satisfactory radiation theory. Now the question is always: how are we to picture something like this? In other words, we demand a mechanical or electrodynamical model of such a resonator. But in mechanics and in present-day electrodynamics, we have no discrete elements of action, and hence we cannot produce a mechanical or electrodynamical model either. It thus seems impossible by mechanical means, and we will have to get used to it. Our attempts to represent the light-ether mechanically also failed completely. We also wanted to picture

9. Albert Einstein, "Über die Entwicklung unserer Anschauungen über das Wesen und die Konstitution der Strahlung," *Phys. Zs.* 10 (1909): 817–25, quotation on 817. Klein, "Einstein and the Wave-Particle Duality," 5–15.

electrical current as mechanical and thought of the comparison with the flow of water, but we had to give that up, too, and as we got used to that we must get used to such a resonator, too.[10]

In the discussion, Planck raised another objection to the light-quantum hypothesis, this one with more intuitive appeal: for a single light quantum to interfere with itself, as experience with light would seem to require, the quantum would have to extend for hundreds of thousands of wavelengths in contradiction with its atomistic character. Einstein did not think this objection was insurmountable. It was only necessary to regard light quanta as interacting with one another: "I would like to compare this with the process of molecularization of the carriers of the electrostatic field. The field produced by atomized electric particles is not essentially different from the previous conception, and it is not impossible that something similar will happen in radiation theory.[11]

A month after the Salzburg meeting, Planck wrote to Wien that he had a new approach to radiation that departed even further from Einstein's and Stark's.[12] Planck continued to correspond with Einstein, Lorentz, and others on the topic of light quanta, convinced that the essential discontinuity in physical processes should be located where it does least harm, in the behavior of resonators—or oscillators, as he now called them—rather than in the ether.[13]

At the time of the Salzburg meeting, and for some time after, Einstein tried repeatedly to determine the new equations to replace Maxwell's. They were to describe both the wave and the quantum aspects of light, and, if possible, they were also to describe electrical quanta, or electrons.[14] But he could not determine these equations, and he began to speak of feeling "powerless" in the face of the basic problems of physical theory.[15]

"Today's physical world picture," he wrote in 1911, rests on two sets of fundamental equations, the mechanical equations of material points and Maxwell's equations of the electromagnetic field, and neither set is valid in general but only for slowly varying periodic processes; the "true theory" accounting for quantum processes remained hidden. To look ahead: in 1916 Einstein supplied yet another argument for light quanta. In connection with a new derivation of Planck's blackbody distribution law, in which he introduced a statistical description of atomic emission and absorption of radiation, he established theoretically the directional, as opposed to the spherical, propagation of light. It was an advance in the understanding of

10. Armin Hermann, "Einstein auf der Salzburger Naturforscherversammlung 1909," *Phys. Bl.* 25 (1969): 433–36, on 435. Quotations from Stark and Planck in the discussion following Einstein's talk, in *Phys. Zs.* 10 (1909): 825–26.

11. Quotation from Einstein in the discussion, *Phys. Zs.* 10 (1909): 826.

12. Planck to Wien, 25 Oct. 1909, Wien Papers, STPK, 1973.110.

13. Planck to Lorentz, 7 Jan. 1910 Lorentz Papers, AR. Planck to Wien, 21 Feb. 1910 and 30 Sept. 1915, Wien Papers, STPK, 1973.110.

14. These strenuous efforts did not lead to publication, but Einstein spoke of them in his letters. They are discussed in Klein, "Thermodynamics in Einstein's Thought," 514; and in McCormmach, "Einstein, Lorentz, and the Electron Theory," 81–83.

15. Einstein to Fischer, 5 Nov. 1910, STPK, Darmst. Coll. 1917.141.

quanta, but it did not come closer to connecting them with the wave theory of light.[16]

Quantum Theory, Other Theories, the *Annalen,* and the Solvay Congress

The year 1911 or thereabouts marks three major developments in theoretical physics. That year Einstein's theory of relativity, then six years old, could be characterized by Sommerfeld as one of the secure possessions of physics.[17] That same year Einstein began his sustained work on the theory of gravitation, which he was to complete together with his general theory of relativity several years later. Finally, in 1911 Sommerfeld, Einstein, and a number of other physical scientists came together at the Solvay Congress, marking a shift in interest in the quantum theory and bringing the first phase of the quantum theory to a close. We begin this section by looking at publications on the quantum and other theories in the *Annalen,* and we conclude it with an account of the Solvay Congress.

When in 1900 Drude became editor of the *Annalen der Physik,* Planck continued on as advisor. Their working relationship was good, even if Planck was not always kept as informed as he wished. When Drude published a worthless paper without his knowledge, for instance, Planck was upset; Drude agreed with Planck that the paper was worthless, but the author had appealed to him personally, and Drude had lacked the heart to refuse him. It meant that Drude then had to face the delicate task of limiting the inevitable polemics that followed.[18] Polemics were always regarded with repugnance, and Planck, Drude, and everyone else connected with the *Annalen* took pains to tone them down.[19]

When Drude died in 1906, the question of his successor at the *Annalen* came up. It "interests all of German physics," Quincke said, since the *Annalen* was the leading journal of physics in Germany. Drude had always spoken of Wien as his successor, a choice Planck approved of. Even before Wien's appointment as editor, Planck discussed with him the working relationship he would like. He accepted that the business of the journal had to be done by one man, Wien, but he wanted to be consulted about all revisions and rejections of papers. In general, he wanted a more responsible position this time; he wanted to appear, as he would, side by side with Wien on the title page.[20]

16. Einstein, "Bemerkungen über eine fundamentale Schwierigkeit in der theoretischen Physik," ms. dated 2 Jan. 1911, STPK; "Zur Quantentheorie der Strahlung," *Phys. Gesellschaft, Zürich, Mitteilungen* 16 (1916): 47–62; discussed in Martin J. Klein, "The First Phase of the Bohr-Einstein Dialogue," *HSPS* 2 (1970): 1–39, on 7–8.

17. Arnold Sommerfeld, "Das Plancksche Wirkungsquantum und seine allgemeine Bedeutung für die Molekularphysik," *Phys. Zs.* 12 (1911): 1057–69, on 1057.

18. Planck to Wien, 12 and 15 Oct. 1906, Wien Papers, STPK, 1973.110.

19. The *Annalen's* policy on polemics was that each opponent was allowed two statements. Since the opponents would not have a chance to refute anything after the last opponent had published his second statement, the two often corresponded, wording their positions so that they did not have to fear further objections. Planck to Stark, 8 Apr. 1912, Stark Papers, STPK.

20. Quincke to J. W. Brühl, 10 July 1906, Heidelberg UB, Hs. 3632,8. Planck to Wien, 28 July and 15 Oct. 1906, Wien Papers, STPK, 1973.110.

Planck and Wien generally saw eye to eye, but at the beginning they differed over the question of fixed guidelines for the rejection of papers. In short, Wien wanted guidelines and Planck did not. The rejection rate of the journal was remarkably low, no higher than five or ten percent, but the consequences of any one rejection—or, for that matter, of an acceptance that should have been a rejection—could be unpleasant. Rules for rejections would obviously help at such times, but the rules had to be the editors' own, Planck insisted. If he and Wien went to the journal's curators for guidelines, their freedom would be lost, and they would live to regret it. To Planck, the highest editorial principle was to insure that the *Annalen* contained the most valuable physical research being done, which required the editors' completely "free hand" to make decisions as they went along.[21]

The danger, as Planck saw it, of acknowledging the curators' competence to judge on the acceptance and revision of papers arose when Bucherer submitted a paper on a new relativity principle. In Planck's judgment, the paper deserved rejection, but because Bucherer was an established physicist, a Privatdocent at Bonn, Planck did not want to treat him so abruptly. He advised Wien to ask Bucherer to shorten the paper. Wien did and Bucherer refused, threatening to publish the paper elsewhere together with Wien's critical letter; he had earlier behaved toward Drude in a similar way, and Planck would have liked to see him go. But this is not what happened; to Planck's alarm, Bucherer appealed to the curators, who took the position that their competence did not extend to judging manuscripts, and the affair only ended with Bucherer's decision not to resubmit his paper to the *Annalen*.[22]

On one occasion Planck reviewed a paper so full of errors that he wanted it rejected emphatically to discourage the author from returning to the *Annalen* soon. But he seldom judged an author to be this hopeless, and he seldom rejected a paper outright. Once in a while the *Annalen* received a paper that had no physical interest, only a mathematical one, but even then Planck was unlikely to recommend rejection unless the mathematics itself was objectionable.[23]

Planck overlooked errors, even serious ones, if the author had good training in physics or even in mathematics or if he had given the journal valuable work in the past or was likely to again in the future. When Stark submitted a paper on a valuable measuring method and included in it a worthless theoretical discussion, Planck advised Wien to accept the paper. Stark's "frequently entirely untenable, almost always arbitrary theoretical ideas" were insufficient reason to alienate this "talented and deserving" experimental physicist. Editors could not guarantee the content of papers, Planck reasoned, so no one would reproach them for publishing a paper by someone of Stark's scientific reputation. All things considered, it did not hurt to publish nonsense now and then.[24] Since the long-term interests of the *Annalen* had to be

21. Planck to Wien, 28 July 1906, and 1 and 3 July 1907, Wien Papers, STPK, 1973.110.

22. Planck to Wien, 29 Nov. and 7 and 21 Dec. 1906, and 5 and 26 Jan. 1907, Wien Papers, STPK, 1973.110.

23. Planck to Wien, 19 Feb. 1912, 2 May 1908, 27 Feb. and 30 Nov. 1909, and 30 Sept. 1915, Wien Papers, STPK, 1973.110.

24. Planck to Wien, 21 Oct. 1910, and 14 Jan. and 9 Feb. 1911, Wien Papers, STPK, 1973.110.

constantly kept in mind, the paper at hand was not the sole consideration. Now and then, the journal published bad papers by good physicists.[25]

Although a paper might be partially redeemed by some secondary value, perhaps a "didactic" one, its "scientific significance" was Planck's primary criterion. Planck did not require a paper to resolve every difficulty, but only to be clear about what the difficulties were. Even if a paper was "strongly fanciful and arbitrarily carried through," it could still have significance if it treated some puzzling phenomena in a stimulating way. Molecular and atomic phenomena were still "very much in the dark," yet they were a worthwhile subject. Planck approved a derivation of an atomic radiation formula, for example; while recognizing that its "speculations are in part very daring" and deficient, he said that in this "maiden area" physicists needed freedom, for otherwise they would never find new and better ways. The principle Planck followed was that it is impossible in advance to know if an idea will prove fruitful for later work, and so it is best to be tolerant toward unfamiliar work. "In general," Planck said, "I shun much more the reproach of having suppressed strange opinions than that of having been too gentle in evaluating them."[26]

While Planck edited others' work, he, of course, did work of his own. He tried to keep his convictions about the desirable directions of research separate from the views of the authors he judged—he accepted a certain paper "although (or perhaps better: because)" he disagreed with its viewpoint[27]—though he naturally took more interest in some papers than in others. Interested always, and above all, in questions of foundations, he regarded Paul Hertz's thorough paper on the mechanical foundations of thermodynamics, which built on the work of Gibbs and Einstein, so important for physicists at the time that he accepted its undue mathematical emphasis, which otherwise would have drawn his criticism. Because he had worked on the "Lorentz-Einstein relativity theory," he found "very pretty" an application of relativity theory that yielded Heinrich Hertz's equations for the electrodynamics of moving bodies. Planck's editing, research, and teaching came together at times, as when he asked Wien to return his student Mosengeil's paper on the relativistic thermodynamics of moving bodies, since in the meantime Planck had discovered an error in it through his own research.[28]

Despite the burgeoning theoretical work, the German periodical physics literature did not divide along the lines of theory and experiment. When in 1910 Waldemar von Ignatowsky and Eugen Jahnke planned a "journal for theoretical physics," Planck acknowledged that it would bring relief to the Annalen, freeing it from the "rush of theoretical papers." But Planck did not approve of the sharp separation of theory and experiment, nor did most of his colleagues.[29]

25. Planck to Wien, 10 Nov. and 21 Dec. 1906, 9 May and 19 June 1907, 13 June 1910, and 28 Nov. 1914, Wien Papers, STPK, 1973.110.

26. Planck to Wien, 5 Jan. 1907, 28 Nov. 1914, 10 Nov. 1906, 12 Apr. 1911, 1 Mar. 1916, and 14 Jan. 1911, Wien Papers, STPK, 1973.110.

27. Planck to Wien, 9 May 1907.

28. Planck to Wien, 1 June 1910, 24 May 1907, 9 Oct. 1908, and 1 Feb. 1907, Wien Papers, STPK, 1973.110.

29. Planck to Wien, 13 June 1910.

Planck's conscientious criticism and practical good sense did much to maintain the standing of the *Annalen* among German physicists. During World War I, the proposal was made to discontinue certain physics journals, but evidently not the *Annalen*. Contributors and readers were sometimes unhappy with editorial decisions, but in letters to his co-editor Wien, Planck never had to speak of "enemies" of the journal (as he would after the war for reasons beyond his control).[30]

Contributors to the *Annalen* continued to be drawn overwhelmingly from the universities and technical institutes. Of the hundred or so contributors in 1911 who were working in Germany, five worked at the Reichsanstalt (a connection Planck valued, which if lost could be "disastrous for the *Annalen* in the future"[31]), three or four taught in secondary schools, one worked in the laboratory of a Frankfurt firm and another in a private laboratory in Berlin. The rest, the great majority, were distributed widely among the universities and technical institutes; all twenty German universities were represented—if unequally, Rostock with one contributor and Berlin with ten or more—and five, about half, of the technical institutes were represented. Within these institutions, the contributors were nearly always affiliated with the physics institute or the institute for theoretical or applied physics. The only other significant affiliation was with a physical chemistry institute, usually Nernst's in Berlin.

Externally, the *Annalen* continued to change. It was becoming mammoth; from the age-old twelve issues per volume still in effect at the turn of this century, the *Annalen* had expanded gradually to sixteen issues in 1912, and the third of the three volumes for 1912 alone ran to over sixteen hundred pages. During the following year, it was decided to expand the journal to eighteen issues—seventeen actually came out—because of the overflow of papers.[32] The *Annalen* looked different inside, too; there were more drawings and more photographs of apparatus and of results and, above all, more graphs. Graphs had commonly entered the *Annalen* from the middle of the nineteenth century, when they were relegated to the section of plates at the end of each issue. As the now preferred—more visual—way of presenting observations, graphs often appeared alongside the accompanying tables and formulas in the texts of the papers. And the notation in the *Annalen* looked different; for example, in recent years, vectors had become so standard that they were used for directed quantities—magnetic fields, electric oscillations, optic axes, and the like—even in nonmathematical experimental papers.[33] The visual aspect of the *Annalen* generally reflected the quantitative nature of experimental work and the experi-

---

30. Unlike the *Verhandlungen* of the Physical Society, which published preliminary communications of research, the *Annalen's* work was complete and definitive, and a journal for that purpose was indispensable. Planck to Wien, 23 June and 1 July 1907. Planck said that if he were asked to decide between the *Physikalische Zeitschrift* and the *Fortschritte der Physik* on the one hand and the Physical Society's *Verhandlungen* and the *Annalen's* abstracting supplement the *Beiblätter* (the two pairs of journals having overlapping functions), he would prefer the former because of the association of the *Fortschritte* with the Physical Society. Planck to Wien, 5 Oct. 1917 and 20 June 1920, Wien Papers, STPK, 1973.110.

31. Planck to Wien, 23 June and 1 July 1907.

32. Planck to Wien, 24 Oct. 1913, Wien Papers, STPK, 1973.110.

33. For example, du Bois and Elias, 35:617–78.

menters' competence in mathematical manipulation, though not every experiment and experimenter could be characterized this way.[34]

If we look at other German journals in and around 1911, our picture of theoretical physics does not markedly change. The *Physikalische Zeitschrift* and the *Verhandlungen* of the German Physical Society published a great many physics papers, most of which were shorter than the *Annalen*'s. Neither journal contained as many pages per year as the *Annalen*, but the *Verhandlungen* was becoming large, its volume for 1911 running to over 1150 pages of text and over 450 pages of bibliography. The *Physikalische Zeitschrift* published a good deal of theoretical physics around 1911, as it had from the beginning. The *Verhandlungen* at this time contained relatively little theoretical physics, only one out of every eight to ten papers; yet five years before, it had not contained any highly mathematical work at all, while only three years later, in 1914, it would be filled with work of that kind. By then, the *Verhandlungen* as well as the *Physikalische Zeitschrift* published many papers that would have been judged too narrowly mathematical in interest by the *Annalen* editors. By contrast, Stark's *Jahrbuch der Radioaktivität und Elektronik* was given over almost entirely to experimental work; the volumes for 1911 and 1912 contained only one paper, another review article on relativity, with any mathematical content to speak of.

The *Annalen* for 1911 and 1912 contained frequent references to the "classical" theory, which could mean a number of things. When Planck spoke of the "classical electrodynamics and electron theory," it was to distinguish them from the physics centered on the quantum concept. When Laue spoke of the "classical elasticity theory" or Gustav Herglotz of "classical mechanics," it was to distinguish these theories from relativity theory. But often the characterization of a theory as "classical" had nothing to do with the recent quantum and relativity physics. Readers of the *Annalen* were not led to think of the recent innovations of Planck and Einstein when they came across an expression such as the "classical theory" of friction. The reference was to what was standard theory and not to the physics that had preceded quanta, which became the later common and narrower meaning.[35]

Much theoretical research entering the *Annalen* in 1911 and 1912 did not touch on quanta and relativity at all. Physicists addressed, for example, long-standing atomic and molecular problems such as the surface tension of fluids and the equation of state of solids.[36] They analyzed the foundations of thermodynamics by the statistical mechanics of atoms and molecules.[37] They speculated on the arrangement, shape, and structure of atoms and molecules and on the law and nature of molecular

34. Lehmann gave many drawings of what he saw in liquid crystals using polarized light, but he gave no graphs, tables, or equations (39:80–110).

35. Planck, 37: 642–56, on 643; Laue, 35: 524–42, on 526; Herglotz, 36: 493–533, on 493; Sackur, 34: 455–68, on 456 and 467; Witte, 34: 543–46, on 543; Reiger, 34: 258–76, on 276.

36. For example, Einstein, 34: 165–69; Grüneisen, 39: 257–306.

37. For example, Otto Sackur applied the "Boltzmann-Planck views" on the relation of entropy and probability to irreversible chemical processes (36: 958–80); Einstein acknowledged the force of Paul Hertz's criticisms of his early papers on the mechanical foundations of thermodynamics, which he would not have published if he had known of Gibbs's book on the subject (34: 175–76); Jan Kroò and Ludwig Silberstein disagreed on the subject of Gibbs's statistical mechanics (34: 907–35; 37: 386–92; 38: 885–87).

force, although they still rarely introduced detailed atomic models and might even be skeptical of them in principle.[38]

Annalen authors showed an ongoing interest in the electron, criticizing, extending, and testing the electron theories of metals, of heat conduction, of thermoelectricity, of magnetic splitting of spectral lines, and of various other magneto-optical phenomena.[39] In addition to their interest in applications like these, they continued to take an interest in the relationship of the electron theory to the electromagnetic foundations of physics. One of their studies, for example, proposed a new electrical fundamental law to eliminate the need for nonelectric forces to account for a finite electron, a basic objection to the "purely electromagnetic foundation of the electron theory." Mie offered a new theory of matter, which was an ambitious extension of the electron theory and which occupied a good deal of space in the *Annalen* in 1912 and 1913. Assuming that experiment had shown that the laws of mechanics and Maxwell's electromagnetic equations are invalid within the interior of atoms, Mie did not look to further experiments to uncover the new laws to put in their place; rather he looked to new foundations of physical theory, which for him meant a new theory of the world-ether. Mie assumed that atoms consist of electrons bound by positive charges and that electrons are not foreign bodies in the world-ether but are special states of it, from which he deduced the fundamental equations of the new ether physics. The equations described all phenomena of the material world in terms of a "world-function," the determination of which Mie regarded as the outstanding problem of the physics of the future.[40]

Along with the electron theory, theoretical optics was extensively investigated and tested in publications in the *Annalen* in 1911–12. The main reason for this was Voigt's continuing interest, which he imparted to students working in his Göttingen institute. One of the researches carried out under his direction applied his method of determining the optical constants of metals, and its author presented it as a continuation of research by three other investigators, all of whom had also worked in Voigt's institute.[41] Proposed or inspired by Voigt or performed by Voigt himself, researches appeared in the *Annalen* during these years on a range of topics, which included metal inflexion, total reflection, dispersion, and emission and absorption of

38. Beckenkamp, 39: 346–76; Schiller, 35: 931–82; W. E. Pauli, 34: 739–79. As Arthur Schidlof did the year before, Peter Lebedew referred to J. J. Thomson's atom model, here to explain the magnetic properties of rotating bodies (35: 90–100; 39: 840–48). In place of Maxwell's inverse fifth-power law of the force of molecular repulsion, Paul Gruner proposed the inverse square law, since it was the law of force of the mutual repulsion of electrons in the electron theory (35: 381–88); building upon Mie's theory, Eduard Grüneisen constructed a "theoretical model" of monatomic solids from interatomic forces of attraction and repulsion obeying the law $-a/r^x + b/r^y$, the exponents of which were to be determined by experiment (39: 257–306); extending van der Waals's theory, Maximilian Reinganum represented molecular forces by electrostatic bipoles on molecules (38: 649–68). "For my part," Voigt said, "I have always been somewhat skeptical of molecular models"; he preferred to set up equations of a theory from certain symmetries and principles, an abstract approach that he justified by the present tendency of physicists to renounce any "insight into the mechanism of the process" (36: 873–906, on 897, 899).

39. Johann Koenigsberger and Weiss, 35: 1–46; Cermak and Schmidt, 36: 575–88; Schneider, 37: 569–93; Baedeker, 35: 75–89; Voigt, 35: 101–8; du Bois and Elias, 35: 617–78.

40. Wolff, 36: 1066–70, on 1066; Mie, 37: 511–34, and 39: 1–40.

41. Erochin, 39: 213–24.

light.[42] The identification of the Göttingen theoretical physics institute with optical researches is suggested by the number of foreign physicists there who carried them out; in 1911–12, at least four of these visitors published on optics in the *Annalen*.[43]

The *Annalen* contained a good many other theoretical researches in optics during these years.[44] Some of them were highly mathematical, such as a study of the time-independent wave equation, which Voigt proposed and Sommerfeld advised. Another study offered a simple solution to a familiar mathematical problem, the determination of the refraction of a plane wave at a cylindrical surface by Huygens's principle. Another gave a rigorous solution to the problem of diffraction of light by a perfectly conducting right-angle wedge. Another, which Sommerfeld and Runge carried out, reformulated the laws of geometrical optics by representing a light ray by a unit vector and by introducing the gradient, divergence, and other operators from vector calculus.[45]

The *Annalen* contained theoretical studies on gases, van der Waals's equation of state, thermodynamic equilibrium,[46] electric discharges in gases, elasticity, applications of physics,[47] and other topics as miscellaneous as gamma rays and sand dunes, the latter of which were analyzed with the help of Helmholtz's theory of vortex motion and experimentally reproduced.[48] As usual, then, in addition to clustering around certain subjects, theoretical researches in the *Annalen* in 1911–12 extended to most parts of physics.

Papers in the *Annalen* in 1911–12 dealing with the quantum theory provided experimental physicists and physical chemists with results they could work with directly. The theory entailed experimental consequences for heat in solids, radiant heat, the interaction of radiation with bodies, and, though still a minor theme, the dynamics of radiation and atoms.

Although the quantum theory promised to be as pervasive, and as fundamental, in physics as relativity theory, researches on the quantum and on relativity were usually distinct. Problems that explicitly engaged both did not appear in the *Annalen* in these years, and the researches appearing there having to do with the quantum theory were less mathematical, more provisional, and more numerous than researches having to do with relativity.

42. Pogány, 37: 257–88; Ignatowsky and Oettinger, 37: 911–22; Ladenburg, 38: 249–318; Voigt, 34: 797–800, and 39: 1381–1407.

43. Fréedericksz, 34: 780–96; Brotherus, 38: 397–433; Erochin, 39: 213–24; Wali-Mohammad, 39: 225–50.

44. Jentzsch, 39: 997–1041; Harnack, 39: 1053–58; Pahlen, 39: 1567–89.

45. Wiegrefe, 39: 449–84; Reiche, 34: 177–81, and 37: 131–56; Sommerfeld and Runge, 35: 277–98.

46. Smoluchowski, 35: 983–1004; Heindlhofer, 37: 247–56; Dieterici, 35: 220–42; Pochhammer, 37: 103–30; Tammann, 36: 1027–54.

47. Greinacher, 37: 561–68; Seeliger, 38: 764–80; Witte, 34: 543–46; Reinstein, 35: 109–44; Möller, 36: 738–78; March, 37: 29–50.

48. Einstein, 34: 165–69; Hertz, 37: 1–28; Lenz, 37: 923–74; Buchwald, 39: 41–52; Bechenkamp, 39: 346–76; Hahmann, 39: 637–76; Merczyng, 39: 1059–69; Koláček, 39: 1491–1539; Esmarch, 39: 1553–66.

Much of the work on the quantum theory in 1911 and 1912 took its starting point in Einstein's theory of specific heats of solids, which he first put forward in 1907. In this theory Einstein argued that the energy not only of resonators and oscillating ions but of all atoms must be quantized. That is, the quantum was not limited to radiation phenomena; it pointed to the need for a new mechanics of ponderable matter on the atomic scale. Little attention was paid to this new direction of Einstein's until around 1911. The reason for the attention then was primarily Nernst's measurements of specific heats, which he believed confirmed Einstein's theory in the "most brilliant way." For Nernst the central point of Einstein's theory was its prediction that at low temperatures the atomic heats of solids fall below the values predicted by the older law of atomic heats, behavior which agreed with Nernst's new heat theorem. In 1911 Nernst and his co-worker F. A. Lindemann improved on Einstein's law, and Debye improved on it further in 1912.[49]

In addition to researches on specific heats that took into account Einstein's theory,[50] other researches recognized a connection between Einstein's theory and the goal of understanding the solid state of matter.[51] Einstein himself was concerned to show the failure of the old "molecular mechanics" not only for the specific heats of solids but also for their heat conductivity. For a full understanding of these and of the optical and elastic phenomena of solids, Einstein looked toward a complete molecular theory of solids.[52] His theory of specific heats was the first step in that direction, "the basis for a *molecular statics*" of solids, according to Arnold Eucken, who was then working in Nernst's laboratory on experiments designed to build a molecular dynamics of solids.[53]

Einstein's other major work in the quantum theory, his light-quantum hypothesis, received scant attention in the *Annalen*. Ehrenfest analyzed the place of light quanta in Planck's radiation theory; Edgar Meyer noted the hypothesis only to say that it did not agree with his experiments on gamma rays; and Abram Joffé said that the attempt to explain the phenomena of photoelectricity, photochemistry, photo-ionization, and fluorescence by light quanta had not been carried through quantitatively and that it was impermissible to speak of "energy quanta" in Einstein's sense.[54]

Planck's law of blackbody radiation continued to enter researches in the *Annalen* in 1911 and 1912. The law was tested for ever shorter wavelengths, a region in which it coincides with Wien's law, and it entered discussions of universal constants

---

49. Nernst, 36: 395–439, on 423; Einstein, 35: 679–94; Debye, 39: 789–839. Working in the Berlin physical chemical laboratory, Fritz Koref made many of the measurements required for a confirmation of the specific heat law of Nernst and Lindemann (36: 49–73).

50. Franz Richarz's "theoretical considerations" on atomic heats, forming a "necessary complement" to Einstein's (39: 1617–24, on 1622–23), were experimentally tested by Oskar Richter (39: 1590–1608).

51. For Sackur, not only did Einstein's theory of specific heats together with the relation between entropy and probability yield Nernst's heat theorem as a "*necessary* consequence," but Einstein's approach promised a "complete kinetic theory of solids" (34: 455–68, on 457, 467); Simon Ratnowsky developed the equation of state of monatomic solids by applying Planck's quantum theory of resonators to the atoms of solids in the spirit of Einstein's theory of specific heats (38: 637–48).

52. Einstein, 35: 679–94, on 680 and 694; 34: 170–74.

53. Eucken, 34: 185–221.

54. Ehrenfest, 36: 91–118; Meyer, 37: 700–720; Joffé, 36: 534–52.

and other matters.[55] The most significant work on Planck's law at this time was Planck's own. He was dissatisfied with his previous derivation of the law; it depended on an electrodynamic part, in which the energy of an oscillator is treated as continuously variable, and a statistical part, in which the energy is treated discontinuously as a multiple of the elementary quantum. He removed this contradiction in a new derivation in which he assumed that linear oscillators in the walls of the blackbody absorb radiation continuously, but that they can emit it only when the energy of oscillation is a multiple of the finite energy element. The new derivation violated the "classical electron theory" but only in the region of the oscillator. The processes causing the emission of radiation Planck assumed to be still "hidden," so that a statistical approach had to be taken in determining their laws.[56]

Planck's constant h entered a considerable range of problems in papers appearing in the *Annalen* in 1911–12.[57] Attributing to the constant a "universal electrodynamic significance," Arthur Schidlof (as A. E. Haas had earlier) related Planck's oscillators to J. J. Thomson's atomic model.[58] Planck himself thought that his constant was probably related to other "atomic constants," and soon quantum research was to be largely directed by atomic theories.[59]

When we turn to the *Physikalische Zeitschrift* and the German Physical Society's *Verhandlungen* around 1911, we find a good deal of theoretical work in the older fields; for example, in classical point-mechanics and even in the mechanical representation of electromagnetic processes. We find there other work that acknowledged, implicitly or explicitly, the difficulties of the "classical" theories and of the "mechanical world picture."[60] We find work on the relativity and quantum theories, again bearing out Sommerfeld's observation that the quantum theory was the less settled of the two. By and large, physicists writing for these journals assumed the validity of the relativity principle and applied it to a variety of dynamical problems, whereas they had as yet no quantum "principle" in which they placed comparable trust. The Berlin physical chemist Fritz Haber, for example, noted that the physical nature of Planck's constant remained unknown since it had not been derived from mechanics or electrodynamics; he went ahead to apply the quantum of energy to the emission of electrons in chemical reactions, drawing testable consequences, and coming away with the "feeling" that h belongs more to the discrete atoms of chemistry than to the continuous processes of physics. Hasenöhrl generalized Planck's earlier quantum hypothesis to allow the frequency of the resonator to be a function

55. Baisch, 35: 543–90; Suchý, 36: 341–82; Paschen, 38: 30–42.
56. Planck, 37: 642–56, on 644.
57. In addition to the problems already mentioned, they included chemical dissociation by light (Einstein, 37: 832–38, 38: 881–84, 888; Stark, 38: 467–69), chemical constants of gases (Tetrode, 38: 434–42), and the production of X rays (Sommerfeld, 38: 473–506).
58. Schidlof, 35: 90–100.
59. Planck, 37:642–56 on 656.
60. Emil Budde, "Zur Theorie des Mitschwingens," *Verh. phys. Ges.* 9 (1911): 121–37; Arthur Korn, "Weiterführung eines mechanischen Bildes der elektromagnetischen Erscheinungen," *Verh. phys. Ges.* 9 (1911): 249–56; Friedrich Hasenöhrl, "Über die Grundlagen der mechanischen Theorie," *Verh. phys. Ges.* 9 (1911): 756–65.

of its energy, which enabled him to discuss the production of line spectra in hydrogen atoms. Eduard Wertheimer replaced Planck's oscillator with an electron rotating about a positive center, a picture that had greater "reality" for him.[61] These and other writers on the quantum theory recognized the rudimentary state of understanding of the physical world on the molecular level and—what was especially important for the development of theoretical physics—the growing number of successes of the quantum theory. In his address to the French Physical Society in 1911, printed in the *Physikalische Zeitschrift*, Planck enumerated the successes, as he saw them, of the quantum hypothesis. It had led not only to the law of blackbody radiation but also to a method for calculating the elementary constants of electricity and magnetism and to an understanding of Nernst's heat theorem, photoelectric and other cathode-ray emissions, and even radioactivity. He acknowledged that physicists knew so "enormously little about the processes within a molecule" that they had a good deal of freedom in choosing hypotheses. Whether or not the quantum hypothesis revealed the "'whole' truth," Planck was confident that it was more successful than the previous explanations.[62]

The Solvay Congress of 1911 pointed up the significance of theoretical questions for physical research in the period we are considering. It was the prototype of the intimate, international meetings that would increasingly attend the latest advances in physics. The Solvay Congress was called into being because of certain troubling theoretical questions raised in recent experimental work in the physics of heat. The instigator and major planner, in short, the "soul," of the congress was Nernst, who had recently taken an interest in the quantum theory, especially in Einstein's quantum theory of specific heats, as we found in our examination of the contents of the *Annalen*. The expenses of the congress were met by the Belgian industrial chemist Ernest Solvay, who took an amateur's interest in physical theory. The members were put up for several days in the elegant Hotel Métropole in Brussels, where the discussions took place.[63]

61. Fritz Haber, "Über den festen Körper sowie über den Zusammenhang ultravioleter und ultraroter Eigenwellenlängen im Absorptionsspektrum fester Stoffe und seine Benutzung zur Verknüpfung der Bildungswärme mit der Quantentheorie," *Verh. phys. Ges.* 13 (1911), 1117–36. Haber to Sommerfeld, 29 Dec. 1911, Sommerfeld Correspondence, Ms. Coll., DM. Hasenöhrl, "Über die Grundlagen der mechanischen Theorie"; Eduard Wertheimer, "Die Plancksche Konstante h und der Ausdruck $h\nu$," *Phys. Zs.* 12 (1911): 408–12.

62. Max Planck, "Energie und Temperatur," *Phys. Zs.* 12 (1911): 681–87, on 686–87. In addition to those enumerated here, there had been other applications of the quantum hypothesis; for example, Einstein's application to Stokes's rule for fluorescent radiation, Wien's to radiation from canal rays and to the determination of wavelengths of X rays, and others we have mentioned, such as Stark's and Haas's.

63. "Soul" was used by Einstein in his acceptance letter to Nernst, 20 June 1911, quoted in the unpublished manuscript by J. Pelseneer, "Historique des Instituts Internationaux de Physique et de Chimie Solvay," 12. A copy of the manuscript is in the AHQP. The following account of the first Solvay Congress is taken largely from Russell McCormmach, "Henri Poincaré and the Quantum Theory," *Isis* 58 (1967): 37–55, on 38–43, which in turn is indebted to Klein, "Einstein, Specific Heats, and the Early Quantum Theory." For further discussion of the congress and its consequences, see Leon Rosenfeld, "La première phase de l'évolution de la Théorie des Quanta," 186–93; Maurice de Broglie, *Les Premièrs Congrès de Physique Solvay* (Paris: Albin Michel, 1951); Max Jammer, *The Conceptual Development of Quantum Mechanics* (New York: McGraw-Hill, 1966), 52–61; and Kuhn, *Black-Body Theory*, 219–20, 226–28, 252.

If the congress had been restricted to only those scientists who had demonstrated a major concern with the quantum theory, it would have been a small gathering. For this reason, Planck had had reservations when Nernst first approached him on the subject in 1910. As Planck could think of only four other well-known physicists who had shown much interest, he encouraged Nernst to wait a couple of years until the older theory had given way and the whole situation had become "*intolerable* for every true theorist." Then, he reasoned, the meeting would have a chance of success.[64] Nernst went ahead anyway. The understanding of the qualifying concerns of the members was broader than Planck's first thoughts on the matter: of the nearly two dozen members, only about half were predominantly theorists, of whom Planck, Einstein, and Lorentz had done the most influential, original, and critical work on the quantum problem. The experimentalists who constituted the other half included leading researchers in atomic phenomena such as Rutherford and Marie Curie and leading researchers in heat phenomena such as the Dutch physicist H. Kamerlingh Onnes and Rubens. Planck was not disappointed; he left the congress speaking of those "scientifically so stimulating days in Brussels."[65]

Some members of the congress were invited to give reports on specific topics, which were circulated in advance of the meeting. Other members were invited to come to discuss the problems. The groundwork for the reports and discussions was well laid by Nernst and those he consulted, notably Planck and Lorentz. There was a thorough review of the older theories and their disagreement with experiment together with a presentation of the several applications and interpretations of the new theory. Altogether, the presentations made a case for a dislocation of physical theory more serious than many of the members of the congress had imagined beforehand.

In the letter inviting the participants, Nernst contended that the acceptance of quanta would require "necessarily and incontestably a vast reform of the current fundamental theories."[66] None of the members took issue with this contention, which the discussions reinforced.[67] The responsibility of the congress, according to

64. Planck to Nernst, 11 June 1910, in Pelseneer, "Historique des Instituts Internationaux de Physique," 6–7. At the time, Planck had good reason to believe in the very limited extent of interest in the quantum theory among leading theorists. Besides Nernst and himself, he could think of only four others with sufficient interest to justify a conference: Einstein, Lorentz, Wien, and Larmor, and he had probably overestimated the interest of Larmor, who turned down his invitation because he had not kept up with recent progress in the quantum theory. Pelseneer, 14.

65. The participants in the "scientific council (a kind of private congress)" were the following: from Germany, Nernst, Planck, Rubens, Sommerfeld, Warburg, and Wien; from England, James Jeans and Ernest Rutherford; from France, Marcel Brillouin, Marie Curie, Paul Langevin, Jean Perrin, and Henri Poincaré; from Austria, Einstein and Hasenöhrl; from Holland, H. Kamerlingh Onnes and Lorentz; from Denmark, Martin Knudsen. Maurice de Broglie from France, F. A. Lindemann from Germany (originally from England), and R. Goldschmidt from Belgium were the congress's scientific secretaries. Rayleigh from England and van der Waals from Holland did not attend, but they were listed as official participants in the council. From the proceedings of the congress, *La théorie du rayonnement et les quanta*, ed. Paul Langevin and Maurice de Broglie (Paris: Gauthier-Villars, 1912). Planck to Wien, 8 Dec. 1911, Autograph Collection I/285, STPK.

66. Nernst's draft together with Ernest Solvay's invitational letter, 15 June 1911, Lorentz Papers, AR.

67. To a few members who joined in a final discussion, however, there appeared to be one unexplored possibility for avoiding a fundamentally discontinuous process in physics. That had to do with

its president Lorentz, was to examine the imperfections of the old theories, to determine the degree of likelihood of the various quantum hypotheses, and to try to
foresee the "future mechanics."[68]

The Solvay Congress discussions naturally dealt at length with the quantum
theories of blackbody radiation, including Planck's new theory, and of specific heats.
At varying lengths, they dealt with a good many other subjects, too, which included
the extension of the quantum to complicated mechanical systems of several degrees
of freedom and to rotational motion, such as atoms and molecules supposedly possess, and to the structure of atoms and of light. These were all subjects that would
become increasingly important to the further development of quantum physics.

It was Lorentz's understanding that most of the work on the quantum theory
still lay ahead. In his introductory address on the "important questions that will
occupy us," he explained that they are *important* because they touch on the very
principles of mechanics and the most intimate properties of matter." Perhaps—he
added, "let us hope it will not be so"—even "the fundamental equations of electrodynamics and our ideas on the nature of the ether, if one is still permitted to use
this word, will be a bit compromised."[69] Einstein in his report at the congress responded to Lorentz's important questions: ". . . our electromagnetism cannot any
more than our mechanics be put in accord with the facts,"[70] an understanding Einstein had reached long before the congress. Others at the congress, for whom the
quantum theory was less familiar, were not ready to go so far. The French physicist
Marcel Brillouin thought that his conclusion might "seem rather timid to the younger among us." Recalling the enormous number of facts that classical physics could
account for so well, he hesitated to "overthrow the foundations of classical electromagnetism and mechanics instead of being content to adapt the new discontinuity
to the old mechanics." But the reports and discussions of the congress had left Brillouin in little doubt that the quantum theory entailed something new and important:
*"It appears certain that from now on it will be necessary to introduce into our physical and
chemical concepts a discontinuity, an element varying by jumps, which we had no idea of
a few years ago."*[71] That was the principal lesson of the quantum theory as of 1911,
as it became known through the French-language publication of the congress's proceedings the following year. That publication did much to acquaint physicists with
the as yet little-known quantum theory, and because it included the discussions as
well as the reports of the congress, it provided a highly stimulating introduction to
the theory at this early stage.[72]

Nernst's proposal of a generalization of the law relating mass to velocity, which might account for the
new facts and still retain a continuous variation of energy. If this had worked—Poincaré determined soon
after the congress that it did not—it would have resolved the current crisis of physical theory by drawing
upon another recent, fundamental change in physical theory. "Conclusions générales," *Théorie du rayonnement,* 451–54.

68. "Discours d'ouverture de M. Lorentz," *Théorie du rayonnement,* 6–9, on 8.

69. "Discours d'ouverture de M. Lorentz," 6.

70. Einstein, "Rapport sur l'état actuel du problème des chaleurs spécifiques," in *Théorie du rayonnement,* 407–35, on 428.

71. "Conclusions générales," in *Théorie du rayonnement,* 451.

72. Born, who had not attended the Solvay Congress, gained a vivid impression of it from the
published proceedings, which he reviewed in *Phys. Zs.* 15 (1914): 166–67:

Two years after the Solvay Congress, Nernst's collaborator Eucken brought out a German translation of the proceedings. To it he appended a history of the quantum theory in the intervening period. The applications of the quantum had greatly expanded, Eucken reported, but there had been little progress in revealing the true foundations of the theory, which made this belated German edition still valuable for physical research. Eucken discussed the basic question of the foundations of the quantum theory: did the theory lie inside mechanics or, by reason of the discontinuities it contained, outside it, or might there be a more general mechanics containing discontinuities and yet be expressible by differential equations? Physicists had not come to agreement on the answer to this question.[73]

Wien, who had attended the Solvay Congress, lectured on recent problems in theoretical physics at Columbia University in the spring of 1913. He spoke of the remarkable success of the quantum theory, and he urged American physicists to take up the subject, which until then had borne the trademark "made in Germany."[74] That trademark could not be attached to the quantum theory much longer. The Solvay Congress with its published proceedings together with the shift of research interest from blackbody radiation to specific heats and other more commonly treated subjects of physics and physical chemistry resulted in a sharp increase, beginning in 1911–12, in the number of authors publishing on quantum topics, and these included an increasing proportion of physicists outside of Germany.[75]

Relativity Theory and the *Annalen*

By 1911 the theory of relativity was widely accepted, but not all of the physicists who accepted it also accepted the implications Einstein drew from it. That is clear from the papers appearing in the *Annalen* around that time; Laue, for example, might follow Einstein in not using the language of the ether, but many other physicists went on speaking of ether waves, ether permeability, and the like. Mie was not alone in assuming the validity of relativity theory in his "ether physics."[76]

The thought of the high-minded Maecenases of physics . . . has remained alive in the published reports. From the full record of the discussions, which concluded the addresses, one acquires insight into the way of thought and work of the men whose minds gave birth to the new revolutionizing views of theoretical physics. In these addresses and discussions, the personalities of the speakers do not retreat behind the subjects, as is otherwise customary in scientific literature. For everyone who takes an active part in the development of science, there is a great appeal in this; for as long as the foundations of the quantum theory are still as unclear as they are today, opinion for or against the new theory depends in no small measure on personal inclinations and views.

73. *Die Theorie der Strahlung und der Quanten. Verhandlungen auf einer von E. Solvay einberufenen Zusammenkunft (30. Oktober bis 3. November 1911). Mit einem Anhange über die Entwicklung der Quantentheorie vom Herbst 1911 bis zum Sommer 1913*, trans. Arnold Eucken (Halle: W. Knapp, 1914), foreword and 372–73.

74. Wilhelm Wien, *Vorlesungen über neuere Probleme der theoretischen Physik, gehalten an der Columbia-Universität in New York im April 1913* (Leipzig and Berlin: B. G. Teubner, 1913), foreword and 76.

75. Kuhn, *Black-Body Theory*, 216, 229.

76. Laue, 38: 370–84; Suchý, 36: 341–82; Weber, 36: 624–46; Beckenkamp, 39: 346–76; Abraham, 38: 1056–58; Mie, 37: 511–34.

The recognition of the theory of relativity did not mean that its development was complete; it continued to be one of the principal themes of the *Annalen*. With Planck's and Minkowski's work in 1907 and 1908, the dynamics of relativity had been largely worked out, but that left a good many problems that still were not.[77] The "new dynamics," as opposed to the "classical," was characterized by the dependence of the mass of a body on its speed; Einstein derived the dependence from relativity and Newton's laws of motion, but by now it had been derived in several other ways: by the electron theory, by the theory of the electromagnetic field, and in 1912 by an energy axiom of Duhem's.[78] Laue and others extended the new dynamics to the treatment of deformable bodies in 1911; they showed, for example, that Helmholtz's and Kirchhoff's laws for fluids of least compressibility, Helmholtz's law of vortex motion, and other results from the older hydrodynamics carried over into relativistic theory.[79] The science of statics, too, was given a relativistic formulation.[80] In addition, statistical mechanics was reformulated: to express the entropy of a gas in terms of probability, a "definite mechanics" had to be chosen, relativistic instead of Newtonian, and just as relativistic mechanics passes over to Newtonian mechanics for "small" velocities, relativistic gas theory passes over to ordinary gas theory for "low" temperatures, below, that is, about a billion degrees.[81]

It was in the nature of relativity theory that work on it tended to have a general character. Relativistic hydrodynamics, for example, was derived without reference to any details from the "old mechanics" beyond Hamilton's principle.[82] Repeatedly the Lorentz transformations were derived kinematically without reference to electrical concepts.[83] The same expression for the four-vector force that Minkowski, Sommerfeld, and Abraham had applied in electrodynamics, Laue extended to "each area of physics."[84] Relativity theory did not belong to electricity or mechanics or any other one area, but to physics as a whole: physicists publishing in the *Annalen* in 1911 and 1912 undertook to cast their theories from whatever source into relativistic form. Despite the highly mathematical nature of this work, they understood that they were addressing the "physical reader," concerned to make the theory and its supporting experiments as clear and visual as possible.[85]

In addition to recasting the older theories of physics to agree with relativity, physicists writing for the *Annalen* further developed the theory of relativity itself. Einstein had begun to speak of the "ordinary" or "old" relativity theory, which he characterized by the constancy of the velocity of light and the invariance of physical

77. One problem was to derive Planck's formulas for the energy and momentum of a moving body from Minkowski's equations of motion (Schaposchnikow, 38: 239–44); another was to express relativistically the laws of motion of rigid bodies (Ignatowsky, 34: 373–75).

78. Frank, 39: 693–703.

79. Laue, 35: 524–42; Herglotz, 36: 493–533; Lamla, 37: 772–96.

80. Epstein, 36: 779–95.

81. Jüttner, 34: 856–82.

82. Lamla, 37: 772–96, on 785.

83. By Ignatowsky in 1910, Wiechert in 1911, and Frank and Rothe in 1912 (Frank, 39: 693–703).

84. In agreement with Planck and Einstein, Laue argued that for each area of physics, there is a world-tensor, the divergence of which yields the corresponding four-vector ponderomotive force (35: 524–42, on 528–29).

85. Frank, 35: 599–606; Laue, 38: 370–84.

laws for observers moving uniformly with respect to one another.[86] In a series of papers in 1911–12, which we discuss below, Einstein extended relativistic considerations to observers moving with uniform relative acceleration; by appealing to Hamilton's principle, he was able to show that the laws of motion acquire a simplicity and a significance in his new theory that they do not have in Newtonian mechanics.[87]

Abraham too recognized that a generalized relativity theory incorporating gravitation was needed if the theory was to lead to a "complete world picture."[88] But Abraham was fundamentally opposed to Einstein's theory, which led to an exchange between them in the *Annalen* in 1912. Abraham conceded that in the six years from 1905 to 1911, "yesterday's relativity theory" had achieved some lasting results, especially the law of the inertia of energy, but he rejected the idea of relativistic spacetime, preferring to view the omnipresent gravitational field as an absolute reference system and an "argument for the 'existence of the ether.'" He believed that until now Einstein's relativity theory had threatened the "sound development of theoretical physics" through its "dogmatic certainty." Einstein defended the theory as an advance in theoretical physics, and he pointed to the coming work of generalizing it.[89]

Relativity theory proved inexhaustibly fascinating to the *Annalen*'s mathematically inclined contributors, and Planck eventually had to call something of a halt to it. In 1911 he told Wien that they should divert papers on relativity dealing with "formulation, illustration, definitions (rigid bodies!)" to the *Physikalische Zeitschrift* or to mathematics journals, which would leave more room in the *Annalen* for "real physical investigations." At this time Planck was tempted to publish a naively critical paper if only to let an opponent of relativity have a say in the *Annalen* for a change.[90]

Einstein on the General Theory of Relativity and Cosmology

Einstein, whose work on the quantum theory had led to the idea of the Solvay Congress, found the Brussels meeting interesting, but he left it feeling he had learned nothing new. He wrote to a friend soon after: "I didn't get any further with the electron theory. In Brussels, too, the failure of the theory was noted with lamentations but without finding a remedy. In general, the congress there was like a lamentation on the ruins of Jerusalem. Nothing positive was accomplished."[91] In another letter at the time, he gave the reason why: ". . . nobody knows anything."[92] From the meeting he returned to Prague to lecture on the "foundations of our poor dead mechanics, which is so beautiful," and to look for its successor, a prob-

86. Einstein, 35: 898–908, on 899.
87. Einstein, 35: 898–908; 38: 355–69; 38: 443–58; 38: 1059–64.
88. Abraham, 38: 1056–58, on 1056.
89. Abraham, 38: 1056–58; 39: 444–48; Einstein, 38: 1059–64.
90. Planck to Wien, 9 Feb. and 30 May 1911, Wien Papers, STPK, 1973.110.
91. Einstein to Besso, 26 Dec. 1911, in *Einstein-Besso Correspondance*, 40–42, on 40.
92. Einstein to Heinrich Zangger, 15 Nov. 1911, EA; this passage is quoted in Banesh Hoffmann with Helen Dukas, *Albert Einstein: Creator and Rebel* (New York: Viking, 1972), 98.

lem he was "slaving over endlessly."[93] Almost everywhere in physics he looked—in mechanics, the electron theory, the theory of radiation—the foundations were shaken and the new ones were yet to be decided.

The Solvay Congress was primarily directed to questions concerning the very small in nature, atoms, electrons, oscillators, quantum elements of energy or action, light quanta, and the like. It was not concerned with gravitation, a subject that was traditionally associated with the very large in nature. By the time of the congress, however, Einstein was hard at work on the problem of gravitation, for him a subject that bore on the foundations of physics no less than did the concerns of the Brussels meeting.

Soon after he had completed his relativity theory of 1905, Einstein looked for a generalization of Newton's gravitational law in keeping with the new kinematical requirements on the mathematical form of natural laws. But he found that his results conflicted with the equality of inertial and gravitational mass, which had received extremely precise experimental confirmation from the Hungarian physicist R. von Eötvös. Gravitation had to find a place within the framework of relativity theory, which was to prove a much more difficult task than Einstein at first envisioned.

"For the time being we do not have a complete world picture that corresponds to the principle of relativity," Einstein said in 1907.[94] By invitation that year, he wrote a long review article on relativity theory for Stark's annual, *Jahrbuch der Radioaktivität und Elektronik*. He concluded it with a "new relativity-theoretical treatment of acceleration and gravitation," in which he extended the principle of relativity to apply to systems moving with uniform relative acceleration. He made an assumption, later called the "equivalence principle," which provided the foundation of his gravitational theory from its inception: from the observation that all bodies fall with the same acceleration regardless of their nature, he assumed the complete physical equivalence of a uniformly accelerated reference system and a uniform gravitational field. The principle says that it is impossible for an observer to tell by experiments carried out in his reference system whether that system is in uniform acceleration or in a uniform gravitational field; the effects are exactly the same in the two cases. The "heuristic" value of the principle, Einstein showed, is that it allows the effects of gravitation to be determined by analyzing uniformly accelerated reference systems, which lend themselves to theoretical treatment. By this procedure, Einstein examined the influence of gravitation on the rates of clocks and on the propagation of light. To a friend, he wrote at the time that he was working on a relativistic theory of gravitation by which he hoped to explain an anomalous motion of the planet Mercury, but that so far he had not succeeded.[95]

93. Einstein to Zangger, 15 Nov. 1911.

94. Albert Einstein, "Über die vom Relativitätsprinzip geforderte Trägheit der Energie," *Ann.* 23 (1907): 371–84, on 371–72. Einstein's observation referred to the limitations of what he called "today's electromechanical world picture" (p. 372). He was not concerned here with gravitation but with a consequence of his relativity theory that would guide his gravitational researches, the inertia of energy.

95. Albert Einstein, "Über das Relativitätsprinzip und die aus demselben gezogenen Folgerungen," *Jahrbuch der Radioaktivität und Elektronik* 4 (1907): 411–62, on 454. Einstein to Habicht, 24 Dec. 1907, quoted in Seelig, *Einstein*, English trans., 76. Our discussion is not intended as a complete history of

In 1911, as we pointed out in our discussion of the *Annalen* for that year, Einstein returned to the gravitational problem of his *Jahrbuch* review. He was dissatisfied with this earlier treatment, and he had changed his mind about the detectability of the influence of gravitation on the propagation of light, which as a form of energy has inertial and therefore gravitational mass. He now believed it would be possible to detect the deflection of the light from stars passing the limb of the sun during a solar eclipse, a question he thought would be a "most desirable thing" for astronomers to take up.[96] Through a colleague at Prague, Einstein was put in touch with the Berlin astronomer Erwin Freundlich, who took an interest in the problem and had photographs of eclipses sent to him for measurements. Einstein wrote to Freundlich that "one thing can be said with certainty: if no such deflection exists, the assumptions of the theory are incorrect."[97] Meanwhile he continued to look for a "scheme of relativity theory" that incorporates the equivalence of inertial and gravitational mass, asking all of his colleagues to "work on this important problem." In 1912 he constructed his first gravitational field equations. For this purpose he regarded the velocity of light as a variable magnitude replacing the Newtonian gravitational potential, but this approach led to a violation of the principle of equality of action and reaction and the need for theoretically puzzling adjustments of the equations.[98]

At this time Einstein was studying the four-dimensional space-time formulation of Minkowski's approach to relativity theory, and he was also looking for a way to

---

Einstein's general theory of relativity and gravitational theory. There are many historical accounts, which include Einstein's own, "Notes on the Origin of the General Theory of Relativity," in *Ideas and Opinions* (New York: Dell, 1973), 279–83; the accounts of Einstein's many biographers such as Hoffmann and Dukas, *Einstein*, 103–33; historical accounts such as Edmund Whittaker, *A History of the Theories of Aether and Electricity*, vol. 2, *The Modern Theories, 1900–1926* (Reprint, New York: Harper and Brothers, 1960), 144–96; J. D. North, *The Measure of the Universe: A History of Modern Cosmology* (Oxford: Clarendon Press, 1965), 52–69; Peter G. Bergmann, *The Riddle of Gravitation* (New York: Scribner's, 1968); M. A. Tonnelat, *Histoire du Principe de Relativité* (Paris: Flammarion, 1971); Jagdish Mehra, "Einstein, Hilbert, and the Theory of Gravitation," in *The Physicist's Conception of Nature*, ed. Jagdish Mehra (Dordrecht: Reidel, 1973), 194–278; A. P. French, "The Story of General Relativity," in *Einstein: A Centenary Volume*, ed. A. P. French (Cambridge, Mass.: Harvard University Press, 1979), 91–112; John Earman and Clark Glymour, "Lost in the Tensors: Einstein's Struggles with Covariance Principles 1912–1916," *Studies in History and Philosophy of Science* 9 (1978): 251–78; John Stachel, "The Genesis of General Relativity," in *Einstein Symposion Berlin*, ed. H. Nelkowski, et al., Lecture Notes in Physics, vol. 100 (Berlin, Heidelberg, and New York: Springer, 1979), 428–42; and historical accounts of the reception of the theory such as Robert Lewis Pyenson, "The Göttingen Reception of Einstein's General Theory of Relativity" (Ph.D. diss., Johns Hopkins University, 1973); John Earman and Clark Glymour, "Relativity and Eclipse: The British Expeditions of 1919 and Their Predecessors," *HSPS* 11 (1980): 49–85; and most of the above references.

96. Albert Einstein, "Über den Einfluss der Schwerkraft auf die Ausbreitung des Lichtes," *Ann.* 35 (1911): 898–908, on 908. In this paper, Einstein discussed another prediction of 1907: a spectral line emitted in a place of high gravitational potential such as the sun should be shifted toward the red; but because of the effects of temperature and pressure, Einstein thought that this test of the gravitational influence on light would be difficult to carry out.

97. Einstein to Besso, 4 Feb. 1912, in *Einstein-Besso Correspondance*, 45–47, on 46. Einstein to Freundlich, 1 Sept. 1911, EA.

98. Albert Einstein, "Relativität und Gravitation. Erwiderung auf eine Bemerkung von M. Abraham," *Ann.* 38 (1912): 1059–64, on 1063–64; "Lichtgeschwindigkeit und Statik des Gravitationsfeldes," *Ann.* 38 (1912): 355–69; "Zur Theorie des statischen Gravitationsfeldes," *Ann.* 38 (1912): 443–58.

introduce into the theory coordinate transformations more general than Lorentz's. He made headway on his theory when he returned to Zurich in 1912 to his new position at the polytechnic: he recognized a crucial connection between Gauss's theory of surfaces and the gravitational problem, and he entered into a collaboration with his former classmate Marcel Grossmann, now professor of mathematics at the polytechnic and an expert on non-Euclidean geometry. Grossmann was familiar with the mathematics Einstein needed to express the equations of his theory in invariant form: it was the absolute differential calculus with its concise notation of tensor quantities, characterized by subscripts and superscripts as running indices.[99] Einstein wrote to Sommerfeld: "On the subject of quanta, I have nothing new to say that could demand anyone's interest. . . . I am now occupying myself exclusively with the problem of gravitation."[100]

The immediate result of Einstein and Grossmann's collaboration was a comprehensive paper in 1913, in which Grossmann gave an exposition of the tensor calculus and Einstein derived the variable gravitational field with it. The theory was developed from the length of a line element in four-dimensional space-time of the ordinary theory of relativity, as stated by Minkowski, $ds = \sqrt{-dx^2 - dy^2 - dz^2 + c^2 dt^2}$, and from the variational equation for the motion of a free particle, a form of Hamilton's principle describing the shortest path through space-time, $\delta\left\{\int ds\right\} = 0$. Einstein retained the line element as the fundamental invariant of the theory and the variational equation as the description of the motion of a particle in an arbitrary gravitational field. He generalized the expression for the line element for this purpose:

$$ds^2 = \sum_{\mu\nu} g_{\mu\nu} dx_\mu dx_\nu.$$

Here the x_μ and x_ν stand for an arbitrary set of coordinates x_1, x_2, x_3, x_4, and the $g_{\mu\nu}$ stand for the ten functions of these coordinates:

$$
\begin{array}{llll}
g_{11} & g_{12} & g_{13} & g_{14} \\
g_{21} & g_{22} & g_{23} & g_{24} \\
g_{31} & g_{32} & g_{33} & g_{34} \\
g_{41} & g_{42} & g_{43} & g_{44}
\end{array}
\qquad (g_{\mu\nu} = g_{\nu\mu})
$$

The $g_{\mu\nu}$ form the complicated mathematical entity that relates the x_μ to measure-

99. The mathematics appropriate to Einstein's theory went back to the nineteenth century, to the work on intrinsic curvature by Gauss and Riemann and to that of E. B. Christoffel on the transformation of quadratic differential forms; that work was incorporated in the absolute differential calculus introduced around the turn of this century by the Italian mathematicians Gregorio Ricci and Tullio Levi-Civita. Our discussion is especially indebted to Stachel, "Genesis of General Relativity," here on 434–36.

100. Einstein to Sommerfeld, 29 Oct. 1912, in *Einstein/Sommerfeld Briefwechsel*, 26–27, on 26.

ments; it is a symmetric "second-rank covariant tensor," in the language of the absolute differential calculus.[101]

The several quantities, g_{11}, g_{12}, . . ., entering the fundamental tensor of the theory replace the single Newtonian gravitational potential of the older theory. Their determination becomes the principal problem of gravitational theory. Einstein looked for a new law of the general form of Newton's, which must reduce to Newton's in the appropriate limit. Newton's law, written not as an action-at-a-distance equation for the force but as a field equation for the gravitational potential ϕ, is $\Delta\phi = 4\pi k\rho$, where Δ is the Laplacian second-order differential operator, k is a constant, and ρ is the mass density; the new gravitational law, Einstein assumed, would take a similar form: $\Gamma_{\mu\nu} = \chi\Theta_{\mu\nu}$, where the field $\Gamma_{\mu\nu}$ is a tensor containing second-order derivatives of the $g_{\mu\nu}$, χ is a constant, and $\Theta_{\mu\nu}$ is a ten-component energy-stress tensor, replacing the Newtonian scalar mass density and assumed to be known. Since Einstein's tensor law summarizes many equations, it is far more complicated than Newton's even if similar in form.[102]

Einstein explained how such a complicated law could ever be discovered in the first place. Suppose, he said, that all that was known about electrodynamics was Coulomb's law of interaction for static electric charges and the requirement that electric actions cannot travel faster than light. Who, he asked, could have discovered Maxwell's equations from such limited information? The gravitational problem was precisely similar. The number of possible generalizations of Coulomb's law is vast, and so are those of Newton's law of gravitation for static masses. It is all but unthinkable that the correct gravitational equations could be guessed at or constructed from hypotheses about the mode of action. Einstein's approach was to reason from empirical principles to the mathematical criteria that the new equations must satisfy. The relativity principle restricts the range of possible gravitational equations by requiring the time coordinate to enter them on the same footing as the spatial coordinates. Further restrictions follow from the principles of the identity of inertial and gravitational masses and the equivalence of gravitation and acceleration and from the conservation principles. These several considerations formed the basis of Einstein's theory of 1913, which guided his thinking in deciding the question of the simplest acceptable gravitational equations.[103]

The new gravitational theory requires a new understanding of measurements, which means that it has implications for the foundations of physics as a whole. The element ds, the distance between two infinitely near space-time points, Einstein called the "naturally measured" distance. Unlike the corresponding element occurring in the ordinary theory of relativity, which is measured by rigid rods and clocks that are independent of the field, the generalized element ds depends on the $g_{\mu\nu}$. Each location in space-time has its own measure of length and time, which depends

101. Albert Einstein and Marcel Grossmann, "Entwurf einer verallgemeinerten Relativitätstheorie und einer Theorie der Gravitation," Zs. f. Math. u. Phys. 62 (1913): 225–61, on 226, 228–30.
102. Einstein and Grossman, "Entwurf," 233.
103. Albert Einstein, "Zum gegenwärtigen Stande des Gravitationsproblems," Phys. Zs. 14 (1913): 1249–62, on 1250.

on the gravitational field there, with the consequence that the special theory of relativity applies only in regions of space-time free from gravitational fields. With Einstein's theory, gravitation moves from the periphery to the center of physics: "The gravitational field ($g_{\mu\nu}$) seems to be, so to speak, the skeleton on which everything hangs," Einstein wrote to Lorentz.[104]

In the "previous conception of the foundations of physics," Einstein observed, the elementary laws of the "distance-action theories" express what happens between finitely separated points. Since the "thorough-going revolution" initiated by Maxwell, however, these theories have been replaced by "contiguous-action theories," to which Einstein's new gravitational theory belongs. This change in foundations has implications for physical geometry. Euclidean geometry, with its rigid measuring rods for determining distances independently of their position, belongs to the older physics; the physical geometry demanded by Einstein's theory represents a "liberation" from a contradictory use of Euclidean geometry in the modern physics of fields.[105]

The goal of Einstein and Grossmann in 1913 was to establish covariant equations for physics. Existing independently of the choice of coordinate systems, such equations express real relationships in nature.[106] If one of two coordinate systems is preferred because, say, physical laws appear simpler that way, that distinction is without a physical cause and so lies outside the relationships of true significance. "A world picture that manages without such arbitrariness is in my opinion preferable," Einstein said, and he regarded his theory of general relativity as a step in the direction of a physics describing a world existing independently of human convenience.[107]

New gravitational laws had been proposed in the nineteenth century, often constructed from, or by analogy with, electrical theory. The early twentieth century witnessed a cluster of gravitational theories, by Poincaré, Minkowski, Lorentz, Sommerfeld, and others; at the time of Einstein and Grossmann's paper in 1913, gravitation was an important topic of current research. Their paper was reviewed that year in the *Physikalische Zeitschrift* by Born, who mentioned other leading researchers who were also currently developing gravitational theories, Abraham, Mie, and Gunnar Nordström.[108] Their theories were all discussed at the 1913 German Association

104. Einstein to Lorentz, 14 Aug. 1913, Lorentz Papers, AR.

105. Albert Einstein, "Die formale Grundlage der allgemeinen Relativitätstheorie," *Sitzungsber. preuss. Akad.*, 1914, 1030–85, on 1079–80.

106. A. D. Fokker, "A Summary of Einstein and Grossmann's Theory of Gravitation," *Philosophical Magazine* 29 (1915): 77–96, on 96. The Dutch physicist Fokker was a collaborator of Einstein on the general theory of relativity.

107. Einstein to Lorentz, 23 Jan. 1915, Lorentz Papers, AR. In correspondence and in publication, Einstein related the absence of any reason for preferring one set of coordinates to another to the idea that all measurements and observations in physics come down to the determination of space-time coincidences, the matching of material points with points of instruments; it should make no difference what set of coordinates describes the point-event. Lorentz accepted this argument. Einstein to Lorentz, 17 Jan. 1916, Lorentz Papers, AR.

108. In his review, Born associated Einstein's theory with the drive toward the unification of natural laws that "characterizes A. Einstein's entire research," calling it bolder than all of his other achievements. *Naturwiss.* 2 (1913): 448–49.

meeting, where, in the combined session for physics, mathematics, and astronomy, Einstein gave an invited talk on the state of the gravitational problem. Einstein discussed briefly Abraham's theory, at length Nordström's, and comprehensively the theory that he had completed with Grossmann's help. Abraham's starting point was Einstein's early theory in which the velocity of light is a variable measure of the gravitational potential, though he rejected Einstein's equivalence principle; Einstein regarded Abraham's theory as self-contradictory. Nordström's theory, which adhered strictly to the older relativity theory in which the velocity of light is constant and which does not contain the equivalence principle, Einstein regarded as a sensible theory, only one which differed from his own and led to some different predictions. Nordström's theory did not predict the deflection of light in a gravitational field, the consequence of Einstein's theory that, to judge from the discussion following his talk, especially interested the audience. Mie's theory Einstein did not take up at all in his talk, an omission Mie called attention to in the discussion. Part of a general electromagnetic theory of matter, Mie's gravitational theory was based on the older relativity theory; it rejected the identity of gravitational and inertial mass, which was the main reason Einstein had not mentioned it and why he thought it had little likelihood of being correct. Mie argued against the principle of general relativity at the meeting and soon after in print, and Einstein responded to his criticisms in both places.[109] The early reaction to Einstein's theory was not all as cool as Mie's, but it was generally guarded.[110]

Planck, as editor of the *Annalen*, said in 1913 that gravitational theories were falling "thick as hail," which made it hard for him to decide where he might expect to find a kernel of truth. Nordström's theory at least held to the constancy of the speed of light, which Planck took to be the "true foundation of relativity theory."[111] Planck's problem can be seen in the paper that the Berlin physicist Alfred Byk sent to the *Annalen* in 1913. Byk combined an atomic hypothesis with non-Euclidean geometry and—the essential point—a force law that for great distances from an atom becomes Newton's gravitational law but for near distances is a different law. Planck commented that in the vicinity of an atom the customary dynamic laws may not apply, so that Byk's bold hypothesis should not be rejected out of hand. Byk was

109. Einstein, "Zum gegenwärtigen Stande des Gravitationsproblems," talk at the 1913 meeting of the German Association in Vienna; discussion following the talk in *Phys. Zs.* 14 (1913): 1262–66. Einstein, in his answer to Mie's published criticism, conceded that his presentation was incomplete and that this contributed to Mie's misunderstanding of the theory. "This incompleteness follows from the fact that in some respects I myself have not yet achieved complete clarity," Einstein said. "Prinzipielles zur verallgemeinerten Relativitätstheorie und Gravitationstheorie," *Phys. Zs.* 15 (1914): 176–80, on 176. Mie came to what he believed was a better understanding of Einstein's theory with the help of Hilbert. Joining Mie's theory to Einstein's, Hilbert made a fundamental contribution to the general theory of relativity, which, Mie told Hilbert, showed him what Einstein really wanted and how close Einstein had come to realizing a general relativity theory. Mie to Hilbert, 13 Feb. 1916, 8 and 16 May, and 2 July 1917, Hilbert Papers, Göttingen UB, Ms. Dept.

110. "Physical humanity is behaving rather passively toward the gravitation paper. Abraham probably understands it best. . . . I'll go to see Lorentz in the spring to discuss the matter with him. He is very interested in it, as is Langevin. Laue is not open to the basic considerations, Planck isn't either, Sommerfeld sooner. In general, free, unclouded vision is not characteristic of the (adult) German (blinker!)." Einstein to Besso, n.d. [late 1913], in *Einstein-Besso Correspondance*, 50.

111. Planck to Wien, 28 Jan. and 29 June 1913, Wien Papers, STPK, 1973.110.

well trained, well read, critically gifted, and he carried through numerical applications of his hypothesis, all of which inclined Planck to urge acceptance of the paper. He came to a contrary decision on another paper submitted to the *Annalen* whose author proposed a connection between gravitation and electricity within the four-dimensional Minkowski world. Planck invoked the criteria of simplicity, clarity or visualizability, and above all experimental consequences; the paper failed to meet them, and the author failed to discuss the theories of others—of Mie, Einstein, and Hilbert—who were then working on a similar subject.[112]

Einstein was dissatisfied with his theory of 1913. To Lorentz he wrote that it "contradicts its own starting point [the equivalence principle]; it then rests on thin air."[113] Although the theory allows physical laws to be expressed in conformity with the requirement of general covariance, the gravitational field equations are not themselves generally covariant. For a time, Einstein thought he had reasons why the $g_{\mu\nu}$ cannot be generally covariant. He made a series of highly technical refinements of the theory, but his dissatisfaction remained. Then, at the end of 1915 (by which time he had moved to Berlin as a member of the academy), prompted by his recognition of an error in the derivation, he returned to the requirement of general covariance of the gravitational field equations, having in the meantime "immortalized in the academy papers" his reversal on this point.[114] With this he returned to the curvature tensor to which Grossmann had earlier introduced him, and in the course of a few weeks he worked his way to the desired result. The new generally covariant field equations for gravitation read:

$$G_{im} = -\kappa \, (T_{im} - \frac{1}{2}g_{im}T),$$

where G_{im} is the so-called Riemann curvature tensor, which is constructed from the g_{im} and their derivatives, κ is a constant, T_{im} is the energy tensor of "matter," and T is the scalar of that tensor. "With this," Einstein said in November 1915 of the paper containing these equations, "the general theory of relativity, as a logical structure, has finally been completed."[115]

To Sommerfeld Einstein wrote that this "is the most valuable discovery I have made in my life."[116] If his German colleagues did not all agree with this assessment, he could report that Planck had now begun to "take the matter more seriously."[117] His colleagues in Holland were more encouraging. Lorentz, with whom he had corresponded extensively as he worked out the theory, wrote to Einstein in the summer of 1916 that he was himself now working on the theory and that he had made a

112. Planck to Wien, 20 Sept. 1913 and 25 Aug. 1916, Wien Papers, STPK, 1973.110.
113. Einstein to Lorentz, 14 Aug. 1913.
114. Einstein to Sommerfeld, 28 Nov. 1915, in *Einstein/Sommerfeld Briefwechsel*, 32–36, on 33.
115. Albert Einstein, "Die Feldgleichungen der Gravitation," *Sitzungsber. preuss. Akad.*, 1915, 844–47, on 845, 847.
116. Einstein to Sommerfeld, 9 Dec. 1915, in *Einstein/Sommerfeld Briefwechsel*, 36–37, on 37.
117. Einstein to Besso, 21 Dec. 1915, in *Einstein-Besso Correspondance*, 61. Einstein added that his experiences with colleagues other than Planck "show a frightening predominance of the all-too-human!"

new contribution to it (which Einstein valued): "In the last months I have been much occupied with your gravitational theory and general relativity theory and have also lectured on it, which was very useful to me. I now believe that I understand the theory in its full beauty; each difficulty I encountered I could overcome upon closer consideration. Also I have succeeded in deriving your field equations . . . from the variational principle."[118]

In 1916 in the *Annalen der Physik*, Einstein published a thorough exposition and analysis of his now completed general theory of relativity. The fundamental requirement of the theory he stated as follows: *"The general laws of nature are to be expressed by equations which hold good for all systems of coordinates, that is, are covariant (generally covariant) with respect to any substitutions whatever."* With this principle, Einstein explained, the foundations of physics had been changed. His earlier principle of the constancy of the velocity of light had departed from classical mechanics, and it together with the principle of relativity had resulted in a "far-reaching" modification of the theory of space and time, by which lengths and time intervals depend on motion. The principle of general relativity now modified the theory even further: lengths and time intervals depend not only on motion but also on position, removing from space and time the "last remnant of physical objectivity."[119]

In advancing to the final form of his theory, in addition to showing Newton's theory to be a first approximation, Einstein calculated the anomalous secular rotation of Mercury's orbit in the direction of the motion. The current astronomical value for the deviation of observation and theory, which had been known since the nineteenth century, was $45'' \pm 5''$ of arc per century. Einstein's theoretical value in 1915 of $43''$ was in complete accord. There were two additional points of contact with direct experience. Einstein reported to the Prussian Academy that the spectral redshift predicted by theory was confirmed in order of magnitude by the light from B- and K-type stars. And by his latest theory, light from stars passing the sun's limb should be bent by $1.7''$, double the value of his earlier theory; the confirmation of this prediction awaited a total solar eclipse.[120] But most of the new consequences of the theory were too small to detect, and at the time all of them were beyond the reach of laboratory experiments. Only the large gravitational fields of astronomical bodies promised observable results. The precision of astronomical measurements, long the ideal of physicists, were now looked to for the confirmation of a theory that entailed a new theoretical understanding of physical measurement.

We have seen that without waiting for the experimental evidence to weigh in its favor, Einstein began to generalize his relativity theory of 1905 beyond its original

118. Lorentz to Einstein, 6 June 1916, Rijksmuseum voor de Geschiedenis der Natuurwetenschappen, Leiden.

119. Albert Einstein, "Die Grundlage der allgemeinen Relativitätstheorie," *Ann.* 49 (1916): 769–822, on 776.

120. Albert Einstein, "Erklärung der Perihelbewegung des Merkur aus der allgemeinen Relativitätstheorie," *Sitzungsber. preuss. Akad.*, 1915, 831–39. In Einstein's report on general relativity and its applications to astronomy on 25 Mar. 1915, his and Nordström's prediction of the gravitational displacement of spectral lines was said to be supported in order of magnitude by a new work by Freundlich. *Sitzungsber. preuss. Akad.*, 1915, 315.

restriction to uniform relative motion. In the course of this work, he discussed the technical questions of the theory with colleagues in Germany and abroad, with physicists, mathematicians, and astronomers; he collaborated with some, and he debated physical principles with others who were developing competing gravitational theories. In a series of researches extending over eight or nine years and involving almost as many preliminary versions, he completed the general theory of relativity together with the theory of gravitation.

The result, the establishment of covariant gravitational equations and the calculation of the anomalous precession of Mercury, left Einstein "speechless with excitement" for days. Unlike the quantum problem, which remained perplexing, the gravitational problem seemed to have yielded a solution. Yet Einstein reported to Lorentz that "this much is clear to me; the quantum difficulties concern the new gravitational theory just as much as Maxwell's theory."[121] Moreover, as he later recalled, he did not regard his achievement as "anything *more* than a theory of the gravitational field, which was somewhat artificially isolated from a total field of as yet unknown structure." His review article in the *Annalen* in 1916 looked to the possibility of combining gravitational and electromagnetic theories to gain insight into the structure of matter. He concluded a lecture on relativity theory at Leiden University in 1920 by characterizing gravitation and electromagnetism as "two realities which are completely separated from each other conceptually." To view them as "one unified conformation" would be a great advance in physics: "Then for the first time the epoch of theoretical physics founded by Faraday and Maxwell would reach a satisfactory conclusion. The contrast between ether and matter would fade away, and, through the general theory of relativity, the whole of physics would become a complete system of thought." Einstein's work of the future was devoted to building a unified theory of gravitation and electromagnetism, freeing physics of its dualisms of particle and field, of mechanics and electricity. Characteristically, he directed his account of the properties of good theories in his late "Autobiographical Notes" to "such theories whose object is the *totality* of all physical appearances."[122]

Einstein's work encouraged other researches in the tradition of total theories of physics. In 1917 the mathematician Hermann Weyl submitted to the *Annalen* a gravitational theory, which Planck characterized as belonging to those researches that try to subsume all physical phenomena under a "single 'world formula.'" Here the formula was Hamilton's principle, containing terms for mechanical and electromagnetic processes. Building on Einstein's theory and non-Euclidean geometry, Weyl's theory dealt with world lines, light signals, and mass points in a gravitational field. Weyl hoped it would yield information about the interior of atoms, but Planck

---

121. Einstein to Lorentz, 17 June 1916, Lorentz Papers, AR.

122. Einstein to Ehrenfest, probably Jan. 1916, quoted in Hoffmann with Dukas, *Einstein*, 125. Einstein, "Ether and the Theory of Relativity," an address delivered on 27 Oct. 1920, translated by G. B. Jeffery and W. Perrett in *Sidelights on Relativity* (New York: E. P. Dutton, 1922), 3–24, on 22–23; "Autobiographical Notes," in *Albert Einstein: Philosopher-Scientist*, ed. P. A. Schilpp (Evanston: The Library of Living Philosophers, 1949), 1–95, on 23, 75.

was doubtful. There could be no question of an experimental test of the theory at present, and Planck thought that theories resolving physical processes into non-Euclidean geometry might best be diverted to mathematical journals. But for now he did not want to decide the question. As far as he was concerned, if a theory stuck closely to gravitation, it should be considered for the *Annalen*. Planck believed that a definite solution to the ultimate problem of the world formula was not to be expected but that progress still could be made toward it.[123]

Einstein soon applied the general theory of relativity to the largest object of all, the totality of physical things, the universe as a whole. Persuaded by Mach's *Science of Mechanics* that the inertia of a body is not a resistance to motion but an interaction of the body with all other bodies of the universe, in 1917 Einstein published a cosmological theory embodying Mach's principle.[124]

The general theory of relativity worked for the planets, accounting for Mercury's anomaly, but it led to difficulties when it was extended to the universe at large. These had to do with Mach's principle and with boundary conditions—the values assumed for the gravitational potentials—at infinite distances. The way around the difficulties was to eliminate such boundary conditions by denying spatial infinity. Einstein assumed a Riemannian universe, *"finite (closed) with respect to its spatial dimensions."* To investigate it mathematically he made the simplifying assumptions that matter is uniformly distributed across enormous spaces and that it is permanently at rest, accepting observational indications that the velocities of stars among themselves are small. Since, as he showed, his field equations of 1915 were incompatible with this cosmology, in the interest of logical consistency he modified them by adding a term:

$$G_{\mu\nu} - \lambda g_{\mu\nu} = -\kappa(T_{\mu\nu} - \tfrac{1}{2}g_{\mu\nu}T),$$

where λ is a small, undetermined, universal constant, which is related to the total mass and positive radius of the universe. Einstein recognized that it would not be easy to verify the new term by observation; according to astronomical estimates of the distribution of stars, the radius of the universe came out to be 10^7 light-years, whereas the most distant visible stars were only 10^4 light-years away. Nevertheless the general theory of relativity could not do without the hypothesis of closed space, Einstein told Sommerfeld, and the modifications of the field equations seemed to be demanded by these cosmological considerations.[125]

Two years later, Einstein showed that the cosmological constant modifying the

123. Planck to Wien, 16 Sept. 1917, Wien Papers, STPK, 1973.110.

124. Albert Einstein, "Cosmological Considerations on the General Theory of Relativity," *Sitzungsber. preuss. Akad.*, 1917, 142–52, translated by W. Perrett and G. B. Jeffery in *The Principle of Relativity: A Collection of Original Memoirs on the Special and General Theory of Relativity by H. A. Lorentz, A. Einstein, H. Minkowski and H. Weyl* (1923; reprint, New York: Dover, n.d.), 177–88, on 180.

125. Einstein, "Cosmological Considerations," 183–86. Einstein to Besso, 9 Mar. 1917, in *Einstein-Besso Correspondance*, 101–3, on 102–3. Einstein to Sommerfeld, 1 Feb. 1918, in *Einstein/Sommerfeld Briefwechsel*, 43–47, on 44.

equations appears as a constant of integration when the equations are applied to the electron. He had introduced it ad hoc as a constant "peculiar to the fundamental law," which he regarded as a defect in the "formal beauty of the theory." He showed that owing to the gravitational curvature of space, a negative pressure obtains in the interior of the electron, accounting for the equilibrium of its charge and thereby its permanence.[126] In this way the problem of the largest object, the universe, and the problem of the smallest, the electron, came together: both problems required the modification of the field equations of 1915. Einstein regarded his general relativistic construction of matter as a preferable alternative to Mie's general theory of matter with its greater need for hypotheses. In time Einstein came to be dissatisfied with his own theory of the electron as he did with his theory of the static universe with its cosmological constant.[127] But Einstein's first general relativistic theory of the universe initiated a powerful theoretical development of cosmology, which eventually merged with observation.[128]

Einstein on the Torch of Mathematics

We began this history with Ohm, who in the early years of the nineteenth century valued mathematics as a "torch" for guiding theoretical understanding. His mathematics came largely from France, as did its impressive applications to physics. The mathematics together with its place of origin changed in various ways after Ohm. But mathematics being the guide, the torch proved enduring, as we will see as we look again at Einstein, the principal physicist with whom we conclude this history.[129]

Throughout his career Einstein held pronounced views on mathematics. Early on he did not value highly the subtler parts of mathematics for the work of the theoretical physicist. This was not owing to any mathematical difficulties of his own, for mathematics was a subject he was drawn to and in which he received early encouragement. In the gymnasium he attended in Munich he was ahead of his classmates in mathematics and apparently in nothing else. Leaving the gymnasium before earning a diploma, he obtained a written statement from his mathematics teacher to

126. Albert Einstein, "Do Gravitational Fields Play an Essential Part in the Structure of the Elementary Particles of Matter?" *Sitzungsber. preuss. Akad.*, 1919, 349–56; in *Principle of Relativity*, 191–98, quotation on 193.

127. At the close of his paper, "Do Gravitational Fields Play an Essential Part?" on 198, Einstein called attention to a limitation of his theory of the electron: it allows *any* symmetrical, spherical charge distribution to be in equilibrium, so that the "problem of the constitution of the elementary quanta cannot yet be solved on the immediate basis of the given field equations." Einstein remained uneasy about the cosmological constant λ, and in 1931 he discarded it. The constant related to the assumption of a static universe, which was replaced by that of an expanding universe. North, *Measure of the Universe*, 84–86, 109–10.

128. "Einstein's basic approach to cosmology [of 1917]—setting up a metrical model and attempting to correlate its features with observations—has dominated theoretical work in the field for over half a century." Stachel, "Genesis of General Relativity," 441.

129. This discussion of Einstein and the relationship of physics and mathematics is based in part on Russell McCormmach, "Editor's Foreword," *HSPS* 7 (1976): xi–xxxv. We thank Princeton University Press for permission to use this material.

the effect that his knowledge of mathematics was uncommon and qualified him for advanced study in the subject. In the entrance examination for the Zurich Polytechnic, he scored ahead of the other candidates in mathematics, and he did well in physics too, while failing in other subjects.[130] His mathematical performance was sufficiently strong that the director of the polytechnic urged him to attend a cantonal school, earn a diploma there, and return, all of which he did. Although in his final diploma examination at the polytechnic he did well in physical subjects, he again scored highest in mathematics.[131]

In the latter half of the curriculum for teachers of mathematics and physics at the polytechnic, students were expected to take seminars in advanced mathematics, but no one could stir Einstein to attend them.[132] He preferred to spend his time in the physics laboratory. According to Hermann Minkowski, one of his mathematics teachers, Einstein "never bothered about mathematics at all," which made his later success in theoretical physics mystifying to Minkowski.[133] Eventually Einstein came to regret that he had passed up an opportunity of gaining a fine mathematical education at the polytechnic from outstanding mathematicians such as Minkowski and Adolf Hurwitz.[134]

Writing in 1905, Minkowski said that in the half century since the death of Dirichlet the development of all parts of mathematics showed Dirichlet's spirit; namely, the desire for "fraternization of the mathematical disciplines, for the unity of our science." David Hilbert, the Göttingen mathematician and colleague of Minkowski after the latter's move from Zurich to Göttingen, believed in the methodological unity of mathematics: "the question is forced upon us whether mathematics is once to face what other sciences have long ago experienced, namely to fall apart into subdivisions whose representatives are hardly able to understand each other and whose connections for this reason will become even looser. I neither believe nor wish this to happen; the science of mathematics as I see it is an indivisible whole, an organism whose ability to survive rests on the connection between its parts." Weyl, who moved to the Zurich Polytechnic as professor of mathematics in 1913, contrasted physics and mathematics in their respective tendencies toward internal specialization and unity: "Whereas physics in its development since the turn of the century resembles a mighty stream rushing on in one direction, mathematics is more like the Nile delta, its waters fanning out in all directions." He recognized as well

130. Philipp Frank, *Einstein. His Life and Times*, trans. G. Rosen, ed. and rev. S. Kusaka (New York: Knopf, 1947), 16, 18.

131. Seelig, *Einstein*, 54.

132. According to Einstein's biographer A. Reiser, cited in Gerald Holton, "Mach, Einstein, and the Search for Reality," *Daedalus* 97 (1968): 636–73, on 638.

133. Seelig, *Einstein*, 33. It should be noted that the Zurich Polytechnic generally and Minkowski in particular did not attract many students in its advanced mathematics courses. When Einstein was in his second year at the polytechnic, Minkowski observed that there was only one student in the school with more than three semesters' mathematics, that the colloquium was sustained chiefly by assistants, and that in each of his three classes he had only about eight students. Minkowski to Hilbert, 23 Nov. 1897, Hilbert Papers, Göttingen UB, Ms. Dept., 258/65.

134. Einstein, "Autobiographical Notes," 15.

that the apparently opposite "tendency of several branches of mathematics to coalesce is another conspicuous feature in the modern development of our science."[135]

Einstein was not a mathematician, and unlike Minkowski, Hilbert, and Weyl, he did not have a sense for the central problems and unifying tendencies of mathematics. In his "Autobiographical Notes," he recalled how the choice of a career in mathematics had looked to him when he was young. He decided against mathematics in part because he was unsure of his way around in it, lacking an intuitive feeling for what was important as opposed to mere erudition. He felt like Buridan's ass, unable to decide which specialty within mathematics he should enter, certain only that each specialty could exhaust a lifetime.[136]

Einstein was drawn instead to the specialty of theoretical physics, to a career that bridged his physical and mathematical interests and one that dealt with problems whose central significance he was reasonably assured of. He recognized that physics, as mathematics, was internally divided into parts, but he soon acquired an intuitive understanding of what was important in physics as he had not in mathematics. He responded strongly to the unifying goals of theoretical physics; after sending his first paper for publication to the *Annalen der Physik* in 1901, he wrote to his former polytechnic classmate Grossmann: "It is a magnificent feeling to recognize the unity of a complex of phenomena which appear to be things quite apart from the direct visible truth." As in mathematics, in physics it was possible to spend a lifetime in any one of its specialties, but it was also possible for a physicist of Einstein's persuasion to spend it seeking general principles for ordering two or more specialties.[137]

The task of bringing unity to physics involved analyzing the mathematics of its various fundamental theories. The debates of the 1890s on the foundations and methods of theoretical physics included much discussion of the intimate relationship of mathematics to physical concepts. Boltzmann, for example, in his lectures on gas theory in 1895, which Einstein read as a student, argued that the mathematical description by differential equations of the inner motions of bodies leads compellingly to the concept of heat as the motion of the smallest particles. Two years later, in his lectures on mechanics, he went further in associating the concepts of physical and mathematical atomicity, countering the phenomenologists' claim for the "picture"-free nature of bare differential equations. With reference to finite time elements, he suggested that the differential equations of physics may represent only average values constructed from elements that are not themselves rigorously differ-

135. Hermann Minkowski, "Peter Gustav Lejeune Dirichlet und seine Bedeutung für die heutige Mathematik," in Minkowski, *Gesammelte Abhandlungen*, ed. David Hilbert, 2 vols. (Leipzig and Berlin: B. G. Teubner, 1911), 2: 447–61, on 450. Hilbert quoted in Hermann Weyl, "Obituary: David Hilbert 1862–1943," *Obituary Notices of Fellows of the Royal Society* 4 (1944): 547–53, reprinted in Weyl, *Gesammelte Abhandlungen*, ed. K. Chandrasekharan (Berlin: Springer, 1968), 4: 121–29, on 123. Weyl, "A Half-Century of Mathematics," *Amer. Math. Monthly* 58 (1951): 523–53, reprinted in Weyl, *Gesammelte Abhandlungen*, 464–94, on 464–65.

136. Einstein, "Autobiographical Notes," 15.

137. Einstein, "Autobiographical Notes," 15. Einstein to Grossmann, 14 Apr. 1901, in Seelig, *Einstein*, English trans., p. 53.

entiable.[138] To take another example, Planck, who was allied to Boltzmann in opposition to the claims for mathematical phenomenology, told the German Mathematical Society in 1899 that the main physical idea underlying Maxwell's electromagnetic equations is that action is contiguous, and that the mathematics appropriate to this idea is fundamentally different from that appropriate to an action-at-a-distance theory. The "critical" contrast between the two theories is that in the former, the space and time coordinates enter as differentials, whereas in the latter, the space coordinates enter as finite intervals.[139]

From the beginning of his work in theoretical physics, Einstein—as Boltzmann and Planck—paid close attention to the mathematical tools of his trade. His 1905 paper on light quanta, for example, was not only about the nature of light but also about the relationship of physical ideas to their mathematical expression and its implication for the unity of physics. That is evident from the way Einstein introduced his paper: There is a "profound formal distinction," he wrote, between the concepts of the atomic theory of matter and those of Maxwell's electromagnetic theory of light. The emphasis is on "formal," for he went on to describe the differences in the formalisms by which physicists presented these two branches of physics. He pointed out that in the atomic theory the energy of a body is given by a discrete "sum" of the energies of a finite number of individual atoms and electrons, whereas in Maxwell's theory the energy is given by a "continuous spatial function." Observing that the continuous functions of Maxwell's theory refer only to time-average values of observations, he suggested that in describing the emission and transformation of light, the "theory of light which operates with continuous spatial functions may lead to contradictions with experience."[140] He thought that physicists would gain a better understanding of such phenomena if they regarded light as behaving like a finite number of spatially localized energy quanta. In this way he accompanied his "very revolutionary"[141] hypothesis of light quanta with the suggestion that the concepts of matter and light may turn out not to be as formally different as physicists had believed up to then. In later writings Einstein characterized the separateness of atomic and field theories by the mathematics used in each: total, or ordinary, differential equations in the former and partial differential equations in the latter.[142] In calling attention to the different formalisms in the two theories, he had a similar kind of distinction in mind as in 1905.

Given the problems of physical theory in 1905, Einstein was able to carry through profound analyses of the foundations of physics using ordinary algebra, differential equations, and the probability calculus, and with that modest mathematical

138. Boltzmann, *Gas Theory*, 27; *Prinzipe der Mechanik* 1: 3–4, 26–27.

139. Max Planck, "Die Maxwell'sche Theorie der Elektricität von der mathematischen Seite betrachtet," *Jahresber. d. Deutsch. Math.-Vereinigung* 7 (1899): 77–89, reprinted in *Phys. Abh.* 1: 601–13, on 603.

140. "Einstein's Proposal of the Photon Concept," 367–68.

141. "Very revolutionary" was Einstein's own judgment at the time on his light quantum hypothesis. Einstein to Konrad Habicht, undated [1905], quoted in Seelig, *Einstein*, 89.

142. Albert Einstein, "Maxwell's Influence on the Evolution of the Idea of Physical Reality" (1931), in *Ideas and Opinions*, 259–63, on 261.

equipment he formulated the kinematics of special relativity and other bodies of conceptually new physical knowledge. The voluminous, mathematically elaborate papers on the electron theory by Sommerfeld, Voigt, Abraham, and others at the time contrast strikingly with Einstein's brief, mathematically uncomplicated papers. The standard electron theory problems entailed intricate calculations of cases: surface and volume electron charge distributions, rigid and deformable electron structures, slowly and rapidly accelerated electrons, electron velocities over and under the velocity of light, and so on. Einstein showed little interest in these problems and in the associated mathematical investigations that occupied many other physicists. He was interested in the dualism of the foundations of physics. The electron theories, for example, were usually built on the dual concepts of discrete particle and continuous field and on their respective formalisms of total and partial differential equations. By a mathematically elementary analysis, in his paper on light quanta Einstein argued that Maxwell's equations had to be revised before the conceptual and mathematical dualism could be resolved.[143]

In his subsequent efforts to build a unified theory comprehending both the continuous and the discrete aspects of the field, Einstein followed a different direction than that of many contemporary theorists in their own efforts to solve the same basic problem. Einstein came to seek field equations that yielded point solutions representing discrete electrons and light quanta. This problem was mathematically daunting in its own right.

Einstein told his students at the University of Zurich around 1910 that "in fact with mathematics one can prove anything" and that all that mattered was the content. At this time he disagreed with a Swiss friend who believed, reasonably enough, that the total length of several matches laid end to end is the sum of the lengths of the individual matches; Einstein gave as his reason: "Car moi, je ne crois pas à la mathématique."[144] After meeting Einstein at the first Solvay Congress in 1911, Lindemann reported that Einstein "says he knows very little mathematics" but added that Einstein "seems to have had a great success with them."[145] Einstein made the same self-deprecating observation in connection with the first mathematical text on relativity theory by Laue in 1911, complaining in a half joking way that he could "hardly understand" it.[146] By 1911 he had begun his sustained work on the general theory of relativity, in the course of which he became aware of mathematical problems he had not confronted before and at the same time of the extent of his own

---

143. McCormmach, "Einstein, Lorentz, and the Electron Theory." The general observation has been made that in both classical and modern physics, theorists sometimes arrived at major results using elementary mathematics. See, for example, J. H. Van Vleck, "Nicht-mathematische theoretische Physik," *Phys. Bl.* 24 (1968): 97–102, on 98.

144. Quoted in Seelig, *Einstein*, 122, 127.

145. Quoted in F. W. F. Smith, Earl of Birkenhead, *The Professor and the Prime Minister: The Official Life of Professor F. A. Lindemann, Viscount Cherwell* (Boston: Houghton, Mifflin, 1962), 43.

146. Quoted in Frank, *Einstein*, 206. Einstein was half joking; as Laue said, his presentation drew on the "usual mathematical equipment of the theoretical physicist," the calculus and vector analysis. Max Laue, *Das Relativitätsprinzip* (Braunschweig: F. Vieweg, 1911), vi.

mathematical limitations. In 1912 he wrote to Sommerfeld that "with the aid of a local mathematician who is a friend of mine [Grossmann] I believe I will now be able to master all difficulties. But one thing is certain that in all my life I have never struggled as hard and that I have become imbued with great respect for mathematics, the subtler parts of which, in my simplemindedness, I had considered pure luxury up to now! Compared with this problem, the original theory of relativity is child's play."[147]

From this time on Einstein never doubted the physical relevance of certain of the subtler parts of mathematics. His correspondence sometimes dealt with mathematical matters that were now at the center of the physical argument; for example, writing to Lorentz in 1913 Einstein remarked that mathematicians had not developed group theory sufficiently for his needs.[148] He selected his closest collaborators—such as Grossmann—in part for their mathematical competence.

"The charm of this theory can hardly be denied to anyone who has really grasped it; it signifies a true triumph of the methods of the general differential calculus as laid down by Gauss, Riemann, Christoffel, Ricci, and Levi-Civita."[149] With these words Einstein expressed his indebtedness to the mathematicians who prepared the way for his general theory of relativity of 1915. An early book on Einstein's new theory by Freundlich was to be recommended, Einstein said, for bringing out the "depth of thought of Riemann, a mathematician so far in advance of his time."[150] Einstein's subsequent work on field theory continually posed new and demanding mathematical problems and occasionally led to great rewards, and in his Herbert Spencer Lecture at Oxford in 1933 Einstein went so far as to claim that in theoretical physics the "creative principle resides in mathematics."[151]

In his "Autobiographical Notes" Einstein took pains to clarify his understanding of the relationship of mathematics and theoretical physics. He said that he had abandoned his early attempts to discover by trial and error the physically unifying field equations accounting for light quanta and electrons. He realized that even if he had hit upon them, he would not have gone deeply into the matter; the equations would remain arbitrary, at best a happy guess. He recognized that the development of physics depended on finding principles applicable to all of physics. The principles took the form of statements about physically permissible equations, so that the development of physics might be seen as the progressive reduction of the ad hoc or empirical element in its mathematical foundations. Einstein used his special and general relativity principles as heuristic guides in seeking the "simplest" equations for describing the total physical world; even the simplest equations are so involved, he had learned, that they "can be found only through the discovery of a logically

147. Einstein to Sommerfeld, 29 Oct. 1912, in *Einstein/Sommerfeld Briefwechsel*, 26–27, on 26.

148. Einstein to Lorentz, 14 Aug. 1913, Lorentz Papers, AR.

149. Einstein, "Zur allgemeinen Relativitätstheorie," *Sitzungsber. preuss. Akad.*, 1915, 778–86, on 779.

150. Einstein, "Preface," to Erwin Freundlich, *The Foundations of Einstein's Theory of Gravitation*, trans. H. L. Brose (London: Methuen, 1924).

151. Albert Einstein, "On the Method of Theoretical Physics," Herbert Spencer Lecture delivered at Oxford, 10 June 1933, in *Ideas and Opinions*, 263–70, on 267–68.

simple mathematical condition which determines the equations completely or [at least] almost completely."[152] So mathematically demanding was this way of building physical theories that he felt constantly inadequate to the task. He likened his mathematical efforts to "a man struggling to climb a mountain without being able to reach its peak." On his deathbed Einstein is said to have lamented: "If only I had more mathematics!"[153]

Einstein was fairly typical of physicists in his understanding that physics asks a good deal of mathematics, even that mathematics may not be up to the needs of physics. On receiving a copy of Wien's book on hydrodynamics in 1900, Planck noted that analysis is powerless to treat some of the simplest physical phenomena, and he hoped that Wien's book would be widely read by mathematicians. For his part, Wien, in an address at the 1905 meeting of the German Association, said that to solve the problems of the electron theory physicists needed everything that analysis could yield. In 1906 Wien complained to Hilbert that he got little help from his present mathematical colleagues at Würzburg and asked Hilbert's advice on candidates for an extraordinary professorship for mathematics there. Wien gave a talk to the German Mathematical Society the following year on partial differential equations in physics, remarking that today theoretical physicists needed the "comprehensive cooperation" of mathematicians.[154]

From the side of mathematics, there was an anticipation of a mathematics oriented more toward applications. The Breslau mathematician Jakob Rosanes (from whom Born learned the matrix calculus, a branch of mathematics which he later used in formulating the new quantum mechanics) observed in 1903 that the estrangement of mathematics and physics of the last several decades was past and that now an "epoch of closer union" had opened.[155] The prospects of closer union appealed to Minkowski, who in 1905 projected that even the most abstract branch of mathematics, number theory, would soon have its "triumph in physics and chemistry."[156] In 1916 Wilhelm Lorey recalled the prediction of the Berlin mathematicians Weierstrass and Leopold Kronecker that the present era of abstract mathematics would be superseded by one of application, and that the problems of theoretical physics, astronomy, and technology were already ripe for a new application of mathematics; Lorey believed that in theoretical physics their prediction was close to realization. In 1914 Klein spoke of the "cry of distress of modern physics which in its

152. Einstein, "Autobiographical Notes," 37, 53, 69, 89.

153. Peter Michelmore, Einstein, Profile of the Man (New York: Dodd, Mead, 1962), 198, 261.

154. Planck to Wien, 24 May 1900, Wien Papers, STPK, 1973.110. Wien, Über Elektronen, 5; "Über die partiellen Differential-Gleichungen der Physik," Jahresber. d. Deutsch. Math.-Vereinigung 15 (1906): 42–51, on 42. Wien to Hilbert, 3 Dec. 1906; also Wien to Hilbert, 2 May 1909, asking Hilbert's advice on how to develop integral equations in physics; Hilbert Papers, Göttingen UB, Ms. Dept.

155. Jakob Rosanes, "Charakteristische Züge in der Entwicklung der Mathematik des 19. Jahrhunderts," Jahresber. d. Deutsch. Math.-Vereinigung 13 (1904): 17–30, on 23.

156. Minkowski, "Dirichlet," 451–52.

stormy, even revolutionary, development calls for the help of mathematicians and threatens to devour a large part of our working energy!"[157]

Together with Klein and another colleague, the Strassburg mathematician Heinrich Weber came up with the idea of what was to become the massive, multi-volume series *Encyklopädie der mathematischen Wissenschaften*, which encompassed both pure and applied mathematics, including the mathematics of mechanics and physics.[158] In 1900–01 Weber brought out another work, related in spirit to the *Encyklopädie*, a new edition of Riemann's lectures on the partial differential equations of physics. Physics had experienced a "sweeping transformation" as a result of Maxwell's electromagnetic theory, Weber said, and in addition to the content of physics, its mathematics had greatly changed as well, which necessitated a complete reworking of the text and a doubling of its length. Weber retained "Riemann" in the title only because the text remained true to Riemann's "purpose and intellectual spirit." Weber brought out five revised editions of Riemann's lectures. What was said of Weber's last edition of 1910–12 could be said of earlier ones: it was indispensable for doing "theoretical physics."[159]

In addition to new aids for solving partial differential equations, new mathematical subjects entered Weber's revision of Riemann's lectures. Prominent among them was vector analysis, which Weber introduced in connection with the concept of a physical "field": associated with each point in space is a directed magnitude, or vector, and the mathematical subject of vector analysis enables the physicist to study the spatial distribution of a vector and its changes over time. Physicists increasingly viewed vector analysis as a tool of research, not just a formal convenience. Abraham said in 1899, for example, that it could yield new laws that would be hard to derive using the customary scalar analysis.[160]

During the 1890s and 1900s, the time of consolidation of vector analysis as a branch of mathematics, the majority of books introducing and using vectors were written by physicists.[161] While there were various "schools" of thought on the subject, for most purposes physicists used the form of vectors introduced by their fellow physicists Heaviside and Gibbs rather than the earlier forms introduced by the mathematicians Hermann Grassmann and William Rowan Hamilton.[162] Physicists found

157. Wilhelm Lorey, *Das Studium der Mathematik an den deutschen Universitäten seit Anfang des 19. Jahrhunderts* (Leipzig and Berlin: B. G. Teubner, 1916), 263. Klein quoted on 307.

158. The origin of the project is described in Walther von Dyck, "Einleitender Bericht über das Unternehmen der Herausgabe der Encyklopädie der mathematischen Wissenschaften," in *Encyklopädie der mathematischen Wissenschaften. Mit Einschluss ihrer Anwendungen*, vol. 1. pt. 1, *Reine Mathematik*, ed. H. Burkhardt and F. Meyer (Leipzig, 1898), v–xx, on vi–vii.

159. The full title of Heinrich Weber's two-volume revision of Riemann's lectures is *Die partiellen Differential-Gleichungen der mathematischen Physik. Nach Riemann's Vorlesungen.* "Vorrede," 1: v–ix, explains the need for the revision. Aurel Voss, "Heinrich Weber," *Jahresber. d. Deutsch. Math.-Vereinigung* 23 (1914): 431–44, on 435, 442.

160. Heinrich Weber, "Vectoren," in *Die partiellen Differential-Gleichungen* 1:207–26. Abraham to Sommerfeld, 7 Jan. 1899, Sommerfeld Correspondence, Ms. Coll., DM.

161. Michael J. Crowe, *A History of Vector Analysis: The Evolution of the Idea of a Vectorial System* (Notre Dame: University of Notre Dame Press, 1967), 242.

162. In preparing the entry dealing with vectors for the *Encyklopädie der mathematischen Wissenschaf-*

vectors useful in their textbooks on mechanics,[163] but even more so in their textbooks and research on electromagnetism.[164] Primarily it was the change in the physical view of nature following the acceptance of Maxwell's theory that prompted an extension of the standard mathematics of physics to include vectors.[165]

To facilitate the teaching and introduction of vectors, physicists wrote books that presented vector analysis as an "independent discipline" followed by examples from physics.[166] They also explained and used vectors in their textbooks on the various branches of physics and on theoretical physics as a whole.[167] Since students would come across both vectors and the older mathematical presentation, physicists might include in their textbooks the main equations in both notations.[168]

It was largely in connection with the development of vector analysis, together with an increasing attention to symmetries, invariances, and transformation properties of physical equations, that theoretical physicists and mathematicians recognized a widening applicability of newer mathematical quantities and calculuses. Four-dimensional vectors were applied in the electron and relativity theories. Minkowski applied them to physics in 1908 in exhibiting the Lorentz invariance of the fundamental electrodynamic equations; he applied as well Arthur Cayley's matrix calculus, explaining at length the elementary properties of matrices and remarking that matrices were preferable for his needs to Hamilton's quaternion calculus.[169] Sommerfeld published a long, two-part paper in 1910 devoted wholly to the new algebraic and analytic methods appropriate to the Minkowski "absolute world" and its character-

---

ten ("Geometrische Grundbegriffe," vol. 4, in 1901), Abraham spoke of the difficult task of uniting the various "schools" of thought on the subject at various times and of striking the right balance, with not too much "mathematical jargon" for the physicists and not too much "modern physics literature" for the mathematicians. Abraham to Sommerfeld, 23 Feb. 1901, Sommerfeld Papers, Ms. Coll., DM. On the form of vectors preferred by physicists: Ludwig Prandtl, "Über die physikalische Richtung in der Vektoranalysis," *Jahresber. d. Deutsch. Math.-Vereinigung* 13 (1904): 436–49, on 436.

163. For example: Voigt's 1889 textbook *Elementare Mechanik*; Emil Budde, *Allgemeine Mechanik der Punkte und starren Systeme. Ein Lehrbuch für Hochschulen*, 2 vols. (Berlin, 1890–91).

164. Crowe, *A History of Vector Analysis*, 225. Alfred M. Bork, "'Vectors Versus Quaternions'—The Letters in *Nature*," *Am. J. Phys.* 34 (1966): 202–11, on 210.

165. A systematic vector presentation of Maxwellian electrodynamics and the electron theory is given in Abraham's revision of Föppl's 1894 textbook, retitled *Theorie der Elektrizität: Einführung in die Maxwellsche Theorie der Elektrizität* (Leipzig: B. G. Teubner, 1904), together with his own companion volume in 1905, *Theorie der Elektrizität: Elektromagnetische Theorie der Strahlung*.

166. Some examples are: Alfred Bucherer, *Elemente der Vektoranalysis mit Beispielen aus der theoretischen Physik*, 2d ed. (Leipzig: B. G. Teubner, 1905); Richard Gans, *Einführung in die Vektoranalysis mit Anwendungen auf die mathematische Physik* (Leipzig: B. G. Teubner, 1905); Waldemar von Ignatowski, *Die Vektoranalysis und ihre Anwendung in der theoretischen Physik*, 2 vols. (Leipzig and Berlin: B. G. Teubner, 1909–10); Eugen Jahnke, *Vorlesungen über die Vektorenrechnung. Mit Anwendungen auf Geometrie, Mechanik und mathematische Physik* (Leipzig: B. G. Teubner, 1905).

167. Because authors of works on the separate branches of physics sometimes use vector analysis exclusively, C. Christiansen and J. J. C. Müller introduced vectors in their text on all of physics, *Elemente der theoretischen Physik*, 3d rev. ed. (Leipzig: J. A. Barth, 1910).

168. This was the plan of Clemens Schaefer, *Einführung in die theoretische Physik*, vol. 1, *Mechanik materieller Punkte, Mechanik starrer Körper und Mechanik der Continua (Elastizität und Hydrodynamik)* (Leipzig: Veit, 1914), as he explains it on p. iv.

169. Hermann Minkowski, "Die Grundgleichungen für die elektromagnetischen Vorgänge in bewegten Körpern," *Gött. Nachr.*, 1908, 53–111, on 78–98. Minkowski's introduction of the four-dimensional space-time element is discussed in Holton, "The Metaphor of Space-Time Events in Science," 68.

istic four-dimensional vector of space and time.[170] Drawing largely on Sommerfeld's paper, Laue devoted a chapter to "world vectors and world tensors" and to associated mathematical concepts in his textbook on relativity theory in 1911.[171] Mie exploited an elaborate apparatus of matrices and higher dimensional vectors in his development of an electromagnetic theory of matter from 1912, as did Born and others who investigated Mie's theory.[172]

Tensors, which Voigt repeatedly urged physicists to adopt,[173] proved especially useful in the physics of fields to express concisely, and to manipulate conveniently, the multiple, related equations for energy, momentum, stress, and even mass; Abraham, for instance, emphasized the tensor character of the electron mass in his efforts to place physics on electromagnetic foundations.[174] Above all, Einstein made fundamental use of tensors in his development of the general theory of relativity. Grossmann devoted the mathematical part of his and Einstein's paper of 1913 to a "general vector analysis," an extension of the vector analysis for four dimensions recently developed by Minkowski, Sommerfeld, Laue, and others.[175] This analysis introduced the compact tensor formulation of Einstein's general theory of relativity.[176]

In its mathematical apparatus, much of theoretical physics changed in the early years of this century. Readers of the theoretical parts of the *Annalen*, for example, now routinely confronted four- and six-vectors, tensor masses, world-points and world-lines, and the rest of four-dimensional physics. They read about the symmetry

170. Arnold Sommerfeld, "Zur Relativitätstheorie. I. Vierdimensionale Vektoralgebra," and ". . . II. Vierdimensionale Vektoranalysis," *Ann.* 32 (1910): 749–76, 33 (1910): 649–89.

171. Laue, *Das Relativitätsprinzip,* chap. 4, "Weltvektoren und -tensoren," 60–76. In the second edition, Laue included the general theory of relativity; he believed that the main reason why physicists kept their distance from this theory was their insufficient familiarity with non-Euclidean geometry and the associated tensor calculus, which deficiency he intended his book to help overcome. "Vorwort," *Die Relativitätstheorie,* vol. 2: *Die allgemeine Relativitätstheorie und Einsteins Lehre von der Schwerkraft* (Braunschweig: F. Vieweg, 1921), v–viii, on vi.

172. Gustav Mie, "Grundlagen einer Theorie der Materie," *Ann.* 37 (1912): 511–34, 39 (1912): 1–40, and 40 (1913): 1–66. Max Jammer remarks on Mie's use of matrices in *The Conceptual Development of Quantum Mechanics,* 206. Max Born, "Der Impuls-Energie-Satz in der Electrodynamik von Gustav Mie," *Gött. Nachr.,* 1914, 23–36. In a footnote Born explained that in matrix multiplication, one multiplies row by column, which points up the expected unfamiliarity of physicists with the properties of matrices at this time (p. 33).

173. Salomon Bochner, "The Significance of Some Basic Mathematical Conceptions for Physics," *Isis* 54 (1963): 179–205, on 193.

174. Max Abraham, "Dynamik des Electrons," *Gött. Nachr.,* 1902, 20–41, on 28.

175. Einstein and Grossmann, "Entwurf einer verallgemeinerten Relativitätstheorie und einer Theorie der Gravitation," 244.

176. The increasing compactness of physical mathematics in this century can be traced through the sequence of Einstein's publications. For example, in 1905 Einstein wrote Maxwell's equations as Hertz had, six equations for the two sets of x, y, and z components, to which two additional equations needed to be added, making *eight* altogether. In 1908 he wrote them using three-dimensional vector notation together with the vector operators of "curl" and "divergence"; the resulting equations were *four* in number, only half as many. In 1916 he wrote the generalization of Maxwell's equations for the vacuum as *two* four-dimensional tensor equations: $\partial F_{\rho\sigma}/\partial x_\tau + \partial F_{\sigma\tau}/\partial x_\rho + \partial F_{\tau\rho}/\partial x_\sigma = 0$, $\partial F^{\mu\nu}/\partial x_\nu = J^\mu$. Here the F components correspond to the electric and magnetic forces and the J to the current and charge. Einstein, "Zur Elektrodynamik bewegter Körper," 907; Einstein and Jakob Laub, "Über die elektromagnetischen Grundgleichungen für bewegter Körper," *Ann.* 26 (1908): 532–40, on 533; Einstein, "Die Grundlage der allgemeinen Relativitätstheorie," 812–13.

properties of sixteen-term matrices or world-tensors and ten-member groups of motions and were directed to the mathematicians' literature on the theory of transformations.[177] In this highly mathematical writing on relativity, Einstein might not even be mentioned, but Minkowski would be for his representation of the Lorentz transformations as an imaginary rotation in space-time, Voigt for his exposition of vector and tensor products, and Sommerfeld for his extension of vector and tensor analysis to four dimensions.[178]

The "ideal" of the Frankfurt theoretical physicist Erwin Madelung was a "theoretical analog" of Kohlrausch's manual for laboratory physics. But Madelung concluded that such a manual could only be prepared by a number of physicists and mathematicians working in collaboration, so he decided to write for the "calculating physicist" a limited, practical book, omitting in most cases proofs of mathematical theorems. The result, *Die mathematischen Hilfsmittel des Physikers*, published in 1922, emphasized vector and tensor analysis, mathematical disciplines which Madelung regarded as having unusual importance for physicists. More than a shorthand, they were an aid to thinking visually, physically, and a knowledge of them led to a familiarity with other valuable mathematical disciplines such as invariance theory and differential geometry. Madelung divided his textbook into a mathematical part and a physical part, the latter consisting of chapters on mechanics, electricity, thermodynamics, and—testifying to the importance he attributed to it—relativity theory.[179]
Madelung's book was the fourth volume of a new series on the mathematical sciences under the general editorship of the Göttingen mathematician Richard Courant. The twelfth volume of the series, appearing two years later in 1924, was written by Courant himself together with his Göttingen colleague Hilbert. Based on Hilbert's lectures, their *Methoden der mathematischen Physik* was intended to unite the interests of mathematicians and physicists. Because the characterization of mathematical quantities by extremum properties leads to "simplification and unification"—a point appreciated by theoretical physicists we have discussed—Courant and Hilbert used the variational calculus throughout, by this approach formulating classical physics as "unified mathematical theories." As an alternative to solving differential equations with boundary conditions, they treated variation problems directly. Using variation methods, they derived fundamental laws of the eigenvalues of vibrating systems, a subject to which they gave much attention and which proved to be of great significance for the next stage of atomic theory.[180] Courant and Hil-

177. Van Alkemade, 38: 1033–40; Frank and Rothe, 34: 825–55, which follows the presentation of group theory in Sophus Lie and G. Scheffers, *Vorlesungen über kontinuierliche Gruppen* (Leipzig, 1893).

178. Laue, 35: 524–42; Epstein, 36: 779–95; Mie, 37: 511–34; Frank, 35: 599–606; Herglotz, 36: 493–533.

179. Erwin Madelung, *Die mathematischen Hilfsmittel des Physikers* (Berlin: Springer, 1922), quotations on v. This is vol. 4 of the series: Die Grundlehren der mathematischen Wissenschaften in Einzeldarstellungen mit besonderer Berücksichtigung der Anwendungsgebiete, under Richard Courant's general editorship.

180. Richard Courant and David Hilbert, *Methoden der mathematischen Physik*, vol. 1 (Berlin: Springer, 1924), quotations on v–vi. Jammer, *The Conceptual Development of Quantum Mechanics*, 207, com-

bert's and Madelung's books were both commended by physicists for the "simplicity" of their mathematical methods, which made them accessible for work in theoretical physics.[181]

Courant and Hilbert's book and the earlier one by Riemann and Weber were the only books with a goal similar to that of the course of lectures at Jena University on the "methodology" of theoretical physics, according to the lecturer Auerbach. Since both of these books were written by mathematicians, Auerbach saw the need for a more elementary textbook written by a physicist such as himself. In 1925 he brought out *Die Methoden der theoretischen Physik,* which dealt primarily with the partial differential equations of physics, still the most important mathematical subject in his opinion. He also included in it vectors, tensors, and Minkowski's form of relativity theory as an example of graphical methods. The "language" of physics is mathematics, Auerbach said, and his textbook and others such as Madelung's and Courant and Hilbert's, originating in university lectures, were intended to aid the student and the researcher to master the expanding vocabulary of this language.[182]

It is clear that by around the time of the Solvay Congress and of Einstein's general theory of relativity, physicists had a wide range of proven mathematical aids at their disposal. Textbooks and handbooks conveniently organized for their needs the essential materials dealing with, for example, differential equations, the series development of arbitrary functions, linear integral equations, linear transformations, the variational calculus, the probability calculus, non-Euclidean and four-dimensional geometries,[183] and "multiple algebras" such as those of complex numbers,[184] dyadics, quaternions, vectors, tensors, matrices, and groups.[185] Moreover, the rapid advance of physics at this time implied that other branches of mathematics, such as the calculus of finite differences, might soon become the source of new standard

---

ments on the importance of this text for quantum mechanics. Volume 2 of Courant and Hilbert's text, treating the classical partial differential equations of physics, came out in 1937, also published by Springer.

181. Born's review of Madelung's book in *Phys. Zs.* 24 (1923): 246–47; P. P. Ewald's review of Courant and Hilbert's book in *Naturwiss.* 13 (1925); 384–87.

182. Felix Auerbach, *Die Methoden der theoretischen Physik* (Leipzig: Akademische Verlagsgesellschaft, 1925), v–vi, 42–43, 180. Physicists commonly spoke of mathematics as their "language," as Auerbach did. Hertz, for example, in writing to Heaviside of his difficulty understanding him, said: "You know mathematical symbols are like a language and your writing is like a very remote dialect of it." Letter of 21 Mar. 1889, quoted in Appleyard, *Pioneers,* 238.

183. Early examples are discussed in Ernst Wölffing, "Die vierte Dimension," *Umschau* 1 (1897): 309–14.

184. Although complex numbers were used in classical physics, especially in relativity theory, "basic conceptualizations and basic formulations continued to be presented and expressed in real variables only"; it was not until quantum mechanics that the "very basic equations" of the theory displayed the symbol i "openly and directly," as we will see in the next chapter. Bochner, "The Significance of Some Basic Mathematical Conceptions for Physics," 196.

185. In his analysis in 1905 of the relativistic composition of velocities, Einstein observed that the transformations of coordinates between parallel, moving coordinate systems form a "group." Einstein, "On the Electrodynamics of Moving Bodies," 51. Henri Poincaré, "Sur la dynamique de l'électron," *Comptes rendus* 140 (1905): 1504–8, and *Palermo, Rend. Circ. Mat.* 21 (1906): 129–75, introduced group theory and four-dimensional vectors into Lorentz's electron theory. This point is discussed in Camillo Cuvaj, "Henri Poincaré's Mathematical Contributions to Relativity and the Poincaré Stresses," *Am. J. Phys.* 36 (1968): 1102–13, especially 1109–11 and 1113.

techniques for the theoretical physicist. For example, as a result of his participation in the Solvay Congress, Poincaré announced in 1912 that the greatest revolution in physics since Newton seemed to be in progress. He characterized this new revolution by the way it upset physicists' assumptions about the branch of mathematics that was appropriate for expressing fundamental physical laws. He explained that ever since Newton the laws of theoretical physics were thought to be inseparable from differential equations, but that the discontinuities associated with Planck's quantum theory now seemed to demand another mathematics.[186]

Experimentalists sometimes felt the need for more mathematics. "Herr Gott," Braun exclaimed to Graetz, "if only I had your mathematics and theory," and Haber wrote to Sommerfeld of his admiration for the "playful ease with which you master the mathematical apparatus." Experimentalists, however, could get by with modest mathematics together with their other skills. It was different for theorists: they knew that to do serious theoretical physics they needed a good mathematical knowledge and ability. Applying to Sommerfeld to come to Munich to work with him, Ehrenfest explained that he could not carry through calculations to the end, and unless he learned he would "be ruined" as a theorist.[187]

For all their need of mathematics, physicists did "physics not mathematics," Helmholtz told his theoretical physics class.[188] Volkmann put it sharper: "Theoretical physics is an independent discipline, which has been enormously served by mathematics, but which will tolerate no mathematical baby halter."[189] Planck, as editor of the *Annalen*, welcomed any interest shown by mathematicians in physics; he admired, for example, Hilbert's derivations from a formal standpoint, but he acknowledged that "physically they bring nothing that is in the least new." Mie explained the difference, as he saw it, between a physicist and a mathematician: they do not mean the same things by terms such as "equal" and "infinity," and so they often misunderstand one another. To the physicist, "equal" means "within the limits of error," and these limits vary from problem to problem; in one problem, two quantities are equal, in another unequal. The physicist sees a problem as numerical, the "modern mathematician" does not.[190]

Physicists sometimes used the expressions "mathematical physics" and "theoretical physics" interchangeably; at other times they used them in consciously contrasting senses. Increasingly they tended to choose their words carefully and to refer to their work as "theoretical physics." Wien explained that mathematicians, such as Hilbert, did mathematical physics, which was concerned with developing mathe-

186. Henri Poincaré, "Sur la théorie des quanta," *Journal de physique théorique et appliquée* 2 (1912): 5–34, on 5. McCormmach, "Henri Poincaré and the Quantum Theory," 43, 49, 55.

187. Braun to Graetz, 10 Apr. 1887, Ms. Coll., DM. Haber to Sommerfeld, 29 Dec. 1911; Ehrenfest to Sommerfeld, 17/30 Sept. 1911, Sommerfeld Correspondence, Ms. Coll., DM.

188. Helmholtz told his audience that in contrast to the pure mathematician who turns to physics for examples of difficult mathematical problems, he was going to treat physics. *Einleitung zu den Vorlesungen über theoretische Physik*, 25.

189. Volkmann to Sommerfeld, 3 Oct. 1899, Sommerfeld Papers, Ms. Coll., DM.

190. Planck to Wien, 4 Oct. 1912, Wien Papers, STPK, 1973.110. Mie to Hilbert, 26 Dec. 1917, Hilbert Papers, Göttingen UB, Ms. Dept.

matical methods applicable to physics. Theoretical physics was done by physicists, who were concerned with developing the most general laws of physical phenomena, and to this end physicists worked with physical hypotheses and approximations. "But what is a theoretical physicist?" Boltzmann asked, and in the answer he gave, he explained why it is inappropriate to characterize the work of the theoretical physicist as "mathematical physics." After all, the experimentalist does complex calculations in evaluating his observations and so does the technical physicist, but neither of them is doing theoretical physics. The theorist tries to grasp the phenomena "qualitatively and quantitatively in their whole course," which means that he must have a sure command of mathematics, but he is not doing mathematics. He is ordering the phenomena, describing them clearly and simply: "theoretical physics has the task of—as we used to say—finding the fundamental causes of phenomena or—as we prefer to say today—it has the task of uniting the experimental results we have obtained under unified points of view."[191]

For Einstein and other theoretical physicists of the twentieth century, as for Helmholtz or, for that matter, Ohm in the nineteenth century, mathematics was an indispensable tool of their work, but it was not their goal. Physical understanding was their goal, to which they brought to bear all kinds of mathematical techniques, which varied with the development of both physics and mathematics. In this chapter we have seen how a physical problem, that of gravitation, was solved by recourse to a recent branch of mathematics, the absolute differential calculus. The result was a fundamentally new understanding: the general theory of relativity entailed the most far-reaching revision of the physical world picture before the creation of an acausal quantum mechanics of the atom, with its own characteristic mathematics, in the mid-1920s.

191. Wien, "Ziele und Methoden der theoretischen Physik," 242. Ludwig Boltzmann, "Josef Stefan," in *Populäre Schriften*, 92–103, on 94.

27

The Coming of the New Masters

In the 1920s the problems of the atom attracted many of the most talented new theoretical physicists, as they had already attracted some of the established masters. Theoretical physics was embarked once again on a test of strength with fundamental problems, this time bringing to it all of the tools of a fully matured field. Its several institutes attracted researchers who joined in the common effort, and the theoretical physics that resulted was of remarkable quality. In this chapter we look at the work of theoretical physics once more. It is not meant to be a conclusion, and in light of the design of this book, it could not be, for the work of theoretical physicists in Germany, of course, continued on after the 1920s and continues today. We offer the following brief account of work in atomic theory *in* conclusion, as a fitting place to end this history of theoretical physics.

Atomic Theory: Relativity and the Quantum Joined

In the midst of his work on the gravitational problem, Einstein wrote to Ehrenfest on 19 August 1914: "Europe, in her insanity, has started something unbelievable. . . . My dear astronomer Freundlich will become a prisoner of war in Russia instead of being able there to observe the eclipse of the sun. I am worried about him."[1]

World War I affected theoretical physicists in a number of ways. Junior teachers and assistants were called up or volunteered to serve in the military. Ordinary professors gave over some of their time to support the war effort while they kept up the activity of their institutes as best they could. Their advanced students were mostly

---

1. Einstein to Ehrenfest, 19 Aug. 1914, quoted in *Einstein on Peace,* ed. O. Nathan and H. Norden (New York: Schocken, 1968), 2. In the event, the measurements of the deflection of light by the sun had to wait for another eclipse; in 1919 two British expeditions were sent out to make the observations, which favored Einstein's predictions.

gone, as were the students who made up their big classes;[2] yet enough students turned up wanting instruction in the understaffed institutes that their directors were burdened and requested assistance.[3] Budget cuts, inflation, and isolation from other countries made it difficult for the directors to keep their institutes running effectively.[4]

Research, as teaching, continued during the war, again under constraints. Even physicists who were called up did research in what spare time and with what resources they could find.[5] Physics journals went on publishing, and even a new journal for atomic physics was proposed, but it was judged irresponsible since existing journals had shrunk in size.[6] The editors of the *Annalen*, anticipating that for the duration of the war and for some time after research would be reduced, agreed to evaluate manuscripts as before, preferring a smaller journal to a lessening of quality.[7]

During the war, German physicists inevitably received less stimulus from physicists abroad than before. At first they occasionally corresponded with colleagues in countries at war with Germany, but they did not keep it up. Publications from these countries were hard to come by or inaccessible. International congresses of physicists, their visits to one another's national meetings, and their easy movements between countries were for the time being a thing of the past. Collegial relations were strained by the war, painfully so by specific actions such as the signing of patriotic manifestos by physicists on both sides. Voigt, who in previous years had met with colleagues at international gatherings in London, Brussels, and St. Petersburg, anticipated that German physicists would not recover their valuable scientific relations with their counterparts in enemy countries for a long time to come.[8]

The war, in addition to causing deprivations such as these, took its toll physically and emotionally. In 1916 the normally prolific researcher Voigt could send a colleague only a small work on acoustics; he explained that in the wake of world events, he lived with political cares day and night, so that physical ideas entered his

2. Des Coudres was told that he could expect about one fifth or one quarter of the normal enrollment in his institute. Wiener to Des Coudres, 23 Oct. 1914, Wiener Papers, Leipzig UB, Ms. Dept. Attendance in Planck's institute fell from two hundred to forty or fifty. Planck to Wien, 8 Nov. 1914; Wien to L. Holborn, 22 Oct. 1915; Wien Papers, STPK, 1973.110.

3. From descriptions of physics institutes early in the war by their directors: Debye to Göttingen U. Curator, 30 Mar. 1915, Göttingen UA, 4/V h/35; Lenard to the Baden Ministry of Justice, Culture, and Education, 9 Nov. 1914, Bad. GLA, Physikalisches Cabinet der Universität Heidelberg; Kayser, "Erinnerungen," 295–96; Riecke to Stark, 2 Jan. 1915, Stark Papers, STPK; Wien, *Aus dem Leben*, 31; Wiener to Wien, 24 Nov. 1914, Ms. Coll., DM.

4. Koenigsberger gave all of these reasons in his request for more support from the Baden government; he especially missed support from abroad, from the Thompson and Solvay foundations, which his institute had relied on. Koenigsberger to the Baden Ministry of Justice, Culture, and Education, 25 Feb. 1915 and 22 Sept. 1916, Bad. GLA, 235/7769.

5. Paul Forman, "Alfred Landé and the Anomalous Zeeman Effect, 1919–1921," *HSPS* 2 (1970): 153–261, on 160.

6. The proposed journal was entitled *Archiv für Radiologie und Atomphysik*. Wien to Stark, 5 Feb. 1916, AHQP.

7. Planck to Wien, 4 May 1915, Wien Papers, STPK, 1973.110.

8. Voigt, "Ansprache gelegentlich der Zusammenkunft der Lehrer der Georgia-Augusta am 31. Oktober 1914," Voigt Papers, Göttingen UB, Ms. Dept.

head only as transient guests. Early in the war, as we have seen, Einstein completed the theory of general relativity, which belongs to the most important researches of his career; but later he told Lorentz that he was "constantly very depressed over the immeasurably sad things which burden our lives" and that it no longer helped, "as it used to, to escape into one's work in physics."[9]

Of the researches done in Germany early in the war, Sommerfeld's on atomic structure was among the most significant for the next major phase of theoretical physics; at the same time it marked the beginning of "Sommerfeld's great period," according to Laue.[10] Before the war, Sommerfeld had been interested in the quantum but not yet in its application to the detailed structure of atoms, and in this regard he was typical of German theorists. At the 1911 Solvay Congress, Sommerfeld had brought together Planck's universal quantum of action with the Maxwell-Lorentz theory, viewing the quantum as compatible with, even as it was foreign to, the "electromagnetic world picture." At this time, Sommerfeld was investigating hypotheses about the quantum, which he expected to account for the existence of atoms.[11]

Following the Danish physicist Niels Bohr's atomic theory of 1913, Sommerfeld moved from his early interest in only the most general properties of atoms to an interest in their detailed structure. Bohr worked with Rutherford's model of the atom, according to which a small positive nucleus is surrounded by negative electrons. Bohr located the electrons in discrete orbits within the atom, and by introducing a frequency condition, which relates energy to frequency through Planck's constant h, he calculated the quantity of radiant energy that an electron emits or absorbs in passing from one orbit to another. To explain the working of this model, Bohr had to restrict the validity of mechanics and electrodynamics.[12]

Bohr's most impressive initial accomplishment was to derive from his atomic theory the Balmer series of the hydrogen spectrum. Sommerfeld, whose interest in spectra was expressed by the regular appearance of topics such as the Zeeman and Stark effects in the Munich colloquium, was drawn to Bohr's theory in the first place because it promised an understanding of the production of spectra.[13] At the same time, Bohr's theory pointed to the vast store of observations of optical spectra as a key to the problem of the structure of atoms. It also pointed to a new branch of spectroscopy that had originated a few years before with Laue's Röntgen-ray work in Sommerfeld's own institute and with the corresponding work of W. H. Bragg and

9. Voigt to Auerbach, 10 Jan. 1916, Auerbach Papers, STPK. Einstein to Lorentz, 18 Dec. 1917, quoted in *Einstein on Peace*, 21.

10. Max Laue, "Sommerfelds Lebenswerk," *Naturwiss.* 38 (1951): 513–18, on 516.

11. Arnold Sommerfeld, "Das Plancksche Wirkungsquantum und seine allgemeine Bedeutung für die Molekularphysik," *Verh. Ges. deutsch. Naturf. u. Ärzte*, pt. 2, 1911, 31–49. Benz, *Sommerfeld*, 77–79. Forman and Hermann, "Sommerfeld," 528.

12. Bohr's atomic theories are analyzed in John L. Heilbron and Thomas S. Kuhn, "The Genesis of the Bohr Atom," *HSPS* 1 (1969): 211–90; Helge Kragh, "Niels Bohr's Second Atomic Theory," *HSPS* 10 (1979): 123–86.

13. Sigeko Nisio, "The Formation of the Sommerfeld Quantum Theory of 1916," *Jap. Stud. Hist. Sci.*, no. 12 (1973): 39–78, on 54–56.

W. L. Bragg in England. Soon after the Braggs's work, H. G. J. Moseley showed experimentally that characteristic Röntgen-ray spectra are, like optical spectra, precisely ordered by simple equations, and that they are governed by the nuclear charge, or atomic number. This work on spectra helped to guide the construction and testing of atomic models.[14]

While still skeptical in principle of atomic models, Sommerfeld tried out Bohr's theory on the Zeeman effect. Encouraged by what he found, he worked intensively on Bohr's theory through 1914, and in 1915 he lectured on it in his institute and began a series of studies on it in relation to spectra.[15] In December 1915 he read a paper to the Bavarian Academy of Sciences in which he analyzed the motions of an electron within an atom in a stationary state by means of the "phase integral":

$$\int p dq = nh,$$

where p is the momentum of the electron corresponding to its conjugate coordinate q, n is a positive integer, and the integration is carried over a complete period of the coordinate. If the system has more than one degree of freedom, each coordinate has a corresponding phase integral. Others, notably Planck, had quantized "phase space," the abstract space of p and q, providing a precedent for this direction of Sommerfeld's thought. Sommerfeld did not prove the phase integrals but began with them, presenting them as a generalization of the quantized treatments of simple systems such as the oscillator and the rotator. Now and for some time, he regarded the quantum condition of the phase integrals as the, perhaps unprovable, foundation of the quantum theory.[16]

In this 1915 paper, Sommerfeld went beyond Bohr, who had examined only the circular motion of an electron about the nucleus. Treating the more realistic Keplerian elliptical motion of an electron, a next step which Bohr had already suggested, Sommerfeld showed that the energy of the electron varies continuously with the eccentricity of the ellipse. This would produce a continuous hydrogen spectrum unless the eccentricity, too, were quantized. To regain the observed discrete spectrum from elliptical motions, Sommerfeld needed two quantum conditions, instead of one, and two associated quantum numbers. For this purpose he introduced a phase integral for the radial coordinate of the electron as well as one for the angular coordinate. The resulting equation for the energy of the Keplerian motion is so "precise and fruitful," Sommerfeld said, that it could not be an "algebraic accident."[17]

14. The bearing of X-ray spectroscopy on Sommerfeld's atomic theory is discussed in John L. Heilbron, "The Kossel-Sommerfeld Theory and the Ring Atom," *Isis* 58 (1967): 451–85.

15. Benz, *Sommerfeld*, 80–84. Forman and Hermann, "Sommerfeld," 528–29.

16. Arnold Sommerfeld, "Zur Theorie der Balmerschen Serie," *Sitzungsber. bay. Akad.*, 1915, 425–58. Independently, and at about the same time, in 1915, the British physicist William Wilson and the Japanese physicist Jun Ishiwara proposed quantum conditions similar to Sommerfeld's phase integrals, and that year Planck introduced a method for quantizing phase space in connection with his generalized quantum theory of systems of any number of degrees of freedom. Sommerfeld and Planck soon entered into a correspondence, which Sommerfeld found useful in further developing his theory. Nisio, "Formation," 69–75. Benz, *Sommerfeld*, 90–91. Kuhn, *Black-Body Theory*, 250–51.

17. Sommerfeld, "Zur Theorie der Balmerschen Serie," 439.

Upon recovering the Balmer formula—a remarkable result, he thought—he compared the theoretical multiplicity of the Balmer lines produced by transitions between elliptical paths with observations of the multiple components of the Stark effect in hydrogen. He pointed out that in more complicated atoms than hydrogen, there is less symmetry, so that to define the motions of electrons in these atoms, a third coordinate must enter and with it a third quantum number. Throughout the study, he made use of the Bohr frequency condition for transitions within the atom, though regarding the condition as provisional only. In the transitions, energy and momentum change in a characteristic way, and since ordinary mechanics deals with processes in which energy and momentum remain constant, Sommerfeld anticipated that "entirely new laws of mechanics must be found."[18]

In January 1916, Sommerfeld read a sequel paper to the Bavarian Academy in which he added to his theory a relativistic treatment of the electron, again following a suggestion by Bohr. Because of the relativistic variation of the electron's mass with velocity, by which the attractive force on the electron deviates slightly from the inverse square law, the elliptical orbit slowly precesses, a new periodicity requiring yet another quantum condition. Sommerfeld compared the theoretical fine splitting and, in certain cases, the magnitudes of spectral lines with observations of characteristic Röntgen rays and of the fine structure of lines of hydrogen, ionized and neutral helium, and lithium. He found excellent agreement in places, and where there was disagreement, as there was in the case of the absolute magnitudes of doublet and triplet lines, he thought that the "foundations of the quantum theory or relativity theory are responsible."[19] The agreement was in any case sufficient for him to claim a direct confirmation of the "reality" of elliptical orbits.[20]

Sommerfeld expanded his two Bavarian Academy papers into a three-part paper for the *Annalen in* 1916. Much of the new paper was simply a republication of the old, but there were significant additions. Sommerfeld generalized the quantum law of the phase integrals to apply to systems of any number of degrees of freedom; he revealed the close connection of the phase integrals with the "concepts and methods of the general Hamilton-Jacobi mechanics"; he quantized the spatial orientation of the elliptical orbits; and in general he made his theory more complete.[21] It was a sign of his confidence in the fundamental, exact nature of the new atomic theory that he proposed in place of Planck's system of "radiation-theoretical universal units" for physics a new system of "universal spectroscopic units."[22]

18. Sommerfeld, "Zur Theorie der Balmerschen Serie," 432. John L. Heilbron, "Lectures on the History of Atomic Physics 1900–1922," in *History of Twentieth Century Physics,* ed. C. Weiner (New York: Academic Press, 1977), 40–108, on 78–80.

19. Arnold Sommerfeld, "Die Feinstruktur der Wasserstoff- und der wasserstoff-ähnlichen Linien," *Sitzungsber. bay. Akad.,* 1915, 459–500, on 459–61. Heilbron, "Lectures," 80–81.

20. Sommerfeld, "Die Feinstruktur," 459.

21. Arnold Sommerfeld, "Zur Quantentheorie der Spektrallinien," *Ann.* 51 (1916): 1–94, 125–67, reprinted in Sommerfeld, *Gesammelte Schriften,* 4 vols., ed. F. Sauter (Braunschweig: F. Vieweg, 1968), 3: 172–308, quotation on 212. The first two parts of the *Annalen* paper bear the same titles as the Bavarian Academy papers. The third, greatly expanded part is entitled "Theorie der Röntgenspektren." The relationship between the academy and the *Annalen* papers is analyzed in Nisio, "Formation," 62–67.

22. Sommerfeld, "Zur Quantentheorie der Spektrallinien," 261–65.

Sommerfeld's work on the atom was generally well received. Planck, who later called Sommerfeld's work a "synthesis of quantum theory and relativity theory," thought that Sommerfeld had improved upon his and Bohr's work by using relativistic instead of "classical" mechanics in quantizing phase space.[23] Planck wrote to Sommerfeld at the time: "That relativistic mechanics not only enters easily into the quantum theory but also takes better account of the facts than classical mechanics is really all that one can demand and wish."[24] Sommerfeld's work came as a "revelation" to Einstein, and he, Planck, and other members of the Prussian Academy of Sciences, in proposing Sommerfeld as a corresponding member, referred to it as one of the strongest supports of Bohr's theory of spectra.[25] Lorentz admired the theory, as did Bohr, who withheld from publication for a time a comprehensive paper of his own after receiving Sommerfeld's paper.[26] In explaining why he did, Bohr said that Sommerfeld's relativistic theory together with Paschen's experimental confirmation removed any doubt about the essential direction of atomic theory, though Bohr preferred different quantum conditions than Sommerfeld's.[27] Toward the end of World War I, in a talk on recent developments in physics in Germany, Sommerfeld had a chance to discuss his own work on the atom. The elucidation of the fine structure of spectral lines was an indication that physics had acquired a "deep view into the interior of atoms," he said; this work united the most significant directions of recent physical theory: the electron theory, the quantum theory, and relativity theory.[28]

Sommerfeld's treatment of the fine structure of spectral lines made him "at once the leading theoretical spectroscopist,"[29] an authority he soon consolidated by writing the standard text on the subject. Based on lectures on atomic models he gave to a general audience at Munich University in the winter of 1916–17, this text, *Atombau und Spektrallinien*, was Sommerfeld's first in theoretical physics. Appearing in 1919, it presented Bohr's atomic theory and Sommerfeld's development of it in 1915 and after.[30] It was the "bible of the modern physicist," Born soon wrote to Sommerfeld.[31]

23. Planck to Wien, 15 Feb. 1916, Wien Papers, STPK, 1973.110. Planck, "Arnold Sommerfeld zum siebzigsten Geburtstag," 370.

24. Planck to Sommerfeld, 11 Feb. 1916, Sommerfeld Correspondence, Ms. Coll., DM.

25. Einstein to Sommerfeld, 8 Feb. 1916, in *Einstein/Sommerfeld Briefwechsel*, 40–41, on 40. Proposal of Sommerfeld as corresponding member of the mathematical-physical class of the Prussian Academy of Sciences, signed by Einstein, Planck, Warburg, Haber, and Rubens [Jan./Feb. 1920], in Kirsten and Treder, eds., *Albert Einstein in Berlin*, 126–27.

26. Nisio, "Formation," 75–76.

27. Niels Bohr's introduction to *Abhandlungen über Atombau aus den Jahren 1913–1916*, trans. H. Stintzing (Braunschweig: F. Vieweg, 1921), iv–v, xii, xv–xvi.

28. Arnold Sommerfeld, "Die Entwicklung der Physik in Deutschland seit Heinrich Hertz," *Deutsche Revue* 43 (1918): 122–32, on 131.

29. Born, "Sommerfeld," 284.

30. Arnold Sommerfeld, *Atombau und Spektrallinien* (Braunschweig: F. Vieweg, 1919). Sommerfeld brought out frequent revisions of this text, which recorded the development of the subject up to the introduction of quantum mechanics, and beyond.

31. Born to Sommerfeld, 13 May 1922, Sommerfeld Correspondence, Ms. Coll., DM. Comments on the second edition of Sommerfeld's text.

Although in *Atombau* Sommerfeld primarily reported the accomplishments of the new field, he also discussed certain unsolved problems; in particular, because the subject was atoms and the light they emit and absorb, he discussed the "most difficult and at the same time most interesting question" of the relationship between the quantum theory of atoms and the wave theory of light, between the "old world of waves" and the "new world of quanta." The energy and momentum of the stationary states of the atom are competently described by atomic theory, but the description of the energy and momentum of the ether runs into the incompatible oppositions of "modern physics." The bridge between the "classical" wave theory of light and Einstein's "extreme standpoint" of light quanta, Sommerfeld noted, had yet to be built.[32]

Sommerfeld's deduction of spectral properties from phase integrals proved useful for hydrogen with its single electron, but it did not for helium with its two electrons. Soon he departed from his earlier deductive approach to seek mathematical relations in the spectroscopic data and to infer from them atomic energy levels and the rules of transition between them. He and his students developed this approach to ordering spectra—atomic, molecular, and Röntgen ray—into a fine art, which inspired similar work outside Sommerfeld's institute.[33]

Sommerfeld's goal was an axiomatic basis for the quantum theory of atoms. He developed mathematical theories of the atom by closely attending to experimental results, in turn drawing consequences from his theories to be tested by further experiments. He was primarily concerned with attaining agreement between theoretical prediction and experimental measurement; his inclination was to make thorough calculations rather than to analyze the internal perfection of theories. He was not the discoverer of new fundamental principles of physics. His work was sharply distinguished from that, say, of Bohr, Einstein, and Planck: "If the distinction between mathematical and theoretical physics has any significance," Born observed, "its application to Sommerfeld ranges him decidedly in the mathematical section."[34]

In 1918 Sommerfeld contributed one of the papers to honor Planck on his sixtieth birthday, to which *Die Naturwissenschaften* devoted an entire issue. Physics in the twentieth century was "in increasing measure a physics of quanta," Sommerfeld wrote, an observation affirmed by the other contributors, who referred repeatedly to the "classical period" in physics that the quantum theory had brought to a close. Sommerfeld predicted that the greatest scientific surprises would come from the connection of the quantum theory with atomism if, that is, the world were ever again to find the peace and quiet for scientific contemplation.[35]

32. Sommerfeld, *Atombau*, vii, 477–78.

33. Laue, "Sommerfelds Lebenswerk," 516–17. Benz, *Sommerfeld*, 122. Forman and Hermann, "Sommerfeld," 529.

34. Born, "Sommerfeld," 282. Born, "Sommerfeld als Begründer," 1035. Laue, "Sommerfelds Lebenswerk," 518. Benz, *Sommerfeld*, 90. Nisio, "Formation," 76–77.

35. Arnold Sommerfeld, "Max Planck zum sechzigsten Geburtstage," *Naturwiss.* 6 (1918): 195–99, on 195, 198. Sommerfeld also gave a talk on Planck at the German Physical Society, which, together with talks on Planck by Einstein, Laue, and Warburg, was separately published that year.

Quantum Mechanics

For some time after the war, German physicists continued to be isolated from colleagues in former enemy countries, and their work was impeded by postwar political and economic dislocations. Yet through it all, including an inflation of runaway proportions, research was not brought to a standstill. Institute budgets were more or less maintained, and physicists got additional research support from a variety of sources, from individuals, societies, foundations, firms, and, most important, the central government.[36]

At the end of the war, the teaching staff for theoretical physics was nearly what it had been at the beginning, with only a few replacements and additions.[37] We have already described the complicated changes at Göttingen brought about by Riecke's and Voigt's retirements and Debye's call there. New universities in Frankfurt am Main, Hamburg, and Cologne and the loss of Strassburg University meant several more changes.[38] But these were about all. The few new men who taught theoretical physics in 1919 did not of themselves point toward an outstanding postwar development of the subject, at least not an immediate one. The hope would seem to lie more with the new students and with the established theorists, with Born and Laue, holders of the new ordinary professorships for theoretical physics in Frankfurt and Berlin, with Sommerfeld in Munich, and with Einstein and perhaps Planck, who was nearing retirement, in Berlin. Experimental physics professors such as Franck and Paschen promised to sustain the theoretical development in atomic physics.

The institutional position of theoretical physics was strengthened after the war. Extraordinary professors were admitted to the faculties, and with them theoretical physics was represented as an independent field. And in the five years after 1921, all but two of the German universities that had not already done so converted their extraordinary professorships for theoretical physics into ordinary ones, either regular

36. Physicists from former enemy nations might not know what to expect when they met. That was so when Planck and several other German physicists met the British physicist W. Barkla in Stockholm at the Nobel prize ceremonies; Barkla proved congenial, so they treated him the same in return. Planck to Wien, 20 June 1920, Wien Papers, STPK, 1973.110. The political and economic conditions under which physics research was carried out in Germany after World War I are analyzed by Paul Forman in, among other places, "The Environment and Practice of Atomic Physics in Weimar Germany: A Study in the History of Science" (Ph.D. diss., University of California, Berkeley, 1967); "Scientific Internationalism and the Weimar Physicists: The Ideology and Its Manipulation in Germany after World War I," *Isis* 64 (1973): 151–80; "The Financial Support and Political Alignment of Physicists in Weimar Germany," *Minerva* 12 (1974): 39–66.

37. Two principal lecturers in theoretical physics died natural deaths during or right after the war: they were Ernst Pringsheim at Breslau, who was replaced by Clemens Schaefer, and Leonhard Weber at Kiel, who after a delay was replaced by Walther Kossel. Mathias Cantor at Würzburg was killed in the war, and he was replaced by Friedrich Harms. Tübingen University's Edgar Meyer returned to Switzerland, his home, to be replaced by Christian Füchtbauer. Gustav Mie left Greifswald to become the experimental physics professor at Halle.

38. Strassburg's theorist Emil Cohn moved to Rostock in 1919 and then, the year following, to Freiburg. In 1919–20 Born taught theoretical physics at Frankfurt am Main, and J. W. Classen taught it at Hamburg. Druxes taught theoretical physics at Cologne in 1921–22.

or honorary.[39] At Münster, for example, the faculty asked the Prussian ministry to name the extraordinary professor for theoretical physics Heinrich Konen "personal ordinary professor." They argued that not only was Konen deserving—he had taught "very successfully" for fourteen years at Münster—but that theoretical physics was deserving as well. At many other Prussian universities, they pointed out, physics was taught by a theoretical physics professor as well as by an experimental one; its teaching needed to be stressed more at Münster, because the "powerful revolution that physics has undergone in the last decade we owe in greatest part to theoretical physics."[40]

When in 1919 Debye received an offer from Zurich, the Göttingen faculty petitioned the minister to try to keep him, even though they recognized the unlikelihood of succeeding. In making their case, they acknowledged the heightened political sensitivity of German universities and ministries after the war by recalling that in the first year of the war Göttingen had won out in the competition with Zurich for Debye. That had counted as a success for Prussia; if now Göttingen were to lose Debye in the same competition in the first year after the war, that would be counted as a failure for Prussia, one for the "whole scientific world" to see. Debye accepted the polytechnic's offer in February 1920; at the same time, he felt anxious about the future of the Göttingen mathematical physics department he would leave behind.[41]

Debye's anxiety proved unnecessary. His replacement was Born, who went on to create at Göttingen an outstanding school of theoretical physics. Born came from Frankfurt, where he had gone right after the war, exchanging positions with Laue who wanted to be near Planck at Berlin.[42] At Frankfurt, Born had directed the theoretical physics institute, which had experimental facilities and a mechanic. He did some experimental work there with others, but he did mainly theoretical work of distinction. When after two years Göttingen asked him to become Debye's successor, his Frankfurt colleague Wachsmuth urged the ministry to keep him at Frankfurt, since none of the "younger theoretical physicists" were in his class.[43]

39. The principal theoretical physics lecturer was elevated at Bonn, Erlangen, Greifswald, Hamburg, Jena, Marburg, Münster, Rostock, Würzburg, and Cologne. That left Freiburg and Heidelberg with only extraordinary professors for the subject.

40. The personal ordinary professorship that Konen received did not affect his salary as extraordinary professor. Dean of Münster U. Philosophical and Scientific Faculty to Prussian Ministry of Culture (draft), 13 Aug. 1919; Minister to Konen, 26 Nov. 1919, Konen Personalakte, Münster UA.

41. Debye to Göttingen U. Curator, 7 Nov. 1919, 12 Feb. and 29 Mar. 1920; Dean of the Philosophical Faculty to Curator, 18 Nov. 1919; Debye Personalakte, Göttingen UA, 4/V b/278.

42. Laue had earlier been considered for Berlin. On 2 August 1914, Born, who was then Privatdocent for theoretical physics at Göttingen, received a letter from Planck informing him that there was to be a second position for theoretical physics at Berlin and asking him if he would accept it. Born said he would, but no official letter came, as World War I began. In December, Planck wrote again to say that Laue wanted the job and would get it, not Born. But Laue irritated the ministry, which offered the job to Born after all. Born took up his duties in 1915. Max Born, My Life: Recollections of a Nobel Laureate (New York: Scribner's, 1978), 161–64.

43. Born to Sommerfeld, 5 Mar. 1920, Sommerfeld Correspondence, Ms. Coll., DM. Prussian Minister of Culture to Frankfurt U. Curator, 4 Jan. 1919; Wachsmuth, as Dean of the Frankfurt U. Science Faculty, to Curator, 27 Apr. 1920; Born Personalakte, Göttingen UA. N. Kemmer and R. Schlapp, "Max Born," Biographical Memoirs of Fellows of the Royal Society 17 (1971): 17–52, on 20–21.

Göttingen's offer to Born of Debye's directorship of the "mathematical department of the physics institute" presented a problem, since Debye's department was equipped for experimental work. Born had never given experimental physics lectures or directed a large physics laboratory, and his attempts at experimental research at Frankfurt had convinced him that he had neither the training nor the patience to be a "true experimentalist." He had scruples about accepting the Göttingen job until he discovered that, on paper, there existed not only Pohl's extraordinary professorship for experimental physics at Göttingen, but also a second extraordinary professorship that had been created out of Voigt's ordinary professorship after Voigt's death. That second extraordinary professorship, by the arrangements made at the time of Debye's call to Göttingen, was to be given to Pohl, and the one Pohl already had was to be discontinued. But the complex plans of the royal ministry of culture of 1915 eluded the postwar ministry officials, and Born was quick to exploit their confusion. He got the ministry to retain both positions and to appoint an experimentalist to the open one, in effect dividing Debye's department into two departments, one for theory and one for experiment. For the latter, Born recommended Franck, who was then hired as ordinary professor for experimental physics.[44] So the arrival in Göttingen of Born, a theoretical specialist, resulted in a multiplication of physics chairs. Instead of the one full professorship that was intended to serve Göttingen after Riecke and Voigt, there were three: Born's, Franck's, and Pohl's.

The old institute building had enough rooms for all three, since Born did not need much space for himself. (Born shared rooms with Franck in the basement and on the ground floor, while Franck had the whole second floor and Pohl the whole first floor and attic.)[45] Born had no mechanic or workshop, and after directing one student in experimental research, he did not try again; Franck and Pohl doubted his "experimental abilities and showed this in a rather discouraging manner," confirming Born's own view of the matter. In any case, there was no need for Born's institute to carry out experimental research; for Born's colleague Franck had an uncommon talent for translating abstract theoretical ideas into experimental ones, which made possible a high degree of cooperation between theory and experiment in Voigt's old, now divided institute.[46]

An admirer of Bohr, Franck helped direct Born's interest toward atomic physics and the quantum theory. Bohr himself visited Göttingen in 1922 and succeeded in generating a good deal of interest in his subject. Born wrote to Sommerfeld that year that he now allowed his people "to quantize [quanteln] to give you a

---

44. Born remembered accurately the essential point; namely, that there were two extraordinary professorships; but his recollection that Voigt had one of the extraordinary professorships originated in a misunderstanding. Odd as the physics appointments at Göttingen must have seemed to Born when he learned of them at the time of his negotiations with the ministry, they contained no error as he and the ministry supposed but were merely incomplete in that the change in Pohl's position had not yet been made. Born, My Life, 200.

45. Prussian Minister of Culture to Born, 30 Nov. 1920, Born Personalakte, Göttingen UA.

46. Born, My Life, 210–11. There was less meeting of minds in the case of Pohl's institute than in the case of Franck's. Born recalled the "permanent petty sniping between the extreme experimentalists of Pohl's school" and his own theorists.

little competition."[47] Göttingen had become a "center of the emerging discipline."[48]

Young theoretical physicists moved readily between Bohr's and Born's institutes and also Sommerfeld's at Munich, the other main center in Germany. Born's first assistant was Wolfgang Pauli, Sommerfeld's student. Although Pauli did not have his degree yet, Born described him to the Göttingen curator as the "greatest talent in the field of physics that has surfaced in recent years." He wanted Pauli himself, but to present the case for him he said that Franck, Hilbert, and Runge all wanted a chance to work with him as well.[49] In ability Pauli was their equal, and although Born got "no great help from him" in his teaching, he learned from him and collaborated with him on research. When he left, Pauli recommended as his successor Heisenberg, another of Sommerfeld's students, who was also then working on his dissertation. Born urged Sommerfeld to let Heisenberg come because of his "unprecedented talent" and temperament and because Born needed help with his teaching. (Although Born had nine doctoral students of his own and four "descendents" of Sommerfeld's, none was far enough along to help him.)[50] So Heisenberg came, and like Pauli before him, he became Born's collaborator. The connections had now been established that were to lead to theoretical collaborations of the greatest significance for atomic physics.

In 1922 Planck assessed the state of theoretical physics in the following terms. He conceded that in the past, physics had had "beautiful theories," and that it had been "simpler, more harmonious, and therefore more satisfying" than it was now. But he insisted that the new, disconcerting theoretical ideas of physics were demanded by the facts, and if German physicists ignored them, they would fall behind physicists in other countries. He thought that over the last ten years the most important developments in physics were the relativity and quantum theories, the former being the more fully worked out of the two. The quantum theory was not yet a "real theory" and would not be until its relationship to the wave theory of light was clarified; it was still undergoing a strong development, as was evidenced by the "astonishing successes of Bohr's atom model."[51]

By the time Planck made this observation, important experimental work had been done in Germany that lent support to major features of Bohr's atom model and its elaboration by Sommerfeld and others. In 1914 Franck and Gustav Hertz, in experiments undertaken for another purpose, demonstrated the existence of stationary states by bringing about transitions between the states through collisions of electrons with atoms. In 1921 Otto Stern and Walter Gerlach demonstrated space quan-

47. Born to Sommerfeld, 13 May 1922, Sommerfeld Correspondence, Ms. Coll., DM.

48. Werner Heisenberg, "Professor Max Born," *Nature* 225 (1970): 669–71.

49. Born to Göttingen U. Curator, 4 July 1921, Göttingen UA, 4/V h/35.

50. Born to Sommerfeld, 5 Jan. 1923, Sommerfeld Correspondence, Ms. Coll., DM. Born, *My Life*, 212.

51. Planck argued against Wien's discouraging assessment of the present condition of theory, which Wien supported by citing some remarks by Lorentz. The immediate question was the choice of speaker at the general session of the 1922 German Association meeting. Planck wanted Einstein to speak on relativity theory, since he thought it was too soon for a general address on the quantum theory. Planck to Wien, 13 June and 9 July 1922, Wien Papers, STPK, 1973.110.

tization, or the quantization of a component of angular momentum in a given direction, by deflecting beams of atomic particles with a magnetic field. Further theoretical development in Germany went hand in hand with experimental spectroscopy. Theoretical physicists anticipated a complete understanding of atomic structure, though they regarded this as a distant goal. For now they had a quantum theory of the hydrogen atom that worked, but not one for the helium atom or, in general, for atoms with more than one electron. Much of their effort went into interpreting complex spectral lines and the splitting of lines by the magnetic interaction of the radiating electron with the rest of the atom; especially useful in the latter connection was Alfred Landé's theoretical work on the anomalous Zeeman effect, an improvement of Sommerfeld's and Debye's explanation of the normal effect. Other work of theirs went into explaining the complex structure of spectral lines as a relativistic effect, still other work into studying the dispersion of light as an approach to the theory of atoms of many electrons. German theorists and theorists abroad developed Bohr's atom model about as far as it could go.[52]

To go beyond, at Göttingen Born urged Heisenberg and others to study classical mechanics; for only by mastering it could they hope to determine the transition to the new "quantum mechanics," the expression Born introduced into the literature in 1924. They applied, for example, to atoms of many electrons the perturbation methods that astronomers applied to heavenly bodies. And where the classical methods failed, they sought new principles, increasingly convinced that the correct understanding of atoms called for a radical reformulation of physical theory.[53]

Born and other atomic physicists questioned the applicability of one after another of the concepts and laws of the classical theory in the atomic domain. For example, Born and Pauli questioned the admissibility of the space and time continuum because it did not deal with "observable things."[54] Bohr and his co-workers questioned the causal connection between transitions in different atoms and the validity of the laws of conservation of energy and momentum for individual atomic processes.[55]

52. Paul Forman, "The Doublet Riddle and Atomic Physics *circa* 1924," *Isis* 59 (1968): 156–74, on 164–65, 171–73.

53. Stimulated by a "private seminar" with Born, Heisenberg studied "thoroughly and intensively" Poincaré's celestial mechanics, which he found "unbelievably" rich; with Born he applied Poincaré's methods to the quantum theory of the atom. Heisenberg to Sommerfeld, 4 Jan. 1923, Sommerfeld Correspondence, Ms. Coll., DM. Heisenberg, "Max Born," 670. Werner Heisenberg, *Physics and Beyond: Encounters and Conversations,* trans. A. J. Pomerans (New York: Harper and Row, 1971), 59. Born, *My Life,* 214–15.

54. Born to Pauli, 23 Dec. 1919, in *Wolfgang Pauli. Wissenschaftlicher Briefwechsel mit Bohr, Einstein, Heisenberg u. a.,* vol. 1: *1919–1929,* ed. A. Hermann, K. v. Meyenn, and V. F. Weisskopf (New York, Heidelberg, and Berlin: Springer, 1979), 9–10, on 10.

55. In 1924 Bohr, H. A. Kramers, and J. C. Slater avoided the need for light quanta through a new theory of the interaction of light and atoms. In their theory, transitions between stationary states are induced by a "virtual" radiation field originating in virtual oscillators with frequencies corresponding to the allowed transitions. The conservation laws are assumed to have only statistical validity, a feature of the theory that aroused the skepticism of Einstein, Born, Pauli, Sommerfeld, and a number of other theorists. The theory was quickly disproved by experiment, but parts of it, notably the virtual oscillators, were retained by Kramers in his theory of dispersion, which Heisenberg extended; this was the first step toward the construction of quantum mechanics. Bohr, Kramers, and Slater, "The Quantum Theory of

To Heisenberg, as to Bohr, the older mechanics came to appear "false";[56] rejecting it for quantum-theoretical problems, he located the essential difficulty in the kinematics underlying it. His kinematical solution in 1925 was to found a new mechanics, which he based on strictly observable quantities, on atomic radiation. In this "quantum mechanics," the picture of the bound electron was given up; the electron could not be associated with a point in space as a function of time.[57] It could not, that is, be interpreted by "circular and elliptical orbits in classical geometry," and Heisenberg devoted all of his efforts to replacing what appeared to him to be without the "least physical sense."[58] In a letter accompanying his paper of 1925 announcing the new mechanics, Heisenberg told Pauli that his opinion on mechanics was becoming "from day to day more radical."[59]

Heisenberg's "completely new idea" in 1925 was to retain the classical equation for the motion of an electron while rejecting its classical kinematical description.[60] According to the latter the position of the periodically moving electron in an atom can be expanded in a Fourier series, each term of which contains a single integer. Heisenberg's new kinematics, by contrast, requires each term to contain a pair of integers defining a quantum transition within the atom, an event known from the radiation accompanying it. Heisenberg drew attention to an unexpected result of the new kinematics; namely, that the multiplication of observable quantities is not always commutative.[61]

Radiation," *Philosophical Magazine* 47 (1924): 785–802; a German version appeared in *Zs. f. Phys.* 24 (1924): 69–87. Pauli to Kramers, 27 July 1925, in *Pauli. Briefwechsel*, 232–34, on 233. Roger H. Stuewer, *The Compton Effect: Turning Point in Physics* (New York: Science History Publications, 1975), 291–305. Klein, "The First Phase of the Bohr-Einstein Dialogue." Arthur I. Miller, "Visualization Lost and Regained: The Genesis of the Quantum Theory in the Period 1913–27," in *On Aesthetics in Science*, ed. J. Wechsler (Cambridge, Mass.: MIT Press, 1978), 73–101.

56. Heisenberg to Sommerfeld, 15 Jan. 1923, in *Pauli. Briefwechsel*, 81, quoted in note to letter 31.

57. Werner Heisenberg, "Über quantentheoretische Umdeutung kinematischer und mechanischer Beziehungen," *Zs. f. Phys.* 33 (1925): 879–93, translated as "Quantum-theoretical Re-interpretation of Kinematic and Mechanical Relations," in B. L. van der Waerden, ed., *Sources of Quantum Mechanics* (1967; reprint, New York: Dover, 1968), 261–76. The main points of Heisenberg's 1925 paper are analyzed by van der Waerden, "Introduction," *Sources*, 1–59, on 28–35. Van der Waerden quotes extensively from the correspondence surrounding this paper, most of which has subsequently been published in full in *Pauli. Briefwechsel*, which is the source we cite here. There is an extensive historical literature on Heisenberg, Born, their associates, and the creation of quantum mechanics. It includes Whittaker, *The Modern Theories*, 253–67; Jammer, *Conceptual Development of Quantum Mechanics*, 196–236; Armin Hermann, *Die Jahrhundertwissenschaft: Werner Heisenberg und die Physik seiner Zeit* (Stuttgart: Deutsche Verlags-Anstalt, 1977), 68–80; Friedrich Hund, *Geschichte der Quantentheorie*, 2d ed. (Mannheim, Vienna, and Zurich: Bibliographisches Institut, 1975), 121–36; Paul Forman, "Weimar Culture, Causality, and Quantum Theory, 1918–1927: Adaptation by German Physicists and Mathematicians to a Hostile Intellectual Environment," *HSPS* 3 (1971): 1–115; Edward MacKinnon, "Heisenberg, Models, and the Rise of Matrix Mechanics," *HSPS* 8 (1977): 137–88; Daniel Serwer, "*Unmechanischer Zwang*: Pauli, Heisenberg, and the Rejection of the Mechanical Atom, 1923–1925," *HSPS* 8 (1977): 189–256; David C. Cassidy, "Heisenberg's First Core Model of the Atom: The Formation of a Professional Style," *HSPS* 10 (1979): 187–224.

58. Heisenberg to Pauli, 9 July 1925, in *Pauli. Briefwechsel*, 231–32, on 231.

59. Heisenberg to Pauli, 9 July 1925, in *Pauli. Briefwechsel*, 231.

60. Heisenberg gave a new kinematical interpretation to x in the standard equation of motion of an electron: $\ddot{x} + f(x) = 0$. "Quantum-theoretical Re-interpretation," 266.

61. That is, $x(t)\, y(t)$ is not necessarily the same as $y(t)\, x(t)$. Heisenberg, "Quantum-theoretical Re-interpretation," 266.

In his first paper on quantum mechanics, where he treated simple mechanical problems of one degree of freedom, Heisenberg developed parallel formulas: one "classical," by which he meant the formulas of the old Bohr-Sommerfeld quantum theory, and the other his own "quantum-theoretical." His object was to "construct a quantum-mechanical formalism corresponding as closely as possible to that of classical mechanics."[62] In this he was guided by Bohr's correspondence principle requiring that classical physics be the limiting case of quantum physics.[63]

Heisenberg gave the manuscript of his paper to Born when he left Göttingen for a short visit to Cambridge. A few days later, after he had had a chance to study the manuscript, Born wrote to Einstein that its contents "appeared very mystical but certainly correct and deep."[64] Born forwarded it to a journal for publication, and as he thought more about it he recognized that Heisenberg's peculiar noncommutative multiplication belonged to the matrix calculus, a branch of mathematics he had become familiar with as a student and which he had made limited use of in electrodynamics. With this calculus and with the collaboration of his student Pascual Jordan, Born set about to clarify the "formal content" of Heisenberg's theory. Their resulting publication presented a "closed mathematical theory of quantum mechanics that displays strikingly close analogies with classical mechanics." They carried the classical equations of motion over to quantum mechanics with the difference that the quantities appearing in them are matrices. They replaced Heisenberg's cumbersome quantum condition by the matrix equation

$$\mathbf{pq} - \mathbf{qp} = h/2\pi i \cdot \mathbf{I},$$

which in form resembles a well-known equation from classical mechanics and reduces to it in the limit in which Planck's constant h vanishes. On this quantum condition, they based all further considerations.[65]

When Heisenberg returned to Göttingen in the fall of 1925, he found that no one there believed any longer in the old mechanics.[66] He regarded Born and Jordan's work as "very great progress" in the new mechanics,[67] and he quickly learned matrix

62. Heisenberg, "Quantum-theoretical Re-interpretation," 267.

63. Werner Heisenberg, "The Development of Quantum Mechanics," Nobel lecture, 11 Dec. 1933, in *Nobel Lectures. Physics 1922–1941*, The Nobel Foundation (Amsterdam, London, and New York: Elsevier, 1965), 290–301, on 290.

64. Born to Einstein, 15 July 1925, in van der Waerden, *Sources*, 36. The letter is published in full in *Albert Einstein-Hedwig und Max Born Briefwechsel 1916–1955*, ed. Max Born (Munich: Nymphenburger Verlag, 1969), 119–22, quotation on 121.

65. Max Born and Pascual Jordan, "Zur Quantenmechanik," *Zs. f. Phys.* 34 (1925): 858–88, translated as "On Quantum Mechanics" in van der Waerden, *Sources*, 277–306, on 292. In the formula for the quantum condition, **p** and **q** are matrices representing the momentum and the spatial coordinates from classical mechanics, and **I** is the unit matrix. As this condition does not select discrete states from a continuous manifold of solutions, it is not a quantum condition in that sense, but Heisenberg pointed out that it is a "fundamental law of this mechanics" all the same. He referred to it as "Born's very clever idea." Heisenberg to Pauli, 18 Sept. 1925, in *Pauli. Briefwechsel*, 236–40, on 237, 240. This and other central ideas of Born and Jordan's paper were worked out independently and with a different approach by Paul Dirac at Cambridge.

66. Heisenberg to Pauli, 18 Sept. 1925, in *Pauli. Briefwechsel*, 236.

67. Heisenberg to Jordan, 13 Sept. 1925, in van der Waerden, *Sources*, 40.

methods to collaborate with them on their "3-man paper." The mathematical methods were the most important contribution of this paper, which contained nearly all of the essential parts of the matrix formulation of quantum mechanics.[68]

Since matrix calculations were not yet familiar to all theoretical physicists, Born and Jordan had prefaced their formulation of quantum mechanics with an elementary exposition of matrices. At Göttingen, Heisenberg found that some people welcomed matrix methods but others did not understand them; he tried in vain to eliminate the mathematical term "matrix" from the new mechanics and replace it by "quantum-theoretical magnitudes." The unusual mathematics of the theory was its evident characteristic, and the mechanics was referred to as "matrix physics."[69]

At the same time that the matrix approach to atomic mechanics was developed in Göttingen, another approach was considered there. In Born's seminar, the student Walter Elsasser heard about recent experiments on electrons scattered from a metallic plate. Elsasser published a note pointing to the unexpected maxima and minima in the angular distribution of the scattered electrons as evidence for the wave nature of matter, which Einstein had advocated in his quantum theory of gases in 1925, after Louis de Broglie had proposed it in his thesis at Paris the year before.[70] While Heisenberg was struggling with the ideas of his initial paper on quantum mechanics, finding everything "still unclear," he asked Pauli if he had seen "Einstein's new work on atoms that move according to the wave theory," which he himself was "very enthusiastic about."[71] Born wrote to Einstein—in the same letter in which he mentioned having just read Heisenberg's manuscript on quantum mechanics—that he had also just read de Broglie's thesis. With Einstein, he now believed that the "'wave theory of matter' can become a very important subject."[72]

It was not, however, Born or anyone else at Göttingen but Schrödinger at Zurich who further developed the mechanics of matter waves into an alternative to the matrix mechanics of the atom. In a series of papers that began appearing in early 1926, Schrödinger examined the consequences of a certain second-order partial differential equation, the "Schrödinger equation," for describing matter waves. The wave equation for the hydrogen atom that Schrödinger introduced in his first paper is:[73]

68. Heisenberg to Pauli, 23 Oct. 1925, in *Pauli. Briefwechsel*, 251–52, on 251. The paper by Born, Heisenberg, and Jordan is "Zur Quantenmechanik II," *Zs. f. Phys.* 35 (1925): 557–615, translated as "On Quantum Mechanics II," in van der Waerden, *Sources*, 321–85.

69. Heisenberg to Pauli, 16 Nov. 1925, in *Pauli. Briefwechsel*, 255–56, on 255.

70. Walter M. Elsasser tells how he came upon the theme of the wave nature of matter at Göttingen in his *Memoirs of a Physicist in the Atomic Age* (New York: Science History Publications, 1978), 59–62. It is described from Born's perspective in *My Life*, 230–31.

71. Heisenberg to Pauli, 24 and 29 June 1925, in *Pauli. Briefwechsel*, 225–28, on 226, and 229–30, on 229.

72. Born to Einstein, 15 July 1925, in *Einstein-Born Briefwechsel*, 119–22, on 120.

73. Erwin Schrödinger, "Quantisierung als Eigenwertproblem," *Ann.* 79 (1926): 361–76, translated as "Quantization as a Problem of Proper Values (Part 1)," in Erwin Schrödinger, *Collected Papers on Wave Mechanics*, trans. J. F. Shearer from the 2d German ed. of 1928 (London and Glasgow: Blackie and Son, 1928), 1–12, equation on 2. Schrödinger's development of wave mechanics is analyzed in many places. These include Whittaker, *The Modern Theories*, 268–307; Jammer, *Conceptual Development of Quantum*

$$\nabla^2 \psi + \frac{2m}{K^2} \left(E + \frac{e^2}{r} \right) \psi = 0,$$

where ψ is the wave function, E is the energy, and m, e, and r characterize the single electron in its Keplerian path; the constant K must equal $h/2\pi$ for the proper values of the energy to correspond to the Balmer frequencies. The equation yields as eigen-frequencies the same frequencies and corresponding energies for the hydrogen atom as does matrix mechanics, an agreement which Schrödinger took to be the clue to the correct physical understanding of Bohr's frequency condition. Schrödinger remarked that "it is hardly necessary to emphasize how much more congenial it would be to imagine that at a quantum transition the energy changes over from one form of vibration to another, than to think of a jumping electron." The stationary states of an atom are simply standing waves, and transitions between the states take place "continuously in space and time."[74] This way of conceiving of the atom appealed to Schrödinger as more intuitive, as closer to the ways of classical physics than that of the other quantum mechanics.

"The idea of your article shows real genius," Einstein wrote to Schrödinger, even though he had not yet penetrated to the physical significance of the wave equation.[75] Planck, who had pointed out Schrödinger's theory to Einstein "with well justified enthusiasm," wrote to Schrödinger that he was "delighted with the beauties that are evident to the eye," though he too had not yet completely grasped the physical meaning of the theory. The representation of an electron by an infinite plane wave seemed to Planck monstrous, but he had no doubt that the theory was "epoch making"; it already appealed to him as more intuitive than the "purely formal" matrix mechanics of Heisenberg, Born, and Jordan.[76]

Despite the "extraordinary differences between the starting-points" of the two theories, Schrödinger was soon able to show from a mathematical standpoint the "very intimate *inner connection*" between matrix and wave mechanics.[77] Pauli, too, recognized a "rather deep connection between the Göttingen mechanics and the Einstein-de Broglie radiation field," convincing himself that matrices can be con-

Mechanics, 236–80; William T. Scott, *Erwin Schrödinger, An Introduction to His Writings* (Amherst: University of Massachusetts Press, 1967); Klein, "Einstein and the Wave-Particle Duality," 38–46; J. U. Gerber, "Geschichte der Wellenmechanik," *Arch. Hist. Ex. Sci.* 5 (1969): 349–416; V. V. Raman and Paul Forman, "Why Was It Schrödinger Who Developed de Broglie's Ideas?" *HSPS* 1 (1969): 291–314; P. A. Hanle, "Erwin Schrödinger's Reaction to Louis de Broglie's Thesis on the Quantum Theory," *Isis* 68 (1977): 606–9.

74. Schrödinger, "Quantization as a Problem of Proper Values (Part I)," 10–11.

75. Einstein to Schrödinger, 16 Apr. 1926, in Erwin Schrödinger, Max Planck, Albert Einstein, and H. A. Lorentz, *Letters on Wave Mechanics,* ed. K. Przibram, trans. Martin J. Klein (New York: Philosophical Library, 1967), 23–24, on 24.

76. Einstein to Schrödinger, 16 Apr. 1926. Planck to Schrödinger, 2 Apr. and 24 May 1926, in *Letters on Wave Mechanics,* 3, 6–7, on 6. Planck to Wien, 19 Feb., 6 and 22 Mar., and 19 Dec. 1926, Wien Papers, STPK, 1973.110.

77. Erwin Schrödinger, "Über das Verhältnis der Heisenberg-Born-Jordanschen Quantenmechanik zu der meinen," *Ann.* 79 (1926): 734–56, translated as "On the Relation between the Quantum Mechanics of Heisenberg, Born, and Jordan, and that of Schrödinger," in *Collected Papers,* 45–61, quotations on 45–46.

structed from wave functions. He quickly found that Schrödinger's wave mechanics was useful for solving atomic problems;[78] Heisenberg found it useful for calculating matrix elements;[79] and Born found that he could describe aperiodic collision processes between electrons by it, which he had not been able to do by matrix mechanics.[80] But Pauli, Heisenberg, and Born did not accept Schrödinger's physical interpretation of wave mechanics. Pauli decided that Schrödinger was wrong to insist on a "continuum field theory of de Broglie radiation"; it was still necessary to introduce "essentially discontinuous elements into the description of quantum phenomena." Heisenberg thought that the physical part of Schrödinger's theory was "horrible" and Schrödinger's claim for its intuitiveness unconvincing.[81] Born rejected Schrödinger's view that de Broglie waves have the same reality as light waves and that in small packets they can represent material particles.

In 1926, in connection with his study of collision processes, Born put forward an alternative to Schrödinger's understanding of the physical meaning of the wave function. Proceeding from Einstein's view of light waves as a "ghost field" that determines the probability of the path of a light quantum, and from the complete analogy between light quanta and electrons, Born argued that the "de Broglie-Schrödinger waves" are also a "ghost field," or "directing field," which determines the probability of the path of an electron. The wave function determines the probability of physical states; the "paradox" of quantum mechanics, Born wrote, is that the "motion of particles follows probability laws, but the probability itself propagates in conformity with the law of causality."[82]

Quantum mechanics did not immediately solve all of the problems of the atom; to understand complex atomic spectra, for example, physicists had to supplement quantum mechanics with the concept of an intrinsic electron spin, which was introduced independently on the basis of Pauli's exclusion principle and an atom model from the older quantum theory. But quantum mechanics enabled physicists to move ahead again. Atomic physics offered a range of problems of compelling interest to many experimental and theoretical physicists. In the mid-1920s, at over half of the German universities, the theoretical physics lecturers did research on spectra, statistics, and other topics bearing on atomic theory, and as these researchers were often former students of Sommerfeld, Planck, and Born, they were equipped to do it expertly.[83]

78. Pauli to Jordan, 12 Apr. 1926, in *Pauli. Briefwechsel*, 315–20, on 316.
79. Heisenberg to Pauli, 8 June 1926, in *Pauli. Briefwechsel*, 328–29, on 328.
80. Max Born, "Zur Quantenmechanik der Stossvorgänge," *Zs. f. Phys.* 37 (1926): 863–67, reprinted in Max Born, *Zur statistischen Deutung der Quantentheorie*, ed. Armin Hermann, Dokumente der Naturwissenschaft, Abteilung Physik, vol. 1 (Stuttgart: E. Battenberg, 1962), 48–52.
81. Pauli to Schrödinger, 24 May 1926; Heisenberg to Pauli, 8 June 1926, in *Pauli. Briefwechsel*, 324–26, on 326, and 328–29, on 328.
82. Max Born, "Quantenmechanik der Stossvorgänge," *Zs. f. Phys.* 38 (1926): 803–27, in Born, *Zur statistischen Deutung*, 53–77, quotation on 53–54.
83. To name some of the former students of leading theoretical professors who were teaching theory and doing significant research in the mid-1920s: among Sommerfeld's graduates, Wentzel, Landé, Wilhelm Lenz, Pauli, and Heisenberg; among Planck's, Laue, Fritz Reiche, Erich Kretschmann, and Walter

While they developed atomic theory, physicists lectured and published texts on it. On the eve of quantum mechanics, Born published his Göttingen lectures on "atomic mechanics," which he recommended as a sequel to Sommerfeld's more elementary text on atomic structure and spectral lines. His treatment differed from Sommerfeld's in its emphasis on the mechanical point of view, offering an extended presentation of the classical mechanics of multiply periodic systems. Between the original German edition of this text, in which Born said that physicists were "still a long way from a 'final' quantum theory," and the English edition in 1927, Born said that the mechanics of the atom had "developed with a vehemence that could scarcely be foreseen"; he thought that the original text had contributed to the "promotion of the new theories, particularly those parts which have been worked out here in Göttingen," and that most of it was still useful.[84] He brought out other texts on the subject: one based on his American lectures of 1925, already complete with matrix mechanics;[85] another the second volume of his Göttingen lectures on atomic mechanics, which he now referred to as "quantum mechanics." The latter he wrote with his former student Jordan. Following the recent "stormy development" of the subject, he and Jordan wrote, Sommerfeld and several other German physicists had treated quantum mechanics from the standpoint of wave mechanics. Born and Jordan proceeded differently, relating quantum to classical mechanics through Bohr's correspondence principle rather than through the formal resemblance between Schrödinger's wave equation and the classical physics of the continuum. Their goal was the "systematic extension into a logically closed system" of Bohr's "conceptual structures (stationary states, quantum jumps, transition probabilities, etc.)." "We believe that the new mechanics signifies no approximation to the classical ideas," they said.[86] They insisted, that is, on the sharpness of the break with classical physics.

Atomic physicists who helped to build quantum mechanics went regularly to and from Göttingen, Munich, Hamburg, Copenhagen, and other places of atomic research inside and outside Germany. They saw their colleagues several times a year, and wherever they went, as visitors, students, assistants, Privatdocenten, or professors, they went as scientific equals; what mattered was talent, knowledge, and determination. On his return to work in Hamburg from his home in Vienna, Pauli

Schottky; and among Born's, Friedrich Hund and Jordan. In some universities, the theoretical lecturers published nothing, or else they published on topics, principally experimental, that had little to do with the interests of the builders of atomic theory.

84. Max Born's Göttingen lectures in 1923–24, prepared for publication with the help of his assistant Friedrich Hund, *Vorlesungen über Atommechanik*, vol. 1 (Berlin: Springer, 1925); translated as *The Mechanics of the Atom* by J. W. Fisher, revised by D. R. Hartree (London: G. Bell, 1927), quotations on vii–xii.

85. Max Born, *Problems of Atomic Dynamics* (Cambridge, Mass.: MIT Press, 1926).

86. Max Born and Pascual Jordan, *Elementare Quantenmechanik. (Zweiter Band der Vorlesungen über Atommechanik)* (Berlin: Springer, 1930), quotation on vi. Later Born recognized that his and Jordan's exclusive attention to matrix methods in this text was a blunder, for at the time Schrödinger's wave and Dirac's operator methods were undergoing great advances. *My Life*, 230. The history of this second volume, then, as of the first, points up the hazards of writing texts about a rapidly developing field.

stopped at all of the places along the way where atomic physics was done, writing ahead to let people know he was coming; while on the train, he went through the whole problem of atomic physics in his head, since he was a theorist and needed no apparatus to aid his thoughts. Each of Pauli's stops was associated with an approach to the problem, with a mentality, a method.[87] When atomic physicists were not visiting one another, they were corresponding about physics problems. They filled their letters with what they called "calculations," instead of the older "derivations," and these became the basis of publications announcing the progress they were making with the new atomic theory.

As Born said, quantum mechanics was "not a product of our group in Göttingen alone but had many sources." But it was in close connection with Born's research program that Heisenberg created quantum mechanics, and the final matrix formulation of it was, as Heisenberg said, the "joint work of Born's school."[88] Quantum mechanics enhanced the recognition of Göttingen's physicists, who attracted a good many research students there to learn and develop the new theory. Born believed that Göttingen had the "most successful school" in quantum theory and that his private theoretical colloquium contained the "most brilliant gathering of young talent then to be found anywhere."[89] Born so liked Göttingen's "scientific atmosphere" and its "ideal collaboration" between physicists and mathematicians that he turned down highly attractive American job offers. He wanted to make Göttingen even stronger by obtaining a second "official chair for theoretical physics," to which he hoped to attract Heisenberg.[90]

In general, in the wake of quantum mechanics, atomic theorists were much in demand when positions became available. One of the first to change positions was Schrödinger; like each of his predecessors at Zurich University, Einstein, Debye, and Laue, he was attracted to Germany, where his atomic theory preceded him and did much to prepare the way.

The appearance of Schrödinger's wave mechanics coincided with Planck's retirement at Berlin University and the search for his successor. As Laue had actually directed the Berlin theoretical physics institute for the past five years, he was a natural choice. But his appointment would not have solved the problem; it would only have displaced it, since the Berlin faculty was persuaded of the continuing need

87. It was customary for physicists to speak of the "Munich school," "Copenhagen physics," "Göttingen mechanics," and the like. For example, Pauli to Kramers, 27 July 1925, and Pauli to Jordan, 12 Apr. 1926, in *Pauli. Briefwechsel*, 232–34, on 234, and 315–20, on 316.

88. Heisenberg, "Max Born," 670. Born, *My Life*, 215. Forman, "Doublet Riddle," 174.

89. Heisenberg, "Max Born," 670. Born, *My Life*, 227, 237.

90. In the face of American job offers following upon his lecture tour there, Born wanted Göttingen to give him good reasons for staying. To counter the appealing prospect of organizing a "new great branch of science," theoretical physics, in America and training there a "generation of researchers and teachers," Born asked that Heisenberg be returned to Göttingen in "any form," preferably to occupy a chair yet to be created. Born was given informal assurances, but nothing came of them, neither the chair nor Heisenberg's return. Born stayed on because he liked Göttingen and because he could not decide to leave Germany. Born to the Göttingen U. Curator, 13 Apr. 1926; R. Courant, then Dean of the Science Faculty, to Curator, 16 Apr. 1926; Curator to the Prussian Minister of Culture, 11 May 1926; entry for 4 July 1926; Ministry to Born, 28 July 1926; Ministry to Curator, 22 Oct. 1928, Born Personalakte, Göttingen UA.

for two ordinary professors for theoretical physics. The "significance and the enormously lively development" of physical research in recent years demanded it, they argued. After determining confidentially that Einstein did not want Planck's position, the faculty considered the remaining leaders among theoretical physicists, Sommerfeld, Schrödinger, Born, Heisenberg, and Debye, and they recommended the first three in that order.[91] The offer went to Sommerfeld, who declined, and then to Schrödinger, who accepted to become Planck's successor in 1927.[92] Schrödinger also became an ordinary member of the Prussian Academy of Sciences; Planck, Laue, and the other physicists recommending him pointed to his "leading role" in theoretical physics since his work on wave mechanics.[93] In Berlin, Schrödinger found, as he put it, an "unparalleled population-density of physicists of the first rank."[94] There he joined Planck, Laue, and Einstein, all of whom shared his skepticism of what came to be called the "Copenhagen interpretation" of quantum mechanics.

To discourage Sommerfeld from moving to Berlin as Planck's successor, Munich improved Sommerfeld's income and the position of theoretical physics there. He received an increased budget for his institute and an extraordinary grant to enlarge it, an assurance that both assistant positions would continue without the prescribed waiting time if one fell vacant, and, most important, the promise of an extraordinary professorship for theoretical physics. That professorship would bring Munich into line with other large German universities, the Munich faculty argued: Leipzig already had an extraordinary professor alongside the ordinary professor for theoretical physics, as Hamburg was soon to have, and Berlin already had two ordinary professors for the subject.[95]

From 1927 Leipzig University came to be associated with atomic and molecular physics alongside Göttingen, Munich, and other places such as Hamburg.[96] In 1926

91. Report of the Philosophical Faculty to the Prussian Ministry of Culture, 4 Dec. 1926, in Kirsten and Treder, eds., *Einstein in Berlin*, 133–34, on 134.

92. Planck became emeritus at the university in October 1926, but his activities did not change much. He still lectured and gave doctoral examinations and remained very much a part of physics in Berlin and, in particular, in the Prussian Academy, where there was no age limit. Planck to Wien, 19 Dec. 1926. Documents related to Sommerfeld's call to Berlin, May–June, 1927, Munich UA, E II-N, Sommerfeld. Scott, *Schrödinger*, 4.

93. "Wahlvorschlag für Erwin Schrödinger (1887–1961) zum OM," drafted by Planck and signed in addition by Laue, Nernst, Warburg, and Paschen. In *Physiker über Physiker*, 259–60.

94. Scott, *Schrödinger*, 4.

95. The Munich faculty referred to the practical need for a second lecturer for theoretical physics. This had to do with the long time it took students to complete the basic course, three years at Munich; since the course was advanced and since students often could take only part of it before they left the university, a second lecturer was needed to give a parallel course with a phase difference of one and a half years. Because of the state's financial straits, Sommerfeld did not receive the second position right away but only the promise that it would be the next extraordinary professorship created at Munich; in the meantime he got an honorary teaching assignment for a Privatdocent for the same purpose. Sommerfeld to Dean of the Munich U. Philosophical Faculty, 16 May 1927; Dean to the Academic Senate, 19 May 1927; Bavarian Ministry of Education and Culture to the Senate, 20 June 1927, Munich UA, E II-N, Sommerfeld.

96. At Hamburg a lively physics seminar was conducted by the atomic physicists Stern and Lenz; since 1921 the latter was the ordinary professor for theoretical physics. Several atomic theorists did their early teaching at Hamburg: as Privatdocenten there, Wentzel moved on to Leipzig, Jordan to Rostock,

Des Coudres and in 1927 Wiener died, making available Leipzig's two main physics positions for physicists working on subjects of greater contemporary interest. To replace them, Leipzig acquired two of Sommerfeld's students, Debye for the experimental chair and Heisenberg for the theoretical one. Leipzig already had another of Sommerfeld's former students, Gregor Wentzel, who had come the year before as extraordinary professor for theoretical physics. The Leipzig theoretical physics institute under Heisenberg's direction was intimate, informal, and intensely occupied with atoms. His seminar on atomic physics soon attracted talented students, assistants, and collaborators, many from abroad. They were joined in the weekly physics colloquium by Debye and other members of the experimental physics institute who were especially interested in applying the quantum theory to molecular chemistry.[97] Heisenberg established a "kind of physicists' exchange" with the Zurich Polytechnic, where Pauli had taken a job, students and assistants moving freely between the two institutions.[98] Heisenberg, Pauli, and Jordan began to develop a quantum electrodynamics corresponding to quantum mechanics and, in general, a quantum field theory. Problems arising from the new fundamental theory of physics were plentiful.[99]

In 1927 Heisenberg deduced what was to be the most famous consequence of quantum mechanics, variously called the "uncertainty" or "indeterminacy" principle. The principle states how the exact knowledge of one variable used to express the result of a physical measurement can preclude the exact knowledge of another such variable. It arises from the inescapable influence on the atomic system of the measurement itself. In "classical" physics, the world is divisible into the observer

---

and Pauli to Zurich, all to theoretical physics positions. Ortwein, "Wilhelm Lenz 60 Jahre," *Phys. Bl.* 4 (1948): 30–31. Pascual Jordan, "Wilhelm Lenz," *Phys. Bl.* 13 (1957): 269–70. E. Brüche, "Pascual Jordan 60 Jahre," *Phys. Bl.* 18 (1962): 513.

97. The Leipzig faculty committee for Des Coudres's replacement recommended, in this order, Debye, Schrödinger, and Born. Wiener wanted Debye, since he would best further Wiener's understanding that theoretical physics had begun to be extraordinarily fruitful for experimental and technical physics. Before the Saxon ministry acted, Wiener died, and the faculty committee immediately recommended Debye for the now vacant experimental physics position rather than for the theoretical. The ministry agreed, and Heisenberg was then offered the theoretical position. Martin Franke, "Zu den Bemühungen Leipziger Physiker um eine Profilierung der physikalischen Institute der Universität Leipzig im zweiten Viertel des 20. Jahrhunderts," *NTM* 19 (1982), 68–76, on 69–70. Heisenberg was offered Schrödinger's position at Zurich University at the same time, but he chose Leipzig because he wanted to work with Debye. Heisenberg became ordinary professor for theoretical physics at Leipzig in October 1927. Heisenberg Personalakte, Leipzig UA. Among the young physicists Heisenberg attracted were Felix Bloch, Rudolf Peierls, Friedrich Hund, C. F. von Weizsäcker, Edward Teller, I. I. Rabi, and L. D. Landau. Nevill Mott and Rudolf Peierls, "Werner Heisenberg. 5 December 1901–1 February 1976," *Biographical Memoirs of Fellows of the Royal Society* 23 (1977): 213–42, on 225–27. Armin Hermann, *Werner Heisenberg in Selbstzeugnissen und Bilddokumenten* (Reinbek b. Hamburg: Rowohlt, 1976), 44–45; Hermann, *Die Jahrhundertwissenschaft*, 100–101. Heisenberg, *Physics and Beyond*, 93, 117. Gerlach, "Debye," 226.

98. Heisenberg to Pauli, 1 Aug. 1929, quoted in Hermann, *Die Jahrhundertwissenschaft*, 101. P. Scherrer, "Wolfgang Pauli," *Phys. Bl.* 15 (1959); 34–35.

99. Mott and Peierls, "Heisenberg," 225–27. Joan Bromberg, "The Concept of Particle Creation before and after Quantum Mechanics," *HSPS* 7 (1976): 161–91, on 181–88. Steven Weinberg, "The Search for Unity: Notes for a History of Quantum Field Theory," *Daedalus* 106 (1977): 17–35, on 21–24.

and the observed, and in principle the interference of the former on the latter can be made negligibly small. In quantum mechanics, because of the discontinuous nature of processes at the atomic level, the interference is not negligible. The formalism of quantum mechanics implies a relationship, Heisenberg showed, between the accuracies with which two classical variables, say, the momentum p and the position q of a particle, can be simultaneously known:

$$\Delta p \, \Delta q \geq h/4\pi.$$

This inequality says that the product of the probable errors in the simultaneous measurements of momentum and position cannot be less than Planck's constant divided by 4π. The uncertainty relationship applies not only to momentum and position but to all so-called canonically conjugated variables, time and energy, to give another example. The whole structure of quantum mechanics is characterized by complementary relationships of one kind or another. The classically complete description in space and time of causal connections between physical phenomena is replaced in quantum mechanics by the "complementarity of space-time description and causality."[100]

In 1927 the fifth Solvay Congress was held in Brussels, the first to which the Germans were invited after World War I.[101] Born, Heisenberg, and Schrödinger came from Germany to give reports on the quantum mechanics they had helped create, and Einstein, Debye, Pauli, and Planck came to discuss it. From America Arthur Compton came to discuss experiments that lent support to the idea of directed quanta of energy, or "photons."[102] From Denmark Bohr came to discuss the complementarity principle, which he had recently formulated, by which the description of the world of atoms must rely on seemingly contradictory pictures, such as those of waves and particles.[103] Paul Dirac and others came, all imbued with the

100. Werner Heisenberg, "Über den anschaulichen Inhalt der quanten-theoretischen Kinematik und Mechanik," Zs. f. Phys. 43 (1927): 172–98. Contrasting the classical and quantum theories, Heisenberg said that by the classical theory the causal connections of phenomena are given descriptions in space and time, but by the quantum theory there are alternatives: either the phenomena are described in space and time but subject to the uncertainty principle, or the phenomena are not described in space and time, in which case a causal relationship exists through exact mathematical laws. The two alternatives are related to one another statistically. Werner Heisenberg, The Physical Principles of the Quantum Theory, trans. C. Eckart and F. C. Hoyt (New York: Dover, 1930), 2–3, 62, 64–65; "The Development of Quantum Mechanics," 297–99. The notion of the indeterminacy of natural law before and after quantum mechanics is discussed in many places; for example, in Jammer, Conceptual Development of Quantum Mechanics, 323–45; Forman, "Weimar Culture, Causality, and Quantum Theory"; Stephen G. Brush, "Irreversibility and Indeterminism: Fourier to Heisenberg," Journ. Hist. of Ideas 37 (1976): 603–30; P. A. Hanle, "Indeterminacy before Heisenberg: The Case of Franz Exner and Erwin Schrödinger," HSPS 10 (1979): 225–69.

101. The fifth Solvay Congress, in 1927, is discussed in Max Jammer, The Philosophy of Quantum Mechanics: The Interpretations of Quantum Mechanics in Historical Perspective (New York: John Wiley, 1974), 121–32.

102. The relationship of Compton's work to the themes of the fifth Solvay Congress is discussed in Stuewer, The Compton Effect, especially in chapter 7, pp. 287–347.

103. Bohr's views on the wave-particle duality and his related principle of complementarity are

sense that physics had just passed through a strong development. They brought with them to the congress the exact laws of atomic behavior, which had been unknown to their predecessors—some of whom were the same people—at the first Solvay Congress sixteen years before. With pride of accomplishment, Born and Heisenberg characterized "*quantum mechanics* as a complete theory, the fundamental physical and mathematical hypotheses of which are not susceptible to modifications." They insisted on the need for indeterminacy in the theory, for probabilities in place of strict causality.[104]

This congress witnessed a rare face-to-face debate between Bohr and Einstein over the fundamentals of physics,[105] with Einstein defending the ideal of a detailed knowledge of atomic processes and Bohr arguing for a limitation of that knowledge imposed by the wave-particle duality. For Einstein, as always, the goal of physical theory was a direct accounting of the world existing independently of us. Upholding causality in the face of the new indeterminism, Einstein allowed that quantum mechanics could be interpreted as a statistical theory giving information about a cloud of particles but not as a complete theory about individual particles.[106] De Broglie, Schrödinger, Pauli, and others disagreed among themselves over the interpretation of quantum mechanics, which lent discussions at this meeting a strongly epistemological flavor.

Lorentz presided over the congress, his last, and his observations and questions pointed up the sharp contrast between the new theory and the old theory that he had helped to build. Lorentz reminded them of what they were being asked to give up. To begin with, the confirmation of light quanta, or photons, was said to call for the elimination of the electromagnetic ether, which no longer seemed to help but only to cause difficulties. But there has to be "something besides photons," Lorentz objected. Something must guide them in their course, and that something is the electromagnetic field, with its waves, as determined by classical theory, leading to the notion of an electromagnetic ether; even relativity theory and certainly Maxwell's equations are compatible with such an ether, which has often been construed, with a certain success, in terms of a mechanical model. For Lorentz, it would be sufficient, and certainly preferable, to retain for physics the ether on which the electron theory was built, one that passes freely through matter, one for which Maxwell's equations are valid. It was not only the way light was being presented at the congress that troubled Lorentz but also the way matter was. Its behavior was de-

---

discussed in Jammer, *Conceptual Development of Quantum Mechanics*, 345–61; Gerald Holton, "The Roots of Complementarity," *Daedalus* 99 (1970): 1015–55.

104. Max Born and Werner Heisenberg, "La mécanique des quanta," in Institut international de physique Solvay, *Electrons et photons. Rapports et discussions du cinquième conseil de physique tenu à Bruxelles du 24 au 29 octobre 1927* (Paris: Gauthier-Villars, 1928), 143–81, on 178.

105. The development of Bohr's and Einstein's positions before the fifth Solvay Congress is discussed in Klein, "The First Phase of the Bohr-Einstein Dialogue."

106. "Discussion générale des idées nouvelles émises," *Electrons et photons*, 248–89, following Bohr's final report on the quantum postulate. This general discussion was divided into three parts: the first on causality, determinism, and probability; the last two on photons and electrons. Einstein's observations referred to here belong to the first part, 253–56.

scribed by the complicated quantities known as matrices, which represented the frequencies of radiation and, in general, whatever was to be observed in the atom. Lorentz was greatly surprised to learn that these matrices satisfy the equations of motion. Although in theory he could see that this worked, it seemed to him a "great mystery," and with it physics had "taken an enormous step down the road of abstraction." The denial of an ether and of the customary meaning of coordinates, potential energy, and other standard magnitudes of physics led physics irresistibly down this road. The new theory succeeded at the cost of the intelligibility of the world as presented by theoretical physics, and for Lorentz the cost was excessive. "We want to make a representation of the phenomena, to form a picture in our mind," he told the members of the fifth Solvay Congress:

> Until now we have always wanted to form these pictures by ordinary notions of time and space. These notions are perhaps innate; in any case, they have developed by our personal experience, by our daily observations. For me, these notions are distinct, and I confess that I cannot have an idea of physics without these notions. The picture that I want to form of phenomena ought to be absolutely distinct and definite, and it seems to me that we cannot form a similar picture without this system of space and time.
>
> For me, an electron is a corpuscle that at a given instant is located at a definite point of space, and if I have the idea that at the following moment this corpuscle is located elsewhere, I should think of its trajectory, which is a line in space. And if this electron meets an atom and penetrates it and after several events it leaves this atom, I forge a theory for myself in which this electron conserves its individuality; that is to say, I imagine a line along which this electron passes through this atom. It is evidently possible that this theory would be very difficult to develop, but it does not appear to me *a priori* impossible.
>
> I picture to myself that in the new theory one still has these electrons. . . . I would like to preserve that former ideal of describing all that occurs in the world by clear images. I am ready to admit other theories on the condition that one can translate them into clear and distinct pictures.[107]

It is evident from Lorentz's remarks, as it is from the entire proceedings, that the participants at the fifth Solvay Congress recognized that they were no longer only questioning, or tentatively departing from, the foundations of "classical" physics. They were now in possession of an axiomatic atomic physics that appeared to be as complete as its predecessor, or nearly so, and that appeared to deny the possibility of Lorentz's "clear and distinct pictures" from classical physics. In the next few years physics would take up the problems of the theory of the atomic nucleus and of

107. Lorentz's observations are taken from the discussions following Compton's, Born and Heisenberg's, and Bohr's reports, in *Electrons et photons*, 86–87, 183, 248–49. The varying views of Bohr, Born, Heisenberg, and Schrödinger on the kind of pictures Lorentz speaks of here are discussed in Miller, "Visualization Lost and Regained."

the elementary particles related to nuclear forces. But that is another stage of theoretical physics. Our account of the development of theoretical physics in Germany closes with the consolidation of the new physics, with Lorentz recalling the classical theories of physics that, in their day, had held the promise of the new.

We have shown how German physicists, from Ohm to Einstein, constructed, extended, criticized, and taught the classical theories, and we have shown how they carried out this work within institutions they created or adapted for the purpose. Several of the German physicists were theorists of the highest originality, and we have discussed their work through many examples. We have also discussed the work of the large body of physicists who sustained and handed on their subject. They all, in varying measure, furthered the "intellectual mastery of nature," to return to Helmholtz's words from our preface. Of the accompaniment of this mastery, Helmholtz's parallel "material mastery of nature," we of the second half of the twentieth century are keenly aware.

Bibliography

Archives

AHQP: Archive for the History of Quantum Physics, American Philosophical Society Library, Philadelphia, and elsewhere
AR: Algemeen Rijksarchief, Den Hague
A Schweiz. Sch., Zurich: Archiv des Schweizerischen Schulrates, ETH Zürich
Bad. GLA: Badisches Generallandesarchiv Karlsruhe
Bay. HSTA: Bayerisches Hauptstaatsarchiv, München
Bay. STB: Bayerische Staatsbibliothek, München
Bonn UA: Archiv der Rheinischen Friedrich-Wilhelms-Universität Bonn
Bonn UB: Universitätsbibliothek Bonn
DM: Bibliothek des Deutschen Museums, München
DZA, Merseburg: Deutsches Zentralarchiv, Merseburg
EA: Einstein Archive, The Hebrew University of Jerusalem
Erlangen UA: Universitäts-Archiv der Friedrich-Alexander-Universität Erlangen
Erlangen UB: Universitätsbibliothek Erlangen-Nürnberg
ETHB: Bibliothek der ETH Zürich
Freiburg SA: Stadtarchiv der Stadt Freiburg im Breisgau
Freiburg UB: Universitäts-Bibliothek Freiburg im Breisgau
Giessen UA: Universitätsarchiv Justus Liebig-Universität Giessen
Göttingen UA: Archiv der Georg-August-Universität Göttingen
Göttingen UB: Niedersächsische Staats- und Universitätsbibliothek Göttingen
Graz UA: Archiv der Universität in Graz
Heidelberg UA: Universitätsarchiv der Ruprecht-Karls-Universität Heidelberg
Heidelberg UB: Universitätsbibliothek Heidelberg
HSTA, Stuttgart: Württembergisches Hauptstaatsarchiv Stuttgart
Jena UA: Universitätsarchiv der Friedrich-Schiller-Universität Jena
LA Schleswig-Holstein: Landesarchiv Schleswig-Holstein, Schleswig

Leipzig UA: Archiv der Karl-Marx-Universität Leipzig
Leipzig UB: Universitätsbibliothek der Karl-Marx-Universität Leipzig
Munich SM: Münchner Stadtmuseum
Munich UA: Archiv der Ludwig-Maximilians-Universität München
Münster UA: Universitäts-Archiv der Westfälischen Wilhelms-Universität Münster
N.-W. HSTA: Nordrhein-Westfälisches Hauptstaatsarchiv Düsseldorf
Öster. STA: Österreichisches Staatsarchiv, Wien
Rijksmuseum voor de Geschiedenis der Natuurwetenschappen, Leiden
STA K Zurich: Staatsarchiv des Kantons Zürich
STA, Ludwigsburg: Staatsarchiv Ludwigsburg
STA, Marburg: Hessisches Staatsarchiv Marburg
STPK: Staatsbibliothek Preussischer Kulturbesitz, Berlin
Tübingen UA: Universitätsarchiv Eberhard-Karls-Universität Tübingen
Tübingen UB: Universitätsbibliothek Tübingen
Wroclaw UB: Biblioteka Uniwersytecka Wroclaw (Breslau)
Würzburg UA: Archiv der Universität Würzburg

We acknowledge special indebtedness to the Archive for the History of Quantum Physics, or AHQP. This invaluable resource for historians of science was a principal starting point for our study of the primary sources of twentieth century physics in Germany. Many of the letters we cite from the rich Sommerfeld and Lorentz Papers we first saw on microfilm at the Library of the American Philosophical Library, one of the depositories of the AHQP. For our work we subsequently obtained photocopies of these letters from the original archives, and these we acknowledge in our sources. The AHQP is described and its contents listed in Thomas S. Kuhn, John L. Heilbron, Paul Forman, and Lini Allen, *Sources for History of Quantum Physics: An Inventory and Report* (Philadelphia: The American Philosophical Society, 1967.

East Germany refuses us permission to consult its state archives. Fortunately, much of what they contain that bears on physics can be learned from other documents and, in some cases, from copies of their documents in collections accessible to us.

Printed Sources

We give scientific articles in the footnotes but because of their great number we do not give them again in this bibliography.

Works on universities and other schools are arranged by location: Berlin, Göttingen, etc.

Aachener Bezirksverein deutscher Ingenieure. "Adolf Wüllner." *Zs. d. Vereins deutsch. Ingenieure* 52 (1908): 1741–42.
Abbe, Ernst. "Gedächtnisrede zur Feier des 50jährigen Bestehens der optischen Werkstätte." In *Gesammelte Abhandlungen*, vol. 3, 60–95. Jena: Fischer, 1906.
Abraham, Max. *Theorie der Elektrizität.* Vol. 1, *Einführung in die Maxwellsche Theorie der Elektrizität*, by August Föppl, revised by Max Abraham. Leipzig: B. G. Teub-

ner, 1904. Vol. 2, *Elektromagnetische Theorie der Strahlung.* Leipzig: B. G. Teubner, 1905.

Allgemeine deutsche Biographie. Vols. 1–56. 1875–1912. Reprint. Leipzig: Duncker und Humblot, 1967–71 (*ADB*).

Angenheister, Gustav. "Emil Wiechert." *Gött. Nachr.*, *Geschäftliche Mitteilungen aus dem Berichtsjahr 1927/28*, 53–62.

Appleyard, Rollo. *Pioneers of Electrical Communication.* London: Macmillan, 1930.

Assmann, Richard, et al. "Vollendung des 50. Jahrganges der 'Fortschritte.'" *Fortschritte der Physik des Aethers im Jahre 1894* 50, pt. 2 (1896): i–xi.

Auerbach, Felix. "Ernst Abbe." *Phys. Zs.* 6 (1905): 65–66.

———. *Ernst Abbe, sein Leben, sein Wirken, seine Persönlichkeit.* Leipzig: Akademische Verlagsgesellschaft, 1918.

———. *Kanon der Physik. Die Begriffe, Principien, Sätze, Formeln, Dimensionsformeln und Konstanten der Physik nach dem neuesten Stande der Wissenschaft systematisch dargestellt.* Leipzig, 1899.

———. *Die Methoden der theoretischen Physik.* Leipzig: Akademische Verlagsgesellschaft, 1925.

———. *Physik in graphischen Darstellungen.* Leipzig and Berlin: B. G. Teubner, 1912.

———. *Das Wesen der Materie nach dem neuesten Stande unserer Kenntnisse und Auffassungen dargestellt.* Leipzig: Dürr, 1918.

———. *The Zeiss Works and the Carl Zeiss Foundation in Jena: Their Scientific, Technical and Sociological Development and Importance.* Translated by R. Kanthack from the 5th German edition, 1925. London: Foyle, n.d.

Baerwald, Hans. "Karl Schering." *Phys. Zs.* 26 (1925): 633–35.

Band, William. *Introduction to Mathematical Physics.* Princeton: Van Nostrand, 1959.

Bandow, F. "August Becker." *Phys. Bl.* 9 (1953): 131.

Baretin, W. "Johann Christian Poggendorff." *Ann.* 160 (1877): v–xxiv.

———. "Ein Rückblick." *Ann.*, Jubelband (1874): ix–xiv.

Baumgarten, Fritz. *Freiburg im Breisgau.* Berlin: Wedekind, 1907.

Becherer, Gerhard. "Die Geschichte der Entwicklung des Physikalischen Instituts der Universität Rostock." *Wiss. Zs. d. U. Rostock, Math.-Naturwiss.* 16 (1967): 825–30.

Benndorf, H. "Philipp Lenard." *Almanach. Österreichische Akad.* 98 (1948): 250–58.

Benz, Ulrich. *Arnold Sommerfeld. Lehrer und Forscher an der Schwelle zum Atomzeitalter, 1868–1951.* Stuttgart: Wissenschaftliche Verlagsgesellschaft, 1975.

Bergmann, Peter G. *The Riddle of Gravitation.* New York: Scribner's, 1968.

Berkson, William. *Fields of Force: The Development of a World View from Faraday to Einstein.* New York: Wiley, 1974.

Berlin. Academy of Sciences. *Max Planck in seinen Akademie-Ansprachen. Erinnerungsschrift.* Berlin: Akademie-Verlag, 1948.

Berlin. Technical Institute. *Die Technische Hochschule zu Berlin 1799–1924. Festschrift.* Berlin: Georg Stilke, 1925.

Berlin. University. *Chronik der Königlichen Friedrich-Wilhelms-Universität zu Berlin.* Berlin.

————. *Forschen und Wirken. Festschrift zur 150-Jahr-Feier der Humboldt-Universität zu Berlin 1810–1960*. Vol. 1. Berlin: VEB Deutscher Verlag der Wissenschaften, 1960.

————. *Idee und Wirklichkeit einer Universität. Dokumente zur Geschichte der Friedrich-Wilhelms-Universität zu Berlin*. Edited by Wilhelm Weischedel. Berlin: Walter de Gruyter, 1960.

————. *Index Lectionum*. Berlin, n.d.

————. See Kurt-R. Biermann; Rudolf Köpke; Max Lenz.

Bertholet, Alfred, et al. "Erinnerungen an Max Planck." *Phys. Bl.* 4 (1948): 161–74.

Besso, Michele. See Albert Einstein.

Bezold, Wilhelm von. "Gedächtnissrede auf August Kundt." *Verh. phys. Ges.* 13 (1894): 61–80.

————. *Gesammelte Abhandlungen aus den Gebieten der Meteorologie und des Erdmagnetismus*. Braunschweig: F. Vieweg, 1906.

Biermann, Kurt-R. *Die Mathematik und ihre Dozenten an der Berliner Universität 1810–1920. Stationen auf dem Wege eines mathematischen Zentrums von Weltgeltung*. Berlin: Akademie-Verlag, 1973.

Biermer, M. "Die Grossherzoglich Hessische Ludwigs-Universität zu Giessen." In *Das Unterrichtswesen im Deutschen Reich*, edited by Wilhelm Lexis, vol. 1, 562–74.

Blackmore, John T. *Ernst Mach. His Work, Life, and Influence*. Berkeley, Los Angeles, and London: University of California Press, 1972.

Bochner, Salomon. "The Significance of Some Basic Mathematical Conceptions for Physics." *Isis* 54 (1963): 179–205.

Böhm, Walter. "Stefan, Josef." *DSB* 13 (1976): 10–11.

Bohr, Niels. *Abhandlungen über Atombau aus den Jahren 1913–1916*. Translated by H. Stintzing. Braunschweig: F. Vieweg, 1921.

————. "The Genesis of Quantum Mechanics." In *Essays 1958–1962 on Atomic Physics and Human Knowledge*, 74–78. New York: Interscience, 1963.

Boltzmann, Ludwig. "Eugen von Lommel." *Jahresber. d. Deutsch. Math.-Vereinigung* 8 (1900): 47–53.

————. *Gesamtausgabe*. Vol. 8, *Ausgewählte Abhandlungen der internationalen Tagung, Wien*. Vienna: Akademische Druck- u. Verlagsanstalt, 1982.

————. *Gustav Robert Kirchhoff*. Leipzig, 1888. Reprinted in *Populäre Schriften*, 51–75.

————. "Josef Stefan." Rede gehalten bei der Enthüllung des Stefan-Denkmals am 8. Dez. 1895. In *Populäre Schriften*, 92–103.

————. "On the Fundamental Principles and Basic Equations of Mechanics." First of Boltzmann's Clark University lectures, 1899. Translated by J. J. Kockelmans, editor of *Philosophy of Science*, 246–60. New York: Free Press, 1968.

————. *Populäre Schriften*. Leipzig: J. A. Barth, 1905.

————. "The Relations of Applied Mathematics." In *International Congress of Arts and Science: Universal Exposition, St. Louis, 1904.* Vol. 1, *Philosophy and Mathematics,* 591–603. Boston and New York: Houghton, Mifflin, 1905.

———. "Über die Entwicklung der Methoden der theoretischen Physik in neuerer Zeit" (1899). In *Populäre Schriften,* 198–227.

———. "Über die Methoden der theoretischen Physik" (1892). In *Populäre Schriften,* 1–10.

———. *Vorlesungen über die Prinzipe der Mechanik.* 2 vols. Leipzig: J. A. Barth, 1897–1904.

———. *Vorlesungen über Gastheorie.* 2 vols. Leipzig, 1896–98. Translated as *Lectures on Gas Theory* by Stephen G. Brush. Berkeley: University of California Press, 1964.

———. *Vorlesungen über Maxwells Theorie der Elektricität und des Lichtes.* Vol. 1, *Ableitung der Grundgleichungen für ruhende, homogene, isotrope Körper.* Leipzig, 1891. Vol. 2, *Verhältniss zur Fernwirkungstheorie; specielle Fälle der Elektrostatik, stationären Strömung und Induction.* Leipzig, 1893.

———. *Wissenschaftliche Abhandlungen.* Edited by Fritz Hasenöhrl. 3 vols. Leipzig: J. A. Barth, 1909. Reprint. New York: Chelsea, 1968.

———. "Zwei Antrittsreden." *Phys. Zs.* 4 (1902–3): 247–56, 274–77. Boltzmann's inaugural lectures at Leipzig University in November 1900 and at the University of Vienna in October 1902.

Bonn. University. *Chronik der Rheinischen Friedrich-Wilhelms-Universität zu Bonn.* Bonn.

———. *Geschichte der Rheinischen Friedrich-Wilhelm-Universität zu Bonn am Rhein.* Edited by A. Dyroff. Vol. 2, *Institute und Seminare, 1818–1933.* Bonn: F. Cohen, 1933.

———. *150 Jahre Rheinische Friedrich-Wilhelms-Universität zu Bonn 1818–1968. Bonner Gelehrte. Beiträge zur Geschichte der Wissenschaften in Bonn. Mathematik und Naturwissenschaften.* Bonn: H. Bouvier, Ludwig Röhrscheid, 1970.

———. *Vorlesungen auf der Rheinischen Friedrich-Wilhelms-Universität zu Bonn.*

Bonnell, E., and H. Kirn. "Preussen. Die höheren Schulen." In *Encyklopädie des gesamten Erziehungs- und Unterrichtswesens,* edited by K. A. Schmid, vol. 6, 180 ff. Leipzig, 1885.

Bopp, Fritz, and Walther Gerlach. "Heinrich Hertz zum hundertsten Geburtstag am 22. 2. 1957." *Naturwiss.* 44 (1957): 49–52.

Bork, Alfred M. "Physics Just Before Einstein." *Science* 152 (1966): 597–603.

———. "'Vectors Versus Quaternions'—The Letters in *Nature*." *Am. J. Phys.* 34 (1966): 202–11.

Born, Max. "Antoon Lorentz." *Gött. Nachr.,* 1927–28, 69–73.

———. "Arnold Johannes Wilhelm Sommerfeld, 1868–1951." *Obituary Notices of Fellows of the Royal Society* 8 (1952): 275–96.

———. *Ausgewählte Abhandlungen.* Edited by the Akademie der Wissenschaften in Göttingen. 2 vols. Göttingen: Vandenhoeck und Ruprecht, 1963.

———. "How I Became a Physicist." In *My Life and My Views,* 15–27.

————. "Max Karl Ernst Ludwig Planck 1858–1947." *Obituary Notices of Fellows of the Royal Society* 6 (1948): 161–88. Reprinted in *Ausgewählte Abhandlungen* 2: 626–46.

————. *My Life and My Views.* New York: Scribner's, 1968.

————. *My Life: Recollections of a Nobel Laureate.* New York: Scribner's, 1978.

————. *Physics in My Generation: A Selection of Papers.* London and New York: Pergamon, 1956.

————. *Problems of Atomic Dynamics.* Cambridge, Mass.: MIT Press, 1926.

————. "Sommerfeld als Begründer einer Schule." *Naturwiss.* 16 (1928): 1035–36.

————. *Vorlesungen über Atommechanik.* Vol. 1. Berlin: Springer, 1925. Translated as *The Mechanics of the Atom* by J. W. Fisher. Revised by D. R. Hartree. London: G. Bell, 1927.

————. *Zur statistischen Deutung der Quantentheorie.* Edited by Armin Hermann. Dokumente der Naturwissenschaft, Abteilung Physik. Vol. 1. Stuttgart: E. Battenberg, 1962.

————. See Albert Einstein.

Born, Max, and Pascual Jordan. *Elementare Quantenmechanik (Zweiter Band der Vorlesungen über Atommechanik).* Berlin: Springer, 1930.

Born, Max, and Max von Laue. "Max Abraham." *Phys. Zs.* 24 (1923): 49–53.

Borscheid, Peter. *Naturwissenschaft, Staat und Industrie in Baden (1848–1914).* Vol. 17 of Industrielle Welt, Schriftenreihe des Arbeitskreises für moderne Sozialgeschichte, edited by Werner Conze. Stuttgart: Ernst Klett, 1976.

Brauer, Ludolph, et al., eds. *Forschungsinstitute, ihre Geschichte, Organisation und Ziele.* Vol. 1. Hamburg: Hartung, 1930.

Braun, Ferdinand. "Hermann Georg Quincke." *Ann.* 15 (1904): i–viii.

Braunmühl, A. v. "Sohncke, Leonhard." *Biographisches Jahrbuch und Deutscher Nekrolog* 2 (1898): 167–70.

Breslau. University. *Chronik der Königlichen Universität zu Breslau.* Breslau.

————. *Festschrift zur Feier des hundertjährigen Bestehens der Universität Breslau.* Pt. 2, *Geschichte der Fächer, Institute und Ämter der Universität Breslau 1811–1911.* Edited by Georg Kaufmann. Breslau: F. Hirt, 1911.

Broda, Engelbert. *Ludwig Boltzmann. Mensch, Physiker, Philosoph.* Vienna: F. Deuticke, 1955.

Broglie, Maurice de. *Les Premièrs Congrès de Physique Solvay.* Paris: Albin Michel, 1951.

Bromberg, Joan. "The Concept of Particle Creation before and after Quantum Mechanics." *HSPS* 7 (1976): 161–91.

Brüche, E. "Aus der Vergangenheit der Physikalischen Gesellschaft." *Phys. Bl.* 16 (1960): 499–505, 616–21; 17 (1961): 27–33, 120–27, 225–32, 400–410.

————. "Ernst Abbe und sein Werk." *Phys. Bl.* 21 (1965): 261–69.

————. "Pascual Jordan 60 Jahre." *Phys. Bl.* 18 (1962): 513.

Brush, Stephen G. "Boltzmann, Ludwig." *DSB* 2 (1970): 260–68.

————. "Irreversibility and Indeterminism: Fourier to Heisenberg." *Journ. Hist. of Ideas* 37 (1976): 603–30.

————. *The Kind of Motion We Call Heat: A History of the Kinetic Theory of Gases in the 19th Century.* Vol. 1, *Physics and the Atomists.* Vol. 2, *Statistical Physics and Irreversible Processes.* Vol. 6 of Studies in Statistical Mechanics. Amsterdam and New York: North-Holland, 1976.

————. *Kinetic Theory.* Vol. 1, *The Nature of Gases and of Heat.* Vol. 2, *Irreversible Processes.* The Commonwealth and International Library; Selected Readings in Physics. Oxford and New York: Pergamon, 1965–66.

————. "Randomness and Irreversibility." *Arch. Hist. Ex. Sci.* 12 (1974): 1–88.

————. "The Wave Theory of Heat." *Brit. Journ. Hist. Sci.* 5 (1970): 145–67.

Bucherer, Alfred. *Elemente der Vektoranalysis mit Beispielen aus der theoretischen Physik.* 2d ed. Leipzig: B. G. Teubner, 1905.

————. *Mathematische Einführung in die Elektronentheorie.* Leipzig: B. G. Teubner, 1904.

Buchheim, Gisela. "Zur Geschichte der Elektrodynamik: Briefe Ludwig Boltzmanns an Hermann von Helmholtz." *NTM* 5 (1968): 125–31.

Budde, Emil. *Allgemeine Mechanik der Punkte und starren Systeme. Ein Lehrbuch für Hochschulen.* 2 vols. Berlin, 1890–91.

Bühring, Friedrich. "Paul Drude." *Zs. für den physikalischen und chemischen Unterricht* 5 (1906): 277–79.

Burchardt, Lothar. *Wissenschaftspolitik im Wilhelminischen Deutschland. Vorgeschichte, Gründung und Aufbau der Kaiser-Wilhelm-Gesellschaft zur Förderung der Wissenschaften.* Göttingen: Vandenhoeck und Ruprecht, 1975.

Cahan, David. "The Physikalisch-Technische Reichsanstalt: A Study in the Relations of Science, Technology and Industry in Imperial Germany." Ph.D. diss., Johns Hopkins University, 1980.

Caneva, Kenneth L. "From Galvanism to Electrodynamics: The Transformation of German Physics and Its Social Context." *HSPS* 9 (1978): 63–159.

Cantor, G. N., and M. J. S. Hodge, eds. *Conceptions of Ether: Studies in the History of Ether Theories 1740–1900.* Cambridge: Cambridge University Press, 1981.

Cassidy, David C. "Heisenberg's First Core Model of the Atom: The Formation of a Professional Style." *HSPS* 10 (1979): 187–224.

Cath, P. G. "Heinrich Hertz (1857–1894)." *Janus* 46 (1957): 141–50.

Cermak, Paul. "Carl Fromme." *Nachrichten der Giessener Hochschulgesellschaft* 19 (1950): 92–93.

Christiansen, C., and J. J. C. Müller. *Elemente der theoretischen Physik.* 3d rev. ed. Leipzig: J. A. Barth, 1910.

Clark, Ronald W. *Einstein: The Life and Times.* New York: World, 1971.

Clausius, Rudolph. *Die mechanische Wärmetheorie.* 2d rev. and completed ed. of *Abhandlungen über die mechanische Wärmetheorie.* Vol. 1. Second title page reads *Entwickelung der Theorie, soweit sie sich aus den beiden Hauptsätzen ableiten lässt, nebst Anwendungen.* Braunschweig, 1876. Vol. 2, *Die mechanische Behandlung der Electricität.* Second title page reads *Anwendung der der mechanischen Wärmetheorie zu Grunde liegenden Principien auf die Electricität.* Braunschweig, 1879.

————. *Ueber den Zusammenhang zwischen den grossen Agentien der Natur.* Rectorats-antritt, 18 Oct. 1884. Bonn, 1885.

————. *Ueber die Energievorräthe in der Natur und ihre Verwerthung zum Nutzen der Menschheit.* Bonn, 1885.

Cohn, Emil. *Physikalisches über Raum und Zeit.* 3d ed. Leipzig and Berlin: B. G. Teubner, 1918.

Conrad, Johannes. *Das Universitätsstudium in Deutschland während der letzten 50 Jahre. Statistische Untersuchungen unter besonderer Berücksichtigung Preussens.* Jena, 1884.

Courant, Richard, and David Hilbert. *Methoden der mathematischen Physik.* Vol. 1. Vol. 12, Die Grundlehren der mathematischen Wissenschaften in Einzeldarstellungen mit besonderer Berücksichtigung der Anwendungsgebiete, edited by Richard Courant. Berlin: Springer, 1924.

Craig, Gordon. *Germany, 1866–1945.* New York: Oxford University Press, 1978.

Crew, Henry. "Heinrich Kayser, 1853–1940." *Astrophys. Journ.* 94 (1941): 5–11.

Crowe, Michael J. *A History of Vector Analysis: The Evolution of the Idea of a Vectorial System.* Notre Dame: University of Notre Dame Press, 1967.

Curry, Charles Emerson. *Theory of Electricity and Magnetism.* London, 1897.

Cuvaj, Camillo. "Henri Poincaré's Mathematical Contributions to Relativity and the Poincaré Stresses." *Am. J. Phys.* 36 (1968): 1102–13.

D'Agostino, Salvo. "Hertz's Researches on Electromagnetic Waves." *HSPS* 6 (1975): 261–323.

Darrow, Karl K. "Peter Debye (1884–1966)." *American Philosophical Society: Yearbook,* 1968, 123–30.

Davies, Mansel. "Peter J. W. Debye (1884–1966)." *Journ. Chem. Ed.* 45 (1968): 467–72.

Debye, Peter. "Antrittsrede." *Sitzungsber. preuss. Akad.,* 1937, cxiii–cxiv.

————. *The Collected Papers of Peter J. W. Debye.* New York: Interscience, 1954.

De Haas-Lorentz, G. L., ed. *H. A. Lorentz. Impressions of His Life and Work.* Amsterdam: North-Holland, 1957.

Des Coudres, Theodor. "Ludwig Boltzmann." *Verh. sächs. Ges. Wiss.* 85 (1906): 615–27.

————. "Das theoretisch-physikalische Institut." In *Festschrift zur Feier des 500jährigen Bestehens der Universität Leipzig,* vol. 4, pt. 2, 60–69.

Deutscher Universitäts-Kalender. Or *Deutsches Hochschulverzeichnis; Lehrkörper, Vorlesungen und Forschungseinrichtungen.* Berlin, 1872–1901. Leipzig, 1902– .

Dictionary of Scientific Biography. Edited by Charles Coulston Gillispie. 15 vols. New York: Scribner's 1970–78 *(DSB).*

Drude, Paul. "Antrittsrede." *Sitzungsber. preuss. Akad.,* 1906, 552–56.

————. *Lehrbuch der Optik.* Leipzig: S. Hirzel, 1900. Translated as *The Theory of Optics* by C. R. Mann and R. A. Millikan. New York: Longmans, Green, 1902.

————. *Physik des Aethers auf elektromagnetischer Grundlage.* Stuttgart, 1894. 2d ed., edited by Walter König. Stuttgart: F. Enke, 1912.

———. *Die Theorie in der Physik.* Antrittsvorlesung gehalten am 5. Dezember 1894 an der Universität Leipzig. Leipzig, 1895.

———. "Wilhelm Gottlieb Hankel." *Verh. sächs. Ges. Wiss.* 51 (1899): lxvii–lxxvi.

Du Bois-Reymond, Emil. *Hermann von Helmholtz. Gedächtnissrede.* Leipzig, 1897.

Dugas, René. *A History of Mechanics.* Translated by J. R. Maddox. New York: Central Book, 1955.

———. *La théorie physique au sens de Boltzmann et ses prolongements modernes.* Neuchâtel-Suisse: Griffon, 1959.

Dukas, Helen. See Banesh Hoffmann.

Earman, John, and Clark Glymour. "Lost in the Tensors: Einstein's Struggles with Covariance Principles 1912–1916." *Stud. Hist. Phil. Sci.* 9 (1978): 251–78.

———. "Relativity and Eclipse: The British Expeditions of 1919 and Their Predecessors." *HSPS* 11 (1980): 49–85.

Ebert, Hermann. *Hermann von Helmholtz.* Stuttgart: Wissenschaftliche Verlagsgesellschaft, 1949.

———. *Lehrbuch der Physik, nach Vorlesungen an der Technischen Hochschule zu München.* Vol. 1, *Mechanik, Wärmelehre.* Leipzig and Berlin: B. G. Teubner, 1912.

———. *Magnetic Fields of Force.* Pt. 1. Translated by C. V. Burton. London, 1897.

Eggert, Hermann. "Universitäten." In *Handbuch der Architektur,* pt. 4, sec. 6, no. 2aI, 54–111.

Einstein, Albert. "Antrittsrede." *Sitzungsber. preuss. Akad.,* 1914. In *Ideas and Opinions,* 216–19.

———. "Autobiographical Notes." In *Albert Einstein: Philosopher-Scientist,* edited by P. A. Schilpp, 1–95. Evanston: The Library of Living Philosophers, 1949.

———. *Einstein on Peace.* Edited by O. Nathan and H. Norden. New York: Schocken, 1968.

———. "Emil Warburg als Forscher." *Naturwiss.* 10 (1922): 824–28.

———. *Ideas and Opinions.* New York: Dell, 1973.

———. "Leo Arons als Physiker." *Sozialistische Monatshefte* 25 (1919): 1055–56.

———. "Max Planck als Forscher." *Naturwiss.* 1 (1913): 1077–79.

———. "Maxwell's Influence on the Evolution of the Idea of Physical Reality." 1931. In *Ideas and Opinions,* 259–63.

———. "Notes on the Origin of the General Theory of Relativity." In *Ideas and Opinions,* 279–83.

———. *Sidelights on Relativity.* Translated by G. B. Jeffery and W. Perrett. New York: E. P. Dutton, 1922.

———. *Ueber die spezielle und die allgemeine Relativitätstheorie (Gemeinverständlich).* 3d rev. ed. Braunschweig: F. Vieweg, 1918.

———. See H. A. Lorentz; Erwin Schrödinger.

Einstein, Albert, and Michele Besso. *Albert Einstein-Michele Besso Correspondance 1903–1955.* Edited by P. Speziali. Paris: Hermann, 1972.

Einstein, Albert, Hedwig Born, and Max Born. *Albert Einstein-Hedwig und Max Born*

Briefwechsel 1916–1955. Edited by Max Born. Munich: Nymphenburger Verlag, 1969.

Einstein, Albert, and Arnold Sommerfeld. *Albert Einstein/Arnold Sommerfeld Briefwechsel.* Edited by Armin Hermann. Basel and Stuttgart: Schwabe, 1968.

Elsasser, Walter M. *Memoirs of a Physicist in the Atomic Age.* New York: Science History Publications, 1978.

Emde, Fritz. "Gustav Mie 80 Jahre." *Phys. Bl.* 4 (1948): 349–50.

———. See Eugen Jahnke.

Encyklopädie der mathematischen Wissenschaften. Mit Einschluss ihrer Anwendungen. Vol. 1, pt. 1, *Reine Mathematik.* Edited by H. Burkhardt and F. Meyer. Leipzig, 1898. Vol. 4, *Mechanik.* Edited by Felix Klein and Conrad Müller. Pt. 1. Leipzig: B. G. Teubner, 1901–08. Vol. 5, *Physik.* Edited by Arnold Sommerfeld. Pt. 2. Leipzig: B. G. Teubner, 1904–22.

Erlangen. University. See Theodor Kolde.

"Ernst Abbe (1840–1905). The Origin of a Great Optical Industry." *Nature,* no. 3664 (20 Jan. 1940): 89–91.

Ernst Mach. Physicist and Philosopher. Vol. 6 of Boston Studies in the Philosophy of Science. Edited by R. S. Cohen and R. J. Seeger. Dordrecht-Holland: Reidel, 1970.

Eulenburg, Franz. *Der akademische Nachwuchs; eine Untersuchung über die Lage und die Aufgaben der Extraordinarien und Privatdozenten.* Leipzig: B. G. Teubner, 1908.

———. "Die Frequenz der deutschen Universitäten." *Abh. sächs. Ges. Wiss.* 24, pt. 2 (1904): 1–323.

Eversheim, Paul. See Heinrich Kayser.

Ewald, P. P. "Ein Buch über mathematische Physik: Courant-Hilbert." *Naturwiss.* 13 (1925): 384–87.

———. "Erinnerungen an die Anfänge des Münchener Physikalischen Kolloquiums." *Phys. Bl.* 24 (1968): 538–42.

———. "Max von Laue 1879–1960." *Biographical Memoirs of Fellows of the Royal Society* 6 (1960): 135–56.

Falkenhagen, Hans. "Zum 100. Geburtstag von Paul Karl Ludwig Drude (1863–1906)." *Forschungen und Fortschritte* 37 (1963): 220–21.

Ferber, Christian von. *Die Entwicklung des Lehrkörpers der deutschen Universitäten und Hochschulen 1864–1954.* Göttingen: Vandenhoeck und Ruprecht, 1956.

Fierz, M. "Pauli, Wolfgang." *DSB* 10 (1974): 422–25.

Fischer, Otto. *Medizinische Physik.* Leipzig: S. Hirzel, 1913.

Föppl, August. *Einführung in die Maxwellsche Theorie der Elektrizität.* Leipzig, 1894.

———. *Vorlesungen über technische Mechanik.* Vol. 1, *Einführung in die Mechanik.* 5th ed. Leipzig: B. G. Teubner, 1917.

———. "Ziele und Methoden der technischen Mechanik." *Jahresber. d. Deutsch. Math.-Vereinigung* 6 (1897): 99–110.

Försterling, Karl. "Woldemar Voigt zum hundertsten Geburtstage." *Naturwiss.* 38 (1951): 217–21.

Folie, F. "R. Clausius. Sa vie, ses travaux et leur portée metaphysique." *Revue des questions scientifiques* 27 (1890): 419–87.

Forman, Paul. "Alfred Landé and the Anomalous Zeeman Effect, 1919–1921." *HSPS* 2 (1970): 153–261.

———. "The Discovery of the Diffraction of X-Rays by Crystals: A Critique of the Myths." *Arch. Hist. Ex. Sci.* 6 (1969): 38–71.

———. "The Doublet Riddle and Atomic Physics *circa* 1924." *Isis* 59 (1968): 156–74.

———. "The Environment and Practice of Atomic Physics in Weimar Germany: A Study in the History of Science." Ph.D. diss., University of California, Berkeley, 1967.

———. "The Financial Support and Political Alignment of Physicists in Weimar Germany." *Minerva* 12 (1974): 39–66.

———. "Paschen, Louis Carl Heinrich Friedrich." *DSB* 10 (1974): 345–50.

———. "Scientific Internationalism and the Weimar Physicists: The Ideology and Its Manipulation in Germany after World War I." *Isis* 64 (1973): 151–80.

———. "Weimar Culture, Causality, and Quantum Theory, 1918–1927: Adaptation by German Physicists and Mathematicians to a Hostile Intellectual Environment." *HSPS* 3 (1971): 1–115.

———. See V. V. Raman.

Forman, Paul, and Armin Hermann. "Sommerfeld, Arnold (Johannes Wilhelm)." *DSB* 12 (1975): 525–32.

Forman, Paul, John L. Heilbron, and Spencer Weart. "Physics *circa* 1900. Personnel, Funding, and Productivity of the Academic Establishments." *HSPS* 5 (1975): 1–185.

Fragstein, C. v. "Clemens Schaefer zum 75. Geburtstag." *Optik* 11 (1954): 253–54.

Franck, James. "Emil Warburg zum Gedächtnis." *Naturwiss.* 19 (1931): 993–97.

———. "Max von Laue (1879–1960)." *American Philosophical Society: Yearbook,* 1960, 155–59.

Franck, James, and Robert Pohl. "Heinrich Rubens." *Phys. Zs.* 23 (1922): 377–82.

Frank, Philipp. *Einstein. His Life and Times.* Translated by G. Rosen. Edited and revised by S. Kusaka. New York: Knopf, 1947.

Franke, Martin. "Zu den Bemühungen Leipziger Physiker um eine Profilierung der physikalischen Institute der Universität Leipzig im zweiten Viertel des 20. Jahrhunderts." *NTM* 19 (1982): 68–76.

Frankfurt am Main. Physical Society. *Jahresbericht des Physikalischen Vereins zu Frankfurt am Main.* Frankfurt am Main, 1831– .

Freiburg i. Br. *Freiburg und seine Universität. Festschrift der Stadt Freiburg im Breisgau zur Fünfhundertjahrfeier der Albert-Ludwigs-Universität.* Edited by Maximilian Kollofrath and Franz Schneller. Freiburg i. Br.: n.p., 1957.

Freiburg. University. *Aus der Geschichte der Naturwissenschaften an der Universität Freiburg i. Br.* Edited by Eduard Zentgraf. Freiburg i. Br.: Albert, 1957.

———. *Die Universität Freiburg seit dem Regierungsantritt Seiner Königlichen Hoheit des Grossherzogs Friedrich von Baden.* Freiburg i. Br. and Tübingen, 1881.

————. See Fritz Baumgarten.

French, A. P., ed. *Einstein: A Centenary Volume*. Cambridge, Mass.: Harvard University Press, 1979.

Freundlich, Erwin. *The Foundations of Einstein's Theory of Gravitation*. Translated by H. L. Brose. London: Methuen, 1924.

Frey-Wyssling, A., and Elsi Häusermann. *Geschichte der Abteilung für Naturwissenschaften an der Eidgenössischen Technischen Hochschule in Zürich 1855–1955*. [Zurich], 1958.

Fricke, Dieter. "Zur Militarisierung des deutschen Geisteslebens im wilhelminischen Kaiserreich. Der Fall Leo Arons." *Zs. f. Geschichtswissenschaft* 8 (1960): 1069–1107.

Fricke, Robert. "Die allgemeinen Abteilungen." In *Das Unterrichtswesen im Deutschen Reich*, edited by Wilhelm Lexis, vol. 4, pt 1, 49–62.

Friedrich, W. "Wilhelm Conrad Röntgen." *Phys. Zs.* 24 (1923): 353–60.

Frommel, Emil. *Johann Christian Poggendorff*. Berlin, 1877.

Fueter, R. "Zum Andenken an Karl VonderMühll (1841–1912)." *Math. Ann.* 73 (1913): i–ii.

Fuoss, Raymond M. "Peter J. W. Debye." In *The Collected Papers of Peter J. W. Debye*, xi–xiv.

Galison, Peter Louis. "Minkowski's Space-Time: From Visual Thinking to the Absolute World." *HSPS* 10 (1979): 85–121.

Gans, Richard. *Einführung in die Vektoranalysis mit Anwendungen auf die mathematische Physik*. Leipzig: B. G. Teubner, 1905.

Gebhardt, Willy. "Die Geschichte der Physikalischen Institute der Universität Halle." *Wiss. Zs. d. Martin-Luther-U. Halle-Wittenberg, Math.-Naturwiss.* 10 (1961): 851–59.

Gehlhoff, Georg. "E. Warburg als Lehrer." *Zs. f. techn. Physik* 3 (1922): 193–94.

Gehlhoff, Georg, Hans Rukop, and Wilhelm Hort. "Zur Einführung." *Zs. f. techn. Physik* 1 (1920): 1–4.

Gehrcke, E. "Otto Lummer." *Zs. f. techn. Physik* 6 (1925): 482–86.

————. "Warburg als Physiker." *Zs. f. techn. Physik* 3 (1922): 186–92.

Gerber, J. U. "Geschichte der Wellenmechanik." *Arch. Hist. Ex. Sci.* 5 (1969): 349–416.

Gerlach, Walther. "Edgar Meyer 80 Jahre." *Phys. Bl.* 15 (1959): 136.

————. "Friedrich Paschen." *Jahrbuch bay. Akad.*, 1944–48, 277–80.

————. "Heinrich Matthias Konen." *Phys. Bl.* 5(1949): 226.

————. "Heinrich Rudolf Hertz 1857–1894." In *150 Jahre Rheinische Friedrich-Wilhelms-Universität zu Bonn 1818–1968. Mathematik und Naturwissenschaften*, 110–16.

————. "Peter Debye." *Jahrbuch bay. Akad.*, 1966, 218–30.

————. See Fritz Bopp.

German Physical Society. Fiftieth anniversary issue. *Verh. phys. Ges.* 15 (1896): 1–40.

————. Foreword. *Die Fortschritte der Physik im Jahre 1845* 1 (1847).

————. "Satzungen der Deutschen Physikalischen Gesellschaft." *Verh. phys. Ges.* 1 (1899): 5–10.

German Society for Technical Physics. "Zur Gründung der Deutschen Gesellschaft für technische Physik." *Zs. f. techn. Physik* 1 (1920): 4–6.

Gibbs, Josiah Willard. "Rudolf Julius Emanuel Clausius." *Proc. Am. Acad.* 16 (1889): 458–65.

Giessen. University. *Ludwigs-Universität, Justus Liebig-Hochschule, 1607–1957. Festschrift zur 350-Jahrfeier.* Giessen, 1957.

————. *Die Universität Giessen von 1607 bis 1907. Beiträge zu ihrer Geschichte. Festschrift zur dritten Jahrhundertfeier.* Edited by Universität Giessen. Vol. 1. Giessen: A. Töpelmann, 1907.

————. See M. Biermer; Wilhelm Lorey.

Glasser, Otto. *Dr. W. C. Röntgen.* Springfield, Ill.: Charles C. Thomas, 1945.

Glymour, Clark. See John Earman.

Göttingen. University. *Chronik der Georg-August-Universität zu Göttingen.* Göttingen.

————. *Die physikalischen Institute der Universität Göttingen.* Edited by Göttinger Vereinigung zur Förderung der angewandten Physik und Mathematik. Leipzig and Berlin: B. G. Teubner, 1906.

————. *Statuten des mathematisch-physikalischen Seminars zu Göttingen.* Göttingen, 1886.

Goldberg, Stanley. "The Abraham Theory of the Electron: The Symbiosis of Experiment and Theory." *Arch. Hist. Ex. Sci.* 7 (1970): 7–25.

————. "Early Response to Einstein's Theory of Relativity, 1905–1911: A Case Study in National Differences." Ph.D. diss., Harvard University, 1969.

————. "The Lorentz Theory of Electrons and Einstein's Theory of Relativity." *Am. J. Phys.* 37 (1969): 982–94.

————. "Max Planck's Philosophy of Nature and His Elaboration of the Special Theory of Relativity." *HSPS* 7 (1976): 125–60.

Goldstein, Eugen. "Aus vergangenen Tagen der Berliner Physikalischen Gesellschaft." *Naturwiss.* 13 (1925): 39–45.

Graetz, Leo. *Der Aether und die Relativitätstheorie. Sechs Vorträge.* Stuttgart: J. Engelhorns Nachf., 1923.

————. *Die Atomtheorie in ihrer neuesten Entwickelung. Sechs Vorträge.* Stuttgart: J. Engelhorns Nachf., 1920.

————. *Lehrbuch der Physik.* 4th rev. ed. Leipzig and Vienna: F. Deuticke, 1917.

Graz. University. *Academische Behörden, Personalstand und Ordnung der öffentlichen Vorlesungen an der K. K. Carl-Franzens-Universität und der K. K. medicinisch-chirurgischen Lehranstalt zu Graz.* Graz. n.d.

————. *Verzeichnis der Vorlesungen an der K. K. Karl-Franzens-Universität in Graz, 1876–1890.* Graz, n.d.

Gregory, Frederick. *Scientific Materialism in Nineteenth Century Germany.* Dordrecht and Boston: Reidel, 1977.

Greifswald. University. *Chronik der Königlichen Universität Greifswald.* Greifswald.

———. *Festschrift zur 500-Jahrfeier der Universität Greifswald.* Vol. 2. Greifswald: Universität, 1956.

Grüneisen, Eduard. "Emil Warburg zum achtzigsten Geburtstage." *Naturwiss.* 14 (1926): 203–7.

Günther, Siegmund. *Handbuch der Geophysik.* 2d rev. ed. 2 vols. Stuttgart, 1897–99.

Guggenbühl, Gottfried. "Geschichte der Eidgenössischen Technischen Hochschule in Zürich." In *Eidgenössische Technische Hochschule 1855–1955,* 3–260.

Guttstadt, Albert, ed. *Die naturwissenschaftlichen und medicinischen Staatsanstalten Berlins. Festschrift für die 59. Versammlung deutscher Naturforscher und Aerzte.* Berlin, 1886.

Gutzmer, A. "Bericht der Unterrichts-Kommission über ihre bisherige Tätigkeit." *Verh. Ges. deutsch. Naturf. u. Ärzte* 77, pt. 1 (1905): 142–200.

Häusermann, Elsi. See A. Frey-Wyssling.

Hahn, Otto. *My Life; the Autobiography of a Scientist.* Translated by E. Kaiser and E. Wilkins. New York: Herder and Herder, 1970.

Halle. University. *Bibliographie der Universitätsschriften von Halle-Wittenberg 1817–1885.* Edited by W. Suchier. Berlin: Deutscher Verlag der Wissenschaften, 1953.

———. *Chronik der Königlichen Vereinigten Friedrichs-Universität Halle-Wittenberg.* Halle.

———. *450 Jahre Martin-Luther-Universität Halle-Wittenberg.* Vol. 2. [Halle, 1953 ?]

———. See Willy Gebhardt; Wilhelm Schrader.

Handbuch der Architektur. Pt. 4, *Entwerfen, Anlage und Einrichtung der Gebäude.* Sec. 6, *Gebäude für Erziehung, Wissenschaft und Kunst.* No. 2a, *Hochschulen, zugehörige und verwandte wissenschaftliche Institute. I. Hochschulen im allgemeinen, Universitäten und Technische Hochschulen. Naturwissenschaftliche Institute.* Edited by H. Eggert, C. Junk, C. Körner, and E. Schmitt. 2d ed. Stuttgart: A. Kröner, 1905.

Handbuch der bayerischen Geschichte. Vol. 4, *Das neue Bayern 1800–1970.* Edited by Max Spindler. Pt. 2. Munich: C. H. Beck, 1975.

Hanle, P. A. "Erwin Schrödinger's Reaction to Louis de Broglie's Thesis on the Quantum Theory." *Isis* 68 (1977): 606–9.

———. "Indeterminacy before Heisenberg: The Case of Franz Exner and Erwin Schrödinger." *HSPS* 10 (1979): 225–69.

Hanle, Wilhelm, and Arthur Scharmann. "Paul Drude (1863–1906)/Physiker." In *Giessener Gelehrte in der ersten Hälfte des 20. Jahrhunderts,* edited by H. G. Gundel, P. Moraw, and V. Press, vol. 2 of *Lebensbilder aus Hessen, Veröffentlichungen der Historischen Kommission für Hessen,* 174–81. Marburg, 1982.

Harig, G., ed. *Bedeutende Gelehrte in Leipzig.* Vol. 2. Leipzig: Karl-Marx-Universität, 1965.

Harman, P. M. *Energy, Force, and Matter: The Conceptual Development of Nineteenth-Century Physics.* Cambridge: Cambridge University Press, 1982.

———. *Metaphysics and Natural Philosophy: The Problem of Substance in Classical Physics.* Brighton: Harvester Press, 1982.

Harnack, Adolf, ed. *Geschichte der Königlich preussischen Akademie der Wissenschaften zu Berlin.* 3 vols. Berlin: Reichsdruckerei, 1900.

Hartmann, H., ed. *Schöpfer des neuen Weltbildes. Grosse Physiker unserer Zeit.* Bonn: Athenäum, 1952.

Havránek, Jan. "Die Ernennung Albert Einsteins zum Professor in Prag." *Acta Universitatis Carolinae—Historia Universitatis Carolinae Pragensis* 17, pt. 2 (1977): 114–30.

Heidelberg. University. *Anzeige der Vorlesungen . . . auf der Grossherzoglich Badischen Ruprecht-Carolinischen Universität zu Heidelberg. . . .* Heidelberg.

————. *Ruperto-Carola. Sonderband. Aus der Geschichte der Universität Heidelberg und ihrer Fakultäten.* Edited by G. Hinz. Heidelberg: Brausdruck, 1961.

————. *Die Ruprecht-Karl-Universität Heidelberg.* Edited by G. Hinz. Berlin and Basel: Länderdienst, 1965.

————. *Zusammenstellung der Vorlesungen, welche vom Sommerhalbjahr 1804 bis 1886 auf der Grossherzoglich Badischen Ruprecht-Karls-Universität zu Heidelberg angekündigt worden sind.* Heidelberg.

————. See Reinhard Riese.

Heilbron, John L. "The Kossel-Sommerfeld Theory and the Ring Atom." *Isis* 58 (1967): 451–85.

————. "Lectures on the History of Atomic Physics 1900–1922." In *History of Twentieth Century Physics,* edited by Charles Weiner, 40–108.

————. See Paul Forman.

Heilbron, John L., and Thomas S. Kuhn. "The Genesis of the Bohr Atom." *HSPS* 1 (1969): 211–90.

Heilbrunn, Ludwig. *Die Gründung der Universität Frankfurt a. M.* Frankfurt a. M.: Joseph Baer, 1915.

Heimann, P. M. "Maxwell, Hertz and the Nature of Electricity." *Isis* 62 (1970): 149–57.

Heisenberg, Werner. "The Development of Quantum Mechanics." In *Nobel Lectures. Physics 1922–1941,* The Nobel Foundation, 290–301. Amsterdam, London, and New York: Elsevier, 1965.

————. *The Physical Principles of the Quantum Theory.* Translated by C. Eckart and F. C. Hoyt. New York: Dover, 1930.

————. *Physics and Beyond: Encounters and Conversations.* Translated by A. J. Pomerans. New York: Harper and Row, 1971.

————. *Physics and Philosophy: The Revolution in Modern Science.* New York: Harper, 1958.

————. "Professor Max Born." *Nature* 225 (1970): 669–71.

————. "Quantenmechanik." *Naturwiss.* 14 (1926): 989–94.

————. "Remarks on the Origin of the Relations of Uncertainty." In *The Uncertainty Principle and Foundations of Quantum Mechanics. A Fifty Years' Survey,* edited by W. C. Price and S. S. Chissick, 3–6. London, New York, Sydney, and Toronto: John Wiley, 1977.

————. *Wandlungen in den Grundlagen der Naturwissenschaft. Zwei Vorträge.* Leipzig: S. Hirzel, 1935.

Heitler, W. "Erwin Schrödinger, 1887–1961." *Biographical Memoirs of Fellows of the Royal Society* 7 (1961): 221–28.

Heller, Karl Daniel. *Ernst Mach: Wegbereiter der modernen Physik.* Vienna and New York: Springer, 1964.

Helm, Georg. *Die Energetik nach ihrer geschichtlichen Entwickelung.* Leipzig, 1898.

———. *Die Lehre von der Energie.* Leipzig, 1887.

———. "Oskar Schlömilch." *Zs. f. Math. u. Phys.* 46 (1901): 1–7.

Helmholtz, Anna von. *Anna von Helmholtz. Ein Lebensbild in Briefen.* Edited by Ellen von Siemens-Helmholtz. Vol. 1. Berlin: Verlag für Kulturpolitik, 1929.

Helmholtz, Hermann (von). "Autobiographical Sketch." In *Popular Lectures on Scientific Subjects,* vol. 2, 266–91.

———. *Epistemological Writings.* Edited by Paul Hertz and Moritz Schlick. Translated by M. F. Lowe. Vol. 37 of Boston Studies in the Philosophy of Science. Dordrecht and Boston: Reidel, 1977.

———. "Gustav Magnus. In Memoriam." In *Popular Lectures on Scientific Subjects,* 1–25.

———. "Gustav Wiedemann." *Ann.* 50 (1893): iii–xi.

———. *Popular Lectures on Scientific Subjects.* Translated by E. Atkinson. London, 1881. New ed. in 2 vols. London: Longmans, Green, 1908–12.

———. "Preface." In Heinrich Hertz's *The Principles of Mechanics.*

———. *Selected Writings of Hermann von Helmholtz.* Edited by R. Kahl. Middletown, Conn.: Wesleyan University Press, 1971.

———. *Vorlesungen über theoretische Physik.* Vol. 1, pt. 1, *Einleitung zu den Vorlesungen über theoretische Physik.* Edited by Arthur König and Carl Runge. Leipzig: J. A. Barth, 1903. Vol. 1, pt. 2, *Vorlesungen über die Dynamik discreter Massenpunkte.* Edited by Otto Krigar-Menzel. Leipzig, 1898. Vol. 2, *Vorlesungen über die Dynamik continuirlich verbreiteter Massen.* Edited by Otto Krigar-Menzel. Leipzig: J. A. Barth, 1902. Vol. 3, *Vorlesungen über die mathematischen Principien der Akustik.* Edited by Arthur König and Carl Runge. Leipzig, 1898. Vol. 4, *Vorlesungen über Elektrodynamik und Theorie des Magnetismus.* Edited by Otto Krigar-Menzel and Max Laue. Leipzig: J. A. Barth, 1907. Vol. 5, *Vorlesungen über die elektromagnetische Theorie des Lichtes.* Edited by Arthur König and Carl Runge. Hamburg and Leipzig, 1897. Vol. 6, *Vorlesungen über die Theorie der Wärme.* Edited by Franz Richarz. Leipzig: J. A. Barth, 1903.

———. *Wissenschaftliche Abhandlungen.* 3 vols. Leipzig, 1882–95.

———. "Zur Erinnerung an Rudolf Clausius." *Verh. phys. Ges.* 8 (1889): 1–7.

Helmholtz, Robert. "A Memoir of Gustav Robert Kirchhoff." Translated by J. de Perott. In *Annual Report of the . . . Smithsonian Institution . . . to July, 1889,* 1890, 527–40.

Henssi, Jacob. *Der physikalische Apparat. Anschaffung, Behandlung und Gebrauch desselben. Für Lehrer und Freunde der Physik.* Leipzig, 1875.

Hermann, Armin. "Albert Einstein und Johannes Stark. Briefwechsel und Verhältnis der beiden Nobelpreisträger." *Sudhoffs Archiv* 50 (1966): 267–85.

———. "Born, Max." *DSB* 15 (1978): 39–44.

―――. "Einstein auf der Salzburger Naturforscherversammlung 1909." *Phys. Bl.* 25 (1969): 433–36.

―――. *The Genesis of the Quantum Theory (1899–1913).* Translated by C. W. Nash. Cambridge, Mass.: MIT Press, 1971.

―――. "Hertz, Heinrich Rudolf." *Neue deutsche Biographie* 8 (1969): 713–14.

―――. *Die Jahrhundertwissenschaft: Werner Heisenberg und die Physik seiner Zeit.* Stuttgart: Deutsche Verlags-Anstalt, 1977.

―――. "Laue, Max von." *DSB* 8 (1973): 50–53.

―――. *Max Planck in Selbstzeugnissen und Bilddokumenten.* Reinbek b. Hamburg: Rowohlt, 1973.

―――. "Schrödinger, Erwin." *DSB* 12 (1975): 217–23.

―――. "Sommerfeld und die Technik." *Technikgeschichte* 34 (1967): 311–22.

―――. "Stark, Johannes." *DSB* 12 (1975): 613–16.

―――. *Werner Heisenberg in Selbstzeugnissen und Bilddokumenten.* Reinbek b. Hamburg: Rowohlt, 1976.

―――. See Paul Forman.

Hermann, L. "Hermann von Helmholtz." *Schriften der Physikalisch-ökonomischen Gesellschaft zu Königsberg* 35 (1894): 63–73.

Herneck, Friedrich. "Max von Laue. Die Entdeckung der Röntgenstrahl-Interferenzen." In *Bahnbrecher des Atomzeitalters; grosse Naturforscher von Maxwell bis Heisenberg,* 273–326. Berlin: Buchverlag Der Morgen, 1965.

Hertz, Heinrich. *Electric Waves, Being Researches on the Propagation of Electric Action with Finite Velocity through Space.* Translated by D. E. Jones. New York, 1893. Reprint. New York: Dover, 1962.

―――. *Erinnerungen, Briefe, Tagebücher.* Edited by J. Hertz. 2d rev. ed. by M. Hertz and Charles Süsskind. San Francisco: San Francisco Press, 1977.

―――. *Gesammelte Werke.* Edited by Philipp Lenard. Vol. 1, *Schriften vermischten Inhalts.* Vol. 2, *Untersuchungen über die Ausbreitung der elektrischen Kraft,* 2d ed. Vol. 3, *Die Prinzipien der Mechanik.* Leipzig, 1894–95.

―――. "Hermann von Helmholtz." In supplement to *Münchener Allgemeine Zeitung,* 31 Aug. 1891. Reprinted and translated by D. E. Jones and G. A. Schott in *Miscellaneous Papers,* 332–40.

―――. *Miscellaneous Papers.* Translation of *Schriften vermischten Inhalts* by D. E. Jones and G. A. Schott. London, 1896.

―――. *Die Prinzipien der Mechanik, in neuem Zusammenhange dargestellt.* Edited by Philipp Lenard. Leipzig, 1894. Translated as *The Principles of Mechanics Presented in a New Form* by D. E. Jones and J. T. Walley. London, 1899. Reprint. New York: Dover, 1956.

―――. *Ueber die Beziehungen zwischen Licht und Elektricität.* Bonn, 1889. Reprinted in *Gesammelte Werke* 1: 339–54.

―――. *Ueber die Induction in rotierenden Kugeln.* Berlin, 1880.

Heydweiller, Adolf. "Friedrich Kohlrausch." In Friedrich Kohlrausch's *Gesammelte Abhandlungen,* vol. 2, xxxv–lxviii.

―――. "Johann Wilhelm Hittorf." *Phys. Zs.* 16 (1915): 161–79.

Hiebert, Erwin N. "The Energetics Controversy and the New Thermodynamics." In *Perspectives in the History of Science and Technology*, edited by D. H. D. Roller, 67–86. Norman: University of Oklahoma Press, 1971.

————. "Ernst Mach." *DSB* 8 (1973): 595–607.

————. "The Genesis of Mach's Early Views on Atomism." In *Ernst Mach. Physicist and Philosopher*, 79–106. Vol. 6, Boston Studies in the Philosophy of Science, ed. R. S. Cohen and R. J. Seeger. Dordrecht: D. Reidel, 1970.

————. "Nernst, Hermann Walther." *DSB*, Supplement, 1978, 432–53.

Hiebert, Erwin N., and Hans-Günther Körber. "Ostwald, Friedrich Wilhelm." *DSB*, Supplement, 1978, 455–69.

Hilbert, David. "Axiomatisches Denken." *Math. Ann.* 78 (1918): 405–15.

————. "Gedächtnisrede auf H. Minkowski." In Hermann Minkowski's *Gesammelte Abhandlungen* 1: v–xxxi.

————. See Richard Courant.

Hirosige, Tetu. "Electrodynamics before the Theory of Relativity, 1890–1905." *Jap. Stud. Hist. Sci.*, no. 5 (1966): 1–49.

————. "The Ether Problem, the Mechanistic Worldview, and the Origins of the Theory of Relativity." *HSPS* 7 (1976): 3–82.

————. "Origins of Lorentz' Theory of Electrons and the Concept of the Electromagnetic Field." *HSPS* 1 (1969): 151–209.

————. "Theory of Relativity and the Ether." *Jap. Stud. Hist. Sci.*, no. 7 (1968): 37–53.

Hölder, O. "Carl Neumann." *Verh. sächs. Ges. Wiss.* 77 (1925): 154–80.

Hönl, H. "Intensitäts- und Quantitätsgrössen. In Memoriam Gustav Mie zu seinem hundertsten Geburtstag." *Phys. Bl.*. 24 (1968): 498–502.

Hoffmann, Banesh, with Helen Dukas. *Albert Einstein: Creator and Rebel.* New York: Viking, 1972.

Hofmann, A. W. "Gustav Kirchhoff," *Berichte der deutschen chemischen Gesellschaft*, vol. 20, pt. 2 (1887): 2771–77.

————. *The Question of a Division of the Philosophical Faculty. Inaugural Address on Assuming the Rectorship of the University of Berlin, Delivered in the Aula of the University on October 15, 1880.* 2d ed. Boston, 1883.

Holborn, Hajo, *A History of Modern Germany, 1840–1945.* New York: Alfred A. Knopf, 1969.

Holt, Niles R. "A Note on Wilhelm Ostwald's Energism." *Isis* 61 (1970): 386–89.

Holton, Gerald. "Einstein's Scientific Program: The Formative Years." In *Some Strangeness in the Proportion*, edited by Harry Woolf, 49–65.

————. "Einstein's Search for the *Weltbild*." *Proc. Am. Phil. Soc.* 125 (1981): 1–15.

————. "Influences on Einstein's Early Work in Relativity Theory." *American Scholar* 37 (1967): 59–79. Reprinted in *Thematic Origins of Scientific Thought: Kepler to Einstein*, 197–217.

————. "Mach, Einstein, and the Search for Reality." *Daedalus* 97 (1968): 636–73. Reprinted in *Thematic Origins of Scientific Thought: Kepler to Einstein*, 219–59.

————. "The Metaphor of Space-Time Events in Science." *Eranos Jahrbuch* 34 (1965): 33–78.

―――. "On the Origins of the Special Theory of Relativity." *Am. J. Phys.* 28 (1960): 627–36. Reprinted in *Thematic Origins of Scientific Thought: Kepler to Einstein*, 165–83.

―――. "On Trying to Understand Scientific Genius." *American Scholar* 41 (1971–72): 95–110.

―――. "The Roots of Complementarity." *Daedalus* 99 (1970): 1015–55.

―――. *Thematic Origins of Scientific Thought: Kepler to Einstein.* Cambridge, Mass.: Harvard University Press, 1973.

Hoppe, Edmund. *Geschichte der Elektrizität.* Leipzig, 1884.

Hort, Wilhelm. "Die technische Physik als Grundlage für Studium und Wissenschaft der Ingenieure." *Zs. f. techn. Physik.* 2 (1921): 132–40.

Hund, Friedrich. *Geschichte der Quantentheorie.* 2d ed. Mannheim, Vienna, and Zurich: Bibliographisches Institut, 1975.

―――. "Höhepunkte der Göttinger Physik." *Phys. Bl.* 25 (1969): 145–53, 210–15.

―――. "Peter Debye." *Jahrbuch der Akademie der Wissenschaften in Göttingen,* 1966, 59–64.

Hunt, Bruce. "Theory Invades Practice: The British Response to Hertz." *Isis* 74 (1983): 341–55.

Ignatowski, Waldemar von. *Die Vektoranalysis und ihre Anwendung in der theoretischen Physik.* 2 vols. Leipzig and Berlin: B. G. Teubner, 1909–10.

Ilberg, Waldemar. "Otto Heinrich Wiener (1862–1927)." In *Bedeutende Gelehrte in Leipzig,* edited by G. Harig, 2: 121–30.

Illy, József. "Albert Einstein in Prague." *Isis* 70 (1979): 76–84.

J., D. E. "Heinrich Hertz." *Nature* 49 (1894): 265–66.

Jaeckel, Barbara, and Wolfgang Paul. "Die Entwicklung der Physik in Bonn 1818–1968." In *150 Jahre Rheinische Friedrich-Wilhelms-Universität zu Bonn 1818–1968,* 91–100.

Jäger, Gustav. "Der Physiker Ludwig Boltzmann." *Monatshefte für Mathematik und Physik* 18 (1907): 3–7.

Jahnke, Eugen. *Vorlesungen über die Vektorenrechnung. Mit Anwendungen auf Geometrie, Mechanik und mathematische Physik.* Leipzig: B. G. Teubner, 1905.

Jahnke, Eugen, and Fritz Emde. *Funktionentafeln mit Formeln und Kurven.* Leipzig and Berlin: B. G. Teubner, 1909.

Jammer, Max. *The Conceptual Development of Quantum Mechanics.* New York: McGraw-Hill, 1966.

―――. *The Philosophy of Quantum Mechanics: The Interpretations of Quantum Mechanics in Historical Perspective.* New York: John Wiley, 1974.

Jena. University. *Beiträge zur Geschichte der Mathematisch-Naturwissenschaftlichen Fakultät der Friedrich-Schiller-Universität Jena anlässlich der 400-Jahr-Feier.* Jena: G. Fischer, 1959.

―――. *Geschichte der Universität Jena 1548/58–1958. Festgabe zum vierhundertjährigen Universitätsjubiläum.* 2 vols. Jena: G. Fischer, 1958.

Jensen, C. "Leonhard Weber." *Meteorologische Zeitschrift* 36 (1919): 269–71.

Jordan, Pascual. "Werner Heisenberg 70 Jahre." *Phys. Bl.* 27 (1971): 559–62.

―――. "Wilhelm Lenz." *Phys. Bl.* 13 (1957): 269–70.

————. See Max Born.

Jost, Walter, "The First 45 Years of Physical Chemistry in Germany." *Annual Review of Physical Chemistry* 17 (1966): 1–14.

Junk, Carl. "Physikalische Institute." In *Handbuch der Architektur*, pt. 4, sec. 6, no. 2aI, 164–236.

Kalähne, Alfred. "Dem Andenken an Georg Quincke." *Phys. Zs.* 25 (1924): 649–59.

————. "Zum Gedächtnis von Rudolf H. Weber." *Phys. Zs.* 23 (1922): 81–83.

Kangro, Hans. "Das Paschen-Wiensche Strahlungsgesetz und seine Abänderung durch Max Planck." *Phys. Bl.* 25 (1969): 216–20.

————. *Vorgeschichte des Planckschen Strahlungsgesetzes.* Wiesbaden: Franz Steiner, 1970.

Karlsruhe. Technical Institute. *Festgabe zum Jubiläum der vierzigjährigen Regierung Seiner Königlichen Hoheit des Grossherzogs Friedrich von Baden.* Karlsruhe, 1892.

————. *Die Grossherzogliche Technische Hochschule Karlsruhe. Festschrift zur Einweihung der Neubauten im Mai 1899.* Stuttgart, 1899.

Kast, W. "Gustav Mie." *Phys. Bl.* 13 (1957): 129–31.

Kaufmann, Walter, "Physik." *Naturwiss.* 7 (1919): 542–48.

Kayser, Heinrich. *Handbuch der Spectroscopie.* Vol. 1. Leipzig: S. Hirzel, 1900.

————. Obituary of Hermann Lorberg in *Chronik der Rheinischen Friedrich-Wilhelms-Universität zu Bonn 1905/06*, 13–14.

Kayser, Heinrich, and Paul Eversheim. "Das physikalische Institut der Universität Bonn." *Phys. Zs.* 14 (1913): 1001–8.

Kelbg, Günter, and Wolf Dietrich Kraeft. "Die Entwicklung der theoretischen Physik in Rostock." *Wiss. Zs. d. U. Rostock* 16 (1967): 839–47.

Kemmer, N., and R. Schlapp. "Max Born." *Biographical Memoirs of Fellows of the Royal Society.* 17 (1971): 17–52.

Ketteler, Eduard. *Theoretische Optik gegründet auf das Bessel-Sellmeier'sche Princip. Zugleich mit den experimentellen Belegen.* Braunschweig, 1885.

Kiebitz, Franz. "Paul Drude." *Naturwiss. Rundschau* 21 (1906): 413–15.

Kiel. University. *Chronik der Universität Kiel.* Kiel.

————. *Geschichte der Christian-Albrechts-Universität Kiel, 1665–1965.* Vol. 6, *Geschichte der Mathematik, der Naturwissenschaften und der Landwirtschaftswissenschaften.* Edited by Karl Jordan. Neumünster: Wachholtz, 1968.

————. See Charlotte Schmidt-Schönbeck.

Kirchhoff, Gustav. *Gesammelte Abhandlungen.* Leipzig, 1882. *Nachtrag.* Edited by Ludwig Boltzmann. Leipzig, 1891.

————. *Vorlesungen über mathematische Physik.* Vol. 1, *Mechanik.* 3d ed. Leipzig, 1883. Vol. 2, *Vorlesungen über mathematische Optik.* Edited by K. Hensel. Leipzig, 1891. Vol. 3, *Vorlesungen über Electricität und Magnetismus.* Edited by Max Planck. Leipzig, 1891. Vol. 4, *Vorlesungen über die Theorie der Wärme.* Edited by Max Planck. Leipzig, 1894.

Kirn, H. See E. Bonnell.

Kirsten, Christa, and Hans-Günther Körber, eds. *Physiker über Physiker.* Berlin: Akademie-Verlag, 1975.

Kirsten, Christa, and H. J. Treder, eds. *Albert Einstein in Berlin 1913–1933*. Pt. 1, *Darstellung und Dokumente*. Berlin: Akademie-Verlag, 1979.

Kistner, Adolf. "Meyer, Oskar Emil." *Biographisches Jahrbuch und Deutscher Nekrolog* 14 (1912): 157–60.

Klein, Felix. "Ernst Schering." *Jahresber. d. Deutsch. Math.-Vereinigung* 6 (1899): 25–27.

———. "Mathematik, Physik, Astronomie an den deutschen Universitäten in den Jahren 1893–1903." *Jahresber. d. Deutsch. Math.-Vereinigung* 13 (1904): 457–75.

———. "Über die Encyklopädie der mathematischen Wissenschaften, mit besonderer Rücksicht auf Band 4 derselben (Mechanik)." *Phys. Zs.* 2 (1900): 90–96.

———. *Vorlesungen über die Entwicklung der Mathematik im 19. Jahrhundert*. Pt. 1 edited by R. Courant and O. Neugebauer. Pt. 2, *Die Grundbegriffe der Invariantentheorie und ihr Eindringen in die mathematische Physik*, edited by R. Courant and St. Cohn-Vossen. Reprint. New York: Chelsea, 1967.

Klein, Martin J. "The Development of Boltzmann's Statistical Ideas." In *The Boltzmann Equation: Theory and Applications*, edited by E. G. D. Cohen and W. Thirring, 53–106. In *Acta Physica Austraica*, Supplement 10. Vienna and New York: Springer, 1973.

———. "Einstein and the Wave-Particle Duality." *The Natural Philosopher*, no. 3 (1964): 3–49.

———. "Einstein, Specific Heats, and the Early Quantum Theory." *Science* 148 (1965): 173–80.

———. "Einstein's First Paper on Quanta." *The Natural Philosopher*, no. 2 (1963): 59–86.

———. "The First Phase of the Bohr-Einstein Dialogue." *HSPS* 2 (1970): 1–39.

———. "Gibbs on Clausius." *HSPS* 1 (1969): 127–49.

———. "Max Planck and the Beginnings of the Quantum Theory." *Arch. Hist. Ex. Sci.* 1 (1962): 459–79.

———. "Maxwell, His Demon, and the Second Law of Thermodynamics." *American Scientist* 58 (1970): 84–97.

———. "Mechanical Explanation at the End of the Nineteenth Century." *Centaurus* 17 (1972): 58–82.

———. "No Firm Foundation: Einstein and the Early Quantum Theory." In *Some Strangeness in the Proportion*, edited by Harry Woolf, 161–85.

———. *Paul Ehrenfest*. Vol. 1, *The Making of a Theoretical Physicist*. Amsterdam and London: North-Holland, 1970.

———. "Planck, Entropy, and Quanta, 1901–1906." *The Natural Philosopher*, no. 1 (1963): 83–108.

———. "Thermodynamics and Quanta in Planck's Work." *Physics Today* 19 (1966): 23–32.

———. "Thermodynamics in Einstein's Thought." *Science* 157 (1967): 509–16.

Klein, Martin J., and Allan Needell. "Some Unnoticed Publications by Einstein." *Isis* 68 (1977): 601–4.

Klemm, Friedrich. "Die Rolle der Mathematik in der Technik des 19. Jahrhunderts." *Technikgeschichte* 33 (1966): 72–91.

Klinckowstroem, Carl von. "Auerbach, Felix." *Neue deutsche Biographie* 1 (1953): 433.

Knapp, Martin. "Prof. Dr. Karl Von der Mühll-His." *Verhandlungen der Schweizerischen Naturforschenden Gesellschaft* 95 (1912), pt. 1, *Nekrologe und Biographien*, 93–105.

Knott, Robert. "Hankel: Wilhelm Gottlieb." *ADB* 49 (1967): 757–59.

———. "Knoblauch: Karl Hermann." *ADB* 51 (1971): 256–58.

———. "Weber: Wilhelm Eduard." *ADB* 41 (1967): 358–61.

König, Walter. "Georg Hermann Quinckes Leben und Schaffen." *Naturwiss.* 12 (1924): 621–27.

König, Walter, and Franz Richarz. *Zur Erinnerung an Paul Drude.* Giessen: A. Töpelmann, 1906.

Königsberg. University. *Chronik der Königlichen Albertus-Universität zu Königsberg i. Pr.* Königsberg.

———. See Hans Prutz.

Koenigsberger, Johann. "F. Pockels." *Centralblatt für Mineralogie, Geologie und Paläontologie,* 1914, 19–21.

Koenigsberger, Leo. *Hermann von Helmholtz.* 3 vols. Braunschweig: F. Vieweg, 1902–3.

———. "The Investigations of Hermann von Helmholtz on the Fundamental Principles of Mathematics and Mechanics." *Annual Report of the . . . Smithsonian Institution . . . to July, 1896,* 1898, 93–124.

———. *Mein Leben.* Heidelberg: Carl Winters, 1919.

Köpke, Rudolf. *Die Gründung der königlichen Friedrich-Wilhelms-Universität zu Berlin.* Berlin, 1860.

Körber, Hans-Günther. "Hankel, Wilhelm Gottlieb." *DSB* 6 (1972): 96–97.

———. "Zur Biographie des jungen Albert Einstein. Mit zwei unbekannten Briefen Einsteins an Wilhelm Ostwald vom Frühjahr 1901." *Forschungen und Fortschritte* 38 (1964): 74–78.

———. See Erwin N. Hiebert.

Körner, Carl. "Technische Hochschulen." *Handbuch der Architektur,* pt. 4, sec. 6, no. 2aI, 112–60.

Kohlrausch, Friedrich. "Antrittsrede." *Sitzungsber. preuss. Akad.,* 1896, pt. 2, 736–43.

———. *Gesammelte Abhandlungen.* Edited by Wilhelm Hallwachs, Adolf Heydweiller, Karl Strecker, and Otto Wiener. 2 vols. Leipzig: J. A. Barth, 1910–11.

———. "Gustav Wiedemann. Nachruf." In *Gesammelte Abhandlungen,* vol. 2, 1064–76.

———. *Leitfaden der praktischen Physik, zunächst für das physikalische Practicum in Göttingen.* Leipzig, 1870. 11th rev. ed., *Lehrbuch der praktischen Physik.* Leipzig: B. G. Teubner, 1910.

———. "Vorwort." In *Lehrbuch der praktischen Physik,* ix–xii. Reprinted in *Gesammelte Abhandlungen,* vol. 1, 1084–88.

————. "Wilhelm v. Beetz. Nekrolog." In *Gesammelte Abhandlungen,* vol. 2, 1048–61.

Kolde, Theodor. *Die Universität Erlangen unter dem Hause Wittelsbach, 1810–1910.* Erlangen and Leipzig: A. Deichert, 1910.

Konen, Heinrich. "Das physikalische Institut." In *Geschichte der Rheinischen Friedrich-Wilhelm-Universität zu Bonn am Rhein,* vol. 2, 345–55.

Korn, Arthur. *Eine Theorie der Gravitation und der elektrischen Erscheinungen auf Grundlage der Hydrodynamik.* 2 vols. in 1. Berlin, 1892, 1894.

"Korn, Arthur." *Reichshandbuch Deutscher Geschichte* 1 (1930): 992–93.

Kossel, Walther. "Walther Kaufmann." *Naturwiss.* 34 (1947): 33–34.

Kraeft, Wolf Dietrich. See Günter Kelbg.

Kragh, Helge. "Niels Bohr's Second Atomic Theory." *HSPS* 10 (1979): 123–86.

Kratzer, A. "Gerhard C. Schmidt." *Phys. Bl.* 6 (1950): 30.

Krause, Martin. "Oscar Schlömilch." *Verh. sächs. Ges. Wiss.* 53 (1901): 509–20.

Küchler, G. W. "Physical Laboratories in Germany." In *Occasional Reports by the Office of the Director-General of Education in India,* no. 4, 181–211. Calcutta: Government Printing, 1906.

Kuhn, K. "Erinnerungen an die Vorlesungen von W. C. Röntgen und L. Grätz." *Phys. Bl.* 18 (1962): 314–16.

————. "Gerhard C. Schmidt." *Naturwiss. Rundschau* 4 (1951): 41.

————. "Johannes Stark." *Phys. Bl.* 13 (1957): 370–71.

Kuhn, Thomas S. *Black-Body Theory and the Quantum Discontinuity 1894–1912.* New York: Oxford University Press, 1978.

————. "Einstein's Critique of Planck." In *Some Strangeness in the Proportion,* edited by Harry Woolf, 186–91.

————. *The Essential Tension: Selected Studies in Scientific Tradition and Change.* Chicago: University of Chicago Press, 1977.

————. "The Function of Measurement in Modern Physical Science." In *Quantification,* edited by Harry Woolf, 31–63. New York: Bobbs-Merrill, 1961.

————. "Mathematical versus Experimental Traditions in the Development of Physical Science." *Journal of Interdisciplinary History* 7 (1976): 1–31. Reprinted in *The Essential Tension,* 31–65.

————. See John L. Heilbron.

Kundt, August. "Antrittsrede." *Sitzungsber. preuss. Akad.,* 1889, pt. 2, 679–83.

————. *Vorlesungen über Experimentalphysik.* Edited by Karl Scheel. Braunschweig: F. Vieweg, 1903.

Kurylo, Friedrich, and Charles Süsskind. *Ferdinand Braun: A Life of the Nobel Prizewinner and Inventor of the Cathode-Ray Oscilloscope.* Cambridge, Mass.: MIT Press, 1981.

Kuznetsov, Boris. *Einstein.* Translated by V. Talmy. Moscow: Progress Publishers, 1965.

Lampa, Anton. "Ludwig Boltzmann." *Biographisches Jahrbuch und Deutscher Nekrolog* 11 (1908): 96–104.

Lampe, Hermann. *Die Entwicklung und Differenzierung von Fachabteilungen auf den Versammlungen von 1828 bis 1913.* Vol. 2 of Schriftenreihe zur Geschichte der Versammlungen deutscher Naturforscher und Ärzte. Hildesheim: Gerstenberg, 1975.

————. *Die Vorträge der allgemeinen Sitzungen auf der 1.–85. Versammlung 1822–1913.* Vol. 1, Schriftenreihe zur Geschichte der Versammlungen deutscher Naturforscher und Ärzte. Hildesheim: Gerstenberg, 1972.

Landolt, Hans Heinrich, and Richard Börnstein, eds. *Physikalisch-chemische Tabellen* Berlin, 1883.

Lang, Victor von. *Einleitung in die theoretische Physik.* 2d rev. ed. Braunschweig, 1891.

————. Obituary of Ludwig Boltzmann. *Almanach. österreichische Akad.* 57 (1907): 307–9.

Laue, Max (von). *Gesammelte Schriften und Vorträge.* 3 vols. Braunschweig: F. Vieweg, 1961.

————. "Heinrich Hertz 1857–1894." In *Gesammelte Schriften und Vorträge* 3: 247–56.

————. "Mein physikalischer Werdegang. Eine Selbstdarstellung." In *Schöpfer des neuen Weltbildes,* edited by H. Hartmann. 178–210.

————. "Paul Drude." *Math.-Naturwiss. Blätter* 3 (1906): 174–75.

————. *Das physikalische Weltbild. Vortrag, gehalten auf der Kieler Herbstwoche 1921.* Karlsruhe: C. F. Müller, 1921.

————. *Das Relativitätsprinzip.* Braunschweig: F. Vieweg, 1911. 2d ed. 2 vols. Vol. 2, *Die allgemeine Relativitätstheorie und Einsteins Lehre von der Schwerkraft.* Braunschweig: F. Vieweg, 1921.

————. "Rubens, Heinrich." *Deutsches biographisches Jahrbuch.* Vol. 4, *Das Jahr 1922* (1929): 228–30.

————. "Sommerfelds Lebenswerk." *Naturwiss.* 38 (1951): 513–18.

————. "Über Hermann von Helmholtz." In *Forschen und Wirken. Festschrift . . . Humboldt-Universität zu Berlin,* vol. 1, 359–66.

————. "Wien, Wilhelm." *Deutsches biographisches Jahrbuch.* Vol. 10, *Das Jahr 1928* (1931): 302–10.

————. See Max Born.

Lehmann, Otto, ed. *Dr. J. Fricks Physikalische Technik; oder, Anleitung zu Experimentalvorträgen sowie zur Selbstherstellung einfacher Demonstrationsapparate.* 7th rev. ed. 2 vols. in 4. Braunschweig: F. Vieweg, 1904–9.

————. "Geschichte des physikalischen Instituts der technischen Hochschule Karlsruhe." In *Festgabe* by the Karlsruhe Technical Institute, 207–65.

————. *Molekularphysik mit besonderer Berücksichtigung mikroskopischer Untersuchungen und Anleitung zu solchen sowie einem Anhang über mikroskopische Analyse.* 2 vols. Leipzig, 1888–89.

————. *Physik und Politik.* Karlsruhe: Braun, 1901.

————. "Vorrede." In *Dr. J. Fricks Physikalische Technik,* vol. 1, pt. 1, v–xx.

Leipzig. University. *Festschrift zur Feier des 500jährigen Bestehens der Universität Leipzig.* Vol. 4, *Die Institute und Seminare der Philosophischen Fakultät.* Pt. 2, *Die mathematisch-naturwissenschaftliche Sektion.* Leipzig: S. Hirzel, 1909.

————. *Die Universität Leipzig, 1409–1909. Gedenkblätter zum 30. Juli 1909.* Leipzig: Press-Ausschuss der Jubiläums-Kommission, 1909.

————. *Verzeichniss der . . . auf der Universität Leipzig zu haltenden Vorlesungen.* Leipzig.

————. See Otto Wiener.

Lemaine, Gerard, Roy Macleod, Michael Mulkay, and Peter Weingart, eds. *Perspectives on the Emergence of Scientific Disciplines.* The Hague: Mouton, 1976.

Lenard, Philipp. "Einleitung." In Heinrich Hertz's *Gesammelte Werke* 1: ix–xxix.

————. *Great Men of Science; A History of Scientific Progress.* Translated by H. Stafford Hatfield. New York: Macmillan, 1933.

————. *Über Relativitätsprinzip, Äther, Gravitation.* Leipzig: S. Hirzel, 1918.

Lenz, Max. *Geschichte der königlichen Friedrich-Wilhelms-Universität zu Berlin.* 4 vols. in 5. Halle a. d. S.: Buchhandlung des Waisenhauses, 1910–18.

Lexis, Wilhelm, ed. *Die deutschen Universitäten.* 2 vols. Berlin, 1893.

————. *Die Reform des höheren Schulwesens in Preussen.* Halle a. d. S.: Buchhandlung des Waisenhauses, 1902.

————, ed. *Das Unterrichtswesen im Deutschen Reich.* Vol. 1, *Die Universitäten im Deutschen Reich.* vol. 4, pt. 1, *Die technischen Hochschulen im Deutschen Reich.* Berlin: A. Asher, 1904.

Lichtenecker, Karl. "Otto Wiener." *Phys. Zs.* 29 (1928): 73–78.

Liebmann, Heinrich. "Zur Erinnerung an Carl Neumann." *Jahresber. d. Deutsch. Math.-Vereinigung* 36 (1927): 174–78.

"Life and Labors of Henry Gustavus Magnus." *Annual Report of the . . . Smithsonian Institution for the Year 1870,* 1872, 223–30.

Lindemann, Frederick Alexander, Lord Cherwell, and Franz Simon. "Walther Hermann Nernst (1864–1941)." *Obituary Notices of Fellows of the Royal Society* 4 (1942): 101–12.

Lommel, Eugen. *Experimental Physics.* Translated from the 3d German ed. of 1896 by G. W. Myers. London, 1899.

Lorentz, H. A. *Collected Papers.* 9 vols. The Hague: M. Nijhoff, 1934–39.

————. "Ludwig Boltzmann." *Verh. phys. Ges.* 9 (1907): 206–38. Reprinted in *Collected Papers,* vol. 9, 359–91.

————. *Versuch einer Theorie der electrischen und optischen Erscheinungen in bewegten Körpern.* Leiden, 1895. Reprinted in *Collected Papers,* vol. 5, 1–137.

————. See Erwin Schrödinger.

Lorentz, H. A., Albert Einstein, Hermann Minkowski, and Hermann Weyl. *The Principle of Relativity: A Collection of Original Memoirs on the Special and General Theory of Relativity by H. A. Lorentz, A. Einstein, H. Minkowski and H. Weyl.* Translated from the 4th German edition of 1922 by W. Perrett and G. B. Jeffery. London: Methuen, 1923. Reprint. New York: Dover, n.d.

Lorenz, Hans. *Technische Mechanik starrer Systeme.* Munich: Oldenbourg, 1902.

————. "Die Theorie in der Technik mit besonderer Berücksichtigung der Entwickelung der Kreiselräder." *Phys. Zs.* 12 (1911): 185–91.

————. "Der Unterricht in angewandter Mathematik und Physik an den deutschen Universitäten." *Jahresber. d. Deutsch. Math.-Vereinigung* 12 (1903): 565–72.

Lorey, Wilhelm. "Paul Drude und Ludwig Boltzmann." *Abhandlungen der Naturforschenden Gesellschaft zu Görlitz* 25 (1907): 217–22.

————. "Die Physik an der Universität Giessen im 19. Jahrhundert." *Nachrichten der Giessener Hochschulgesellschaft* 15 (1941): 80–132.

————. *Das Studium der Mathematik an den deutschen Universitäten seit Anfang des 19. Jahrhunderts.* Leipzig and Berlin: B. G. Teubner, 1916.

Losch, P. "Melde, Franz Emil." *Biographisches Jahrbuch und Deutscher Nekrolog* 6 (1901): 338–40.

Ludwig, Hubert. *Worte am Sarge von Heinrich Rudolf Hertz am 4. Januar 1894 im Auftrage der Universität gesprochen.* Bonn, 1894.

Lüdicke, Reinhard. *Die preussischen Kultusminister und ihre Beamten im ersten Jahrhundert des Ministeriums, 1817–1917.* Stuttgart: J. G. Cotta, 1918.

Lummer, Otto. "Physik." In *Festschrift . . . Universität Breslau,* vol. 2, 440–48.

McCormmach, Russell. "Editor's Foreword." *HSPS* 7 (1976): xi–xxxv.

————. "Einstein, Lorentz, and the Electron Theory." *HSPS* 2 (1970): 41–87.

————. "H. A. Lorentz and the Electromagnetic View of Nature." *Isis* 61 (1970): 459–97.

————. "Henri Poincaré and the Quantum Theory." *Isis* 58 (1967): 37–55.

————. "J. J. Thomson and the Structure of Light." *Brit. Journ. Hist. Sci.* 3 (1967): 362–87.

————. "Lorentz, Hendrik Antoon." *DSB* 8 (1973): 487–500.

————. *Night Thoughts of a Classical Physicist.* Cambridge, Mass.: Harvard University Press, 1982.

McGucken, William. *Nineteenth-Century Spectroscopy.* Baltimore: Johns Hopkins University Press, 1969.

McGuire, J. E. "Forces, Powers, Aethers and Fields." *Boston Studies in the Philosophy of Science* 14 (1974): 119–59.

Mach, Ernst. *Die Mechanik in ihrer Entwickelung. Historisch-kritisch dargestellt.* Leipzig, 1883. 2d rev. ed. Leipzig, 1889. Translated as *The Science of Mechanics. A Critical and Historical Exposition of Its Principles* by T. J. McCormack. Chicago, 1893.

————. *Populär-wissenschaftliche Vorlesungen.* Leipzig, 1896. Translated as *Popular Scientific Lectures* by T. J. McCormack. Chicago, 1895.

————. *Die Principien der Wärmelehre. Historisch-kritisch entwickelt.* Leipzig, 1896.

MacKinnon, Edward. "Heisenberg, Models, and the Rise of Matrix Mechanics." *HSPS* 8 (1977): 137–88.

Macleod, Roy. See Gerard Lemaine.

Madelung, Erwin. *Die mathematischen Hilfsmittel des Physikers.* Vol. 4, Die Grundlehren der mathematischen Wissenschaften in Einzeldarstellungen mit besonderer Berücksichtigung der Anwendungsgebiete, edited by Richard Courant. Berlin: Springer, 1922.

Manegold, Karl-Heinz. *Universität, Technische Hochschule und Industrie.* Vol. 16,

Schriften zur Wirtschafts- und Sozialgeschichte, edited by W. Fischer. Berlin: Duncker und Humblot, 1970.

Marburg. University. *Catalogus professorum academiae Marburgensis; die akademischen Lehrer der Philipps-Universität in Marburg von 1527 bis 1910.* Edited by Franz Gundlach. Marburg: Elwert, 1927.

————. *Chronik der Königlich Preussischen Universität Marburg.* Marburg.

————. *Die Philipps-Universität zu Marburg 1527–1927.* Edited by H. Hermelink and S. A. Kaehler. Marburg: Elwert, 1927.

Max-Planck-Gesellschaft. *50 Jahre Kaiser-Wilhelm-Gesellschaft und Max-Planck–Gesellschaft zur Förderung der Wissenschaften 1911–1961.* Göttingen: Max-Planck-Gesellschaft, 1961.

Maxwell, James Clerk. "Hermann Ludwig Ferdinand Helmholtz." *Nature* 15 (1877): 389–91.

————. *Lehrbuch der Electricität und des Magnetismus.* Translated by Bernhard Weinstein. 2 vols. Berlin, 1883.

Mehra, Jagdish. "Albert Einsteins erste wissenschaftliche Arbeit." *Phys. Bl.* 27 (1971): 386–91.

————. "Einstein, Hilbert, and the Theory of Gravitation." In *The Physicist's Conception of Nature,* edited by Jagdish Mehra, 194–278.

————, ed. *The Physicist's Conception of Nature.* Dordrecht: D. Reidel, 1973.

————, ed. *The Solvay Conferences on Physics: Aspects of the Development of Physics since 1911.* Dordrecht: D. Reidel, 1975.

Meissner, Walther. "Max von Laue als Wissenschaftler und Mensch." *Sitzungsber. bay. Akad.,* 1960, 101–21.

M[elde, Franz]. "Der Erweiterungs- und Umbau des mathematisch-physikalischen Instituts der Universität Marburg." *Hessenland. Zeitschrift für Hessische Geschichte und Literatur* 5 (1891): 141–42.

Mendelsohn, Kurt. *The World of Walther Nernst. The Rise and Fall of German Science, 1864–1941.* Pittsburgh: University of Pittsburgh Press, 1973.

Merz, John Theodore. *A History of European Thought in the Nineteenth Century.* 4 vols. 1904–12. Reprint. New York: Dover, 1965.

Meyer, Oskar Emil. "Das physikalische Institut der Universität zu Breslau." *Phys. Zs.* 6 (1905): 194–96.

Meyer, Stefan. "Friedrich Hasenöhrl." *Phys. Zs.* 16 (1915): 429–33.

————, ed. *Festschrift Ludwig Boltzmann gewidmet zum sechzigsten Geburtstage/20. Februar 1904.* Leipzig: J. A. Barth, 1904.

Michelmore, Peter. *Einstein, Profile of the Man.* New York: Dodd, Mead, 1962.

Mie, Gustav. "Aus meinem Leben." *Zeitwende* 19 (1948): 733–43.

————. *Die Einsteinsche Gravitationstheorie. Versuch einer allgemein verständlichen Darstellung der Theorie.* Leipzig: S. Hirzel, 1921.

————. *Entwurf einer allgemeinen Theorie der Energieübertragung.* Vienna, 1898.

————. *Lehrbuch der Elektrizität und des Magnetismus. Eine Experimentalphysik des Weltäthers für Physiker, Chemiker, Elektrotechniker.* Stuttgart: F. Enke, 1910.

―――. *Die Materie. Vortrag gehalten am 27. Januar 1912 (Kaisers Geburtstag) in der Aula der Universität Greifswald.* Stuttgart: F. Enke, 1912.

―――. "Die mechanische Erklärbarkeit der Naturerscheinungen. Maxwell. ― Helmholtz. ― Hertz." *Verhandlungen des naturwissenschaftlichen Vereins in Karlsruhe* 13 (1895–1900): 402–20.

―――. *Moleküle, Atome, Weltäther.* Leipzig: B. G. Teubner, 1904.

―――. *Die neueren Forschungen über Ionen und Elektronen.* Stuttgart: F. Enke, 1903.

Miller, Arthur I. *Albert Einstein's Special Theory of Relativity.* Reading, Mass.: Addison-Wesley, 1981.

―――. "A Study of Henri Poincaré's 'Sur la Dynamique de l'Electron,' " *Arch. Hist. Ex. Sci.* 10 (1973): 207–328.

―――. "Visualization Lost and Regained: The Genesis of the Quantum Theory in the Period 1913–27." In *On Aesthetics in Science,* edited by J. Wechsler, 73–101. Cambridge, Mass.: MIT Press, 1978.

Minkowski, Hermann. *Briefe an David Hilbert/Hermann Minkowski.* Edited by L. Rüdenberg and H. Zassenhaus. Berlin, Heidelberg, and New York: Springer, 1973.

―――. *Gesammelte Abhandlungen.* Edited by David Hilbert. 2 vols. Leipzig and Berlin: B. G. Teubner, 1911.

―――. "Peter Gustav Lejeune Dirichlet und seine Bedeutung für die heutige Mathematik." In Minkowski's *Gesammelte Abhandlungen* 2: 447–61.

―――. See H. A. Lorentz.

Mohl, Robert von. *Lebens-Erinnerungen.* Vol. 1. Stuttgart and Leipzig: Deutsche Verlags-Anstalt, 1902.

Mott, Nevill, and Rudolf Peierls. "Werner Heisenberg. 5 December 1901–1 February 1976." *Biographical Memoirs of Fellows of the Royal Society* 23 (1977): 213–42.

Mrowka, B. "Richard Gans." *Phys. Bl.* 10 (1954): 512–13.

Müller, J. A. von. "Das physikalisch-metronomische Institut." In *Die wissenschaftlichen Anstalten der Ludwig-Maximilians-Universität zu München,* 278–79.

Müller, J. J. C. See C. Christiansen.

Mulkay, Michael. See Gerard Lemaine.

Munich Technical Institute, ed. *Darstellungen aus der Geschichte der Technik, der Industrie und Landwirtschaft in Bayern.* Munich: R. Oldenbourg, 1906.

Munich. University. *Die Ludwig-Maximilians-Universität in ihren Fakultäten.* Vol. 1. Edited by L. Boehm and J. Spörl. Berlin: Duncker und Humblot, 1972.

―――. *Ludwig-Maximilians-Universität, Ingolstadt, Landshut, München, 1472–1972.* Edited by L. Boehm and J. Spörl. Berlin: Duncker und Humblot, 1972.

―――. *Die wissenschaftlichen Anstalten der Ludwig-Maximilians-Universität zu München.* Edited by Karl Alexander von Müller. Munich: R. Oldenbourg und Dr. C. Wolf, 1926.

―――. See Clara Wallenreiter.

Narr, Friedrich. *Ueber die Erkaltung und Wärmeleitung in Gasen.* Munich, 1870.

Needell, Allan A. "Irreversibility and the Failure of Classical Dynamics: Max

Planck's Work on the Quantum Theory 1900–1915." Ph.D. diss., Yale University, 1980.

———. See Martin J. Klein.

Nernst, Walther. "Antrittsrede." *Sitzungsber. preuss. Akad.*, 1906, 549–52.

———. "Development of General and Physical Chemistry during the Last Forty Years." *Annual Report of the Smithsonian Institution*, 1908, 245–53.

———. "Rudolf Clausius 1822–1888." In *150 Jahre Rheinische Friedrich-Wilhelms-Universität zu Bonn 1818–1968*, 101–9.

———. *Theoretische Chemie vom Standpunkte der Avogadroschen Regel und der Thermodynamik*. Stuttgart, 1893.

———. *Das Weltgebäude im Lichte der neueren Forschung*. Berlin: Springer, 1921.

Neuer Nekrolog der Deutschen.

Neuerer, Karl. *Das höhere Lehramt in Bayern im 19. Jahrhundert*. Berlin: Duncker und Humblot, 1978.

Neumann, Carl. *Beiträge zu einzelnen Theilen der mathematischen Physik, insbesondere zur Elektrodynamik und Hydrodynamik, Elektrostatik und magnetischen Induction*. Leipzig, 1893.

———. *Untersuchungen über das logarithmische und Newton'sche Potential*. Leipzig, 1877.

———. "Worte zum Gedächtniss an Wilhelm Hankel." *Verh. sächs. Ges. Wiss.* 51 (1899): lxii–lxvi.

Neumann, Franz. *Vorlesungen über mathematische Physik, gehalten an der Universität Königsberg*. Edited by his students. Leipzig, 1881–94. The individual volumes are as follows. *Einleitung in die theoretische Physik*. Edited by Carl Pape. Leipzig, 1883. *Vorlesungen über die Theorie der Capillarität*. Edited by Albert Wangerin. Leipzig, 1894. *Vorlesungen über die Theorie der Elasticität der festen Körper und des Lichtäthers*. Edited by Oskar Emil Meyer. Leipzig, 1885. *Vorlesungen über die Theorie des Magnetismus, namentlich über die Theorie der magnetischen Induktion*. Edited by Carl Neumann. Leipzig, 1881. *Vorlesungen über die Theorie des Potentials und der Kugelfunctionen*. Edited by Carl Neumann. Leipzig, 1887. *Vorlesungen über elektrische Ströme*. Edited by Karl Von der Mühll. Leipzig, 1884. *Vorlesungen über theoretische Optik*. Edited by Ernst Dorn. Leipzig, 1885.

Neumann, Luise. *Franz Neumann, Erinnerungsblätter von seiner Tochter*. 2d ed. Tübingen: J. C. B. Mohr (P. Siebeck), 1907.

Nisio, Sigeko. "The Formation of the Sommerfeld Quantum Theory of 1916." *Jap. Stud. Hist. Sci.*, no. 12 (1973): 39–78.

Nitske, W. Robert. *The Life of Wilhelm Conrad Röntgen: Discoverer of the X Ray*. Tucson: University of Arizona Press, 1971.

North, J. D. *The Measure of the Universe: A History of Modern Cosmology*. Oxford: Clarendon Press, 1965.

Obituary of Hermann Ebert. *Leopoldina* 18 (1913): 38.

Obituary of Hermann von Helmholtz. *Nature* 50 (1894): 479–80.

Obituary of Eduard Ketteler. *Leopoldina* 37 (1901): 35–36.

Obituary of Ludwig Matthiessen. *Leopoldina* 42 (1906): 158.

Obituary of Franz Melde. *Leopoldina* 37 (1901): 46–47.

Olesko, Kathryn Mary. "The Emergence of Theoretical Physics in Germany: Franz Neumann and the Königsberg School of Physics, 1830–1890." Ph.D. diss., Cornell University, 1980.

Oppenheim, A. "Heinrich Gustav Magnus." *Nature* 2 (1870): 143–45.

Ortwein. "Wilhelm Lenz 60 Jahre." *Phys. Bl.* 4 (1948): 30–31.

Ostwald, Wilhelm. *Aus dem wissenschaftlichen Briefwechsel Wilhelm Ostwalds.* Vol. 1, *Briefwechsel mit Ludwig Boltzmann, Max Planck, Georg Helm und Josiah Willard Gibbs.* Edited by Hans-Günther Körber. Berlin: Akademie-Verlag, 1961.

———. "Gustav Wiedemann." *Verh. sächs. Ges. Wiss.* 51 (1899): lxxvii–lxxxiii.

———. *Lehrbuch der allgemeinen Chemie.* 2d ed. 2 vols. in 4. Leipzig: W. Engelmann, 1891–1906.

———. "Recent Advances in Physical Chemistry." *Nature* 45 (1892): 590–93.

Paalzow, Adolph. "Stiftungsfeier am 4. Januar 1896." *Verh. phys. Ges.* 15 (1896): 36–37.

Paschen, Friedrich. "Gedächtnisrede des Hrn. Paschen auf Emil Warburg." *Sitzungsber. preuss. Akad.*, 1932, cxv–cxxiii.

———. "Heinrich Kayser." *Phys. Zs.* 41 (1940): 429–33.

Pasler, M. "Leben und wissenschaftliches Werk Max von Laues." *Phys. Bl.* 16 (1960): 552–67.

Paul, Wolfgang. See Barbara Jaeckel.

Pauli, Wolfgang. "Albert Einstein in der Entwicklung der Physik." *Phys. Bl.* 15 (1959): 241–45.

———. *Wolfgang Pauli. Wissenschaftlicher Briefwechsel mit Bohr, Einstein, Heisenberg u. a.* Vol. 1, *1919–1929.* Edited by Armin Hermann, K. v. Meyenn, and V. F. Weisskopf. New York, Heidelberg, and Berlin: Springer, 1979.

Paulsen, Friedrich. *Die deutschen Universitäten und das Universitätsstudium.* Berlin: A. Asher, 1902.

Peierls, Rudolf. See Nevill Mott.

Perron, Oskar, Constantin Carathéodory, and Heinrich Tietze. "Das Mathematische Seminar." In *Die wissenschaftlichen Anstalten . . . zu München,* 206.

Pfannenstiel, Max, ed. *Kleines Quellenbuch zur Geschichte der Gesellschaft Deutscher Naturforscher und Ärzte.* Berlin, Göttingen, and Heidelberg: Springer, 1958.

Pfaundler, Leopold, ed. *Müller-Pouillet's Lehrbuch der Physik und Meteorologie.* 9th rev. ed. 3 vols. Braunschweig, 1886–98. 10th rev. ed. 4 vols. Braunschweig: F. Vieweg, 1905–14.

Pfetsch, Frank. "Scientific Organization and Science Policy in Imperial Germany, 1871–1914: The Foundations of the Imperial Institute of Physics and Technology." *Minerva* 8 (1970): 557–80.

Philippovich, Eugen von. *Der badische Staatshaushalt in den Jahren 1868–1889.* Freiburg i. Br., 1889.

Planck, Max. *Acht Vorlesungen über theoretische Physik.* Leipzig: S. Hirzel, 1910. Translated as *Eight Lectures on Theoretical Physics Delivered at Columbia University in 1909* by A. P. Wills. New York: Columbia University Press, 1915.

————. "Antrittsrede." *Sitzungsber. preuss. Akad.*, 1894, 641–44. Reprinted in *Physikalische Abhandlungen und Vorträge* 3: 1–5.

————. "Arnold Sommerfeld zum siebzigsten Geburtstag." *Naturwiss.* 26 (1938): 777–79. Reprinted in *Physikalische Abhandlungen und Vorträge* 3: 368–71.

————. *Die Entstehung und bisherige Entwicklung der Quantentheorie.* Leipzig: J. A. Barth, 1920. Reprinted in *Physikalische Abhandlungen und Vorträge* 3: 121–34.

————. "Gedächtnissrede auf Heinrich Hertz." *Verh. phys. Ges.* 13 (1894): 9–29. Reprinted in *Physikalische Abhandlungen und Vorträge* 3: 268–88.

————. "Gedächtnisrede des Hrn. Planck auf Heinrich Rubens." *Sitzungsber. preuss. Akad.*, 1923, cviii–cxiii.

————. *Grundriss der allgemeinen Thermochemie.* Breslau, 1893.

————. "Helmholtz's Leistungen auf dem Gebiete der theoretischen Physik." *ADB* 51 (1906): 470–72. Reprinted in *Physikalische Abhandlungen und Vorträge* 3: 321–23.

————. "Das Institut für theoretische Physik." In *Geschichte der . . . Universität zu Berlin,* edited by Max Lenz, vol. 3, 276–78.

————. "James Clerk Maxwell in seiner Bedeutung für die theoretische Physik in Deutschland." *Naturwiss.* 19 (1931): 889–94. Reprinted in *Physikalische Abhandlungen und Vorträge* 3: 352–57.

————. "Max von Laue. Zum 9. Oktober 1929." *Naturwiss.* 17 (1929): 787–88. Reprinted in *Physikalische Abhandlungen und Vorträge* 3: 350–51.

————. "Paul Drude." *Ann.* 20 (1906): i–iv.

————. "Paul Drude." *Verh. phys. Ges.* 8 (1906): 599–630. Reprinted in *Physikalische Abhandlungen und Vorträge* 3: 289–320.

————. *Physikalische Abhandlungen und Vorträge.* 3 vols. Braunschweig: F. Vieweg, 1958.

————. *Das Princip der Erhaltung der Energie.* Leipzig, 1887.

————. "Theoretische Physik." In *Aus fünfzig Jahren deutscher Wissenschaft,* edited by Gustav Abb, 300–309. Reprinted in *Physikalische Abhandlungen und Vorträge* 3: 209–18.

————. *Über den zweiten Hauptsatz der mechanischen Wärmetheorie.* Munich, 1879.

————. "Verhältnis der Theorien zueinander." In *Physik,* edited by Emil Warburg, 732–37.

————. *Vorlesungen über die Theorie der Wärmestrahlung.* Leipzig: J. A. Barth, 1906.

————. *Vorlesungen über Thermodynamik.* Leipzig, 1897. Translated as *Treatise on Thermodynamics* by A. Ogg. London, New York, and Bombay: Longmans, Green, 1903.

————. *Wissenschaftliche Selbstbiographie.* Leipzig: J. A. Barth, 1948. Reprinted in *Physikalische Abhandlungen und Vorträge* 3: 374–401.

————. See Erwin Schrödinger.

Plücker, Julius. *Gesammelte physikalische Abhandlungen.* Edited by Friedrich Pockels. Leipzig, 1896.

Pockels, Friedrich. "Gustav Robert Kirchhoff." In *Heidelberger Professoren aus dem 19. Jahrhundert,* vol. 2, 243–63.

————. *Lehrbuch der Kristalloptik.* Leipzig and Berlin: B. G. Teubner, 1906.

————. *Über die partielle Differentialgleichung* $\Delta u + k^2 u = 0$ *und deren Auftreten in der mathematischen Physik.* Leipzig, 1891.

Poggendorff, Johann Christian. *J. C. Poggendorff's biographisch-literarisches Handwörterbuch zur Geschichte der exacten Wissenschaften.* Leipzig, 1863– .

————. "Meine Rede zur Jubelfeier am 28. Februar 1874." In *Johann Christian Poggendorff* by Emil Frommel, 68–72.

Preston, David Lawrence. "Science, Society, and the German Jews: 1870–1933." Ph.D. diss., University of Illinois, 1971.

Pringsheim, Peter. "Gustav Magnus." *Naturwiss.* 13 (1925): 49–52.

Prutz, Hans. *Die Königliche Albertus-Universität zu Königsberg i. Pr. im neunzehnten Jahrhundert. Zur Feier ihres 350jährigen Bestehens.* Königsberg, 1894.

Pyenson, Robert Lewis. "The Göttingen Reception of Einstein's General Theory of Relativity." Ph.D. diss., Johns Hopkins University, 1973.

————. "Hermann Minkowski and Einstein's Special Theory of Relativity." *Arch. Hist. Ex. Sci.* 17 (1977): 71–95.

————. "Mathematics, Education, and the Göttingen Approach to Physical Reality, 1890–1914." *Europa* 2 (1979): 91–127.

————. "Physics in the Shadow of Mathematics: The Göttingen Electron-Theory Seminar of 1905." *Arch. Hist. Ex. Sci.* 21 (1979): 55–89.

————. "La réception de la relativité généralisée: disciplinarité et institutionalisation en physique." *Revue d'histoire des sciences* 28 (1975): 61–73.

R., D. "Jolly: Philipp Johann Gustav von." *ADB* 55 (1971): 807–10.

Raman, V. V., and Paul Forman. "Why Was It Schrödinger Who Developed de Broglie's Ideas?" *HSPS* 1 (1969): 291–314.

Ramsauer, Carl. "Zum zehnten Todestag. Philipp Lenard 1862–1957." *Phys. Bl.* 13 (1957): 219–22.

Rees, J. K. "German Scientific Apparatus." *Science* 12 (1900): 777–85.

Reiche, F. "Otto Lummer." *Phys. Zs.* 27 (1926): 459–67.

Reid, Constance. *Courant in Göttingen and New York: The Story of an Improbable Mathematician.* New York: Springer, 1976.

————. *Hilbert, With an Appreciation of Hilbert's Mathematical Work by Hermann Weyl.* Berlin and New York: Springer, 1970.

Reindl, Maria. *Lehre und Forschung in Mathematik und Naturwissenschaften, insbesondere Astronomie, an der Universität Würzburg von der Gründung bis zum Beginn des 20. Jahrhunderts.* Neustadt an der Aisch: Degener, 1966.

Reinganum, Max. "Clausius: Rudolf Julius Emanuel." *ADB* 55 (1971): 720–29.

Richarz, Franz. See Walter König.

Riebesell, P. "Die neueren Ergebnisse der theoretischen Physik und ihre Beziehungen zur Mathematik." *Naturwiss.* 6 (1918): 61–65.

Riecke, Eduard. "Friedrich Kohlrausch." *Gött. Nachr.,* 1910, 71–85.

————. *Lehrbuch der Experimental-Physik zu eigenem Studium und zum Gebrauch bei Vorlesungen.* 2 vols. Leipzig, 1896.

————. *Die Principien der Physik und der Kreis ihrer Anwendung.* Festrede. Göttingen, 1897.

————. "Rede." In *Die physikalischen Institute der Universität Göttingen,* 20–37.

————. "Rudolf Clausius." *Abh. Ges. Wiss. Göttingen* 35 (1888): appendix, 1–39.

————. "Wilhelm Weber." *Abh. Ges. Wiss. Göttingen* 38 (1892): 1–44.

Riese, Reinhard. *Die Hochschule auf dem Wege zum wissenschaftlichen Grossbetrieb. Die Universität Heidelberg und das badische Hochschulwesen 1860–1914.* Vol. 19 of Industrielle Welt, Schriftenreihe des Arbeitskreises für moderne Sozialgeschichte, edited by Werner Conze. Stuttgart: Ernst Klett, 1977.

Riewe, K. H. *120 Jahre Deutsche Physikalische Gesellschaft.* N.p., 1965.

Ringer, Fritz K. *The Decline of the German Mandarins: The German Academic Community, 1890–1933.* Cambridge, Mass.: Harvard University Press, 1969.

Röntgen, W. C. *W. C. Röntgen. Briefe an L. Zehnder.* Edited by Ludwig Zehnder. Zurich, Leipzig, and Stuttgart: Rascher, 1935.

Rohmann, H. "Ferdinand Braun." *Phys. Zs.* 19 (1918): 537–39.

Rosanes, Jakob. "Charakteristische Züge in der Entwicklung der Mathematik des 19. Jahrhunderts." *Jahresber. d. Deutsch. Math.-Vereinigung* 13 (1904): 17–30.

Roscoe, Henry. *The Life and Experiences of Sir Henry Enfield Roscoe, D.C.L., LL.D., F.R.S.* London and New York: Macmillan, 1906.

Rosenberg, Charles E. "Toward an Ecology of Knowledge: On Discipline, Context and History." In *The Organization of Knowledge in Modern America 1860–1920,* edited by A. Oleson and J. Voss, 440–55. Baltimore: Johns Hopkins University Press, 1979.

Rosenberger, Ferdinand. *Die Geschichte der Physik.* Vol. 3, *Geschichte der Physik in den letzten hundert Jahren.* Braunschweig, 1890. Reprint. Hildesheim: G. Olms, 1965.

Rosenfeld, Léon. "Kirchhoff, Gustav Robert." *DSB* 7 (1973): 379–83.

————. "La première phase de l'évolution de la Théorie des Quanta." *Osiris* 2 (1936): 149–96.

————. "The Velocity of Light and the Evolution of Electrodynamics." *Nuovo Cimento,* supplement to vol. 4 (1957): 1630–69.

Rostock. University. See Günter Kelbg.

Rubens, Heinrich. "Antrittsrede." *Sitzungsber. preuss. Akad.,* 1908, 714–17.

————. "Das physikalische Institut." In *Geschichte der . . . Universität zu Berlin,* edited by Max Lenz, vol. 3, 278–96.

Runge, Carl. "Woldemar Voigt." *Gött. Nachr.,* 1920, 46–52.

Runge, Iris. *Carl Runge und sein wissenschaftliches Werk.* Göttingen: Vandenhoeck und Ruprecht, 1949.

Salié, Hans. "Carl Neumann." In *Bedeutende Gelehrte in Leipzig,* vol. 2, edited by G. Harig, 13–23.

Schachenmeier, R. See A. Schleiermacher.

Schaefer, Clemens. *Einführung in die theoretische Physik.* Vol. 1, *Mechanik materieller Punkte, Mechanik starrer Körper und Mechanik der Continua (Elastizität und Hydrodynamik).* Leipzig: Veit, 1914.

————. "Ernst Pringsheim." *Phys. Zs.* 18 (1917): 557–60.

Schaffner, K. F. "The Lorentz Electron Theory [and] Relativity." *Am. J. Phys.* 37 (1969): 498–513.

Scharmann, Arthur. See Wilhelm Hanle.

Scheel, Karl. "Bericht über den internationalen Katalog der wissenschaftlichen Literatur." *Verh. phys. Ges.*, 1903, 83–86.

————. "Die literarischen Hilfsmittel der Physik." *Naturwiss.* 16 (1925): 45–48.

————. "Physikalische Forschungsstätten." In *Forschungsinstitute, ihre Geschichte, Organisation und Ziele*, edited by Ludolph Brauer, et al., 175–208.

Scherrer, P. "Wolfgang Pauli." *Phys. Bl.* 15 (1959): 34–35.

Schilpp, P. A., ed. *Albert Einstein: Philosopher-Scientist.* Evanston: The Library of Living Philosophers, 1949.

Schlapp, R. See N. Kemmer.

Schleiermacher, A., and R. Schachenmeier. "Otto Lehmann." *Phys. Zs.* 24 (1923): 289–91.

Schmidt, Gerhard C. "Eilhard Wiedemann." *Phys. Zs.* 29 (1928): 185–90.

————. "Wilhelm Hittorf." *Phys. Bl.* 4 (1948): 64–68.

Schmidt, Karl. "Carl Hermann Knoblauch." *Leopoldina* 31 (1895): 116–22.

Schmidt-Ott, Friedrich. *Erlebtes und Erstrebtes, 1860–1950.* Wiesbaden: Franz Steiner, 1952.

Schmidt-Schönbeck, Charlotte. *300 Jahre Physik und Astronomie an der Kieler Universität.* Kiel: F. Hirt, 1965.

Schmitt, Eduard. "Hochschulen im allgemeinen." In *Handbuch der Architektur*, pt. 4, sec. 6, no. 2aI, 4–53.

Schnabel, Franz. "Althoff, Friedrich Theodor." *Neue deutsche Biographie* 1(1953): 222–24.

Schrader, Wilhelm. *Geschichte der Friedrichs-Universität zu Halle.* 2 vols. Berlin, 1894.

Schröder, Brigitte. "Caractéristiques des relations scientifiques internationales, 1870–1914." *Journal of World History* 10 (1966): 161–77.

Schrödinger, Erwin. *Collected Papers on Wave Mechanics.* Translated by J. F. Shearer from the 2d German ed. of 1928. London and Glasgow: Blackie and Son, 1928.

————. *Science and the Human Temperament.* Translated by J. Murphy and W. H. Johnston. New York: W. W. Norton, 1935.

Schrödinger, Erwin, Max Planck, Albert Einstein, and H. A. Lorentz. *Letters on Wave Mechanics.* Edited by K. Przibram. Translated by Martin J. Klein. New York: Philosophical Library, 1967.

Schroeter, Joachim. "Johann Georg Koenigsberger (1874–1946)." *Schweizerische Mineralogische und Petrographische Mitteilungen* 27 (1947): 236–46.

Schuler, H. "Friedrich Paschen." *Phys. Bl.* 3 (1947): 232–33.

Schulz, H. "Otto Lummer." *Zs. für Instrumentenkunde* 45 (1925): 465–67.

Schulze, F. A. "Franz Richarz." *Phys. Zs.* 22 (1921): 33–36.

————. "Wilhelm Feussner." *Phys. Zs.* 31 (1930): 513–14.

Schulze, Friedrich. *B. G. Teubner 1811–1911. Geschichte der Firma in deren Auftrag.* Leipzig, 1911.

Schulze, O. F. A. "Zur Geschichte des Physikalischen Instituts." In *Die Philipps-Universität zu Marburg 1527–1927*, 756–63.

Schuster, Arthur. "International Science." *Annual Report of the Smithsonian Institution*, 1906, 493–514.

————. *The Progress of Physics During 33 Years (1875–1908).* Cambridge: Cambridge University Press, 1911.

Schwalbe, B. "Nachruf auf G. Karsten." *Verh. phys. Ges.* 2 (1900): 147–59.

Schwarzschild, Karl. "Antrittsrede." *Sitzungsber. preuss. Akad.*, 1913, 596–600.

Scott, William T. *Erwin Schrödinger, An Introduction to His Writings.* Amherst: University of Massachusetts Press, 1967.

Seelig, Carl. *Albert Einstein: Eine dokumentarische Biographie.* Zurich: Europa-Verlag, 1952. Translated as *Einstein: A Documentary Biography* by M. Savill. London: Staples, 1956.

Segré, Emilio. *From X-Rays to Quarks: Modern Physicists and Their Discoveries.* San Francisco: W. H. Freeman, 1980

Serwer, Daniel. "*Unmechanischer Zwang:* Pauli, Heisenberg, and the Rejection of the Mechanical Atom, 1923–1925." *HSPS* 8 (1977): 189–256.

Siemens, Werner von. *Personal Recollections.* Translated by W. C. Coupland. New York, 1893.

Simpson, Thomas K. "Maxwell and the Direct Experimental Test of His Electromagnetic Theory." *Isis* 57 (1966): 411–32.

Skalweit, Stephan. "Gossler, Gustav Konrad Heinrich." *Neue deutsche Biographie* 6 (1964): 650–51.

Smith, F. W. F., Earl of Birkenhead. *The Professor and the Prime Minister: The Official Life of Professor F. A. Lindemann, Viscount Cherwell.* Boston: Houghton, Mifflin, 1962.

Solvay Congress. Instituts Solvay. Institut international de physique. *Electrons et photons. Rapports et discussions du cinquième conseil de physique tenu à Bruxelles du 24 au 29 octobre 1927.* Paris: Gauthier-Villars, 1928.

————. *La théorie du rayonnement et les quanta. Rapports et discussions de la réunion tenue à Bruxelles, du 30 octobre au 3 novembre 1911.* Edited by Paul Langevin and Maurice de Broglie. Paris: Gauthier-Villars, 1912. Translated as *Die Theorie der Strahlung und der Quanten. Verhandlungen auf einer von E. Solvay einberufenen Zusammenkunft (30. Oktober bis 3. November 1911). Mit einem Anhange über die Entwicklung der Quantentheorie vom Herbst 1911 bis zum Sommer 1913* by Arnold Eucken. Halle: Wilhelm Knapp, 1914.

Sommerfeld, Arnold. "Abraham, Max." *Neue deutsche Biographie* 1: 23–24.

————. *Atombau und Spektrallinien.* Braunschweig: F. Vieweg, 1919.

————. "Die Entwicklung der Physik in Deutschland seit Heinrich Hertz." *Deutsche Revue* 43 (1918): 122–32.

————. *Gesammelte Schriften.* 4 vols. Edited by F. Sauter. Braunschweig: F. Vieweg, 1968.

———. "Das Institut für Theoretische Physik." In *Die wissenschaftlichen Anstalten . . . zu München,* 290–91.

———. "Max Planck zum sechzigsten Geburtstage." *Naturwiss.* 6 (1918): 195–99.

———. "Max von Laue zum 70. Geburtstag." *Phys. Bl.* 5 (1949): 443.

———. "Oskar Emil Meyer." *Sitzungsber. bay. Akad.* 39 (1909): 17.

———. "Some Reminiscences of My Teaching Career." *Am. J. Phys.* 17 (1949): 315–16.

———. "Überreichung der Planck-Medaille für Peter Debye." *Phys. Bl.* 6 (1950): 509–12.

———. "Woldemar Voigt." *Jahrbuch bay. Akad.,* 1919 (1920): 83–84.

———. See Albert Einstein.

Stachel, John. "The Genesis of General Relativity." In *Einstein Symposion Berlin,* edited by H. Nelkowski, et al., 428–42. Lecture Notes in Physics, vol. 100. Berlin, Heidelberg, and New York: Springer, 1979.

Stäckel, Paul. "Angewandte Mathematik und Physik an den deutschen Universitäten." *Jahresber. d. Deutsch. Math.-Vereinigung* 13 (1904): 313–41.

Stein, Howard. "'Subtler Forms of Matter' in the Period Following Maxwell." In *Conceptions of Ether: Studies in the History of Ether Theories 1740–1900,* edited by G. N. Cantor and M. J. S. Hodge, 309–40.

Stevens, E. H. "The Heidelberg Physical Laboratory." *Nature* 65 (1902): 587–90.

Strassburg. University. *Festschrift zur Einweihung der Neubauten der Kaiser-Wilhelms-Universität Strassburg.* Strassburg, 1884.

Stuewer, Roger H. *The Compton Effect: Turning Point in Physics.* New York: Science History Publications, 1975.

Sturm, Rudolf. "Mathematik." In *Festschrift . . . Breslau,* vol. 2, 434–40.

Süss, Eduard. Obituary of Josef Stefan. *Almanach. Österreichische Akad.* 43 (1893): 252–57.

Süsskind, Charles. "Hertz and the Technological Significance of Electromagnetic Waves." *Isis* 56 (1965): 342–55.

———. "Observations of Electromagnetic-Wave Radiation before Hertz." *Isis* 55 (1964): 32–42.

———. See Friedrich Kurylo.

Swenson, Loyd S., Jr. *The Ethereal Ether: A History of the Michelson-Morley-Miller Aether-Drift Experiments, 1880–1930.* Austin and London: University of Texas Press, 1972.

Täschner, Constantin. "Ferdinand Reich, 1799–1884. Ein Beitrag zur Freiberger Gelehrten- und Akademiegeschichte." *Mitteilungen des Freiberger Altertumsvereins,* no. 51 (1916): 23–59.

Tammann, G. "Wilhelm Hittorf." *Gött. Nachr.,* 1915, 74–78.

Thiele, Joachim. "Einige zeitgenössische Urteile über Schriften Ernst Machs. Briefe von Johannes Reinke, Paul Volkmann, Max Verworn, Carl Menger und Jakob von Uexküll." *Philosophia Naturalis* 11 (1969): 474–89.

———. "Ernst Mach und Heinrich Hertz. Zwei unveröffentlichte Briefe aus dem Jahre 1890." *NTM* 5 (1968): 132–34.

————. " 'Naturphilosophie' und 'Monismus' um 1900 (Briefe von Wilhelm Ostwald, Ernst Mach, Ernst Haeckel und Hans Driesch)." *Philosophia Naturalis* 10 (1968): 295–315.

Todhunter, Isaac. *A History of the Theory of Elasticity and of the Strength of Materials from Galilei to the Present Time.* Vol. 2, *Saint-Venant to Lord Kelvin.* Pt. 2. Cambridge, 1893.

Tomascheck, R. "Zur Erinnerung an Alfred Heinrich Bucherer." *Phys. Zs.* 30 (1929): 1–8.

Tonnelat, M. A. *Histoire du Principe de Relativité.* Paris: Flammarion, 1971.

Truesdell, C. "History of Classical Mechanics. Part II, the 19th and 20th Centuries." *Naturwiss.* 63 (1976): 119–30.

Tübingen. University. *Festgabe zum 25. Regierungs-Jubiläum seiner Majestät des Königs Karl von Württemberg.* Tübingen, 1889.

Turner, R. Steven. "Helmholtz, Hermann von." *DSB* 6 (1972): 241–53.

Van der Waerden, B. L., ed. *Sources of Quantum Mechanics.* Amsterdam: North-Holland, 1967. Reprint. New York: Dover, 1968.

Van't Hoff, J. H. *Acht Vorträge über physikalische Chemie gehalten auf Einladung der Universität Chicago 20. bis 24. Juni 1901.* Braunschweig: F. Vieweg, 1902.

————. "Friedrich Wilhelm Ostwald." *Zs. f. phys. Chem.* 46 (1903): v–xv.

————. *Vorlesungen über theoretische und physikalische Chemie.* 3 vols. Braunschweig: F. Vieweg, 1898–1900.

Van Vleck, J. H. "Nicht-mathematische theoretische Physik." *Phys. Bl.* 24 (1968): 97–102.

Vienna. University. *Geschichte der Wiener Universität von 1848 bis 1898.* Edited by the Akademischer Senat der Wiener Universität. Vienna, 1898.

Voigt, Woldemar. "Eduard Riecke als Physiker." *Phys. Zs.* 16 (1915): 219–221.

————. *Elementare Mechanik als Einleitung in das Studium der theoretischen Physik.* 2d rev. ed. Leipzig: Veit, 1901.

————. *Erinnerungsblätter aus dem deutsch-französischen Kriege 1870/71.* Göttingen: Dietrich, 1914.

————. *Die fundamentalen physikalischen Eigenschaften der Krystalle.* Leipzig, 1898.

————. "Der Kampf um die Dezimale in der Physik." *Deutsche Revue* 34 (1909): 71–85.

————. *Kompendium der theoretischen Physik.* Vol. 1, *Mechanik starrer und nichtstarrer Körper. Wärmelehre.* Leipzig, 1895. Vol. 2, *Elektricität und Magnetismus. Optik.* Leipzig, 1896.

————. *Lehrbuch der Kristallphysik (mit Ausschluss der Kristalloptik).* Leipzig and Berlin: B. G. Teubner, 1910.

————. "Ludwig Boltzmann." *Gött. Nachr.,* 1907, 69–82.

————. *Magneto- und Elektrooptik.* Leipzig: B. G. Teubner, 1908.

————. "Paul Drude." *Phys. Zs.* 7 (1906): 481–82.

————. *Physikalische Forschung und Lehre in Deutschland während der letzten hundert Jahre. Festrede im Namen der Georg-August-Universität zur Jahresfeier der Universität am 5. Juni 1912.* Göttingen, 1912.

———. "Rede." In *Die physikalischen Institute der Universität Göttingen*, 37–43.

———. *Thermodynamik*. 2 vols. Leipzig: G. J. Göschen, 1903–4.

———. "Zum Gedächtniss von G. Kirchhoff." *Abh. Ges. Wiss. Göttingen* 35 (1888): 3–10.

———. "Zur Erinnerung an F. E. Neumann, gestorben am 23. Mai 1895 zu Königsberg i/Pr." *Gött. Nachr.*, 1895, 248–65. Reprinted as "Gedächtnissrede auf Franz Neumann" in *Franz Neumanns Gesammelte Werke*, vol. 1, 3–19.

Voit, C. "August Kundt." *Sitzungsber. bay. Akad.* 25 (1895): 177–79.

———. "Eugen v. Lommel." *Sitzungsber. bay. Akad.* 30 (1900): 324–39.

———. "Leonhard Sohncke." *Sitzungsber. bay. Akad.* 28 (1898): 440–49.

———. "Philipp Johann Gustav von Jolly." *Sitzungsber. bay. Akad.* 15 (1885): 119–36.

———. "Wilhelm von Beetz." *Sitzungsber. bay. Akad.* 16 (1886): 10–31.

———. "Wilhelm von Bezold." *Sitzungsber. bay. Akad.* 37 (1907): 268–71.

Volkmann, H. "Ernst Abbe and His Work." *Applied Optics* 5 (1966): 1720–31.

Volkmann, Paul. *Einführung in das Studium der theoretischen Physik insbesondere in das der analytischen Mechanik mit einer Einleitung in die Theorie der physikalischen Erkenntniss*. Leipzig: B. G. Teubner, 1900.

———. *Erkenntnistheoretische Grundzüge der Naturwissenschaften und ihre Beziehungen zum Geistesleben der Gegenwart. Allgemein wissenschaftliche Vorträge*. Leipzig, 1896. 2d ed. Leipzig and Berlin: B. G. Teubner, 1910.

———. "Franz Neumann als Experimentator." *Phys. Zs.* 11 (1910): 932–37.

———. *Franz Neumann. 11. September 1798, 23. Mai 1895*. Leipzig, 1896.

———. "Hermann von Helmholtz." *Schriften der Physikalisch-ökonomischen Gesellschaft zu Königsberg* 35 (1894): 73–81.

———. *Die materialistische Epoche des neunzehnten Jahrhunderts und die phänomenologisch-monistische Bewegung der Gegenwart. Rede am Krönungstage, 18. Januar 1909* . . . Leipzig and Berlin: B. G. Teubner, 1909.

———. *Vorlesungen über die Theorie des Lichtes. Unter Rücksicht auf die elastische und die elektromagnetische Anschauung*. Leipzig, 1891.

Voss, A. "Heinrich Weber." *Jahresber. d. Deutsch. Math.-Vereinigung* 23 (1914): 431–44.

Wachsmuth, Richard. *Die Gründung der Universität Frankfurt*. Frankfurt a. M.: Englert und Schlosser, 1929.

Wallenreiter, Clara. *Die Vermögensverwaltung der Universität Landshut-München: Ein Beitrag zur Geschichte des bayerischen Hochschultyps vom 18. zum 20. Jahrhundert*. Berlin: Duncker und Humblot, 1971.

Wangerin, Albert. *Franz Neumann und sein Wirken als Forscher und Lehrer*. Braunschweig: F. Vieweg, 1907.

Warburg, Emil. "Friedrich Kohlrausch." *Verh. phys. Ges.* 12 (1910): 911–38.

———. *Lehrbuch der Experimentalphysik für Studirende*. Freiburg i. Br. and Leipzig, 1893.

———. "Das physikalische Institut." In *Die Universität Freiburg*, 91–96.

———. "Die technische Physik und die Physikalisch-Technische Reichsanstalt." *Zs. f. techn. Physik* 2 (1921): 225–27.

———. "Über Plancks Verdienste um die Experimentalphysik." *Naturwiss.* 6 (1918): 202–3.

———. "Verhältnis der Präzisionsmessungen zu den allgemeinen Zielen der Physik." In *Physik*, edited by Emil Warburg, 653–60.

———. "Zur Erinnerung an Gustav Kirchhoff." *Naturwiss.* 13 (1925): 205–12.

———. "Zur Geschichte der Physikalischen Gesellschaft." *Naturwiss.* 13 (1925): 35–39.

———, ed. *Physik*. Kultur der Gegenwart, ser. 3, vol. 3, pt. 1. Berlin: B. G. Teubner, 1915.

Weart, Spencer. See Paul Forman.

Weber, Heinrich. *Die partiellen Differential-Gleichungen der mathematischen Physik. Nach Riemann's Vorlesungen.* 4th rev. ed. 2 vols. Braunschweig: F. Vieweg, 1900–1901.

Weber, Heinrich. *Wilhelm Weber. Eine Lebensskizze.* Breslau, 1893.

Weber, Leonhard. "Gustav Karsten." *Schriften d. Naturwiss. Vereins f. Schleswig-Holstein* 12 (1901): 63–68.

Weber, Wilhelm. *Wilhelm Weber's Werke.* Edited by Königliche Gesellschaft der Wissenschaften zu Göttingen. Vol. 1, *Akustik, Mechanik, Optik und Wärmelehre.* Edited by Woldemar Voigt. Berlin, 1892. Vol. 2, *Magnetismus.* Edited by Eduard Riecke. Berlin, 1892. Vol. 3, *Galvanismus und Elektrodynamik, erster Theil.* Edited by Heinrich Weber. Berlin, 1893. Vol. 4, *Galvanismus und Elektrodynamik, zweiter Theil.* Edited by Heinrich Weber. Berlin, 1894. Vol. 5, with E. H. Weber, *Wellenlehre auf Experimente gegründet oder über die Wellen tropfbarer Flüssigkeiten mit Anwendung auf die Schall- und Lichtwellen.* Edited by Eduard Riecke. Berlin, 1893.

Weickmann, Ludwig. "Nachruf auf Otto Wiener." *Verh. sächs. Ges. Wiss.* 79 (1927): 107–18.

Weinberg, Steven. "The Search for Unity: Notes for a History of Quantum Field Theory." *Daedalus* 106 (1977): 17–35.

Weiner, Charles, ed. *History of Twentieth Century Physics.* New York: Academic Press, 1977.

Weiner, K. L. "Otto Lehmann, 1855–1922." In vol. 3, *Geschichte der Mikroskopie*, edited by H. Freund and A. Berg, 261–71. Frankfurt a. M.: Umschau, 1966.

Weingart, Peter. See Gerard Lemaine.

Weinstein, Bernhard. *Einleitung in die höhere mathematische Physik.* Berlin: F. Dümmler, 1901.

Weis, E. "Bayerns Beitrag zur Wissenschaftsentwicklung im 19. und 20. Jahrhundert." In *Handbuch der bayerischen Geschichte*, vol. 4, pt. 2, 1034–88.

Weyl, Hermann. "A Half-Century of Mathematics." *Amer. Math. Monthly* 58 (1951): 523–53.

———. "Obituary: David Hilbert 1862–1943." *Obituary Notices of Fellows of the*

Royal Society 4 (1944): 547–53. Reprinted in Weyl's *Gesammelte Abhandlungen*, edited by K. Chandrasekharan, vol. 4, 121–29. Berlin: Springer, 1968.

———. See H. A. Lorentz.

Wheaton, Bruce R. "Philipp Lenard and the Photoelectric Effect, 1889–1911." *HSPS* 9 (1978): 299–322.

Whittaker, Edmund. *A History of the Theories of Aether and Electricity.* Vol. 1, *The Classical Theories.* Vol. 2, *The Modern Theories, 1900–1926.* Reprint. New York: Harper and Brothers, 1960.

Wiechert, Emil. "Eduard Riecke." *Gött. Nachr.*, 1916, 45–56.

———. "Das Institut für Geophysik." In *Die physikalischen Institute der Universität Göttingen*, 119–88.

Wiedemann, Eilhard. "Die Wechselbeziehungen zwischen dem physikalischen Hochschulunterricht und dem physikalischen Unterricht an höheren Lehranstalten." *Zs. f. math. u. naturwiss. Unterricht* 26 (1895): 127–40.

Wiedemann, Gustav. *Ein Erinnerungsblatt.* Leipzig, 1893.

———. "Hermann von Helmholtz' wissenschaftliche Abhandlungen." In Helmholtz's *Wissenschaftliche Abhandlungen*, vol. 3, xi–xxxvi.

———. *Die Lehre von der Elektricität.* 2d rev. ed. 4 vols. Braunschweig, 1893–98.

———. "Stiftungsfeier am 4. Januar 1896." *Verh. phys. Ges.* 15 (1896): 32–36.

———. "Vorwort." *Ann.* 39 (1890): i–iv.

Wiederkehr, K. H. *Wilhelm Eduard Weber. Erforscher der Wellenbewegung und der Elektrizität 1804–1891.* Vol. 32, *Grosse Naturforscher.* Stuttgart: Wissenschaftliche Verlagsgesellschaft, 1967.

Wien, Wilhelm. *Aus dem Leben und Wirken eines Physikers.* Edited by K. Wien. Leipzig: J. A. Barth, 1930.

———. "Helmholtz als Physiker." *Naturwiss.* 9 (1921): 694–99.

———. "Mathias Cantor." *Phys. Zs.* 17 (1916): 265–67.

———. "Das neue physikalische Institut der Universität Giessen." *Phys. Zs.* 1 (1899): 155–60.

———. *Die neuere Entwicklung unserer Universitäten und ihre Stellung im deutschen Geistesleben.* Rede für den Festakt in der neuen Universität am 29. Juni 1914 . . . Würzburg: Stürtz, 1915.

———. "Das physikalische Institut und das physikalische Seminar." In *Die wissenschaftlichen Anstalten . . . zu München*, 207–11.

———. *Die Relativitätstheorie vom Standpunkte der Physik und Erkenntnislehre.* Leipzig: J. A. Barth, 1921.

———. "Ein Rückblick." In *Aus dem Leben und Wirken eines Physikers*, 1–76.

———. "Theodor Des Coudres." *Phys. Zs.* 28 (1927): 129–35.

———. "Über die partiellen Differential-Gleichungen der Physik." *Jahresber. d. Deutsch. Math.-Vereinigung* 15 (1906): 42–51.

———. *Über Elektronen. Vortrag gehalten auf der 77. Versammlung deutscher Naturforscher und Ärzte in Meran.* 2d rev. ed. Leipzig and Berlin: B. G. Teubner, 1909.

————. *Universalität und Einzelforschung. Rektorats-Antrittsrede, gehalten am 28. November 1925.* Munich: Max Hueber, 1926.

————. *Vergangenheit, Gegenwart und Zukunft der Physik. Rede gehalten beim Stiftungsfest der Universität München am 19. Juni 1926.* Munich: Max Hueber, 1926.

————. *Vorlesungen über neuere Probleme der theoretischen Physik, gehalten an der Columbia-Universität in New York im April 1913.* Leipzig and Berlin: B. G. Teubner, 1913.

————. *Vorträge über die neuere Entwicklung der Physik und ihrer Anwendungen. Gehalten im Baltenland im Frühjahr 1918 auf Veranlassung des Oberkommandos der achten Armee.* Leipzig: J. A. Barth, 1919.

————. "Ziele und Methoden der theoretischen Physik." *Jahrbuch der Radioaktivität und Elektronik* 12 (1915): 241–59.

Wiener, Otto. "Die Erweiterung unsrer Sinne." *Deutsche Revue* 25 (1900): 25–41.

————. "Nachruf auf Theodor Des Coudres." *Verh. sächs. Ges. Wiss.* 78 (1926): 358–70.

————. "Nachruf auf Wilhelm Hallwachs." *Verh. sächs. Ges. Wiss.* 74 (1922): 293–313.

————. "Das neue physikalische Institut der Universität Leipzig und Geschichtliches." *Phys. Zs.* 7 (1906): 1–14.

Wigand, Albert. "Ernst Dorn." *Phys. Zs.* 17 (1916): 297–99.

Winkelmann, Adolph, ed. *Handbuch der Physik.* 2d ed. 6 vols. Leipzig: J. A. Barth, 1903–9.

Wise, M. Norton. "German Concepts of Force, Energy, and the Electromagnetic Ether: 1845–1880." In *Conceptions of Ether: Studies in the History of Ether Theories 1740–1900,* edited by G. N. Cantor and M. J. S. Hodge, 269–307.

Witte, H. "Die Ablehnung der Materialismus-Hypothese durch die heutige Physik." *Annalen der Naturphilosophie* 8 (1909): 95–130.

————. "Die Monismusfrage in der Physik." *Annalen der Naturphilosophie* 8 (1909): 131–36.

Wolf, Franz. "Aus der Geschichte der Physik in Karlsruhe." *Phys. Bl.* 24 (1968): 388–400.

————. "Philipp Lenard zum 100. Geburtstag." *Phys. Bl.* 18 (1962): 271–75.

Wolkenhauer, W. "Karsten, Gustav." *Biographisches Jahrbuch und Deutscher Nekrolog* 5 (1900): 76–78.

Woodruff, A. E. "The Contributions of Hermann von Helmholtz to Electrodynamics." *Isis* 59 (1968): 300–311.

Woolf, Harry, ed. *Some Strangeness in the Proportion.* Reading, Mass.: Addison-Wesley, 1980.

Wüllner, Adolph. *Lehrbuch der Experimentalphysik.* 4th rev. ed. 4 vols. Leipzig, 1882–86.

Württemberg. Statistisches Landesamt. *Statistik der Universität Tübingen.* Edited by the K. Statistisch-Topographisches Bureau. Stuttgart, 1877.

Würzburg. University. *Verzeichniss der Vorlesungen welche an der Königlich-Bayerischen*

Julius-Maximilians-Universität zu Würzburg . . . gehalten werden. Würzburg, n.d.

——. See Maria Reindl; W. C. Röntgen.

Zehnder, Ludwig. See W. C. Röntgen.

Zenneck, J. "Ferdinand Braun (1850–1918)/Professor der Physik." In *Lebensbilder aus Kurhessen und Waldeck 1830–1930,* edited by Ingeborg Schnack, vol. 2, 51–62. Marburg: Elwert, 1940.

Ziegenfuss, Werner. "Helmholtz, Hermann von." In *Philosophen-Lexikon,* 1: 498–501. Berlin: de Gruyter, 1949.

Zöllner, J. C. F. *Erklärung der universellen Gravitation aus den statischen Wirkungen der Elektricität und die allgemeine Bedeutung des Weber'schen Gesetzes. Mit Beiträgen von Wilhelm Weber.* 2d ed. Leipzig, 1886.

——. *Principien einer elektrodynamischen Theorie der Materie.* Vol. 1, *Abhandlungen zur atomistischen Theorie der Elektrodynamik.* Leipzig, 1876.

Zurich. ETH. *Eidgenössische Technische Hochschule 1855–1955.* Zurich: Buchverlag der Neuen Zürcher Zeitung, 1955.

——. *Festschrift zur Feier des fünfzigjährigen Bestehens des Eidg. Polytechnikums.* Pt. 1, *Geschichte der Gründung des Eidg. Polytechnikums mit einer Übersicht seiner Entwicklung 1855–1905* by Wilhelm Oechsli. Frauenfeld: Huber, 1905. Pt. 2, *Die bauliche Entwicklung Zürichs in Einzeldarstellungen.* Zurich: Zürcher & Furrer, 1905.

——. *100 Jahre Eidgenössische Technische Hochschule. Sonderheft der Schweizerischen Hochschulzeitung* 28 (1955).

——. See A. Frey-Wyssling.

Acknowledgments

We are grateful to the National Science Foundation for the award of two major grants to support our research and to the many historians of science who supported it through their evaluations of the proposal that resulted in this book. For their encouragement and help, we are especially grateful to Martin J. Klein and Thomas S. Kuhn. We want to thank the owners of manuscripts and photographs who granted us permission to use them, and we want to thank the archivists and librarians of the many manuscript collections here and abroad who assisted us in our work.

415

Index